Embryonic Stem Cell Protocols, Vol. 1
SECOND EDITION

METHODS IN MOLECULAR BIOLOGY™

John M. Walker, SERIES EDITOR

346. **Dictyostelium discoideum Protocols**, edited by *Ludwig Eichinger and Francisco Rivero-Crespo, 2006*
345. **Diagnostic Bacteriology Protocols**, *Second Edition*, edited by *Louise O'Connor, 2006*
344. **Agrobacterium Protocols, Second Edition:** *Volume 2*, edited by *Kan Wang, 2006*
343. **Agrobacterium Protocols, Second Edition:** *Volume 1*, edited by *Kan Wang, 2006*
342. **MicroRNA Protocols**, edited by *Shao-Yao Ying, 2006*
341. **Cell–Cell Interactions:** *Methods and Protocols*, edited by *Sean P. Colgan, 2006*
340. **Protein Design:** *Methods and Applications*, edited by *Raphael Guerois and Manuela López de la Paz, 2006*
339. **Microchip Capillary Electrophoresis:** *Methods and Protocols*, edited by *Charles Henry, 2006*
338. **Gene Mapping, Discovery, and Expression:** *Methods and Protocols*, edited by *M. Bina, 2006*
337. **Ion Channels:** *Methods and Protocols*, edited by *J. D. Stockand and Mark S. Shapiro, 2006*
336. **Clinical Applications of PCR:** *Second Edition*, edited by *Y. M. Dennis Lo, Rossa W. K. Chiu, and K. C. Allen Chan, 2006*
335. **Fluorescent Energy Transfer Nucleic Acid Probes:** *Designs and Protocols*, edited by *Vladimir V. Didenko, 2006*
334. **PRINS and In Situ PCR Protocols:** *Second Edition*, edited by *Franck Pellestor, 2006*
333. **Transplantation Immunology:** *Methods and Protocols*, edited by *Philip Hornick and Marlene Rose, 2006*
332. **Transmembrane Signaling Protocols:** *Second Edition*, edited by *Hydar Ali and Haribabu Bodduluri, 2006*
331. **Human Embryonic Stem Cell Protocols**, edited by *Kursad Turksen, 2006*
330. **Embryonic Stem Cell Protocols, Second Edition,** *Vol. II: Differentiation Models*, edited by *Kursad Turksen, 2006*
329. **Embryonic Stem Cell Protocols, Second Edition,** *Vol. I: Isolation and Characterization*, edited by *Kursad Turksen, 2006*
328. **New and Emerging Proteomic Techniques**, edited by *Dobrin Nedelkov and Randall W. Nelson, 2006*
327. **Epidermal Growth Factor:** *Methods and Protocols*, edited by *Tarun B. Patel and Paul J. Bertics, 2006*
326. **In Situ Hybridization Protocols,** *Third Edition*, edited by *Ian A. Darby and Tim D. Hewitson, 2006*
325. **Nuclear Reprogramming:** *Methods and Protocols*, edited by *Steve Pells, 2006*
324. **Hormone Assays in Biological Fluids**, edited by *Michael J. Wheeler and J. S. Morley Hutchinson, 2006*
323. **Arabidopsis Protocols,** *Second Edition*, edited by *Julio Salinas and Jose J. Sanchez-Serrano, 2006*
322. **Xenopus Protocols:** *Cell Biology and Signal Transduction*, edited by *X. Johné Liu, 2006*
321. **Microfluidic Techniques:** *Reviews and Protocols*, edited by *Shelley D. Minteer, 2006*
320. **Cytochrome P450 Protocols**, *Second Edition*, edited by *Ian R. Phillips and Elizabeth A. Shephard, 2006*
319. **Cell Imaging Techniques**, *Methods and Protocols*, edited by *Douglas J. Taatjes and Brooke T. Mossman, 2006*
318. **Plant Cell Culture Protocols**, *Second Edition*, edited by *Victor M. Loyola-Vargas and Felipe Vázquez-Flota, 2005*
317. **Differential Display Methods and Protocols**, *Second Edition*, edited by *Peng Liang, Jonathan Meade, and Arthur B. Pardee, 2005*
316. **Bioinformatics and Drug Discovery**, edited by *Richard S. Larson, 2005*
315. **Mast Cells:** *Methods and Protocols*, edited by *Guha Krishnaswamy and David S. Chi, 2005*
314. **DNA Repair Protocols.** *Mammalian Systems, Second Edition*, edited by *Daryl S. Henderson, 2006*
313. **Yeast Protocols:** *Second Edition*, edited by *Wei Xiao, 2005*
312. **Calcium Signaling Protocols:** *Second Edition*, edited by *David G. Lambert, 2005*
311. **Pharmacogenomics:** *Methods and Protocols*, edited by *Federico Innocenti, 2005*
310. **Chemical Genomics:** *Reviews and Protocols*, edited by *Edward D. Zanders, 2005*
309. **RNA Silencing:** *Methods and Protocols*, edited by *Gordon Carmichael, 2005*
308. **Therapeutic Proteins:** *Methods and Protocols*, edited by *C. Mark Smales and David C. James, 2005*
307. **Phosphodiesterase Methods and Protocols,** edited by *Claire Lugnier, 2005*
306. **Receptor Binding Techniques:** *Second Edition*, edited by *Anthony P. Davenport, 2005*
305. **Protein–Ligand Interactions:** *Methods and Applications*, edited by *G. Ulrich Nienhaus, 2005*
304. **Human Retrovirus Protocols:** *Virology and Molecular Biology*, edited by *Tuofu Zhu, 2005*
303. **NanoBiotechnology Protocols,** edited by *Sandra J. Rosenthal and David W. Wright, 2005*
302. **Handbook of ELISPOT:** *Methods and Protocols*, edited by *Alexander E. Kalyuzhny, 2005*
301. **Ubiquitin–Proteasome Protocols,** edited by *Cam Patterson and Douglas M. Cyr, 2005*
300. **Protein Nanotechnology:** *Protocols, Instrumentation, and Applications*, edited by *Tuan Vo-Dinh, 2005*
299. **Amyloid Proteins:** *Methods and Protocols*, edited by *Einar M. Sigurdsson, 2005*

METHODS IN MOLECULAR BIOLOGY™

Embryonic Stem Cell Protocols

Volume 1: Isolation and Characterization

SECOND EDITION

Edited by

Kursad Turksen

Ottawa Health Research Institute
Ottawa, Ontario, Canada

HUMANA PRESS ✳ TOTOWA, NEW JERSEY

© 2006 Humana Press Inc.
999 Riverview Drive, Suite 208
Totowa, New Jersey 07512

www.humanapress.com

All rights reserved. No part of this book may be reproduced, stored in a retrieval system, or transmitted in any form or by any means, electronic, mechanical, photocopying, microfilming, recording, or otherwise without written permission from the Publisher. Methods in Molecular Biology™ is a trademark of The Humana Press Inc.

All papers, comments, opinions, conclusions, or recommendations are those of the author(s), and do not necessarily reflect the views of the publisher.

This publication is printed on acid-free paper. ∞
ANSI Z39.48-1984 (American Standards Institute)

Permanence of Paper for Printed Library Materials.

Production Editor: Jennifer Hackworth

Cover design by Patricia F. Cleary

Cover illustration: Figure 2D from Chapter 7 in Vol. 1, "Derivation and Propagation of Embryonic Stem Cells in Serum- and Feeder-Free Culture," by Jennifer Nichols and Qi-Long Ying.

For additional copies, pricing for bulk purchases, and/or information about other Humana titles, contact Humana at the above address or at any of the following numbers: Tel.: 973-256-1699; Fax: 973-256-8341; E-mail: orders@humanapr.com; or visit our Website: www.humanapress.com

Photocopy Authorization Policy:
Authorization to photocopy items for internal or personal use, or the internal or personal use of specific clients, is granted by Humana Press Inc., provided that the base fee of US $30.00 per copy is paid directly to the Copyright Clearance Center at 222 Rosewood Drive, Danvers, MA 01923. For those organizations that have been granted a photocopy license from the CCC, a separate system of payment has been arranged and is acceptable to Humana Press Inc. The fee code for users of the Transactional Reporting Service is: [1-58829-498-6/06 $30.00].

Printed in the United States of America. 10 9 8 7 6 5 4 3 2 1

eISBN: 1-59745-037-5
ISSN: 1064-3745

Library of Congress Cataloging-in-Publication Data

Embryonic stem cell protocols / edited by Kursad Turksen.-- 2nd ed.
    v. ; cm. -- (Methods in molecular biology ; v. 329-330)
  Includes bibliographical references and index.
  Contents: v. 1. Isolation and characterization -- v. 2. Differentiation models.
  ISBN 1-58829-498-6 (v. 1 : alk. paper) -- ISBN 1-58829-784-5 (v. 2 : alk. paper)
  1. Stem cells--Laboratory manuals. 2. Embryos--Laboratory manuals.
  [DNLM: 1. Stem Cells. 2. Embryo Research. 3. Models, Animal. QU 325 N812 2006] I. Turksen, Kursad. II. Series: Methods in molecular biology (Clifton, N.J.) ; v. 329-330.
    QH588.S83E46 2006
    616'.02774--dc22

2005018023

Preface

The potentials, and hence popularity, of assessing embryonic stem (ES) cells in regenerative medicine applications is no longer a surprise to either scientists or the general public. This is clearly reflected in the ever-increasing publications in which ES cell biology and differentiation along diverse lineages appear in the academic as well as the popular press. It is also reflected in the intense interest in the isolation and characterization of ES cells from other species for preclinical studies. It therefore seemed timely to capture important advances in the field since the publication of the *Embryonic Stem Cells: Methods and Protocols* volume four years ago.

To provide an update and complement the original mouse ES cell book, I have focused the initial part of the first volume of the new series on ES cells recently isolated from other/nonmouse species. Second, the volumes contain numerous updates, more advanced approaches, and completely new protocols for the use of ES cells in studies of diverse cell lineages. I believe that these two volumes will complement and expand the experimental repertoires of both experts and novices in the field. I would therefore like to take this opportunity to thank all of the contributors for their generosity and dedication in putting together their protocols. Without them, these volumes would not exist.

I am grateful to Dr. John Walker for his support and encouragement during the process. I would also like to thank several others at the Humana Press for their support: initially Elyse O'Grady and Craig Adams, and more recently Damien DeFrances. Also, I am grateful to Jennifer Hackworth for her wonderful support during the production of this volume.

I would also like to thank Jane Aubin and N. Urfe for their continuous support and encouragement, as well as Tammy Troy who has once again been fantastic in helping to put together these volumes.

Kursad Turksen

Contents

Preface v
Contents of the Companion Volume xi
Contributors xv
Guide to the Companion CD xxi

PART I. ISOLATION AND MAINTENANCE
  1 Isolation and Differentiation of Medaka Embryonic Stem Cells
     Yunhan Hong and Manfred Schartl 3
  2 Maintenance of Chicken Embryonic Stem Cells In Vitro
     Hiroyuki Horiuchi, Shuichi Furusawa,
     and Haruo Matsuda 17
  3 Derivation and Culture of Mouse Trophoblast
     Stem Cells In Vitro
     Satoshi Tanaka 35
  4 Derivation, Maintenance, and Characterization
     of Rat Embryonic Stem Cells In Vitro
     Maren Schulze, Hendrik Ungefroren, Michael Bader,
     and Fred Fändrich 45
  5 Derivation, Maintenance, and Induction of the Differentiation
     In Vitro of Equine Embryonic Stem Cells
     Shigeo Saito, Ken Sawai, Arika Minamihashi, Hideyo Ugai,
     Takehide Murata, and Kazunari K. Yokoyama 59
  6 Generation and Characterization of Monkey
     Embryonic Stem Cells
     Hirofumi Suemori and Norio Nakatsuji 81
  7 Derivation and Propagation of Embryonic Stem Cells
     in Serum- and Feeder-Free Culture
     Jennifer Nichols and Qi-Long Ying 91

PART II. SIGNALING IN EMBRYONIC STEM CELL DIFFERENTIATION
  8 Internal Standards in Differentiating
     Embryonic Stem Cells In Vitro
     Christopher L. Murphy 101

9 Matrix Assembly, Cell Polarization, and Cell Survival: Analysis of Peri-Implantation Development With Cultured Embryonic Stem Cells
Shaohua Li and Peter D. Yurchenco 113

10 Phosphoinositides, Inositol Phosphates, and Phospholipase C in Embryonic Stem Cells
Leo R. Quinlan 127

11 Cripto Signaling in Differentiating Embryonic Stem Cells
Gabriella Minchiotti, Silvia Parisi, and M. Graziella Persico 151

12 The Use of Embryonic Stem Cells to Study Hedgehog Signaling
Sandy Becker and Laura Grabel 171

13 Transfection and Promoter Analysis in Embryonic Stem Cells
Sangmi Chung and Kwang-Soo Kim 187

14 SAGE Analysis to Identify Embryonic Stem Cell-Predominant Transcripts
Kenneth R. Boheler and Kirill V. Tarasov 195

15 Utilization of Digital Differential Display to Identify Novel Targets of Oct3/4
Yoshimi Tokuzawa, Masayoshi Maruyama, and Shinya Yamanaka 223

16 Gene Silencing Using RNA Interference in Embryonic Stem Cells
J. Matthew Velkey, Nicole A. Slawny, Theresa E. Gratsch, and K. Sue O'Shea 233

PART III. GENETIC MANIPULATION OF EMBRYONIC STEM CELLS

17 Efficient Transfer of HSV-1 Amplicon Vectors Into Embryonic Stem Cells and Their Derivatives
Dieter Riethmacher, Filip Lim, and Thomas Schimmang 265

18 Lentiviral Vector-Mediated Gene Transfer in Embryonic Stem Cells
Masahiro Oka, Lung-Ji Chang, Frank Costantini, and Naohiro Terada 273

19 Use of the Cytomegalovirus Promoter for Transient and Stable Transgene Expression in Mouse Embryonic Stem Cells
Katie M. Barrow, Flor M. Perez-Campo, and Christopher M. Ward 283

Contents

20 Use of Simian Immunodeficiency Virus Vectors
for Simian Embryonic Stem Cells
Takayuki Asano, Hiroaki Shibata, and Yutaka Hanazono 295

21 Generation of Green Fluorescent Protein-Expressing
Monkey Embryonic Stem Cells
*Tatsuyuki Takada, Yutaka Suzuki, Nae Kadota,
Yasushi Kondo, and Ryuzo Torii* 305

22 DNA Damage Response and Mutagenesis
in Mouse Embryonic Stem Cells
Yiling Hong, Rachel B. Cervantes, and Peter J. Stambrook 313

23 Ultraviolet-Induced Apoptosis in Embryonic Stem Cells In Vitro
Dakang Xu, Trevor J. Wilson, and Paul J. Hertzog 327

PART IV. USE OF EMBRYONIC STEM CELLS IN PHARMACOLOGICAL
AND TOXICOLOGICAL SCREENS

24 Use of Differentiating Embryonic Stem Cells
in Pharmacological Studies
*Brigitte Wdziekonski, Phi Villageois, Cécile Vernochet,
Blaine Phillips, and Christian Dani* 341

25 Embryonic Stem Cells as a Source of Differentiated Neural
Cells for Pharmacological Screens
*Patrick J. Mee, Carmel M. O'Brien, Hazel Thomson,
Sjaak van der Sar, Viktor Lakics,
and Timothy E. Allsopp* 353

26 Use of Murine Embryonic Stem Cells in Embryotoxicity Assays:
The Embryonic Stem Cell Test
*Andrea E. M. Seiler, Roland Buesen, Anke Visan,
and Horst Spielmann* 371

27 Use of Chemical Mutagenesis in Mouse Embryonic Stem Cells
*Sonja Becker, Martin Hrabé de Angelis,
and Johannes Beckers* 397

PART V. EPIGENETIC ANALYSIS OF EMBRYONIC STEM CELLS

28 Nuclear Reprogramming of Somatic Nucleus Hybridized
With Embryonic Stem Cells by Electrofusion
Masako Tada and Takashi Tada 411

29 Methylation in Embryonic Stem Cells In Vitro
*Koichiro Nishino, Jun Ohgane, Masako Suzuki,
Naka Hattori, and Kunio Shiota* 421

PART VI. TUMOR-LIKE PROPERTIES

30 Identification of Genes Involved in Tumor-Like Properties of Embryonic Stem Cells
   Kazutoshi Takahashi, Tomoko Ichisaka, and Shinya Yamanaka 449

31 In Vivo Tumor Formation From Primate Embryonic Stem Cells
   Takayuki Asano, Kyoko Sasaki, Yoshihiro Kitano, Keiji Terao, and Yutaka Hanazono 459

PART VII. ANIMAL MODELS AND THERAPY

32 Directed Differentiation and Characterization of Genetically Modified Embryonic Stem Cells for Therapy
   Adeline A. Lau, Kim M. Hemsley, Adrian Meedeniya, Aaron J. Robinson, and John J. Hopwood 471

33 Use of Differentiating Embryonic Stem Cells in the Parkinsonian Mouse Model
   Fumihiko Nishimura, Hayato Toriumi, Shigeaki Ishizaka, Toshisuke Sakaki, and Masahide Yoshikawa 485

Index 495

CONTENTS OF THE COMPANION VOLUME
Volume 2: Differentiation Models

1. Neural Stem Sphere Method: *Induction of Neural Stem Cells and Neurons by Astrocyte-Derived Factors in Embryonic Stem Cells In Vitro*
   Takashi Nakayama and Nobuo Inoue
2. Generation and Characterization of Oligodendrocytes From Lineage-Selectable Embryonic Stem Cells In Vitro
   Nathalie Billon, Christine Jolicoeur, and Martin Raff
3. Derivation and Characterization of Neural Cells From Embryonic Stem Cells Using Nestin Enhancer
   Nibedita Lenka
4. Optimized Neuronal Differentiation of Murine Embryonic Stem Cells: *Role of Cell Density*
   Matthew T. Lorincz
5. Generation of Inner Ear Cell Types From Embryonic Stem Cells
   Marcelo N. Rivolta, Huawei Li, and Stefan Heller
6. Derivation of Epidermal Colony-Forming Progenitors From Embryonic Stem Cell Cultures
   Tammy-Claire Troy and Kursad Turksen
7. Directing Epidermal Fate Selection by a Novel Co-Culture System
   Tammy-Claire Troy and Kursad Turksen
8. In Vitro Generation of T Lymphocytes From Embryonic Stem Cells
   Renée F. de Pooter, Thomas M. Schmitt, and Juan Carlos Zúñiga-Pflücker
9. The Role of *Hex* in Hemangioblast and Hematopoietic Development
   Rebecca J. Chan, Robert Hromas, and Mervin C. Yoder
10. Generation of Osteoblasts and Chondrocytes From Embryonic Stem Cells
    Jitsutaro Kawaguchi
11. Analysis of Embryonic Stem Cell-Derived Osteogenic Cultures
    Nicole L. Woll and Sarah K. Bronson
12. Generation of Chondrocytes From Embryonic Stem Cells
    Jaspal Singh Khillan

13 Derivation and Characterization of Chondrocytes From Embryonic Stem Cells In Vitro
    Jan Kramer, Gunnar Hargus, and Jürgen Rohwedel

14 Generation and Characterization of Cardiomyocytes Under Serum-Free Conditions
    Cornelia Gissel, Michael Xavier Doss, Rita Hippler-Altenburg, Jürgen Hescheler, and Agapios Sachinidis

15 Analysis of Arrhythmic Potential of Embryonic Stem Cell-Derived Cardiomyocytes
    Lijuan L. Shang, Samuel C. Dudley, Jr., and Arnold E. Pfahnl

16 Derivation and Characterization of Alveolar Epithelial Cells From Murine Embryonic Stem Cells In Vitro
    Ali Samadikuchaksaraei and Anne E. Bishop

17 Derivation and Characterization of Thyrocyte-Like Cells From Embryonic Stem Cells In Vitro
    Reigh-Yi Lin and Terry F. Davies

18 Derivation and Characterization of Gut-Like Structures From Embryonic Stem Cells
    Takatsugu Yamada and Yoshiyuki Nakajima

19 Formation of Gut-Like Structures In Vitro From Mouse Embryonic Stem Cells
    Shigeko Torihashi

20 In Vitro Derivation and Expansion of Endothelial Cells From Embryonic Stem Cells
    Kara E. McCloskey, Steven L. Stice, and Robert M. Nerem

21 Differentiation of Mouse Embryonic Stem Cells Into Endothelial Cells: *Genetic Selection and Potential Use In Vivo*
    Clotilde Gimond, Sandrine Marchetti, and Gilles Pagès

22 Integrins and Vascular Development in Differentiated Embryonic Stem Cells In Vitro
    Sheila E. Francis

23 TGF-β Signaling in Embryonic Stem Cell-Derived Endothelial Cells
    Tetsuro Watabe, Jun K. Yamashita, Koichi Mishima, and Kohei Miyazono

24 The Role of the Adapter Protein SHB in Embryonic Stem Cell Differentiation Into the Pancreatic β-cell and Endothelial Lineages
    Johan Saldeen, Nina Ågren, Björn Åkerblom, Lingge Lu, Bashir Adem, Łukasz Sędek, and Vitezslav Kriz

Contents of the Companion Volume

25 In Vitro Differentiation of Embryonic Stem Cells Into the Pancreatic Lineage
    Przemyslaw Blyszczuk and Anna M. Wobus

26 Derivation and Characterization of Hepatocytes From Embryonic Stem Cells In Vitro
    Shigeaki Ishizaka, Yukiteru Ouji, Masahide Yoshikawa, and Kazuki Nakatani

27 Differentiation of Embryonic Stem Cells to Retinal Cells In Vitro
    Xing Zhao, Jianuo Liu, and Iqbal Ahmad

28 Derivation and Characterization of Lentoid Bodies and Retinal Pigment Epithelial Cells From Monkey Embryonic Stem Cells In Vitro
    Masayo Takahashi and Masatoshi Haruta

29 Differentiation of Rhesus Monkey Embryonic Stem Cells in Three-Dimensional Collagen Matrix
    Silvia Sihui Chen, Roberto P. Revoltella, Joshua Zimmerberg, and Leonid Margolis

Contributors

TIMOTHY E. ALLSOPP • *Stem Cell Sciences Ltd., University of Edinburgh, Edinburgh, UK*
TAKAYUKI ASANO • *Division of Regenerative Medicine, Center for Molecular Medicine, Jichi Medical School, Tochigi, Japan*
MICHAEL BADER • *Department of General and Thoracic Surgery, University Hospital of Kiel, Kiel, Germany*
KATIE M. BARROW • *Centre for Molecular Medicine, Faculty of Medical and Human Sciences, The University of Manchester, Manchester, UK*
SANDY BECKER • *Biology Department, Wesleyan University, Middletown, CT*
SONJA BECKER • *Institute of Experimental Genetics, Ingolstaedter Landstr, Neuherberg, Germany*
JOHANNES BECKERS • *Institute of Experimental Genetics, Ingolstaedter Landstr, Neuherberg, Germany*
KENNETH R. BOHELER • *Molecular Cardiology Unit, Laboratory of Cardiovascular Science, NIH, National Institute on Aging, Baltimore, MD*
ROLAND BUESEN • *National Center for Documentation and Evaluation of Alternative Methods to Animal Experiments (ZEBET), Federal Institute for Risk Assessment (BfR), Berlin, Germany*
RACHEL B. CERVANTES • *Department of Cell Biology, Neurobiology and Anatomy, University of Cincinnati, Vontz Center for Molecular Studies, Cincinnati, OH*
LUNG-JI CHANG • *Program in Stem Cell Biology, College of Medicine, Department of Pathology, University of Florida, Gainesville, FL*
SANGMI CHUNG • *Molecular Neurobiology Laboratory, McLean Hospital, Harvard Medical School, Belmont, MA*
FRANK COSTANTINI • *Program in Stem Cell Biology, Department of Pathology, College of Medicine, University of Florida, Gainesville, FL*
CHRISTIAN DANI • *DR2 INSERM Directeur du laboratoire Cellules Souches et Différenciation Centre de Biochimie, Institut de Recherches Signalisation, Biologie du Développement et Cancer, Nice, France*
FRED FÄNDRICH • *Department of General and Thoracic Surgery, University Hospital of Kiel, Kiel, Germany*
SHUICHI FURUSAWA • *Laboratory of Immunobiology, Department of Molecular and Applied Biosciences, Graduate School of Biosphere Science, Hiroshima University, Higashi-Hiroshima, Japan*

LAURA GRABEL • *Biology Department, Wesleyan University, Middletown, CT*
THERESA E. GRATSCH • *Department of Cell and Developmental Biology, University of Michigan Medical School, Ann Arbor, MI*
YUTAKA HANAZONO • *Division of Regenerative Medicine, Center for Molecular Medicine, Jichi Medical School, Tochigi, Japan*
NAKA HATTORI • *Laboratory of Cellular Biochemistry Animal Resource, Sciences/Veterinary Medical Sciences, Graduate School of Agriculture and Life Sciences, The University of Tokyo, Tokyo, Japan*
KIM M. HEMSLEY • *Lysosomal Diseases Research Unit, Department of Genetic Medicine, Women's and Children's Hospital, North Adelaide, Australia*
PAUL J. HERTZOG • *Center for Functional Genomics and Human Disease, Monash Institute of Reproduction and Development, Monash University, Victoria, Australia*
YUNHAN HONG • *Department of Biological Sciences, National University of Singapore, Crescent, Singapore*
YILING HONG • *Department of Cell Biology, Neurobiology and Anatomy, University of Cincinnati College of Medicine, Cincinnati, OH*
JOHN J. HOPWOOD • *Lysosomal Diseases Research Unit, Department of Genetic Medicine, Women's and Children's Hospital, North Adelaide, Australia*
HIROYUKI HORIUCHI • *Laboratory of Immunobiology, Department of Molecular and Applied Biosciences, Graduate School of Biosphere Science, Hiroshima University, Higashi-Hiroshima, Japan*
MARTIN HRABÉ DE ANGELIS • *Institute of Experimental Genetics, Ingolstaedter Landstr, Neuherberg, Germany*
TOMOKO ICHISAKA • *Laboratory of Animal Molecular Technology, Research and Education Center for Genetic Information, Nara Institute of Science and Technology, Nara, Japan*
SHIGEAKI ISHIZAKA • *Program in Tissue Engineering, Department of Parasitology, Nara Medical University, Nara, Japan*
NAE KADOTA • *Discovery Research Laboratory, Tanabe Seiyaku Co. Ltd., Osaka, Japan*
KWANG-SOO KIM • *Molecular Neurobiology Laboratory, Harvard Medical School, Belmont, MA*
YOSHIHIRO KITANO • *Division of General Surgery, National Center for Child Health and Development, Tokyo, Japan*
YASUSHI KONDO • *Discovery Research Laboratory, Tanabe Seiyaku Co. Ltd., Osaka, Japan*
VIKTOR LAKICS • *Eli Lilly and Company Ltd., Erl Wood Manor, Surrey, UK*

Contributors

ADELINE A. LAU • *Lysosomal Diseases Research Unit, Department of Genetic Medicine, Women's and Children's Hospital, North Adelaide, Australia*

SHAOHUA LI • *Department of Pathology and Laboratory Medicine, UMDNJ-Robert Wood Johnson Medical School, Piscataway, NJ*

FILIP LIM • *Center for Molecular Neurobiology, University of Hamburg, Hamburg, Germany*

MASAYOSHI MARUYAMA • *Department of Stem Cell Biology, Institute for Frontier Medical Sciences, Kyoto University, Kyoto, Japan*

HARUO MATSUDA • *Laboratory of Immunobiology, Department of Molecular and Applied Biosciences, Graduate School of Biosphere Science, Hiroshima University, Higashi-Hiroshima, Japan*

PATRICK J. MEE • *Stem Cell Sciences Ltd., University of Edinburgh, Edinburgh, UK*

ADRIAN MEEDENIYA • *Eskitis Institute for Cell and Molecular Therapies, Griffith University, Australia*

ARIKA MINAMIHASHI • *Hokkaido Animal Research, Shintoku, Hokkaido, Japan*

GABRIELLA MINCHIOTTI • *Institute of Genetics and Biophysics Adriano Buzzati-Traverso, CNR, Naples, Italy*

TAKEHIDE MURATA • *Gene Engineering Division, BioResource Centre, RIKEN (The Institute of Physical and Chemical Research), Ibaraki, Japan*

CHRISTOPHER L. MURPHY • *Kennedy Institute of Rheumatology, Imperial College, London, UK*

NORIO NAKATSUJI • *Department of Development and Differentiation, Institute for Frontier Medical Sciences, Kyoto University, Kyoto, Japan*

JENNIFER NICHOLS • *Centre Development in Stem Cell Biology, Institute for Stem Cell Research, University of Edinburgh, Edinburgh, UK*

FUMIHIKO NISHIMURA • *Division of Developmental Biology, Department of Parasitology, Nara Medical University, Nara, Japan*

KOICHIRO NISHINO • *Laboratory of Cellular Biochemistry, Animal Resource Sciences/Veterinary Medical Sciences, Graduate School of Agriculture and Life Sciences, The University of Tokyo, Tokyo, Japan*

CARMEL M. O'BRIEN • *Stem Cell Sciences Ltd., Melbourne, Victoria, Australia*

K. SUE O'SHEA • *Department of Cell and Developmental Biology, University of Michigan Medical School, Ann Arbor, MI*

JUN OHGANE • *Laboratory of Cellular Biochemistry, Animal Resource Sciences/Veterinary Medical Sciences, Graduate School of Agriculture and Life Sciences, The University of Tokyo, Tokyo, Japan*

MASAHIRO OKA • *Program in Stem Cell Biology, Department of Pathology, University of Florida College of Medicine, Gainesville, FL*

SILVIA PARISI • *Institute of Genetics and Biophysics Adriano Buzzati-Traverso, CNR, Naples, Italy*

FLOR M. PEREZ-CAMPO • *Faculty of Medical and Human Sciences, Centre for Molecular Medicine, The University of Manchester, Manchester, UK*

M. GRAZIELLA PERSICO • *Institute of Genetics and Biophysics, Adriano Buzzati-Traverso, CNR, Naples, Italy*

BLAINE PHILLIPS • *DR2 INSERM Cellules Souches et Différenciation, Centre de Biochimie, CNRS Institut de Recherches Signalisation Biologie du Développement et Cancer, Nice, France*

LEO R. QUINLAN • *Physiology Department, National University of Ireland, Galway, Ireland*

DIETER RIETHMACHER • *Center for Molecular Neurobiology, University of Hamburg, Hamburg, Germany*

AARON J. ROBINSON • *Lysosomal Diseases Research Unit, Department of Genetic Medicine, Women's and Children's Hospital, North Adelaide, Australia*

SHIGEO SAITO • *Saito Laboratory of Cell Technology, Kataoku, Yuita, Techigi, Japan*

TOSHISUKE SAKAKI • *Department of Neurosurgery, Nara Medical University, Nara, Japan*

KYOKO SASAKI • *Division of Regenerative Medicine, Center for Molecular Medicine, Jichi Medical School, Tochigi, Japan*

KEN SAWAI • *Hokkaido Animal Research, Shintoku, Hokkaido, Japan*

MANFRED SCHARTL • *Physiological Chemistry, Biocenter of the University of Würzburg, Würzburg, Germany*

THOMAS SCHIMMANG • *Institute for Biology and Molecular Genetics, University of Valladolid, Valladolid, Spain*

MAREN SCHULZE • *Department of General and Thoracic Surgery, University Hospital of Kiel, Kiel, Germany*

ANDREA E. M. SEILER • *National Center for Documentation and Evaluation of Alternative Methods to Animal Experiments (ZEBET), Federal Institute for Risk Assessment (BfR), Berlin, Germany*

HIROAKI SHIBATA • *Division of Regenerative Medicine, Center for Molecular Medicine, Jichi Medical School, Tochigi, Japan*

KUNIO SHIOTA • *Laboratory of Cellular Biochemistry, Animal Resource Sciences/Veterinary Medical Sciences, Graduate School of Agriculture and Life Sciences, The University of Tokyo, Tokyo, Japan*

NICOLE A. SLAWNY • *Department of Cell and Developmental Biology, University of Michigan Medical School, Ann Arbor, MI*

Contributors   xix

HORST SPIELMANN • *National Center for Documentation and Evaluation of Alternative Methods to Animal Experiments (ZEBET), Federal Institute for Risk Assessment (BfR), Berlin, Germany*
PETER J. STAMBROOK • *Department of Cell Biology, Neurobiology, and Anatomy, Vontz Center for Molecular Studies, University of Cincinnati, Cincinnati, OH*
HIROFUMI SUEMORI • *Laboratory of Embryonic Stem Cell Research, Stem Cell Research Center, Institute for Frontier Medical Sciences, Kyoto University, Kyoto, Japan*
MASAKO SUZUKI • *Laboratory of Cellular Biochemistry, Animal Resource Sciences/Veterinary Medical Sciences, Graduate School of Agriculture and Life Sciences, The University of Tokyo, Tokyo, Japan*
YUTAKA SUZUKI • *Research Center for Animal Life Science, Shiga University of Medical Science, Shiga, Japan*
MASAKO TADA • *ReproCELL Inc., Tokyo and Laboratory of Stem Cell Engineering, Institute for Frontier Medical Sciences, Kyoto University, Kyoto, Japan*
TAKASHI TADA • *Laboratory of Stem Cell Engineering, Institute for Frontier Medical Sciences, Kyoto University, Kyoto, Japan*
TATSUYUKI TAKADA • *Discovery Research Laboratory, Tanabe Seiyaku Co., Ltd., Osaka, Japan*
KAZUTOSHI TAKAHASHI • *Laboratory of Animal Molecular Technology Research and Education, Center for Genetic Information, Nara Institute of Science and Technology, Nara, Japan*
SATOSHI TANAKA • *Laboratory of Cellular Biochemistry, Animal Resource Sciences/Veterinary Medical Sciences, Graduate School of Agricultural and Life Sciences, University of Tokyo, Tokyo, Japan*
KIRILL V. TARASOV • *Laboratory of Cardiovascular Science, NIH National Institute on Aging, Baltimore, MD*
NAOHIRO TERADA • *Program in Stem Cell Biology, Department of Pathology, University of Florida College of Medicine, Gainesville, FL*
KEIJI TERAO • *Tsukuba Primate Research Center, National Institute of Biomedical Innovation, Ibayaki, Japan*
HAZEL THOMSON • *Stem Cell Sciences Ltd., University of Edinburgh, Edinburgh, UK*
YOSHIMI TOKUZAWA • *Department of Stem Cell Biology, Institute for Frontier Medical Sciences, Kyoto University, Kyoto, Japan*
RYUZO TORII • *Research Center for Animal Life Science, Shiga University of Medical Science, Shiga, Japan*
HAYATO TORIUMI • *Division of Developmental Biology, Department of Parasitology, Nara Medical University, Nara, Japan*

KURSAD TURKSEN • *Development Program, Ottawa Health Research Institute, Ottawa, Ontario, Canada*
HIDEYO UGAI • *Gene Engineering Division, BioResource, RIKEN (The Institute of Physical and Chemical Research), Ibaraki, Japan*
HENDRIK UNGEFROREN • *Department of General and Thoracic Surgery, University Hospital of Kiel, Kiel, Germany*
SJAAK VAN DER SAR • *Stem Cell Sciences Ltd., University of Edinburgh, Edinburgh, UK*
J. MATTHEW VELKEY • *Department of Cell and Developmental Biology, University of Michigan Medical School, Ann Arbor, MI*
CÉCILE VERNOCHET • *Cellules Souches et Différenciation, Centre de Biochimie, Institut de Recherches Signalisation Biologie du Développement et Cancer, Nice, France*
PHI VILLAGEOIS • *Cellules Souches et Différenciation, Centre de Biochimie, Institut de Recherches Signalisation Biologie du Développement et Cancer, Nice, France*
ANKE VISAN • *National Center for Documentation and Evaluation of Alternative Methods to Animal Experiments (ZEBET), Federal Institute for Risk Assessment (BfR), Berlin, Germany*
CHRISTOPHER M. WARD • *Faculty of Medical and Human Sciences, Centre for Molecular Medicine, The University of Manchester, UK*
BRIGITTE WDZIEKONSKI • *Cellules Souches et Différenciation, Centre de Biochimie, Institut de Recherches Signalisation Biologie du Développement et Cancer, Nice, France*
TREVOR J. WILSON • *Center for Functional Genomics and Human Disease, Monash Institute of Reproduction and Development, Monash University, Victoria, Australia*
DAKANG XU • *Center for Functional Genomics and Human Disease, Monash Institute of Reproduction and Development, Monash University, Victoria, Australia*
SHINYA YAMANAKA • *Department of Stem Cell Biology, Institute for Frontier Medical Sciences, Kyoto University, Kyoto, Japan*
QI-LONG YING • *Centre Development in Stem Cell Biology, Institute for Stem Cell Research, University of Edinburgh, Edinburgh, UK*
KAZUNARI K. YOKOYAMA • *Gene Engineering Division, BioResource, RIKEN (The Institute of Physical and Chemical Research), Ibaraki, Japan*
MASAHIDE YOSHIKAWA • *Division of Developmental Biology, Department of Parasitology, Nara Medical University, Nara, Japan*
PETER D. YURCHENCO • *Department of Pathology and Laboratory Medicine, UMDNJ-Robert Wood Johnson Medical School, Piscataway, NJ*

COMPANION CD

for *Embryonic Stem Cell Protocols, Volume 1, SECOND EDITION*

All of the electronic versions of illustrations in this book may be found on the Companion CD attached to the inside back cover. The image files are organized into folders by chapter number and are viewable in most Web browsers. The number following "f" at the end of the file name identifies the corresponding figure in the text. The CD is compatible with both Mac and PC operating systems.

COLOR FIGURES

CHAPTER 1   FIG. 1   CHAPTER 18 FIGS. 2, 3, 4

CHAPTER 4   FIGS. 1–5   CHAPTER 19 FIGS. 1, 2

CHAPTER 5   FIGS. 1–4   CHAPTER 24 FIG. 2

CHAPTER 6   FIG. 2   CHAPTER 25 FIG. 2

CHAPTER 9   FIG. 1   CHAPTER 26 FIGS. 2, 6

CHAPTER 12 FIGS. 1–4   CHAPTER 29 FIGS. 1, 2, 4

CHAPTER 15 FIGS. 1–5   CHAPTER 33 FIGS. 1, 2

CHAPTER 16 FIG. 5

I

ISOLATION AND MAINTENANCE

1

Isolation and Differentiation of Medaka Embryonic Stem Cells

Yunhan Hong and Manfred Schartl

Summary

Medaka is a small laboratory fish that daily produces eggs easily controllable by light cycles. This fish represents a unique lower vertebrate compared to mammals, in which embryonic stem (ES) cell lines can be derived from midblastula embryos (MBEs). Like mouse ES cells, medaka ES cells most resemble the totipotent embryonic cells at the blastula stage. Medaka ES cells retain a diploid karyotype, pluripotency in vitro, and chimera competence in vivo. They give rise to high efficiencies of transient and stable gene transfer and maintain their pluripotency after long-term drug selection for transgene integration. They can also be directed to differentiate into particular cell types. Medaka is the most distantly related vertebrate to mammals, and its ES cell lines provide an ideal reference to mammalian ES cells for the molecular analysis of stemness. More important, medaka ES cell lines on their own offer an excellent tool for studying stem cell biology in vitro and in vivo because production and observation of ES-derived chimeras as well as phenotypic analyses are very easy because of its external, transparent, and temperature-adjustable embryology.

Key Words: Chimera formation; ES cells; medaka.

1. Introduction

Embryonic stem (ES) cells are pluripotent cell cultures from early developing embryos and represent a powerful system in many fields of biomedical research. The medaka is a complementary system to the zebrafish, with which it shares many characteristics, but it also some peculiarities. Medaka is a small teleost fish well suited for the development and exploitation of the ES cell technology because of its many advantageous features *(1)*.

It is easy to maintain under laboratory conditions. It has a small genome of less than 800 Mb, which is only a quarter the size of the human genome and less than half that of zebrafish. The short generation time is only 10 wk. Medaka embryos develop externally and are transparent, making them easily accessible for manipulation and imaging. Many inbred strains and stocks derived from a plethora of wild fish collected over the

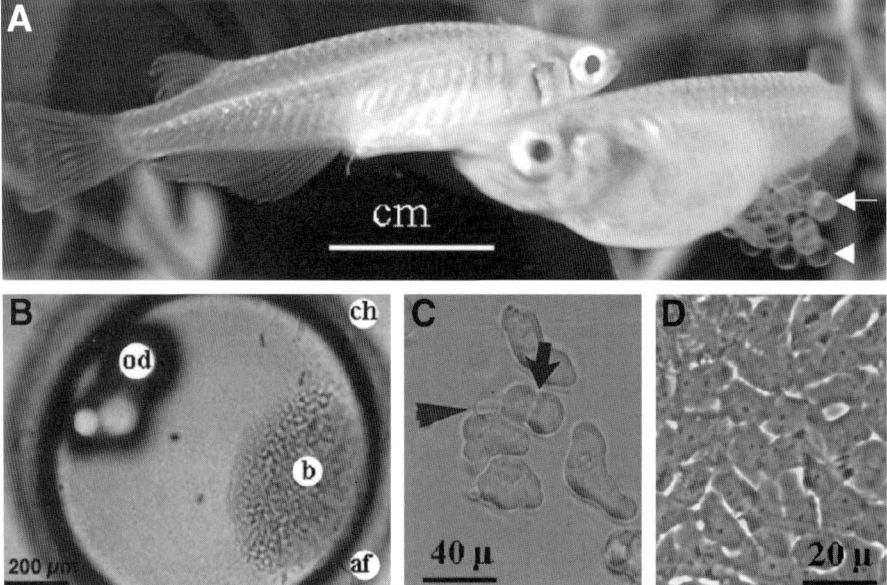

Fig. 1. Derivation of medaka embryonic stem (ES) cells. (**A**) Adult male (left) and female (right) medaka with eggs hanging in the belly. Live eggs are transparent (arrowhead), easily distinguishable from dead embryos (arrow). (**B**) MBE showing the blastoderm (b) of 1000 cells opposite to oil droplets (od), surrounded by the chorion (ch) with remaining attachment filaments (af). (**C**) Freshly seeded MBE cells are 30–40 µm in diameter, active in division (arrow), and have amoebalike movement by forming pseudopodia (arrowhead). (**D**) ES cells at 90% confluence ready for subculture. Large nuclei and prominent nucleoli are evident.

whole natural range of medaka are available. They provide an extremely rich genetic diversity for ES cell derivation and phenotypic analysis. A single female medaka produces 20–40 eggs every day (**Fig. 1A**). Embryonic development can be slowed and even stopped at lower temperatures without affecting normal development. This allows the manipulation of a large number of embryos every day and detailed observation of the highly dynamic processes during early embryonic development.

In medaka, Wakamatsu et al. *(2)* adopted the feeder layer technique and reported the first medaka ES-like cell line OLES1. We developed a feeder-free culture system, in which blastula-derived stem cells are cultured on gelatin-coated substrata *(3)*. One of several medaka ES cell lines, MES1, has been characterized as pluripotent in vitro, including stable growth, normal karyotype, and the ability to differentiate into functional cell types of the three germ layers *(4)*. MES1 cells, following transplantation to the host blastulae, participate in chimeric development *(5)*. This cell line displays pluripotency-specific gene expression, as shown by the ability to activate the mouse Oct4 regulatory sequence *(6,7)*. This feeder-free protocol is also applicable even in marine fish species such as gilthead seabream *(8)*, red seabream *(9)*, and sea perch *(10)*.

In mouse, ES cell lines are able to colonize the host germline and thus are used as a cellular vehicle for germline transmission. Medaka ES lines are capable of contribution to all somatic cell lineages and organs of chimeric embryos, although germline transmission has not been achieved. Germline contribution may depend on the use of proper donor/recipient combinations and optimized protocols for chimera production. For gene targeting, drug selection procedures have been developed that can be used to enrich for rare homologous recombination (HR) events. Vector cassettes and the appropriate selection schemes for HR with medaka ES cells have been developed *(11)*.

ES cells offer an ideal system for studying stem cell self-renewal and differentiation and have promise for generating specific cell types in larger quantities for cell replacement therapy. Differentiation of ES cells in vitro faces major challenges: the difficulty to establish complicated culture conditions that, in mice, usually involve a step of embryoid body formation; the combinatorial use of growth factors; and the heterogeneity of induced differentiation in terms of cell states and types. Often, a uniform population of cells differentiating along a particular cell lineage is difficult to obtain. Instead, the cell population is a mixture of undifferentiated ES cells and of cells differentiating along various lineages. Uncontrolled proliferation of undifferentiated ES cells leads to teratoma formation following transplantation into a host *(12)*. Using medaka ES cells, we have established a system of directed differentiation in which the melanocyte-specific microphthalmia-associated transcription factor (mitf) was used as a master regulator to direct differentiation predominantly into pigment cells *(13)*. This chapter describes protocols for the derivation, maintenance, transfection, drug selection, directed differentiation, and transplantation of medaka ES cells.

2. Materials

2.1. Equipment

1. Borosilicate glass capillaries, 1-mm diameter (World Precision Instrument, Sarasota, FL; cat. no. MTW100-4).
2. Centrifuge (MIRKRO 22R, Hettich, Tullingen, Germany; Centrifuge 5415, Eppendorf, Hamburg, Germany).
3. Flaming/brown micropipet puller (Sutter Instrument, Novato, CA; model P-87).
4. Forceps (World Precision Instrument, Dumont 5 and Dumont 500085).
5. Homogenizer (Ultra-Turrax T25, Jank and Kunkel, Staufen, Germany).
6. Microinjector, self-built.
7. Micromanipulator (Leica, Wetzlar, Germany; cat. no. 11520137).
8. Microscopes: Zeiss (Thornwood, NY) Stemi 2000-C and Leica (Heerbrugg, Switzerland) MZFLIII stereomicroscope equipped with a Fluo III ultraviolet light system, filters for green fluorescent protein (GFP) and red fluorescent protein (RFP) and a Nikon (Tokyo, Japan) E4500 digital camera for photography; Zeiss Axiovert inverted microscope and Zeiss Axioskop 2 plus upright microscope, both equipped with optics for phase contrast, DIC, and fluorescent microscopy; and a Zeiss AxioCam MRc digital camera with AxioVision 4 software.
9. Tissue culture plasticware.
10. Corex tubes.
11. Gel loading pipet tip (Scientific, Boca Raton, FL; cat. no. 1022-0000).

2.2. Reagents and Solutions

1. 1/15 M phosphate buffer, pH 6.8: for 1000 mL, dissolve 9.46 g Na_2HPO_4 and 5.60 g NaH_2PO_4 to 900 mL H_2O, bring final volume to 1000 mL with water.
2. 2X freezing medium: ESM4 that contains 25% fetal bovine serum and 20% dimethylsulfoxide as cryoprotectant.
3. 4% paraformaldehyde: for 100 mL, add 4 g paraformaldehyde to 60 mL H_2O at 60°C in a fume hood and stir. Add 60 µL 1 N NaOH and stir until the solution is clear; remove from heat. Add 10 mL 10X phosphate-buffered saline (PBS) and bring final volume to 100 mL with H_2O. Mix well and pour into 50-mL Falcon tubes; cool on ice for at least 20 min and store at 4°C in darkness.
4. Anticoagulation solution (15% ethylenediaminetetraacetic acid [EDTA], 0.9% NaCl, 1.6% NaOH). For 100 mL, dissolve 15.0 g EDTA, 0.9 g NaCl, and 1.6 g NaOH into 80 mL H_2O and bring final volume to 100 mL with water.
5. BCIP/NBT substrate solution for alkaline phosphatase (AP) staining: freshly made by adding 66 µL of NBT stock (Sigma, St. Louis, MO; cat. no. N-6876; 0.5 g/10 mL of 70% dimethylformamide; store at 4°C in darkness) and 33 µL of BCIP stock (Sigma, cat. no. B-8503; 0.5 g/10 mL of 100% dimethyformamide; store at 4°C in darkness) in 10 mL of AP buffer (100 mM NaCl, 5 mM $MgCl_2$, 100 mM Tris-HCl, pH 9.5) to the solution and then bring final volume to 1000 mL with water.
6. BSS (balanced salt solution): for 500 mL, add 25 mL of solution A and 1 mL of solution B in water and bring final volume to 500 mL. Solution A (for 500 mL): 65 g NaCl, 4 g KCl, 2 g $MgSO_4·7H_2O$, 2 g $CaCl_2·H_2O$, 5 mg phenol red; autoclave. Solution B: 5% $NaHCO_3$; sterilize by filtration through 0.2-µm filter.
7. BSS-1% polyethylene glycol (PEG): BSS that contains 1% PEG 2000, 100 µg/mL streptomycin, 100 U/mL penicillin. Sterilize by filtration through 0.2-µm filter.
8. Colchicine solution (10 ng/mL; Sigma, cat. no. C9754).
9. EM (embryo medium; 0.1% NaCl, 0.003% KCl, 0.004% $CaCl_2$-H_2O, 0.016% $MgSO_4$-$7H_2O$, 0.00001% methylene blue). We make and autoclave a 100X stock without methylene blue. For a working solution, dilute 10 mL of the stock to 1000 mL of autoclaved water and add 1 mL 0.01% methylene blue.
10. Enzol enzymatic detergent (World Precision Instrument, cat. no. 2254).
11. ESM4 is comprised of four components: (1) Pure medium: Dulbecco's modified Eagle's medium (DMEM), 4.5 g/L high glucose (cat. no. 12100); 20 mM HEPES, pH 7.7, filtrate through 0.2-µm filter; (2) antibiotics: 100X stock penicillin-streptomycin, (cat. no. 15070-063); (3) nonprotein supplements: 100X stock L-glutamine (cat. no. 25030-081), 100X stock nonessential amino acids (cat. no. 11140-050), 100X stock Na-pyruvate (cat. no. 11360-070), Na-selenite (Sigma, cat. no. S-9133), self-made 2 µM 1000X stock, 2-mercaptoethanol (Sigma, cat. no. M7522), self-made 50 mM 500X stock; (3) protein supplements: fetal bovine serum (cat. no. 1131776), final 15%; basic fibroblast growth factor (PeproTech, Roch Hill, NJ; cat. no. 100-18B), final 10 ng/mL, fish serum, self-made from rainbow trout, final 1%; fish embryo extract (FEE), self-made from medaka, final 4 U/mL. Reagents are from Invitrogen Life Technologies (Carlsbad, CA) unless otherwise indicated.
12. Fugene reagent (10%; Roche, Basel, Switzerland; cat. no. 1814443).
13. G418, 50 mg/mL stock Geneticin solution (Sigma, cat. no. G7034).
14. Gelatin, type A, porcine skin, 300 Bloom (Sigma, cat. no. G2500). Working solution: 0.1% in water; autoclave.
15. GeneJuice Transfection Reagent (Novagen, San Diego, CA, cat. no. 70967-3).

16. Hygromycin B (Sigma, cat. no. H3274).
17. Light mineral oil (Sigma, cat. no. M3516).
18. pEGFP-N1 (Clontech, cat. no. 632318).
19. pHygEGFP (Clontech, cat. no. 632305).
20. PBS: 108 mM NaCl, 252 mM Na$_2$HPO$_4$, 22.5 mM KH$_2$PO$_4$, pH 7.3; diluted from 10X solution. For 1000 mL 10X solution: dissolve 63.1 g NaCl, 35.77 g Na$_2$HPO$_4$, 18.9 g KH$_2$PO$_4$ to 900 mL H$_2$O; adjust pH to 7.3 by adding 1 N HCl; bring final volume to 1000 mL with water; autoclave.
21. PEG 2000 (Merck, Darmstadt, Germany; cat. no. S35263-210).
22. Proteinase K (Sigma, cat. no. P6556): dilute to 10 mg/mL in water, aliquot, and store at $-20°C$.
23. Puromycin (Sigma, cat. no. P8833). Stock solution: 1 mg/mL in water.
24. Transplantation medium (TM): 100 mM NaCl, 5 mM KCl, 5 mM HEPES (pH 7.1), 0.1% phenol red as a tracing dye. Sterilize by filtration through 0.2-μm filter.
25. Transfection solution A: for 50 μL, gently mix 46 μL of DMEM medium (pure medium; *see* **Subheading 2.2.1.**) and 4 μL of GeneJuice Transfection Reagent.
26. Transfection solution B: for 50 μL, gently mix 49 μL of DMEM medium and 1 μL of plasmid DNA (1 μg/μL).
27. Trypsin/EDTA (diluted from 10X solution; Invitrogen Life Technologies; cat. no. 15400-054). For a working solution (0.05% trypsin/0.53 mM EDTA), dilute 10 mL 10X stock with 90 mL PBS.
28. Yamamoto Ringer's solution (diluted from 10X solution): for 10X solution, dissolve 75 g NaCl, 2 g KCl, 2 g CaCl$_2$, 0.2 g NaHCO$_3$ in 900 mL H$_2$O; adjust pH to 7.3; bring final volume to 1000 mL with water; autoclave.

2.3. Media

Mammalian ES cells are usually maintained in the presence of leukemia inhibitory factor (LIF) or on a layer of mitotically inactivated feeder cells. Medaka ES cells can be initiated and maintained under feeder-free conditions on gelatin-coated substrates. The culture media used for medaka ES cells are prepared from a powder of DMEM (4.5 g/L high glucose). We use ESM3 for primary cultures and early passages (up to 120 d). ESM4 is for ES cell maintenance. ESM3 is the same as ESM4 but contains FEE at 10 U/mL. ESM3 and ESM4 are derived from ESM1 and ESM2 by omitting mammalian LIF as it has no effect on medaka ES cells *(2)*. The basic medium is dispensed into glass bottles and stored for up to 1 yr at 4°C in darkness. The complete media ESM3 and ESM4 should be used within 6 mo. Cells are cultured under humidified air without CO$_2$, usually at 28°C.

3. Methods

3.1. Preparation of FEE From Medaka

1. Collect embryos 3–6 h after spawning (*see* **Note 1**). Mechanically remove the attachment filaments by putting clusters of eggs in a fine net and rolling them in the net with your fingers so that eggs slide against the net wall.
2. Incubate eggs in a Petri dish containing EM at 26°C for 7 d.
3. Remove dead (blue) and abnormal embryos and change EM every day.
4. Wash 7-d-old embryos with PBS, transfer into a 2-mL tube, remove PBS, and store at –20°C until more than 5000 embryos have been collected.
5. Thaw and combine the stored embryos in a 50-mL Falcon tube. Measure the volume and estimate the number of embryos: 400 embryos/mL.

6. Remove PBS as much as possible. Homogenize in an ice-water bath.
7. Freeze/thaw three times alternatively in liquid nitrogen and a 37°C water bath for 5 min each.
8. Spin at 4°C for 30 min at 3200g. There should be three phases: the upper lipid layer, the middle green supernatant, and the black bottom debris. Pipet the green supernatant into Corex tubes.
9. Spin at 4°C for 60 min at 15,000g.
10. Collect and filtrate the supernatant through a 0.2-µm filter. Add PBS to adjust volume to 2.5 µL/embryo.
11. Make 1-mL aliquots in 1.5-mL tubes and store at −20°C (*see* **Notes 2** and **3**).

3.2. Preparation of Fish Serum

Fish serum is mitogenic for medaka ES cells in culture. We have successfully used sera from the rainbow trout (*Oncorhynchus mykiss*), common carp (*Cyprinus carpio*), and seabass (*Lates calcarifer*).

1. Soak a 10-mL syringe with anticoagulation solution.
2. Shock fish by a blow on the head. Collect blood from the tail vein into 50-mL Falcon tubes on ice.
3. Spin blood at 4°C for 30 min at 3200g and transfer the upper clear supernatant to 50-mL tubes.
4. Incubate overnight at 4°C. Spin again.
5. Sterilize the serum by filtration through a 0.2-µm filter and store 5-mL aliquots in 15-mL Falcon tubes at −20°C.
6. Shortly before use, spin at 3200g for 30 min and add the supernatant to media.

3.3. Derivation of ES Cell Cultures

Cell cultures can only be initiated from permissive medaka strains such as HB32C *(2,11)*. The blastoderm, the cell mass of an MBE, consists of 1000 cells opposite to the oil droplets (**Fig. 1B**). Freshly seeded MBEs are 30–40 µm in size, irregular in shape, and active in amoebalike movement by forming pseudopodia (**Fig. 1C**). They divide quickly and begin to attach 12–18 h after seeding at 28°C. An even distribution and a proper cell density are important to prevent differentiation into neural and fibroblastlike cells. A too high cell density favors the appearance of muscle cells. A density that results in a subconfluent (90%) (**Fig. 1D**) monolayer at d 3–4 of culture is appropriate. Appearance of pigment cells at d 3 of culture from some strains/lines seems to be independent of cell density. Cells in primary and early passage (30 d) culture are sensitive to drying out. We use partial medium change during this stage to remove any residual debris, dead cells, and oil droplets.

1. Collect eggs at 2 h postfertilization (hpf) in a 6-cm Petri dish with EM.
2. Roll the eggs with fingers in a net to remove filaments.
3. Wash and transfer fertilized eggs to a new dish with EM. Fertilized eggs are easily recognized because they are transparent (**Fig. 1A**).
4. Incubate at 26°C until 6 hpf (*see* **Note 4**).
5. Under sterile conditions, transfer embryos to a six-well plate containing 12 mL of 0.5% bleach in PBS and incubate for 1 min.

Isolation and Differentiation of Medaka ES Cells

6. Wash embryos four or five times by transferring between wells containing PBS.
7. Transfer 20–30 MBEs (at 6.5 hpf) in a drop (300–400 µL) of PBS using a wide-mouth plastic Pasteur pipet to the middle of a cover of a 6-cm dish.
8. Under a dissecting microscope (from this step to **step 14**), poke the chorion through the oil droplet (opposite the cell mass) with fine forceps without damaging the cell mass of the blastoderm. Let stand for 5 min.
9. While the embryos shrink and release yolk material, the cell mass moves to the hole. Replace half of the volume of the drop three to five times with fresh PBS to remove debris, oil droplets, and yolk using a fine tip.
10. Quickly press the chorion (opposite the hole) using fine forceps to release cells through the hole. Discard chorions.
11. Make a cell suspension by pipetting using the fine pipet tip; swirl the suspension by pipetting in one direction at the marginal region of the drop. Let stand for 5 min for the cells to settle down in the middle of the drop.
12. Aspirate debris and yolk membrane fragments in 200–300 µL of PBS using the fine pipet. Add 200–300 µL of PBS by blowing along the peripheral margin, so that cells move to the center (*see* **Note 5**).
13. Repeat three to five times until the majority of cells are collected at the center, and no fragments or debris are visible.
14. After the final washing step, collect the cells with 100 µL of PBS and seed them evenly in 1 mL of ESM3 medium in a gelatin-coated 24-well plate (*see* **Subheading 3.3.1.**). Culture the cells at 28°C without CO_2 (*see* **Note 6**).
15. On the following day, the majority of cells should be attached. Aspirate approx 800 µL of the medium from the surface to remove oil droplets and add 800 µL fresh medium. Monitor cell growth and morphology daily under an inverted microscope. Culture for 2–4 d until a monolayer is formed.
16. For subculture, add PBS continuously and gently along the wall while aspirating the medium. This prevents the cells from detaching and drying. In the same way, add trypsin/EDTA to replace PBS. Incubate in 200 µl trypsin/EDTA at room temperature for 2 min. Add 200 µL ESM3 to stop trypsinization. Make single-cell suspension by pipetting up and down using a yellow tip. Add 0.6 mL ESM3 and culture for 1 d. Partially change the medium daily as in **step 15** during first 30 d.
17. After 2–4 d of culture, a monolayer is formed. Trypsinize and seed the cells into two wells of the same 24-well plate.
18. Subculture at confluence and expand to 12- and 6-well plates and eventually to a 10-cm dish. During the initial 30 d, we subculture at high density to prevent differentiation into neural and fibroblastlike cells CO_2 (*see* **Note 7**).
19. After 20 d of culture, the majority of cells are propagated as described in **Subheading 3.3., step 16**, for expansion or storage in liquid nitrogen; an aliquot is examined for the ES cell phenotype, including AP staining, chromosome number, and differentiation potential (*see* **Subheading 3.7.**).
20. After 120 d of culture, cells are most probably established and usually maintained in ESM4.

3.3.1. Gelatin Coating

1. Under sterile conditions, add 0.1% gelatin solution to cover multiwell plates and 10-cm dishes; incubate at room temperature for not less than 30 min.
2. Aspirate gelatin solution and allow plates to air-dry for not less than 2 h. Gelatin-coated plates/dishes can be stored at room temperature for up to 6 mo.

3.4. Subculture

Medaka ES cells can be maintained at high density without overt differentiation. Just before forming a monolayer, they are subcultured (generally every 3–7 d) at a splitting ratio of 1:3 to 1:6. At low density, enhanced differentiation (mostly into neural cells and fibroblasts) takes place. However, when they reach a high density, morphologically undifferentiated ES cells predominate, as differentiated derivatives stop division and disappear (see **Note 8**).

1. Wash exponentially growing cells in a six-well plate with 2 mL PBS.
2. Aspirate PBS and add 1 mL 1X trypsin-EDTA. Trypsinize at room temperature for 5 min.
3. Aspirate trypsin-EDTA and add 1 mL ESM4. Make a single-cell suspension by pipetting up and down three to five times.
4. Transfer single cells into three to six wells of a gelatin-coated six-well plate or a 10-cm tissue culture dish.

3.5. Freezing Cells

1. Trypsinize exponentially growing cells in a 10-cm dish and make single cells in 3 mL ESM4 medium as described for subculture in **Subheading 3.4.**
2. Transfer the 3 mL of cell suspension into a 15-mL tube containing 3 mL of 2X freezing medium precooled on ice.
3. Incubate the cell suspension on ice for 5–10 min. Meanwhile, label 2-mL cryovials with the name of cell line, passage number, and days of culture.
4. Aliquot 1.5 mL of the cell suspension into cryovials.
5. Transfer the cryovials into a styrofoam box. Immediately close and put the box into a −80°C freezer for 1–5 d.
6. Transfer the cryovials to liquid nitrogen.

3.6. Thawing Cells (see Note 9)

1. Warm a cryovial in a 37°C water bath until ice is completely thawed.
2. Spin at 3500g for 2 min.
3. Aspirate the freezing medium and resuspend the cell pellet in 0.5 mL of fresh ESM4 medium.
4. Transfer the resuspended cells into six-well plates containing 2 mL ESM3.

3.7. Characterization of Medaka ES Cells

Medaka ES cells are polygonal at low density and oval with a diameter of 12–15 μm at high density. They have sparse cytoplasm and large nuclei with prominent nucleoli. They divide rapidly, with a population doubling time of 48 h in ESM2 *(4)*. The pluripotency-specific mouse Oct4 promoter is active in medaka ES cells *(6)*. In mice, undifferentiated ES cells show AP activity, and euploidy is a prerequisite for germline transmission. We use both these criteria to monitor the undifferentiated state of medaka ES cells. Growth and the ES cell phenotype are monitored daily; AP staining and chromosome analysis are performed every 5–10 passages.

3.7.1. AP Staining and Chromosome Analysis (see Note 10)

1. Wash cells gently in PBS in culture plates or dishes.
2. Fix for 10 min at room temperature in methanol:acetone (1:1).
3. Wash the cells twice with PBS.

Isolation and Differentiation of Medaka ES Cells

4. Stain with BCIP/NPT by adding fresh BCIP/NPT substrate solution to cover the cells.
5. Develop in darkness at 28°C for 1–24 h. Positive cells are stained red to purple.

3.7.2. Chromosome Preparation From Medaka ES Cells

1. Culture actively growing ES cells in 12- or 6-well plates in the presence of colchicine for 2–4 h at 28°C to accumulate metaphases.
2. Remove the medium, wash with PBS, add trypsin, and incubate for 3–5 min.
3. Aspirate trypsin, add 1 mL ESM4, make cell suspensions by pipetting, and transfer the cells into a 1.5-mL tube.
4. Spin for 5 min at 3500g. Aspirate the medium, wash once with PBS, spin, and resuspend the cell pellet in 0.1 mL PBS.
5. Add 1 mL 40 mM KCl for hypotonic treatment, mix by gentle inversion, and incubate at room temperature for 30 min (*see* **Note 11**).
6. Add 0.2 mL fresh fixative (3:1 methanol:acetic acid), mix by inversion, and spin.
7. Aspirate the supernatant, leaving approx 0.3 mL above the cells, resuspend the cells by blowing air bubbles through the pellet and add 1 mL of fixative drop by drop. Repeat at least twice. After the last spin, resuspend the cells in 0.2–0.5 mL fixative.
8. Several slides can be made from each sample (10–50 µL; per slide). Drop the cell suspension onto a wet (immersed in water), cold glass slide at a distance of 5 cm. Air-dry at room temperature. Monitor the first slide with a ×20 objective for cell density under phase contrast optics. If necessary, dilute the suspension or pellet the cells and resuspend them in a smaller volume of fixative. Prepare slides and air-dry (*see* **Note 12**).
9. Stain in 5% Giemsa for 10 min at room temperature in 1/15 M phosphate buffer (pH 6.8). Wash in water; air-dry at room temperature.
10. Examine under a microscope. Nuclei and chromosomes stain dark red; the cytoplasm stains pale blue. Scan the slides with a ×10 or ×20 objective for well-spread metaphases. Count the chromosomes using a ×100 oil immersion objective (*see* **Note 13**).

3.8. Chimera Formation

Chimera formation provides a system to examine the in vivo pluripotency of putative ES cells. It is also the ultimate step to bring the ES cells after gene targeting into the germline for transmission into the next generation. Wakamatsu et al. *(14)* reported the production of medaka germline chimeras from noncultivated blastula cells. We obtained chimeras from medaka ES cell lines *(5,11)*. The frequency of chimera formation from medaka ES cell lines is 6% if pigmentation is monitored *(5)* and 100% by fluorescent microscopy of GFP-labeled donor cells *(11)*. Chimera formation in medaka strongly depends on the genetic compatibility between donor and host strains and on the physiological compatibility between donor ES cells and host embryos *(15)*.

The medaka embryo has a tough chorion that consists of a thin outer layer and a thick inner layer. The outer layer is selectively digested by proteinase K; the inner layer is specifically digested by medaka hatching enzymes (HEs). Dechorionated embryos are fragile but can be manipulated in BSS-1% PEG. On average, up to 100 cultured MBE cells are transplantable to each recipient embryo at the midblastula stage without affecting normal development (*see* **Note 14**).

1. Separate spawning females from males 1 d before the experiment. On the day of experiment, mix both sexes. Spawning takes place within 30 min. Collect egg clusters 30 min

after egg deposition. Roll them gently but thoroughly with fingers in a net to remove attachment filaments and hairs.
2. Sort out transparent, healthy eggs. Wash in autoclaved water in a 6-cm Petri dish.
3. Treat with proteinase K (5 μg/mL in water) for 2 h to partially digest the outer layer of the chorion.
4. Wash in autoclaved water in a 6-cm Petri dish.
5. Transfer cleavage embryos to a drop of water on a cover of a 6-cm dish in such a way that embryos are arranged in a drop no larger than 1 cm in diameter. Remove excessive water so that all embryos are just immersed but never exposed to air.
6. Add HE prepared from medaka embryos at hatching (*see* **Subheading 3.7.1.**) to the drop to cover all embryos (total volume 300–400 μL).
7. Arrange embryos in a single layer. Be sure that all embryos are submerged.
8. Close the cover to prevent water loss and incubate for 1–3 h under a stereomicroscope for regular observation. While dechorionation is in progress, prepare the transplantation system and donor ES cells (**steps 11–14**).
9. When approx 50% of embryos are completely free of chorion and approx 40% have fragmented chorion (after incubation for 2 h), stop digestion by gently pipetting BSS-1%PEG to the drop.
10. Remove chorion fragments manually using a pair of fine forceps.
11. Transfer dechorionated embryos with a wide-mouth plastic Pasteur pipet into a 6-cm dish with 12 mL BSS-1%PEG. Put the dish at 8–12°C to slow the developmental speed (*see* **Note 15**).
12. Prepare microinjection agarose plates in Yamamoto Ringer's solution in a 6-cm dish (*see* **Subheading 3.7.2.**) that has a V-shaped ramp. Immerse in BSS-1%PEG precooled at 8–12°C. Arrange embryos in a single row along the V-shaped ramp.
13. Wash the microinjector thoroughly with ethanol and rinse with water (*see* **Note 16**). Fill the air-dried device with light mineral oil.
14. Pull transplantation needles from 1-mm borosilicate glass capillaries using a flaming/brown micropipet puller. Clip the needles using a fine forceps to form an opening of 5–10 μm in inner diameter at the tip. Immerse the needle inversely (tips up) in a 2-mL tube filled with TM for loading by capillary action. Connect the TM-loaded needle to the microinjector mounted on a Leica micromanipulator. Spot three drops of TM of 10 μL each on a cover of a 2-cm dish, with a distance of 0.5 cm between drops. While monitoring under a stereoscope, move the tip into one drop and adjust the tip opening to a slope with diameter of 20–25 μm (approximately double the size of a donor cell) using a fine and clean forceps. Move the tip into a new drop and press out air and excessive TM in the needle until the TM-oil interphase is 5 mm distant from the tip.
15. Prepare single donor cell suspension by trypsinizing cells at the log phase of growth. Wash twice, resuspend donor cells in TM (20 μL for cells from a 12-well plate), put on ice, and use within 6 h.
16. Pipet 2 μL donor cell suspension into a drop at the middle of the three TM-drops on the 2-cm dish cover (**step 13**). Move the tip into the cell drop and collect the cell into the needle by gentle negative pressure.
17. Carefully position the embryos (**step 11**) using an orientation needle (*see* **Note 17**), so that the milky blastoderm faces the needle tip at an angle of approx 40°.
18. Insert needle into the blastoderm and quickly inject 50–100 cells into the deep cells of each midblastula recipient. Withdraw the tip quickly to prevent cells from escaping. Transfer the operated embryos into a new dish containing BSS-1%PEG and incubate at 18°C for 1 d and then at 25–28°C until hatching (*see* **Notes 18** and **19**).

3.8.1. Preparation of HE

Large-scale preparation of HE is similar to that for FEE (*see* **Subheading 3.1.1.**). Embryos at hatching (at 10 d post fertilization [dpf]) are used. Aliquots (200 µL) are stored in 0.2-mL tubes at −20°C.

1. Transfer embryos into a 5- to 10-mL glass tissue homogenizer.
2. Homogenize in ice water bath.
3. Transfer the homogenate into 1.5-mL tubes and spin at 4°C for 30 min at 20,000g.
4. Collect, filtrate through a 0.2-µm filter, and aliquot the clear supernatant for storage at −20°C.

3.8.2. Preparation of Microinjection Agarose Plates

1. Boil 1.5% agarose in Yamamoto Ringer's solution in a microwave.
2. Pour 5 mL into a 6-cm Petri dish and place a glass microscope slide onto the agarose at a 20° angle. Leave for more than 10 min for gelling.
3. Remove the slide. Multiple plates are made for direct use or storage in Ringer's solution at 4°C.

3.9. Transfection and Drug Selection

Mouse ES cells are widely used for gene targeting. We have established vector systems and drug selection conditions for gene targeting in medaka ES cells. Several protocols are effective for gene transfer into medaka ES cells. These include the Fugene reagent (10% transfection efficiency), the GeneJuice Transfection Reagent (20%), and electroporation (up to 40%) *(11)*. We usually use the GeneJuice for transient and stable transfection.

We use pSTneo *(11)* or pHygEGFP in combination with G418 or hygromycin for positive drug selection in stable gene transfer and pSTk to express the herpes simplex virus thymidine kinase for negative selection with ganciclovir. The positive–negative selection procedure widely used for gene targeting in mouse ES cells works well in medaka ES cells *(11)*. We use high concentrations of drugs (500 µg/mL G418 or hygromycin) to select for and low concentrations (200 µg/mL either drug) to propagate stable resistant cell clones. G418 and hygromycin selection does not compromise the pluripotency of medaka ES cells *(11)*.

1. Seed 1 mL of cell suspension in ESM4 (10^5 cells/mL) per well of a gelatin-coated 12-well plate 1 d before transfection so that the cell density will be roughly 70% confluence at the day of transfection.
2. Aspirate ESM4, wash with PBS, and add 1 mL of pure medium while the transfection mix is incubating.
3. Combine transfection solutions A and B, mix, and incubate at room temperature for 5 min.
4. Add the mix dropwise to the cells. Incubate at 28°C for 6 h.
5. Aspirate the pure medium and add 1 mL ESM4. On the following day, 10–40% of cells display green fluorescence if pEGFP-N1 is used for transfection.
6. For stable gene transfer, subculture ES cells into 10-cm dishes at 2 dpt. After 1 d more of culture, start drug selection with G418 (500 µg/mL) or hygromycin (500 µg/mL) after transfection with pSTneo or pHygEGFP *(11)*. Single-cell colonies are formed after 15–18 d of culture.

3.10. Directed Differentiation

Directed differentiation of ES cells is a useful approach for analyzing the mechanisms of cell differentiation. The ability to induce ES cells to differentiate into certain cell types is highly desirable for cell replacement therapy. Differentiation in vitro faces two major challenges: the difficulty to establish complicated culture conditions that, in mammals, usually involve a step of embryoid body formation and the combinatorial use of several growth factors. Furthermore, such procedures result in heterogeneous mixtures of different cell types. Often, a uniform population of cells differentiating along a particular cell lineage is impossible to obtain. Like mouse ES cells, medaka ES cells undergo spontaneous differentiation into many different types of cells, including fibroblasts, neural cells, muscles, and pigment cells *(3,4,11)*. To establish a model of directed differentiation, we made use of the melanocyte-specific mitf as a master regulator for differentiation of medaka ES cells into pigment cells. After transient transfection with pCMVmitf as described above, medaka ES cells are directed to differentiation into melanocytes at as early as 3 dpt. On d 5, 10–30% of transfected cells are melanin-synthesizing melanocytes *(13)*.

3.11. Perspectives

For the development and exploitation of ES cell technology in medaka, future work will focus on the gene targeting by HR and germline transmission. Germline transmission may be achieved by cell transplantation for producing germline chimeras *(14)* and nuclear transfer *(16)*. Most recently, we have succeeded in the establishment of a normal medaka spermatogonial cell line capable of sperm production in vitro *(17)*. It will be intriguing to determine whether in vitro spermatogenesis from this cell line can be used for germline transmission by artificial insemination. Molecular signature and analyses of ES cells and germ cells will provide tools to manipulate ES cell contribution to the germline.

4. Notes

1. For medaka strains and maintenance, *see* http://biol1.bio.nagoya-u.ac.jp:8000.
2. One tube is sufficient to supplement 500 mL of ESM4 (*see* **Subheading 2.2.1.**). FEE from a single embryo is arbitrarily defined as 10 U. Components are added sequentially to pure medium.
3. Test each batch of FEE preparation for activity by cell culture for at least 2 wk.
4. Perform the following steps in a biological safety cabinet.
5. To avoid cell loss, try to keep the drop diameter as small as possible.
6. Healthy blastomeres are recognized because they undergo cell division and show amoeba-like movement by forming pseudopodia (**Fig. 1C**).
7. Appearance of pigment cells at d 3 of culture from some strains/lines seems to be independent of cell density.
8. Subculture by trypsinization, freezing, and thawing of medaka ES cells is done essentially according to standard protocols for animal cell culture. Care should be taken to ensure that a single cell suspension is obtained to prevent differentiation. Freshly trypsinized single cells display pseudopodia before they attach to gelatin-coated substrata within 2 h, a feature reminiscent of their counterparts at the midblastula stage.
9. Thawing cells should be done as quickly as possible to maintain the integrity of cells.
10. BCIP/NBT substrates generate an intense red-purple precipitate by AP.

11. The time of hypotonic treatment is crucial. If it is too short, the chromosomes will be tightly packed with cytoplasm and will be spread insufficiently, preventing counting; if it is too long, some of the chromosomes will not remain in their metaphases.
12. Slides are best made within 2 h of the final fixation. They can also be made several days later if the cells are kept in fixative at 4°C and resuspended in fresh fixative before slide preparation.
13. A total of 20–50 good metaphases give reasonable sampling. Approximately 80% of medaka ES cells properly maintained have a diploid number of 48 chromosomes, and 15% are polyploid. ES cell cultures with a minimum 70% of diploid counts are acceptable. If not, subclones can be established from the original culture with a euploid count (5). Optionally, karyotype analysis after chromosome banding can be performed (5).
14. Materials used for chimera production should be very clean and sterile, and all solutions are sterilized by filtration and, shortly before use, centrifuged at maximal speed. Forceps are presoaked in Enzol (detergent enzymatic) followed by washing with water.
15. Dechorionated embryos are very fragile. During embryo transfer, avoid production of air bubbles and direct contact between embryos and solution surface, which result in collapse of the embryo. During transfer, embryos should float in the middle of the narrow part of a plastic pipet fully filled with BSS-1% PEG. Insert the pipet tip down to the solution in a dish; release embryos slowly and gently out from the pipet.
16. We use a self-built microinjector mounted on a Leica micromanipulator for cell transplantation.
17. Do not contact embryos with the sharp tip but with the side of the tip.
18. Rear fry for 30 d in green algae-containing aquarium water that has previously been adapted by maintaining a few adult males for more than 10 d before transfer into glass tanks.
19. We use donor cells from wild-type pigmentation strains and the albino strain i[1] as the host. Black pigment cells begin at d 3 or 5 of development from donor cells of noncultivated and short-term cultivated (less than 12 d of culture) or long-term cultured MBE cells and ES cell lines, respectively.

Acknowledgments

We thank T. Wagner and L. J. Qin for critically reading the manuscript. This work was supported by the National University of Singapore (R-154-000-153-720); the Biomedical Research Council of Singapore (BMRC R-154-000-204-305); the Chinese Academy of Sciences and Technology (High Tech-973 Program, 2004 CB 117406); and the Deutsche Forschungsgemeinschaft (SFB465).

References

1. Wittbrodt, J., Shima, A., and Schartl, M. (2002) Medaka—a model organism from the Far East. *Nat. Rev. Genet.* **3**, 53–64.
2. Wakamatsu, Y., Ozato, K., and Sasado, T. (1994) Establishment of a pluripotent cell line derived from a medaka (*Oryzias latipes*) blastula embryo. *Mol. Mar. Biol. Biotechnol.* **3**, 185–191.
3. Hong, Y. and Schartl, M. (1996) Establishment and growth responses of early medakafish (*Oryzias latipes*) embryonic cells in feeder layer-free cultures. *Mol. Mar. Biol. Biotechnol.* **5**, 93–104.
4. Hong, Y., Winkler, C., and Schartl, M. (1996) Pluripotency and differentiation of embryonic stem cell lines from the medakafish (*Oryzias latipes*). *Mech. Dev.* **60**, 33–44.
5. Hong, Y., Winkler, C., and Schartl, M. (1998) Production of medakafish chimeras from a stable embryonic stem cell line. *Proc. Natl. Acad. Sci. USA* **95**, 3679–3684.

6. Hong, Y., Winkler, C., Liu, T., Chai, G., and Schartl, M. (2004) Activation of the mouse Oct4 promoter in medaka embryonic stem cells and its use for ablation of spontaneous differentiation. *Mech. Dev.* **121,** 933–943.
7. Hong, Y., Gui, J., Chen, S., Deng, J., and Schartl, M. (2003) Embryonic stem cells in fish. *Acta Zool. Sinica* **49,** 281–294.
8. Bejar, J., Hong, Y., and Alvarez, M.C. (2002) An ES-like cell line from the marine fish *Sparus aurata*: characterization and chimaera production. *Transgenic Res.* **11,** 279–289.
9. Chen, S. L., Ye, H., Sha, Z., and Hong, Y. (2003) Derivation of a pluripotent embryonic cell line from red sea bream blastulae (*Lateolabrax japonicus*) embryos. *J. Fish Biol.* **63,** 795–805
10. Chen, S. L., Sha, Z., and Ye, H. (2003) Establishment of a pluripotent embryonic cell line from sea perch (*Lateolabrax japonicus*) embryos. *Aquaculture* **218,** 141–151.
11. Hong, Y., Chen, S., Gui, J., and Schartl, M. (2004) Retention of the developmental potential in medakafish embryonic stem cells after gene transfer and long-term drug selection for gene targeting in fish. *Transgenic Res.* **13,** 41–50.
12. Foley, A. and Mercola, M. (2004). Heart induction: embryology to cardiomyocyte regeneration. *Trends Caridiovasc. Med.* **14,** 121–125.
13. Béjar, J., Hong, Y., and Schartl, M. (2003) Mitf expression is sufficient to direct differentiation of medaka blastula derived stem cells to melanocytes. *Development* **130,** 6545–6553.
14. Wakamatsu, Y., Ozato, K., Hashimoto, H., et al. (1993) Generation of germline chimeras in medaka (*Oryzias latipes*). *Mol. Mar. Biol. Biotechnol.* **2,** 325–332.
15. Hong, Y., Winkler, C., and Schartl, M. (1998) Efficiency of cell culture derivation from blastula embryos and of chimera formation in the medaka (*Oryzias latipes*) depends on donor genotype and passage number. *Dev. Genes Evol.* **208,** 595–602.
16. Wakamatsu, Y., Ju, B., Pristyaznhyuk, I., et al. (2001) Fertile and diploid nuclear transplants derived from embryonic cells of a small laboratory fish, medaka (*Oryzias latipes*). *Proc. Natl. Acad. Sci. USA* **98,** 1071–1076.
17. Hong, Y., Liu, T., Zhao, H., et al. (2004) Establishment of a normal medakafish spermatogonial cell line capable of sperm production in vitro. *Proc. Natl. Acad. Sci. USA* **101,** 8011–8016.

2

Maintenance of Chicken Embryonic Stem Cells In Vitro

Hiroyuki Horiuchi, Shuichi Furusawa, and Haruo Matsuda

Summary

In this chapter, we describe the methods we have used to show that chicken leukemia inhibitory factor (LIF) maintains chicken embryonic stem (ES) cells in an undifferentiated state in culture. Recombinant chicken LIF (rchLIF) was expressed as a fusion protein linked to glutathione *S*-transferase (GST) and purified to greater than 90% purity in two chromatography stages, the first an affinity step using the GST tail, which was cleaved before further purification by gel chromatography. Chicken ES cells were obtained by culturing chicken blastodermal cells isolated from stage X embryos of freshly laid chicken eggs. These cells can be maintained in media containing rchLIF for at least 9 d without any other cytokines or feeder cells. Chicken ES cells were characterized by the expression of alkaline phosphatase activity, stage-specific embryonic antigen (SSEA)-1 and embryonal carcinoma cell monoclonal antibody-1. In addition, the phosphorylation of signal transducers and activators of transcription-3 by LIF, which is sufficient to maintain the undifferentiated state of ES cells, was detected by Western blotting analysis.

Key Words: Alkaline phosphatase (AP); chicken; chicken blastodermal cells (CBC); embryonic stem (ES) cells; leukemia inhibitory factor (LIF); stage-specific embryonic antigen (SSEA)-1; signal transducers and activators of transcription-3 (STAT3).

1. Introduction

The avian embryo represents an important model system in developmental and cell biology because of its ease of manipulation and its similarity to mammalian development. In addition, the production of transgenic birds using early embryos is an important technology in both fundamental and applied avian biology. Transgenic chickens are particularly in the spotlight at present because of their potential as bioreactors. For this research to progress, the development of transgene technology for use in the early embryo is necessary. Different methods have been employed to introduce transgenes into avian embryos, including microinjection, retroviruses, and transfection of avian blastodermal or primordial germ cells. However, which technique will be most generally applicable is not yet clear.

From: *Methods in Molecular Biology, vol. 329: Embryonic Stem Cell Protocols: Second Edition: Volume 1*
Edited by: K. Turksen © Humana Press Inc., Totowa, NJ

Mouse embryonic stem (ES) cells can be maintained in culture without feeder cells in the presence of fetal bovine serum and mouse leukemia inhibitory factor (LIF) *(1)*. The injection of ES cells into blastocysts can give rise to chimeric mice, and the ES cells can contribute to all tissues, including the germ cells. Mutant mouse ES cells have been used to produce mouse strains with disrupted genes. The relative ease with which one can manipulate ES cells in vitro has made them a powerful tool for the targeted disruption of endogenous genes, and this led to a dramatic increase in the number of transgenic ("knockout") mice during the 1990s.

In chickens, the existence of pluripotent ES-like cells in the stage X *(2)* chicken blastoderm has been demonstrated directly by immunostaining *(3)* and indirectly by the ability of chicken blastodermal cells to contribute to the germline as well as somatic ectodermal, mesodermal, and endodermal lineages *(4–7)*. Pain et al. *(8)* first reported that pluripotent avian stem cells could be produced and maintained by long-term culture of stage X blastodermal cells with several cytokines: chicken stem cell factor (SCF), bovine basic fibroblast growth factor (bFGF), mouse interleukin (IL)-6, human IL-11, and mouse LIF. However, there are no reports of chicken ES cell lines contributing to the germline after extended culture, although there have been some promising developments.

More recently, the chicken homologue of mammalian LIF was cloned, and the effects of recombinant chicken LIF (rchLIF) and mouse LIF on chicken ES-like cells, chicken blastodermal cells (CBCs), were compared *(9)*. In these experiments, rchLIF was found to maintain chicken ES cells in an undifferentiated state, whereas mouse LIF did not. This finding should contribute to the establishment of chicken ES cell lines and the development of transgenic chicken technology. In this chapter, we describe protocols for the expression and purification of rchLIF in *Escherichia coli*, the culture of CBCs, and the detection in these cells of markers characteristic of an undifferentiated state.

2. Materials
2.1. Cloning, Expression, and Purification of rchLIF
2.1.1. Reverse Transcription Polymerase Chain Reaction

1. Thermal cycler (Applied Biosystems, Foster City, CA; model 9700).
2. Agarose gel apparatus and reagents.
3. Isogen-LS (100 mL; Wako, Osaka, Japan; cat. no. 311-02621) (*see* **Note 1**).
4. Chloroform (500 mL; Sigma, St. Louis, MO; cat. no. 05-3400-5).
5. 2-Propanol (500 mL; Sigma, cat. no. 15-2320-5).
6. Diethyl pyrocarbonate (DEPC)-H_2O.
7. Oligotex-dT30 <Super> Kit (Roche, Mannheim, Germany; cat. no. 489991).
8. 70 and 100% ethanol.
9. SuperScript First-Strand Synthesis System for reverse transcriptase polymerase chain reaction (RT-PCR) (Invitrogen, Carlsbad, CA; cat. no. 11904-018). Package contents: 0.5 µg/µL Oligo (dT)$_{12-18}$, 10 mM deoxyribonucleoside triphosphate (dNTP) mix, 10X first-strand buffer, 0.1 M dithiothreitol (DTT), 40 U/µL RNase OUT (inhibitor), 50 U/µL SuperScript II reverse transcriptase.

10. PCR primers diluted to 10 pmol/μL with sterile distilled water (dH$_2$O).
11. Takara Ex Taq (5 U/μL; Takara, Shiga, Japan; cat. no. RR001A). Supplied reagents: 10X Ex Taq buffer, 2.5 mM dNTP mix.

2.1.2. Ligating Plasmid and Target DNA and Transformation

1. pGEX-6P-1 plasmid (Amersham Pharmacia Biotech, Piscataway, NJ; cat. no. 27-4597-01).
2. Competent cells, *E. coli* BL21 strain (Novagen, Madison, WI; cat. no. 70232-3).
3. *Bam* HI (30–60 U/μL; Takara, cat. no. 1010AH). Supplied reagents: 10X K buffer.
4. *Xho* I (30–60 U/μL; Takara, cat. no. 1094AH). Supplied reagents: 10X K buffer.
5. T4 DNA ligase (1 U/μL; Invitrogen, cat. no. 15224-017). Supplied reagents: 5X ligase reaction buffer.
6. 100 × 15-mm plastic Petri dishes (Becton Dickinson, Bedford, MA; cat. no. 351999).
7. Super optimal catabolite (SOC) medium (Invitrogen, cat. no. 15544-034).
8. LB agar (500 g; Invitrogen, cat. no. 22700-025).
9. Ampicillin solution, 100 mg/mL stock: dissolve 200 mg Ampicillin sodium salt in 2 mL sterile H$_2$O.
10. Ampicillin-LB plate. For 100 mL: mix 3.2 g LB agar in 100 mL dH$_2$O and autoclave. Remove the medium from the autoclave and swirl it gently (*see* **Note 2**). Allow the medium to cool to 50–60°C and then add 100 μL ampicillin solution (100 mg/mL stock) and swirl. Pour 10–15 mL of medium per 100 × 15-mm plastic Petri dish. When the medium has completely set, invert the dishes and store them at 4°C until needed.

2.1.3. Expression and Purification

1. LB broth base, powder (500 g; Invitrogen, cat. no. 12780-052). For 1 L: mix 20 g LB broth base in 1 L dH$_2$O and autoclave. Remove the medium from the autoclave and swirl it gently (*see* **Note 2**). Store the medium at room temperature until needed.
2. Isopropyl β-D-thiogalactoside (IPTG) (100 mM IPTG stock): dissolve 0.5 mg IPTG in 20 mL dH$_2$O and sterilize by passing it through a 0.22-μM disposable filter. Dispense the solution into 1-mL aliquots and store at −20°C.
3. Ampicillin solution (100 mg/mL stock) (*see* **Subheading 2.1.2., item 9**).
4. Protease inhibitor (20 tablets; Roche, cat. no. 1-697-498).
5. PreScission protease (2 U/μL; Amersham Pharmacia Biotech, cat. no. 27-0843-01).
6. PreScission cleavage buffer: 50 mM Tris-HCl, pH 7.0, 150 mM NaCl, 1 mM ethylenediaminetetraacetic acid (EDTA)·2Na, 1 mM DTT. For 100 mL: combine 33 mL 5 M NaCl, 0.2 mL 500 mM EDTA·2Na, 1 mL 100 mM DTT, and 5 mL 1 M Tris-HCl, pH 7.0.
7. Glutathione-Sepharose 4B (100 mL; Amersham Pharmacia Biotech, cat. no. 27-4574-01).
8. PD-10 empty column (Amersham Pharmacia Biotech, cat. no. 17-0435-01).
9. HiLoad 26/60 Superdex 75 pg (Amersham Pharmacia Biotech, cat. no. 17-1070-01).
10. Dulbecco's phosphate-buffered saline (PBS), pH 7.4 (Nissui, Tokyo, Japan; cat. no. 08190). For 1 L: dissolve 9.6 g Dulbecco's PBS powder in 1 L dH$_2$O and autoclave.

2.2. Cell Culture

1. Iscove's modified Dulbecco's medium (IMDM) (1 L; Gibco BRL, Carlsbad, CA; cat. no.12440-046).
2. Dulbecco's modified Eagle medium (DMEM) (500 mL; Gibco BRL, cat. no.11965-092) (*see* **Note 3**).
3. Fetal bovine serum (FBS) (500 mL; Hyclone, Logan, UT; cat. no. SH30070.03) (*see* **Note 4**).

4. Chicken serum (500 mL; Gibco BRL, cat. no.16110-082) (*see* **Note 4**).
5. 24-well culture plates (Becton Dickinson, cat. no. 353047).
6. 10 mM minimum essential medium (MEM) nonessential amino acid (NEAA) solution (100 mL; Gibco BRL, cat. no. 11140-050).
7. 100 mM MEM sodium pyruvate solution (100 mL; Gibco BRL, cat. no. 11360-070).
8. 2-Mercaptoethanol (100 mL; Sigma, cat. no. M7522).
9. Adenosine (5 g; Sigma, cat. no. A4036).
10. Cytidine (1 g; Sigma, cat. no. A4036).
11. Guanosine (5 g; Sigma, cat. no. G6264).
12. Thymidine (1 g; Sigma, cat. no. T1895).
13. Uridine (5 g; Sigma, cat. no. U3003).
14. 100 μM nucleotide stock solution. For 100 mL: mix 80 mg adenosine, 73 mg cytidine, 85 mg guanosine, 24 mg thymidine, and 73 mg uridine in 100 mL dH$_2$O, and dissolve the nucleotide mixture at 37°C. Sterilize by passing it through a 0.22-μm disposable filter, dispense the solution into 1-mL aliquots, and store at −20°C. Thaw nucleotide stock at 37°C before use.
15. Fertilized, freshly laid chicken eggs.
16. Lipopolysaccharide (LPS) from *E. coli* serotype O127:B8 (100 mg; Sigma, cat. no. L3880).
17. Penicillin-streptomycin 100X (100 mL; Gibco BRL, cat. no. 15140-122).
18. 100 × 20-mm tissue culture dish (Becton Dickinson, cat. no. 353003).
19. 15-mL centrifuge tube (Sumilon, Tokyo, Japan; cat. no. MS-56150).
20. Filter paper rings (*see* **Note 5**).
21. Glass or plastic tube (*see* **Note 6**).

2.2.1. Media

1. Tissue culture medium for IN24 cell line: IN24 cells (chicken monocytic leukemia cells) *(10)* are maintained in IMDM supplemented with 10% heat-inactivated FBS and 1% penicillin-streptomycin (10% FBS-IMDM). To prepare 100 mL 10% FBS-IMDM, combine 10 mL heat-inactivated FBS, 1 mL penicillin-streptomycin, and 89 mL IMDM.
2. Tissue culture medium for CBCs: CBCs are maintained in DMEM supplemented with 10% heat-inactivated FBS, 2% heat-inactivated chicken serum, 0.1 mM MEM sodium pyruvate solution, 0.05 mM MEM NEAA, 1 μM nucleotide stock solution, and 1% penicillin-streptomycin. This medium is called cytokine-free chicken embryonic stem cell medium (CESM). To prepare 100 mL CESM, combine 10 mL heat-inactivated FBS, 2 mL heat-inactivated chicken serum, 0.1 mL MEM sodium pyruvate solution, 0.5 mL MEM NEAA, 1 mL nucleotide stock solution, 1 mL penicillin-streptomycin, and 85.4 mL DMEM. When CBC are cultured, 100 mL CESM is added with 0.7 μL 2-mercaptoethanol and 2 mL purified rchLIF (1 μg/mL).

2.3. Detection of Markers of Undifferentiated CBC

2.3.1. Alkaline Phosphatase Reaction

1. Naphthol AS-BI alkaline solution (10 mL; Sigma, cat. no. 86-1).
2. Fast Red violet-alkaline solution (10 mL; Sigma, cat. no. 86-2).
3. Sodium nitrite solution (10 mL; Sigma, cat. no. 91-4).
4. Hematoxylin solution, Gill no. 3 (50 mL; Sigma, cat. no. GHS-3).
5. Citrate solution (50 mL; Sigma, cat. no. 91-5).
6. Acetone (500 mL; Wako, cat. no. 016-00346).
7. 37% formaldehyde solution (500 mL; Nacalai tesque, Kyoto, Japan; cat. no. 16223-55).

8. Citrate-acetone-formaldehyde fixative solution: to 25 mL citrate solution, add 65 mL acetone, 8 mL 37% formaldehyde solution. Place in a glass bottle and cap tightly. Store in a refrigerator (2–8°C). Warm to 18–26°C prior to use.

2.3.2. Immunofluorescence

1. PBS at pH 7.4 (see **Subheading 2.1.3., item 10**).
2. 10% formaldehyde neutral buffer, pH 7.0 (1 L; Nacalai tesque, cat. no. 37152-51).
3. Block Ace (250-mL; Dainippon Pharmaceutical Co., Osaka, Japan; cat. no. UK-B25).
4. Anti-stage-specific embryonic antigen (anti-SSEA)-1 mouse monoclonal antibody (ascites) (Developmental Studies Hybridoma Bank, Iowa City, IA; cat. no. MC-480).
5. Embryonal carcinoma cell monoclonal antibody (EMA)-1 mouse monoclonal antibody (partially purified immunoglobulin [Ig]) (Developmental Studies Hybridoma Bank, cat. no. EMA-1).
6. Fluorescein isothiocyanate-conjugated sheep antimouse Ig F(ab′)$_2$ fragment (1-mL vial; Silenus, Victoria, Australia; cat. no. DDAF).

2.3.3. Western Blotting Analysis

1. Cell extraction buffer: 10 mM Tris-HCl, pH 7.4, 150 mM NaCl, 1 mM EDTA·2Na, 1 mM Na$_3$VO$_4$, 0.1% sodium deoxycholate, 0.1% sodium dodecyl sulfate (SDS), 1% Nonidet P-40. For 100 mL: combine 33 mL 5 M NaCl, 0.2 mL 500 mM EDTA·2Na, 1 mL 0.1 M Na$_3$VO$_4$, 1 mL 10% sodium deoxycholate, 1 mL 10% SDS, 1 mL Nonidet P-40, and 1 mL 1 M Tris-HCl, pH 7.4. Add one tablet of protease inhibitor (see **Subheading 2.1.3., item 4**) to 25 mL prior to use.
2. PVDF membrane (Bio-Rad, Hercules, CA; cat. no. 13236).
3. Precision plus protein standards (protein marker) (1 mL; Bio-Rad, cat. no. 161-0363).
4. Blocking buffer: 20 mM Tris-HCl, pH 7.4, 140 mM NaCl, 25 mM EDTA·2Na, 0.2% Tween-20, 3% bovine serum albumin. For 100 mL: combine 3 mL 5 M NaCl, 5 mL 0.5 M EDTA·2Na, 0.2 mL Tween-20, 3 g bovine serum albumin, and 2 mL 1 M Tris-HCl, pH 7.4.
5. Anti-STAT3 (749-769) rabbit antibody (1 mg/mL; Calbiochem, San Diego, CA; cat. no. 57609).
6. Anti-Phospho-STAT3 (Tyr705) antibody (100 µL; Cell Signaling Technology, Beverly, MA; cat. no. 9131).
7. Horseradish peroxidase-labeled antirabbit IgG (1 mg/mL; KPL, Gaithersburg, MD; cat. no. 474-1516).
8. ECL plus Western blotting detection reagents (Amersham Pharmacia Biotech, cat. no. RPN2132).

3. Methods

3.1. Cloning, Expression, and Purification of rchLIF

3.1.1. Isogen RNA Extraction

Chicken LIF cDNA was cloned using mRNA from LPS-stimulated IN24 cells, and chicken LIF mRNA was most abundant in adult chicken liver and thymus (**9**). The following procedure of RNA extraction was derived from the manufacturer's protocol (see **Note 7**).

1. Aspirate the culture medium from IN24 culture dishes and rinse the dishes three times with IMDM.
2. Add 10 mL of 10 µg/mL LPS-IMDM and culture for 16–24 h at 38.5°C in a CO$_2$ incubator.

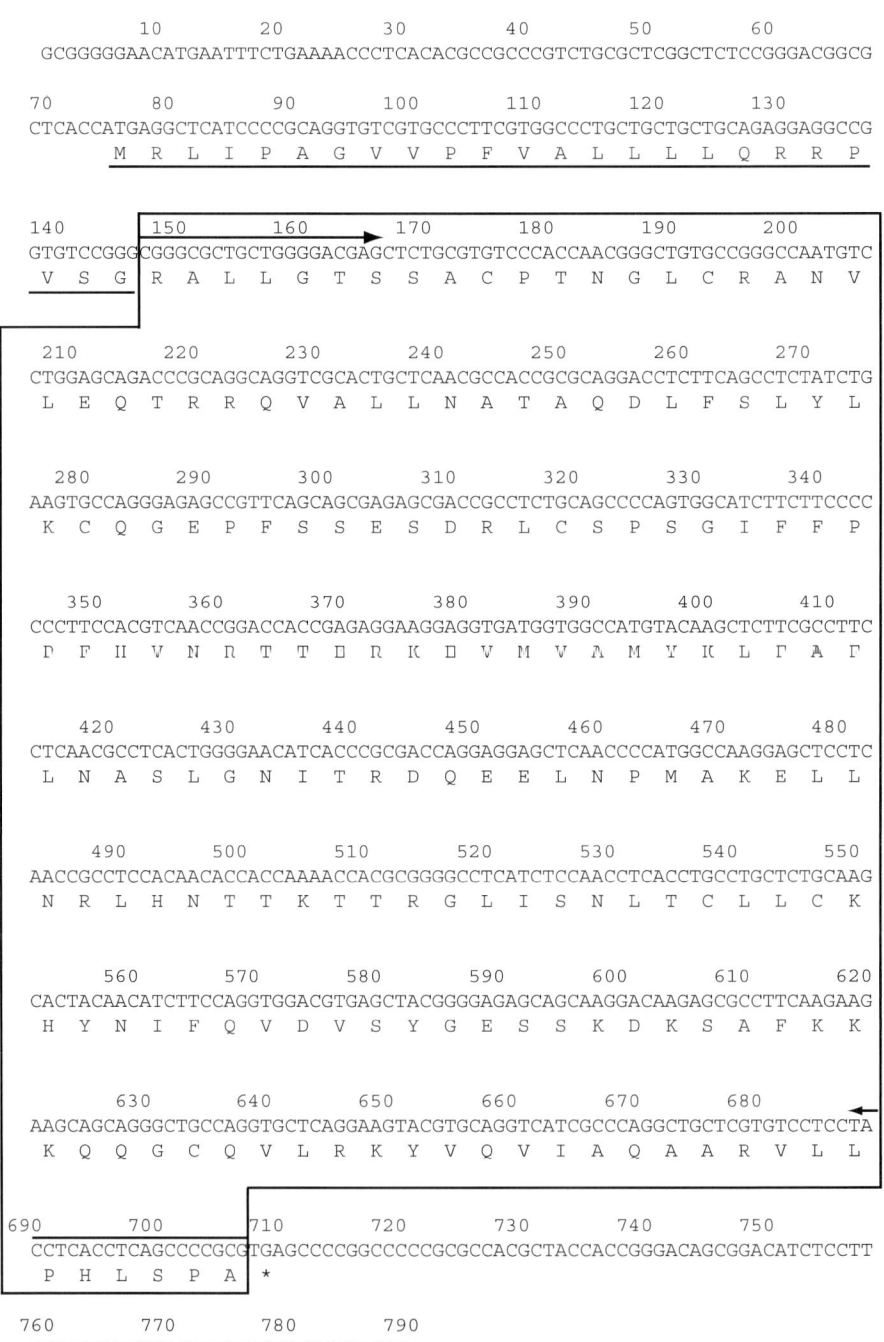

Fig. 1.

Chicken ES Cells

3. Lyse the cells in the culture dish by adding the appropriate amount of Isogen (1 mL per 10 cm^2).
4. Transfer 1-mL samples of the homogenate to microcentrifuge tubes and incubate them for 5 min at room temperature.
5. Add 0.2 mL chloroform per 1 mL of Isogen.
6. Cap the samples securely and shake the microcentrifuge tubes vigorously by hand for 15 s, then incubate at room temperature for 3 min.
7. Centrifuge at 12,000g for 15 min at 4°C. Following centrifugation, the mixture separates into a lower red, phenol-chloroform phase, an interphase, and a colorless upper aqueous phase. The RNA remains exclusively in the aqueous phase.
8. Transfer the aqueous phase to a fresh tube. Precipitate the RNA by adding 0.5 mL RNase-free 2-propanol per 1 mL of Isogen used for the initial homogenization. Incubate the samples for 5–10 min at room temperature.
9. Centrifuge at 12,000g for 10 min at 4°C. The RNA forms a pellet on the bottom and the sides of the tube.
10. Remove the supernatant and wash the pellet with 1 mL of RNase-free 70% ethanol. Mix the sample by vortexing and centrifuge at 12,000g for 5 min at 4°C.
11. Remove the supernatant and air-dry or vacuum-dry the pellet and resuspend it in DEPC-H$_2$O.
12. Measure the optical density (OD) at A$_{260}$ and A$_{280}$. The ratio of ODs at A$_{260}$:A$_{280}$ should be between 1.6 and 2. Determine the concentration (*see* **Note 8**).

3.1.2. Reverse Transcription

The following procedure was derived from the manufacturer's protocol for SuperScript II reverse transcriptase:

1. Add the following components to a microcentrifuge tube: 1.0 µL Oligo (dT)$_{12-18}$ (0.5 g/mL), 1.0 µL 1–5 µg/µL total RNA, 1.0 µL 10 mM dNTP mix, and 11.0 µL dH$_2$O.
2. Heat the mixture to 65°C for 5 min, then quickly chill it on ice. Collect the contents of the tube by brief centrifugation and add 2.0 µL 10X first-strand buffer, 2.0 µL 0.1 M DTT, and 1.0 µL RNase inhibitor (40 U/µL).
3. Mix contents of the tube gently and incubate at 42°C for 2 min.
4. Add 1.0 µL (50 U) of SuperScript II reverse transcriptase, mix by pipetting gently up and down, and incubate for 50 min at 42°C.
5. Inactivate the reaction by heating at 70°C for 15 min and then store at 4°C.

3.1.3. Polymerase Chain Reaction

1. Design and synthesize the appropriate oligonucleotide primer for amplification of mature chicken LIF (*see* **Fig. 1**). For cloning into the expression vector, modify the 5′-end sequence of the primers: forward primer 5′-CG*GGATCC*CGGGCGCTGCTGGGGACGAG-3′ (italics, *Bam* HI restriction site); reverse primer 5′-CC*CTCGAG*TTATCACGCGGGGCTGAGGTGAGGTA-3′ (italics, *Xho* I restriction site; underline, double stop codon).

Fig. 1. Nucleotide and deduced amino acid sequences of chicken leukemia inhibitory factor (LIF) cDNA and primer positions for reverse transcriptase polymerase chain reaction (RT-PCR). The termination codon (TGA) is marked with an asterisk. The numbers above refer to the nucleotide sequence, and the putative signal peptide is underlined. The mature chicken LIF sequence is boxed, and primer positions used for RT-PCR are marked with arrows. The nucleotide sequence has been submitted to the GeneBank/EBI Data Bank with accession number BD187371.

2. Prepare the PCR mixture (50 μL final volume) into the amplification tube: 1.0 μL template DNA (approx 100 ng), 5.0 μL 2.5 mM dNTPs, 5.0 μL 10X Ex Taq buffer, 1.0 μL forward primer (10 pmol/μL), 1.0 μL reverse primer (10 pmol/μL), 0.5 μL Takara Ex Taq DNA polymerase (5 U/μL), and 36.5 μL dH$_2$O.
3. In the thermal cycler, heat the samples to 98°C for 6 min and then run 30 amplification cycles in the linear range of 10 s at 98°C (denaturation), 30 s at 60°C (annealing), and 1 min at 72°C (polymerization). Finally, hold for 10 min at 72°C as an extension step and then store at 4°C.
4. Analyze the amplification products on a 1.0–1.5% agarose gel. The correct amplification product should be approx 600 bp.
5. Dilute the amplification product to 100 μL with distilled water.
6. Add 100 μL of phenol-chloroform (1:1 v/v) and vortex.
7. Centrifuge at 12,000g for 5 min at room temperature and transfer the upper aqueous layer, containing the cDNA, to a new tube. Do not remove any of the interface with the aqueous layer.
8. Add an equal volume of chloroform and vortex.
9. Centrifuge at 12,000g for 3 min at room temperature and transfer the upper aqueous layer, containing the cDNA, to a new tube.
10. Precipitate the cDNA from the aqueous layer by adding the following: 10 μL 3 M sodium acetate and 250 μL 100% (v/v) ethanol.
11. Vortex and leave to precipitate for 15 min at −80°C.
12. Centrifuge at 12,000g for 10 min at 4°C. The cDNA forms a pellet on the bottom and the sides of the tube.
13. Remove the supernatant and wash the pellet with 1 mL 70% ethanol. Mix the sample by vortexing and centrifuge at 12,000g for 5 min at 4°C.
14. Remove the supernatant and air-dry or vacuum-dry the pellet and resuspend it in TE buffer.
15. Measure the OD at A_{260} and A_{280}. The A_{260}:A_{280} ratio should be between 1.6 and 2. Determine the concentration of the cDNA (*see* **Note 9**).

3.1.4. Expression of rchLIF

We succeeded in producing rchLIF using a GST gene fusion system. The following procedure was derived from the manufacturer's protocol.

3.1.5. Ligating cDNA Into the Expression Vector

1. For restriction digestion of pGEX-6P-1 vector (*see* **Fig. 2**) and inserting cDNA, prepare the following reaction mixtures separately for the vector and insert: 5.0 μg pGEX-6P-1 vector or 0.3 pmol/μL insert cDNA, add 5.0 μL 10X K buffer, 1.0 μL *Bam* HI, 1.0 μL *Xho* I, and make up to 50 μL with distilled water.
2. Incubate at 37°C for 16 h.
3. Electrophorese the digested DNAs on 1.0% agarose gels and purify the DNAs using a commercially available gel extraction kit.
4. Determine the concentration of the DNAs using the spectrophotometer (*see* **Note 9**).
5. To anneal the linearized vector and insert DNA, they should be mixed at a vector:insert molar ratio of 1:5. Mix the components listed in a microcentrifuge tube: 100 ng plasmid vector, 61 ng insert DNA (chicken mature LIF cDNA), 4.0 μL 5X T4 DNA ligase buffer, 0.1 μL T4 DNA ligase (100 U/μL). Make up to 20 μL with distilled water.
6. Mix gently and incubate at 23–26°C for 1 h and store the unused portion of the cDNA at −20°C.

Chicken ES Cells

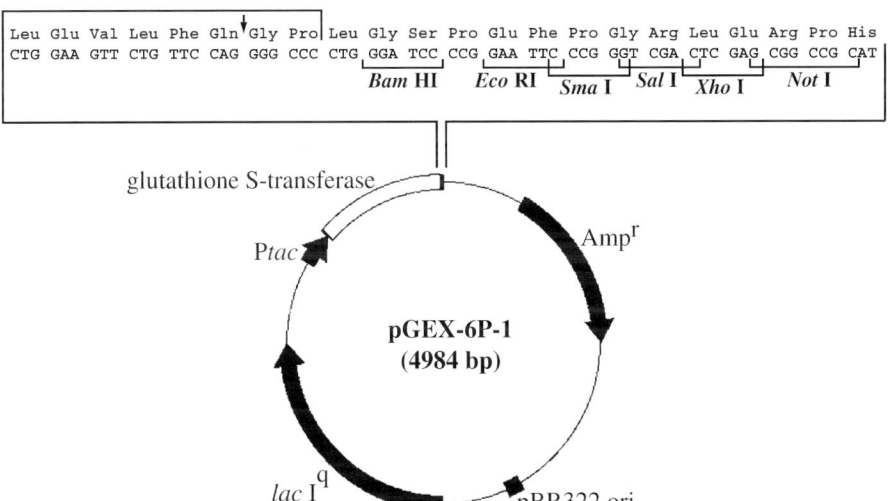

Fig. 2. Map of pGEX-6P-1 vector showing the reading frames and main features. The vector contains a glutathione-S-transferase (GST) coding sequence, an ampicillin resistance marker (Amp^r), a *Tac* promoter (P*tac*), and a PreScission protease recognition site for cleaving the desired product from the fusion protein. Mature chicken leukemia inhibitory factor nucleotides were cloned into the *Bam* HI-*Xho* I sites.

3.1.6. Transformation of Competent Cells

1. Place a clean microcentrifuge tube and a tube of BL21-competent cells on ice.
2. Let the competent cells thaw on ice and mix gently to ensure that the cells are evenly suspended.
3. Pipet 100-µL aliquots of cells into the prechilled microcentrifuge tube.
4. Add 5 µL of the ligated DNA solution to the cells immediately. Stir gently to mix.
5. Leave the tube on ice for 20 min.
6. Heat the tube for exactly 30 s in a 42°C water bath, do not shake.
7. Place on ice for 2 min.
8. Add 400 µL of room temperature SOC medium to the tube.
9. Culture at 37°C for 1 h with vigorous shaking.
10. Plate 100-µL aliquots of the diluted, transformed cells onto ampicillin-LB plate.
11. Incubate the plates at 37°C overnight.
12. Select colonies of BL21 cells containing a recombinant plasmid (*see* **Note 10**).

3.1.7. Preparation of Large-Scale Bacterial Sonicates

1. Use a single colony of BL21 cells containing a recombinant plasmid to inoculate 10 mL ampicillin-LB medium.
2. Incubate for 12–15 h at 37°C with vigorous shaking.
3. Dilute the culture 1:100 into 1 L of fresh ampicillin-LB medium (approx 330 mL in a 1-L flask) and grow in a 20°C shaking incubator until the A_{600} reaches 0.5.

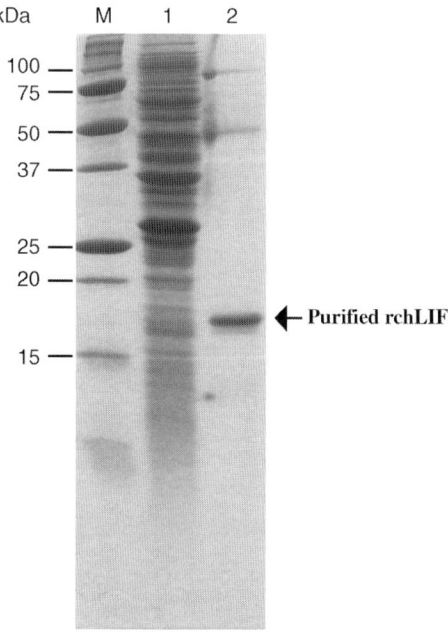

Fig. 3. Sodium dodecyl sulfate polyacrylamide-gel electrophoresis (SDS-PAGE) analysis of bacterial lysate (**lane 1**) and purified recombinant chicken leukemia inhibitory factor (rchLIF) (**lane 2**). The samples were analyzed on a 15% SDS-PAGE stained with Coomassie blue. The molecular sizes of marker proteins (**lane M**) are shown on the left.

4. Add 100 mM IPTG to a final concentration of 0.1 mM and continue incubation for an additional 10–15 h.
5. Transfer the culture to appropriate centrifuge containers and centrifuge at 7700g for 10 min at 4°C.
6. Discard the supernatant and drain the pellet. Place on ice.
7. Using a pipet, completely suspend the cell pellet by adding 50 mL ice-cold 1X PBS in which one tablet of protease inhibitor has been dissolved.
8. Disrupt the resuspended cells using a sonicator, on ice, in short bursts.
9. Centrifuge at 12,000g for 10 min at 4°C. Transfer the supernatant to a fresh container. Save aliquots of the supernatant for analysis by SDS-PAGE (polyacrylamide gel electrophoresis) (*see* **Fig. 3, lane 1**).

3.1.8. Two-Step Purification of rchLIF

rchLIF is expressed as a fusion protein linked to GST and purified to greater than 90% purity in a two-stage (affinity and gel) chromatography method, which includes removal of the GST affinity tail. Using this method, 1–5 µg purified rchLIF can be obtained from 1 L of *E. coli* culture medium.

3.1.9. Affinity Purification Using Glutathione-Sepharose 4B

To facilitate the removal of the GST affinity tail from rchLIF, the GST fusion protein used contains the PreScission protease recognition site. This site is recognized by human rhinovirus 3C protease, which is used to cleave rchLIF from the GST affinity

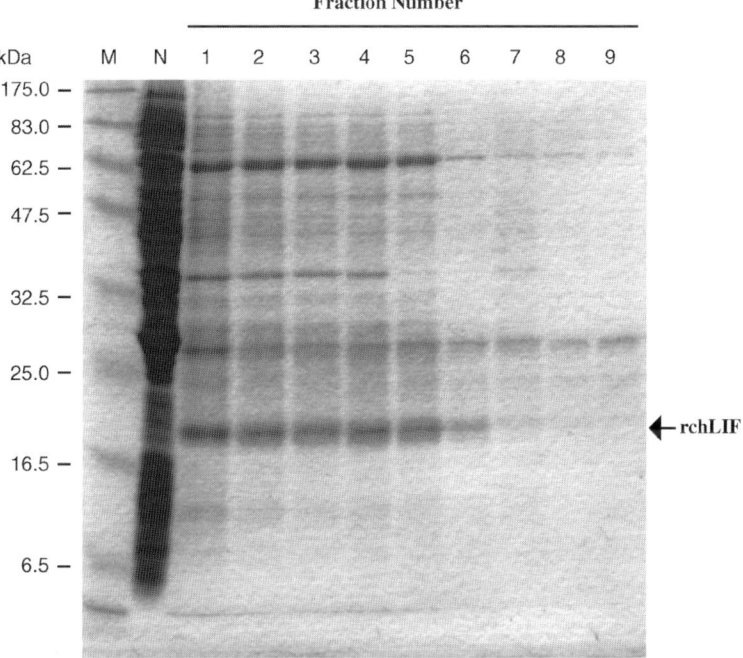

Fig. 4. Sodium dodecyl sulfate-polyacrylamide gel electrophoresis (SDS-PAGE) analysis of eluted fractions from affinity chromatography with glutathione-Sepharose 4B. The samples were analyzed on a 15% SDS-PAGE stained with Coomassie blue. The molecular sizes of marker proteins (**lane M**) are shown on the left. **Lane N** is the flow-through fraction.

tail and is eluted from the column, leaving the GST affinity tail and the protease bound to the glutathione-Sepharose 4B (*see* **Note 11**).

1. Gently shake the bottle of glutathione-Sepharose 4B to resuspend the matrix.
2. Use a pipet to remove sufficient slurry to form a 5-mL bed volume in the disposable column supplied with the purification kit.
3. Remove the bottom cap and allow the column to drain.
4. Wash the glutathione-Sepharose 4B by adding 50 mL cold (4°C) PBS and allow the column to drain.
5. Use a pipet to apply the bacterial sonicate (*see* **Subheading 3.1.7.**) to the column.
6. Remove the bottom cap and allow the sonicate to flow through the column.
7. Wash the matrix by adding 150 mL cold (4°C) 1X PBS. Allow the column to drain. The GST-rchLIF fusion protein will have bound to the glutathione-Sepharose matrix.
8. Replace the bottom cap on the washed column and add 5 mL PreScission protease reaction mixture; mix 200 µL (400 U) of PreScission protease with 4.8 mL of ice-cold PreScission cleavage buffer.
9. Replace the top cap and gently rotate (10 cycles/min) the suspension at 4°C for 14 h.
10. Remove the bottom cap and collect the eluate in 1-mL fraction in new tubes.
11. Analyze by SDS-PAGE to estimate the yield and purity of every fraction (*see* **Fig. 4**).

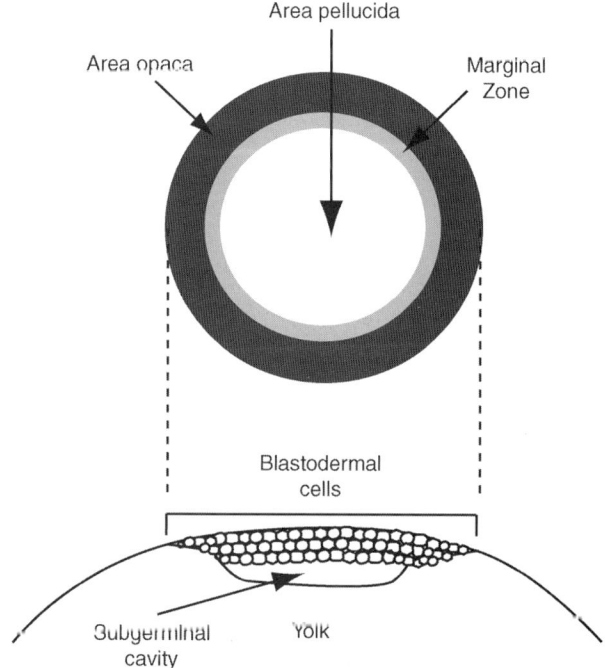

Fig. 5. Schematic illustration of stage X chicken embryo. The upper figure shows the ventral view, and the lower figure shows the cross section. Two distinct regions of blastoderm can be identified, the area pellucida and the area opaca, consisting of the darker cells at the margin of the blastoderm and yolk. Between the area pellucida and the yolk is a space called the subgerminal cavity.

3.1.10. Gel Chromatography

As shown in **Fig. 4**, several bands in addition to rchLIF are present on gels following elution from glutathione-Sepharose 4B. Highly purified rchLIF is obtained after a final gel chromatography step using the following procedure.

1. Pool the major fractions containing rchLIF in a clean tube.
2. Dialyze against cold PBS at 4°C overnight.
3. Equilibrate a Superdex 75 column with 1 L PBS.
4. Apply 5 mL of the dialyzed sample filtered through a 0.45-µm filter to the Superdex 75 column and elute with PBS at a flow rate of 1.0 mL/min.
5. Monitor by measuring the absorbance at 280 nm and collect 1.5-mL fractions.
6. Fraction eluted between 200 and 215 mL should contain rchLIF (*see* **Fig. 3, lane 2**).

3.2. Cell Culture

3.2.1. Method for Isolation of CBCs

Freshly laid fertilized chicken eggs are used to isolate CBCs from stage X embryos, as defined by Eyal-Giladi and Kochav *(2)* and illustrated in **Fig. 5**. Stage X blastodermal cells are pluripotent cells containing primordial germ cells or their immediate precursor cells. We use the stage X blastodermal cells in the area pellucida to isolate and cultivate CBCs.

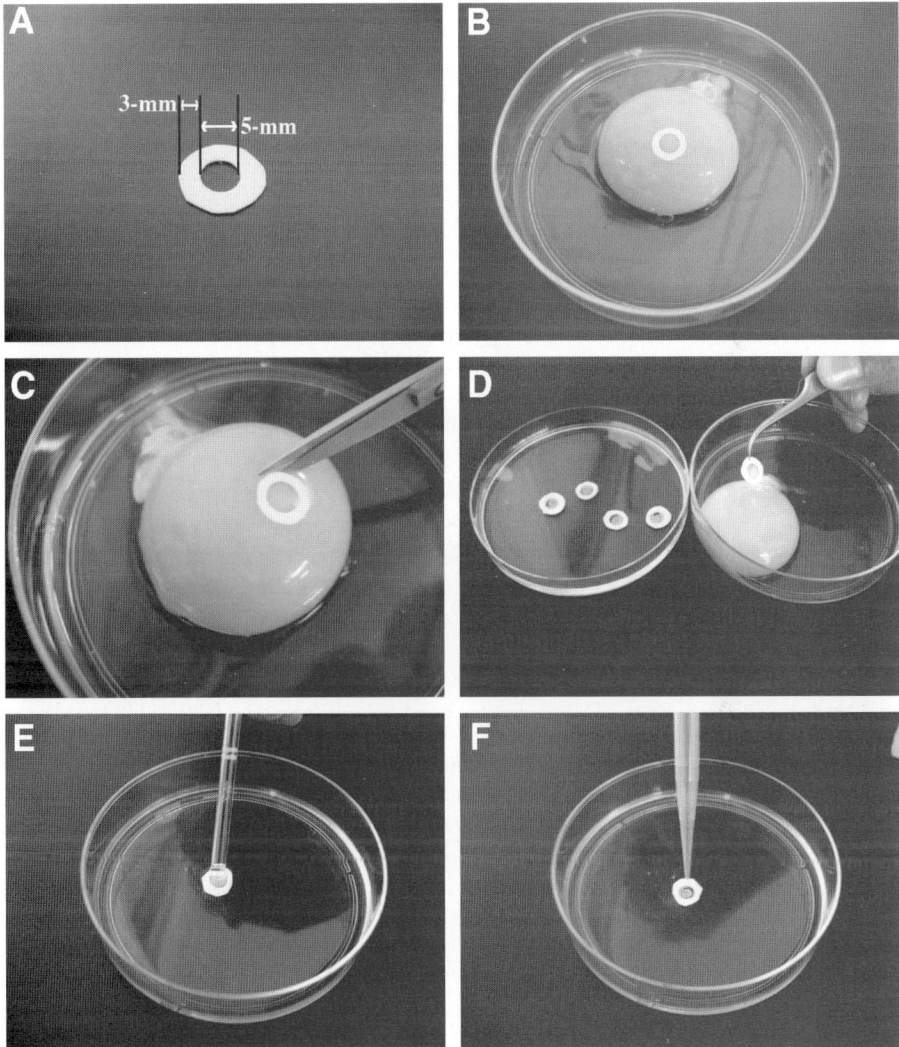

Fig. 6. Method for isolation of chicken blastoderm. (**A**) Dimensions of a filter paper ring. (**B**) Yolk positioned with embryo at the top with a filter paper ring placed centrally over the blastoderm. (**C**) Cutting the blastoderm along the periphery of the ring with scissors. (**D**) Lifting and inverting the ring and transferring it to a dish with PBS. (**E**) Punching out the area pellucida with the bottom of a glass pipet. (**F**) Collecting the area pellucida with a micropipet.

1. Prepare filter paper rings (*see* **Note 5** and **Fig. 6A**).
2. Autoclave and dry the prepared rings.
3. Break the fresh fertilized egg and put the contents into a 100-mm culture dish.
4. Rotate the yolk so the blastoderm is on the top surface.
5. Place a filter paper ring centrally over the blastoderm (*see* **Fig. 6B**).
6. Cut along the periphery of the paper ring with scissors to separate the blastoderm from the yolk (*see* **Fig. 6C**).

7. Pick up the blastoderm and filter paper together using a pair of tweezers, invert them, and transfer them to a 100-mm culture dish containing 10 mL PBS (*see* **Fig. 6D**).
8. Remove any yolk granules by washing carefully twice with PBS.
9. Cut the area pellucida from the blastoderm using a glass or plastic tube, such as the bottom of a pipet, as a punch (*see* **Note 6** and **Fig. 6E**).
10. Collect the area pellucida with a micropipet (*see* **Fig. 6F**) and transfer it to a 15-mL tube with 5 mL of serum-free CESM.
11. Dissociate the area pellucida by pipetting with a Pasteur pipet and then centrifuge at 120g for 10 min to pellet the cells.
12. Remove the supernatant, add 10 mL serum-free CESM, and gently resuspend.
13. Estimate the cell number with a hemocytometer. Approximately 2×10^4 cells should be obtained from each embryo.

3.2.2. Maintenance of CBCs

CBCs can be maintained in CESM containing 20 ng/mL rchLIF for at least 9 d without the addition of other cytokines or feeder cells.

1. After counting the cells, centrifuge the cell suspension (*see* **Subheading 3.2.1.**) at 120g for 10 min.
2. Remove the supernatant and gently resuspend the cell pellet at 4×10^4 cells/mL in 20 ng/mL rchLIF-CESM.
3. Distribute 1 mL to each well of a 24-well culture plate and incubate at 38.5°C in 5% CO_2 and 90% humidity.
4. The medium is partially replaced (by 50%) on every third day.

3.3. Detection of Markers of Undifferentiated CBCs

Alkaline phosphatase (AP) activity, the differentiation antigens recognized by EMA-1 and SSEA-1 are useful markers of chicken ES cells *(6)*. As another method for identifying undifferentiated cells, we measured the level of phosphorylation of STAT3. The advantage of this method is that it directly reflects LIF activity because LIF is responsible for this phosphorylation in ES cells. In fact, the activation of STAT3 by LIF is sufficient to maintain the undifferentiated state of both chicken and mouse ES cells *(9,11,12)*.

3.3.1. AP Reaction of CBCs

1. Measure 45 mL dH_2O and adjust temperature to 18–26°C.
2. Prepare diazonium salt solution by adding 1 mL sodium nitrite solution to 1 mL of Fast Red violet-alkaline solution and mixing gently by inversion. Allow to stand for 2 min.
3. Add diazonium salt solution (**step 2**) to the 45 mL dH_2O (**step 1**).
4. Add 1 mL naphthol AS-BI alkaline solution to diluted diazonium salt solution (**step 3**) and mix thoroughly.
5. Remove the medium from the CBC culture plate (*see* **Subheading 3.2.2.**) and carefully rinse twice with PBS.
6. Add 500 µL alkaline-dye mixture (**step 4**) per well to the washed culture plate and incubate at 18–26°C for 15 min. Protect the culture plate from direct light.
7. Discard the alkaline-dye mixture and rinse three times for 2 min with 1 mL dH_2O per well.
8. Counterstain with 500 mL hematoxylin solution for 2 min and rinse three times for 2 min with 1 mL dH_2O.
9. View and photograph the cultures in distilled water (*see* **Fig. 7**).

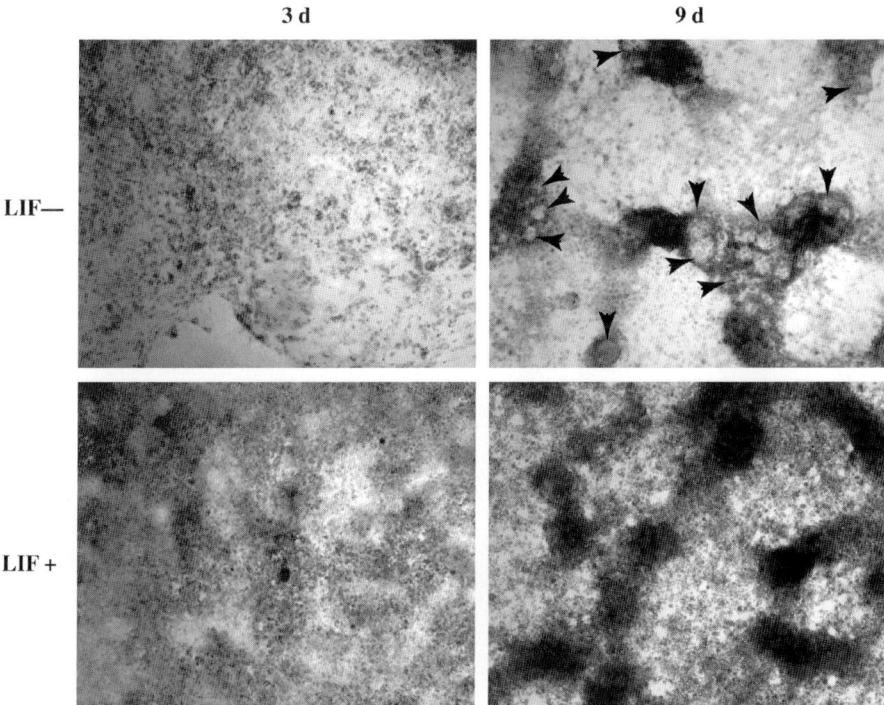

Fig. 7. AP staining of chicken embryonic stem (ES) cells after 3 and 9 d in culture. When ES cells stain positive for alkaline phosphatase (AP), this indicates that they are in an undifferentiated state. Cultures with rchLIF (LIF+) contained a high number of AP-positive cells compared to cultures without rchLIF (LIF−). After 9 d in culture, almost all of the ES cells formed cystlike embryoid bodies (arrowheads) in the absence of LIF, but very few appeared in the presence of LIF.

3.3.2. Immunofluorescence of CBCs

1. Remove the medium from the CBC culture plate (*see* **Subheading 3.2.2.**) and carefully rinse three times with cold (4°C) PBS.
2. Fix the cells with cold (4°C) 10% formaldehyde neutral buffer for 30 min at 4°C.
3. For blocking, add 500 µL/well of cold (4°C) 1% skimmed milk and 10% BlockAce-PBS and incubate for 30 min at 4°C.
4. Remove the blocking solution and add 300 µL/well of diluted antibody (anti-EMA-1 [1:50] or anti-SSEA-1 [1:30]) in 10% BlockAce-PBS. Incubate for 1 h at 4°C.
5. Rinse the wells three times with PBS to remove unbound primary antibody.
6. Add 300 µL/well of diluted fluorescein isothiocyanate-conjugated sheep antimouse Ig (1:50) in 10% BlockAce-PBS. Incubate for 30 min at 4°C.
7. Rinse the wells three times with PBS to remove unbound secondary antibody.
8. View and photograph the cultures in PBS (*see* **Fig. 8**).

3.3.3. Detection of Activated STAT3 Using Western Blotting

We have detected activated STAT3 in both CBCs immediately after isolation (*see* **Subheading 3.2.1., step 13**) and cultured CBCs (*see* **Subheading 3.2.2.**). However,

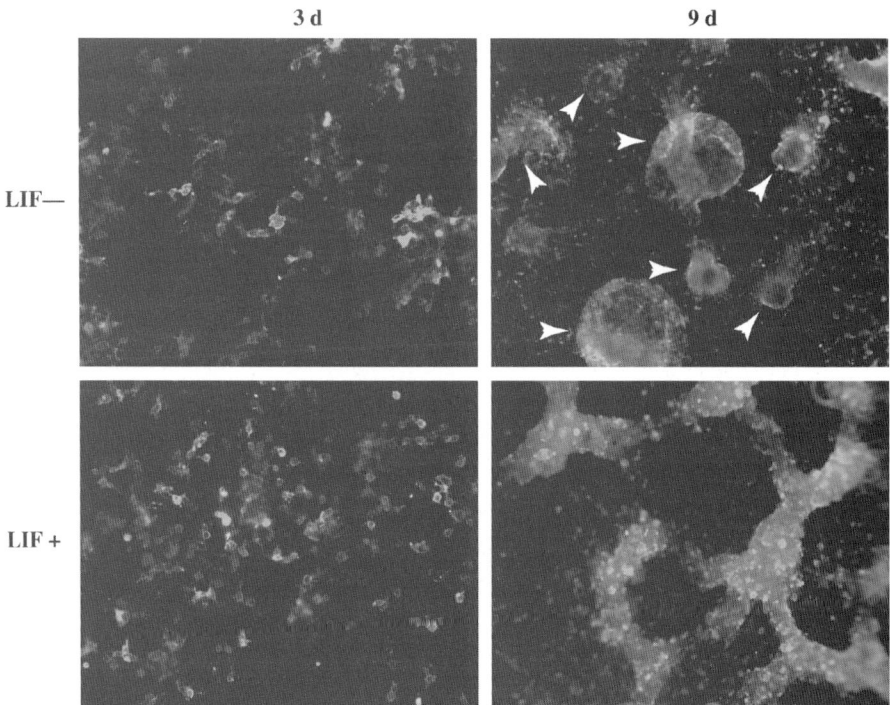

Fig. 8. Immunofluorescence, after labeling with anti-SSEA-1 antibody, of chicken embryonic stem (ES) cells after 3 and 9 d in culture. ES cells that are anti-SSEA-1-positive are in an undifferentiated state. Cultures with recombinant chicken leukemia inhibitory factor (rchLIF) (LIF+) contained a large number of anti-SSEA-1-positive cells compared to cultures without rchLIF (LIF–). After 9 d of culture, almost all of the ES cells formed cystlike embryoid bodies (arrowheads) in the absence of LIF, but very few appeared in the presence of LIF.

when CBCs are cultured without chicken LIF for 24 h or more, the levels of endogenous activated STAT3 decrease. We describe here the method used to detect activated STAT3 in fresh CBCs.

1. Incubate isolated CBC (*see* **Subheading 3.2.1., step 13**) with serum-free CESM containing rchLIF for 15 min at 38.5°C.
2. Lyse the cells ($1-2 \times 10^5$ cells) with 400 μL cold (4°C) cell extraction buffer in a microcentrifuge tube and rotate (10 cycles/min) for 16 h at 4°C.
3. Centrifuge the tube at 14,000*g* for 10 min and transfer the supernatant to a new tube on ice.
4. Add 300 μL cell lysate to 60 μL of 6X SDS loading buffer, vortex briefly, and heat for 5 min at 90–100°C.
5. Centrifuge briefly, then load 10 μL of the sample onto a 7.5% SDS-polyacrylamide gel.
6. Transfer the separated proteins from the electrophoresis gel to PVDF membrane.
7. Transfer the membrane onto which the protein has been blotted into an appropriate container, such as a Petri dish.
8. Add 50–100 mL of blocking buffer to the container and incubate for 2 h at 37°C with gentle shaking.

Chicken ES Cells

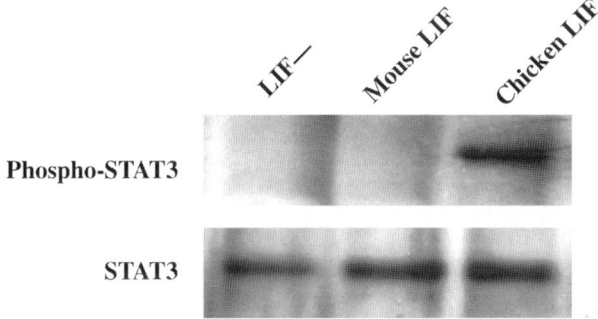

Fig. 9. Detection of activated STAT3 using Western blotting. The CBCs were incubated in serum-free CESM with or without 20 ng/mL recombinant chicken LIF or 20 ng/mL mouse LIF for 15 min at 38.5°C. The application of equivalent amounts of proteins was confirmed by determining nonphosphorylated STAT3 (STAT3). Only treatment with recombinant chicken LIF resulted in the phosphorylation of STAT3 (Phospho-STAT3).

9. Decant and discard the blocking buffer and add 50 mL of diluted anti-STAT3 antibody (1:1000) or diluted anti-Phospho-STAT3 antibody (1:1000) in fresh blocking buffer. Incubate for 10–16 h at 4°C with gentle shaking.
10. Decant and discard the first antibody solution.
11. Rinse the membrane twice by shaking for 15 min in 50 mL fresh blocking buffer.
12. Decant and discard the blocking buffer and add 50 mL diluted horseradish peroxidase-labeled antirabbit antibody (1:3000) in fresh blocking buffer. Incubate for 1 h at 37°C with gentle shaking.
13. Decant and discard the second antibody solution.
14. Rinse the membrane twice by shaking for 15 min in 50 mL fresh blocking buffer.
15. Develop the blot with ECL plus Western blotting detection reagent (see **Fig. 9**).

4. Notes

1. Isogen-LS reagent is a ready-to-use reagent for the isolation of total RNA from cells and tissue. The reagent, a monophasic solution of phenol and guanidine isothiocyanate, is an improvement on the single-step RNA isolation method, similar to Invitrogen's Trizol reagent. In this method, Trizol reagent should work equally well, but we have used Isogen-LS as it is made in Japan.
2. Be careful in case it boils over.
3. This media is high in glucose (4500 mg/L D-glucose) and contains L-glutamine and pyridoxine hydrochloride.
4. Both FBS and chicken serum are heat inactivated by thawing the bottle of serum and incubating at 56°C for 30 min, with agitation every 10 min.
5. To isolate the chicken blastoderm from the yolk, we use filter paper rings (see **Fig. 6A**). To make these, we use a paper punch to make many holes (5-mm diameter) in the filter paper (Advantec, cat. no. 023634007), then we cut around each hole to make rings approx 3 mm wide.
6. To isolate the area pellucida from the chicken blastoderm, we use the bottom of a glass or plastic pipet with a 2- or 3-mm diameter opening (see **Fig. 6E**).
7. The key to successful purification of intact mRNA is speed to avoid degradation by endogenous RNase in the cells. It is crucial during this extraction procedure to use RNase-free instruments and solutions and to change gloves often to minimize the risk of RNase contamination from other sources.

8. Determine the concentration of the RNA using the following equation:

   RNA concentration (μg/μL) = [A_{260} × 40 × Dilution factor] / 1000

9. Determine the concentration of the cDNA using the following equation:

   DNA concentration (μg/μL) = [A_{260} × 50 × Dilution factor] / 1000

10. The 5′ pGEX sequencing primer and the 3′ pGEX sequencing primer can be used to screen transformants rapidly by colony PCR. We recommend determining the sequence of the clones obtained using these sequencing primers.
11. We have used the bulk GST purification module, but the appropriate module should be selected for the scale of your experiment.

References

1. Nichols, J., Evans, E. P., and Smith, A. G. (1990) Establishment of germ-line-competent embryonic stem (ES) cells using differentiation inhibiting activity. *Development* **110,** 1341–1348.
2. Eyal-Giladi, H. and Kochav, S. (1976) From cleavage to primitive streak formation: a complementary normal table and a new look at the first stage of the development of the chick. I. General morphology. *Dev. Biol.* **95,** 321–337.
3. Karagenc, L., Cinnamon, Y., Ginsburg, M., and Petitte, J. N. (1996) Origin of primordial germ cells in the prestreak chick embryo. *Dev. Genet.* **19,** 290–301.
4. Petitte, J. N., Clark, M. E., Liu, G., Verrinder Gibbins, A. M., and Etches, R. J. (1990) The production of somatic and germline chimeras in the chick by transfer of early blastodermal cells. *Development* **108,** 185–190.
5. Fraser, R. A., Carsience, R. S., Clarke, M. E., Etches, R. J., and Gibbins, A. M. V. (1993) Efficient incorporation of transfected blastodermal cells into chimeric chicken embryos. *Int. J. Dev. Biol.* **37,** 381–385.
6. Etches, R. J., Clarke, M. E., Zajchowski, L., and Speksnijder, G. (1996) Manipulation of blastodermal cells. *Poultry Sci.* **76,** 1075–1083.
7. Kagami, H., Tagami, T., Matsubara, Y., et al. (1997) The developmental origin of primordial germ cells and the transmission of the donor-derived gametes in mixed-sex germline chimeras to the offspring in the chicken. *Mol. Reprod. Dev.* **48,** 1–10.
8. Pain, B., Clarke, M. E., Shen, M., et al. (1996) Long-term in vitro culture and characterization of avian embryonic stem cells with multiple morphogenetic potentialities. *Development* **122,** 2339–2348.
9. Horiuchi, H., Tategaki, A., Yamashita, Y., et al. (2004) Chicken leukemia inhibitory factor maintains chicken embryonic stem cells in undifferentiated state. *J. Biol. Chem.* **279,** 24,514–24,520.
10. Inoue, M. and Sato, A. (1988) Establishment and in vitro differentiation of a chicken monocytic leukemia cell line. *Jpn. J. Vet. Sci.* **50,** 648–653.
11. Niwa, H., Burdon, T., Chambers, I., and Smith, A. (1998) Self-renewal of pluripotent embryonic stem cells is mediated via activation of STAT3. *Genes Dev.* **12,** 2048–2060.
12. Matsuda, T., Nakamura, T., Nakao, K., et al. (1999) STAT3 activation is sufficient to maintain an undifferentiated state of mouse embryonic stem cells. *EMBO J.* **18,** 4261–4269.

3

Derivation and Culture of Mouse Trophoblast Stem Cells In Vitro

Satoshi Tanaka

Summary

In the mouse preimplantation embryo, the first cell fate determination segregates two morphologically and functionally distinct cell lineages. One is the inner cell mass, and the other is the trophectoderm. A subset of the trophectoderm maintains a proliferative capacity and forms the extraembryonic ectoderm, the ectoplacental cone, and the secondary giant cells of the early conceptus after implantation. A stem cell population of the trophectoderm lineage can be isolated and maintained in vitro under the presence of fibroblast growth factor 4, heparin, and a feeder layer of mouse embryonic fibroblast cells. Such apparently immortal stem cells, trophoblast stem (TS) cells, exhibit the potential to differentiate to multiple cell types in vitro. TS cells also have the ability to contribute to normal development in chimeras. However, TS cells exclusively contribute to the trophoblastic component of the placenta and of the parietal yolk sac, making a striking contrast with embryonic stem cells, which never contribute to these tissues in chimeras. In this chapter, detailed protocols for the isolation and establishment of TS cell lines from blastocysts and their maintenance are described.

Key Words: Blastocyst; fibroblast growth factor; heparin; mouse embryonic fibroblast cell; trophoblast; TS cell.

1. Introduction

During blastocyst formation, irreversible segregation of two morphologically and functionally distinct cell lineages takes place in the mammalian embryo. One is the inner cell mass (ICM), and the other is the trophectoderm. The trophectoderm, a monolayer of polarized epithelial cells, forms a shape of a ball encasing the ICM at one end of its inner surface. As it is also mentioned in other chapters, the ICM gives rise to germ cells and all somatic cells of the fetus as well as some extraembryonic membranes. On the other hand, the polar trophectoderm (the subset of the trophectoderm that is in direct contact with the ICM) maintains a proliferative capacity and forms the extraembryonic ectoderm, the ectoplacental cone, and the secondary giant cells of the early conceptus after implantation *(1)*.

It was found that a stem cell population of the trophoblast lineage can be isolated from mouse embryos and maintained in a proliferative, undifferentiated state in the presence of fibroblast growth factor (FGF)4 with its cofactor, heparin, and a feeder layer of mouse embryonic fibroblast (MEF) cells *(2)*. Such stem cells, trophoblast stem (TS) cells, mostly differentiate into giant cells on the removal of FGF4/heparin or MEF cells. Expression analysis of marker genes, however, has indicated that subpopulations of TS cells also differentiate into spongiotrophoblast and labyrinthine trophoblast cell fates in vitro *(2–4)*. TS cells have the ability to contribute to normal development in chimeras. However, TS cells exclusively contribute to the trophoblastic component of the placenta and of the parietal yolk sac, making a striking contrast with ES cells, which never contribute to these tissues in chimeras.

TS cell lines have been utilized as a powerful system not only for examining gene functions or signal pathways controlling trophoblast cell development (*see* **ref. 5** for review) but also for addressing an attractive, yet challenging, issue: what "stemness" is. Ko and colleagues, for example, performed transcriptome analyses of mouse preimplantation embryos, embryo-derived stem cell lines (ES) (including TS cells), and adult stem cells as an approach toward a possible definition of a molecular scale of cellular potency *(6,7)*. We recently analyzed the genomewide DNA methylation status of TS cells and compared it with that of ES and embryonic germ (EG) cells to approach the same issue from an epigenetic point of view *(8)*.

TS cell lines can be derived from preimplantation blastocysts or the trophoblastic tissue of postimplantation embryos *(2,9)*. This chapter focuses on the protocols for isolation and establishment of TS cell lines from blastocysts and their maintenance.

2. Materials

2.1. Equipment

1. Tissue culture dishes (Nalge Nunc International, Tokyo, Japan; cat. no. 168381, 150 mm; 172958, 100 mm; 150288, 60 mm; 153066, 30 mm).
2. Four-well multidish (Nalge Nunc International, cat. no. 176740).
3. Nontissue culture-treated, U-bottom, 96-well plate (BD Biosciences, Tokyo, Japan; cat. no. 351177).
4. 50-mL centrifuge tube (Corning, Tokyo, Japan; cat. no. 430291).
5. Cryovial (Nalge Nunc International, cat. no. 377224).
6. Cell-freezing container (e.g., Cryo 1°C freezing container, Nalge Nunc International, cat. no. 5100-0001).

2.2. Reagents

1. Phosphate-buffered saline (PBS) (−): calcium/magnesium-free PBS: dissolve one tablet (Sigma-Aldrich, Tokyo, Japan; cat. no. P4417) per 200 mL water. Autoclave to sterilize and store at 4°C.
2. 0.05% trypsin/0.2 mM ethylenediaminetetraacetic acid (EDTA): dilute 0.25% trypsin/1 mM EDTA·4Na (Invitrogen, Tokyo, Japan; cat. no. 27250-018) with 4X volume of PBS(−). Store at 4°C.

Table 1
Numbers of MMC-Treated MEF Cells to Plate

		Diameter of Dish (mm)			
	4-well plate	35	60	100	150
Coculture	4×10^4/well	2×10^5	4×10^5	1.2×10^6	3×10^6
Conditioned medium				2.4×10^6	6×10^6

3. 0.1% trypsin/0.4 mM EDTA: dilute 0.25% trypsin/1 mM EDTA·4Na (Invitrogen, cat. no. 27250-018) with 1.5X volume of PBS(−).
4. FGF4 1000X stock solution (25 µg/mL): add 1 mL of PBS(−) containing 0.1% (w/v) bovine serum albumin (BSA) to a vial of lyophilized human recombinant FGF4 (25 µg; PeproTech, London, UK; cat. no. 100-31). Mix well by gentle pipetting and freeze in 100-µL aliquots at −80°C. Sterilization of FGF4 solution is not necessary as long as sterile PBS(−) containing BSA (passed through 0.45-µm filter) is used. Thaw each aliquot as needed and store at 4°C; do not refreeze.
5. Heparin 1000X stock solution (1 mg/mL): resuspend heparin (Sigma-Aldrich, cat. no. H3149) in PBS(−) to a concentration of 1.0 mg/mL and store at −80°C in 100-µL aliquots. Thaw each aliquot as needed and store at 4°C; do not refreeze.
6. Mitomycin C (MMC) (2 mg; Sigma-Aldrich, cat. no. M0503).
7. 10 mM β-mercaptoethanol stock solution: dilute pure liquid of 14.3 M β-mercaptoethanol (Sigma-Aldrich, cat. no. M7522) with water. Store at −20°C.

2.3. Culture Media

1. Dulbecco's modified Eagle medium (DMEM) plus 10% fetal bovine serum (FBS): DMEM (pH 7.2) supplemented with 10% FBS, 100 µM β-mercaptoethanol, 2 mM L-glutamine, 50 U/mL penicillin, and 50 µg/mL streptomycin. For 1 L, add 100 mL FBS to 870 mL DMEM (Invitrogen, cat. no. 12800-017). Add 10 mL each of 10 mM β-mercaptoethanol stock solution, 200 mM L-glutamine (Invitrogen, cat. no. 25030-081), and 100X penicillin-streptomycin solution (Invitrogen, cat. no. 15070-063). Store at 4°C.
2. TS medium: RPMI1640 (pH 7.2) supplemented with 20% FBS, 100 µM β-mercaptoethanol, 2 mM L-glutamine, 1 mM sodium pyruvate, 50 U/mL penicillin, and 50 µg/mL streptomycin. For 1 L, add 200 mL FBS to 760 mL RPMI1640 (pH 7.2; Invitrogen, cat. no. 31800-022). Add 10 mL each of 10 mM β-mercaptoethanol stock solution, 200 mM L-glutamine (Invitrogen, cat. no. 25030-081), 100 mM sodium pyruvate (Invitrogen, cat. no. 11360-070), and 100X penicillin-streptomycin solution (Invitrogen, cat. no. 15070-063). Store at 4°C.
3. TS + F4H: TS medium containing 25 ng/mL FGF4 and 1 µg/mL heparin. Add 10 µL each of FGF4 and heparin 1000X stock solutions to 10 mL TS medium.
4. TS + 1.5X F4H: TS medium containing 37.5 ng/mL FGF4 and 1.5 µg/mL heparin. Add 15 µL each of FGF4 and heparin stock solutions to 10 mL TS medium.
5. MEF-CM: TS medium conditioned by feeder cells. Culture MMC-treated MEF cells (*see* **Subheading 3.1.**) in TS medium (without FGF4 and heparin) for 3 d (*see* **Table 1** for adequate number of MEF to seed). Collect the medium and store at −20°C while preparing additional batches. Prepare two more batches with the same dish of MMC-treated MEF cells. Combine three batches of conditioned media and spin at approx 2300g, 4°C, for 20 min to remove debris. Filter (0.45 µm) and store at −20°C in aliquots. Thaw each aliquot as needed and store at 4°C; do not refreeze.

6. 70CM + F4H: add 3 mL TS medium and 10 µL each FGF4 and heparin 1000X stock solutions to 7 mL MEF-CM.
7. 70CM + 1.5X F4H: add 3 mL TS medium and 15 µL each FGF4 and heparin 1000X stock solutions to 7 mL MEF-CM.
8. 70CM + 1.8X F4H: add 3 mL TS medium and 18 µL each FGF4 and heparin 1000X stock solutions to 7 mL MEF-CM.
9. 2X freezing medium for MEF: 50% FBS, 20% dimethylsulfoxide in DMEM.
10. 2X freezing medium for TS: 50% FBS, 20% dimethylsulfoxide in TS medium.

3. Methods (see Note 1)

3.1. Preparation of MMC-Treated Feeder Cell Stocks (see Note 2)

1. Quickly thaw a frozen vial of MEF (see **Note 3**) at 37°C.
2. Transfer entire contents to 10 mL DMEM/10% FBS in a 50-mL tube and centrifuge at 200g for 3 min.
3. Discard supernatant, loosen the cell pellet by gently tapping bottom of the tube, then resuspend the cells in 25 mL DMEM/10% FBS. Split cells onto five 150-mm dishes, each containing 20 mL DMEM/10% FBS.
4. Incubate cells at 37°C, 5% CO_2/95% air.
5. Change medium the next day to remove cell debris.
6. When the cells become approx 90% confluent (approx 3 d after thawing), remove the medium, rinse cells with 10 mL PBS(−) twice, and then add 3 mL 0.05% trypsin/0.2 mM EDTA/PBS to each dish.
7. Incubate at 37°C, 5% CO_2/95% air for 3–5 min.
8. Add 3 mL DMEM/10% FBS to each dish.
9. Gather cell suspensions in a single 50-mL tube, spin down (200g, 3 min), and discard supernatant.
10. Loosen the cell pellet by gently tapping bottom of the tube, then resuspend the cells in 40 mL DMEM/10% FBS.
11. Split cells onto 20 150-mm dishes, each containing 23 mL DMEM/10% FBS.
12. Incubate cells at 37°C, 5% CO_2/95% air and change medium the next day.
13. When the cells become almost confluent (2–3 d after the passage), remove the medium and add 10 mL DMEM/10% FBS containing 10 µg/mL MMC (see **Note 4**).
14. Incubate cells at 37°C, 5% CO_2/95% air for 2 h.
15. Remove the medium and rinse cells twice with 10 mL PBS(−).
16. Trypsinize cells as described in **steps 6–10**. Finally, resuspend the cell pellet in 20 mL DMEM/10% FBS.
17. Count viable cells and dilute to desired density with DMEM/10% FBS. Another set of spinning down/resuspending steps may be required to accomplish high density (see **Note 5**).
18. Add equal volume of 2X freezing medium for MEF and mix well by gentle pipetting.
19. Add 1 mL cell suspension to each cryovial, store at –80°C in a cell-freezing container overnight.
20. The next day, transfer the vials to liquid nitrogen tank (see **Note 6**).

3.2. Establishment of TS Cell Lines From Blastocyst

As the culture conditions for establishing TS cell lines are not truly exclusive, isolation of TS cells from blastocysts often results in a contamination of rapidly growing unidentified cell types. The following protocol describes the case when there is such contamination (see **Note 7**).

Trophoblast Stem Cell Lines

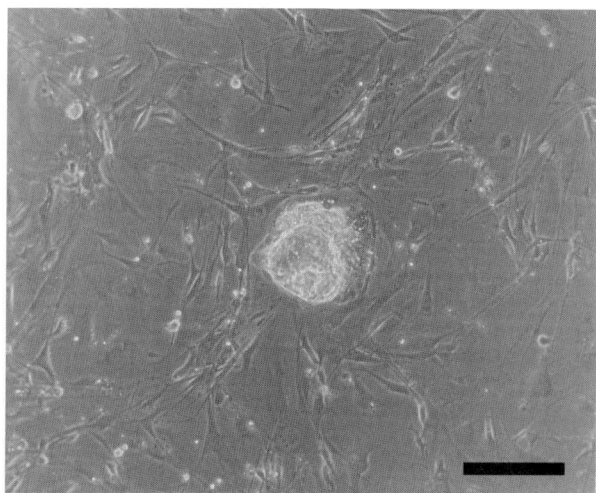

Fig. 1. A blastocyst outgrowth on feeder cells (d 3). Scale bar = 200 µm.

1. Day 0: thaw frozen stocks of MMC-treated MEF cells and seed on four-well plates (4×10^4 cells per well) in TS medium the day before collecting blastocysts.
2. Day 1: replace the medium with TS + F4H medium before flushing blastocysts.
3. Flush 3.5 *dpc* blastocysts from the uterine horns *(10)* and place one blastocyst per well in the four-well plates containing feeder cells; culture at 37°C, 5% CO_2/95% air.
4. Day 2: the blastocysts should hatch and attach to the wells. Embryos of certain strains may take more time and may still be floating at this point (*see* **Note 8**).
5. Day 3: a small outgrowth should be formed from each embryo (**Fig. 1**). Change the medium with fresh 500 µL TS+F4H medium gently or the attached embryo could detach.
6. Day 4: once the outgrowth becomes a suitable size (**Fig. 2**), remove the medium by aspiration and wash the cells twice with 500 µL PBS(−). Discard the PBS(−), add 100 µL 0.1% trypsin/0.4 m*M* EDTA, and incubate for 5 min at 37°C, 5% CO_2/95% air. Disaggregate the cell clump by pipetting through a "yellow tip" vigorously (*see* **Note 9**). Add 500 µL 70CM + 1.8X F4H and incubate at 37°C, 5% CO_2/95% air.
7. Change the medium 8 h after **step 5** (500 µL 70CM + 1.5X F4H). This step is optional but recommended.
8. Day 6: change the medium (500 µL 70CM + 1.5X F4H) and continue to refeed every other day.
9. Days 7–12 (highly variable, *see* **Note 10**): TS cell colonies will begin to appear. They look like flat, epithelial sheets with a distinctive colony boundary (**Fig. 3**). In many cases, colonies of morphologically distinct cells will also appear (**Fig. 4**). TS cell colonies should be picked up in as follows in such cases.
10. Prepare feeder cells in four-well plates as previously described. Refeed cells with 300 µL TS + 1.5X F4H just before picking TS cells.
11. Add 50 µL 0.1% trypsin/0.4 m*M* EDTA into each well of nontissue culture-treated, U-bottom, 96-well plate before picking TS colonies.
12. Remove the medium from four-well plates containing TS cell colonies and wash the cells twice with 500 µL PBS(−). Do not aspirate the PBS(−) after second wash.

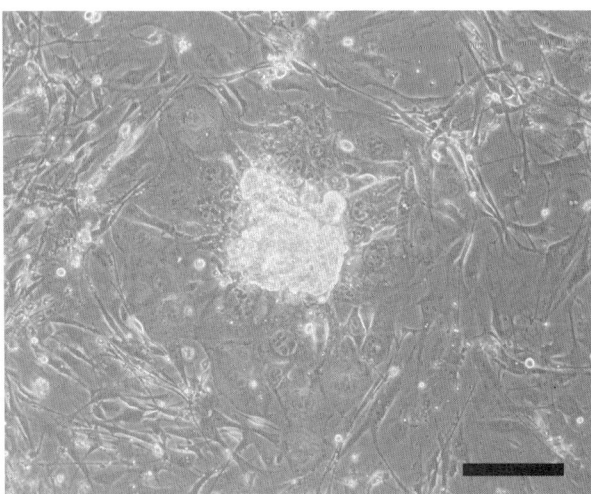

Fig. 2. A blastocyst outgrowth on feeder cells (d 4) before trypsinization. Scale bar = 200 µm.

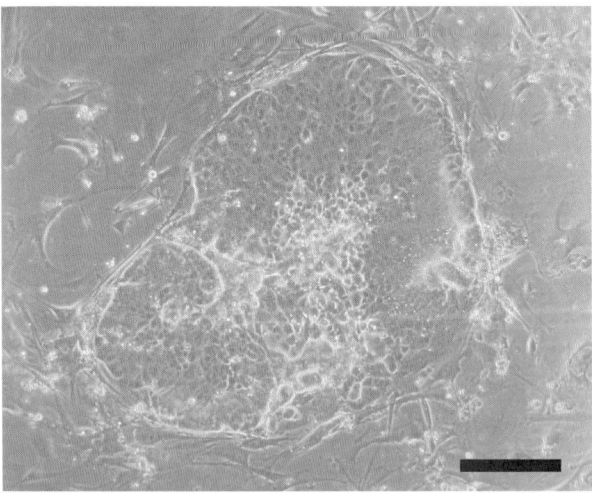

Fig. 3. A typical colony of trophoblast stem cells (d 12). Note an epithelial cell-like appearance with a distinctive colony boundary. Scale bar = 200 µm.

13. Under a dissection microscope in a tissue culture hood, pick TS cell colonies with Pipetteman set at 1 µL. Put lifted TS cells into the 96-well plate containing trypsin, then incubate at 37°C, 5% CO_2/95% air for 5 min.
14. Disaggregate the cell clumps by pipetting. Add 150 µL TS + 1.5X F4H to stop trypsinization and then transfer cells to four-well plates containing fresh feeder cells (**step 10**).
15. Change the medium 8 h after **step 14** (500 µL TS + 1.5X F4H). This step is optional but recommended.

Trophoblast Stem Cell Lines

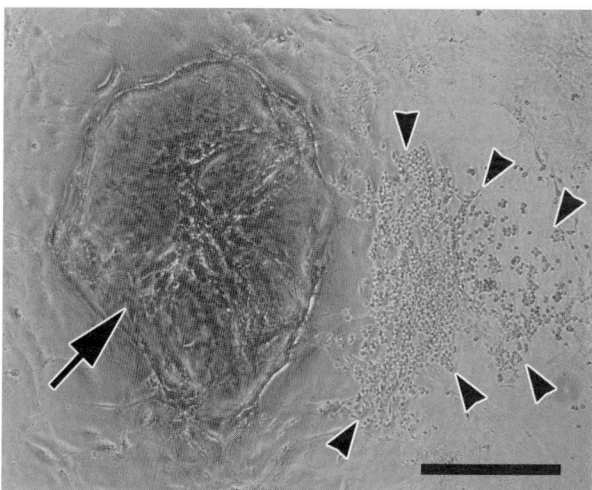

Fig. 4. An example of contaminating unidentified cells (arrowheads) with a trophoblast stem (TS) cell colony (arrow). Note a morphological distinction of these cells from either the TS cells or the feeder cells. Scale bar = 500 μm.

16. Continue to refeed cells every 2 d. TS cell colonies should reappear 3–4 d after **step 14**. If the TS cells reach half-confluency within 1 wk after **step 14**, proceed to next step. If only a few TS cells reappear, **steps 6–8** should be repeated.
17. Passage the half-confluent well of TS cells to a 35-mm dish of preplated feeders (2×10^5 cells per dish). After this step, the cells can be cultured in TS + F4H. Change the medium every other day. Culture of TS cells should gradually be expanded up to a 100-mm dish, keeping a 1.10 dilution ratio at each passage before the cells are frozen for storage.

3.3. Passage of TS Cells

1. When the TS cells reach approx 80% confluency (normally 4 d from previous passage), aspirate the medium, wash the cells twice with PBS(−), and add 0.05% trypsin/0.2 mM EDTA (2 mL per 100-mm dish) (*see* **Note 11**).
2. Incubate for 3–5 min at 37°C, 5% CO_2/95% air.
3. Add TS medium (equal volume to trypsin) to stop trypsinization.
4. Disaggregate cell clumps by pipetting up and down with glass pipet.
5. Transfer the cell suspension into 50-mL tube and spin down (200g, 3 min). Discard supernatant.
6. Loosen the cell pellet by gently tapping bottom of the tube, then resuspend the cells in TS+F4H (*see* **Note 12**).
7. Seed the TS cells onto preplated MMC-treated MEF cells in the ratio of 1:10–1:20 (approx 5×10^4 cells/35-mm dish). Refer to **Table 1** for cell numbers of MEF cells to plate (*see* **Note 13**).
8. Incubate at 37°C, 5% CO_2/95% air. Change the medium every 2 d. Repeat **steps 1–7**.

3.4. Removal of MEF From the Culture of TS Cells

Removal of feeder cells from culture will be desired when DNA/RNA are extracted from TS cells. This can be achieved by culturing TS cells in a feeder-free condition (*see* **Subheading 3.5.**), gradually diluting mitotically inactive MEF cells through several passages. Immediate removal of MEF cells can also be done by taking advantage of differences in adherence to the dish surface between MEF and TS cells.

1. Trypsinize and pellet the TS cells co-cultured with MMC-treated MEF cells as described in **Subheading 3.3.**
2. Resuspend the cells in 70CM + F4H and seed on a new dish containing no MEF cells.
3. Incubate for 30 min at 37°C, 5% CO_2/95% air.
4. Transfer supernatant to a new dish and incubate for another 30 min at 37°C, 5% CO_2/95% air.
5. Remove the supernatant, which should consist almost entirely of TS cells. These cells can be cultured further in a feeder-free condition (*see* **Subheading 3.5.**).

3.5. Feeder-Free Culture of TS Cells

1. Culture the TS cells in 70CM + F4H on standard tissue culture dish. The dishes do not need to be gelatin coated.
2. Passage the cells as described in **Subheading 3.3.**

3.6. Freezing and Thawing of TS Cells (see Note 14)

3.6.1. Freezing

1. Prepare 2X freezing medium for TS.
2. Harvest TS cells from an approx 80% confluent 100-mm dish by trypsinization and pellet them (*see* **Subheading 3.3.**).
3. Resuspend the cells in 1.5 mL TS medium.
4. Add 1.5 mL 2X freezing medium for TS; mix gently.
5. Add the cell suspension into cryovials (1 mL per vial; three vials from a 100-mm dish) (*see* **Note 15**).
6. Put the cryovials into a cell-freezing container and immediately place the container in −80°C freezer overnight.
7. The next day, transfer the cryovials to a liquid nitrogen tank.

3.6.2. Thawing

1. Quickly thaw a frozen vial of TS cells (containing one-third of a 100-mm culture) at 37°C.
2. Transfer entire contents to 10 mL TS + F4H in a 50-mL tube and centrifuge at 200g for 3 min.
3. Discard supernatant, loosen the cell pellet by gently tapping bottom of the tube, and resuspend the cells in 30 mL TS + F4H. Split cells onto three 100-mm dishes, each containing preplated MMC-treated MEF cells.
4. Incubate cells at 37°C, 5% CO_2/95% air.
5. Change medium the next day to remove cell debris.
6. Feed the cells every other day. Passage the cells when the cells reach approx 80% confluency (normally 3–4 d after thawing the cells).

4. Notes

1. As a general rule, all of the culture media and buffers must be prewarmed in 37°C water bath or air incubator before use but do not leave trypsin-containing solutions too long.
2. We found it more convenient to prepare stocks of MMC-treated feeder cells frozen at several different densities rather than treating cells with MMC just before use to flexibly and quickly meet various requirements.
3. A vial contains cells harvested from an 80–90% confluent 150-mm dish.

4. A vial available from Sigma-Aldrich contains 2 mg MMC, making it convenient to prepare an exact volume of MMC-containing medium for 20 150-mm dishes.
5. **Caution:** the density is 2X final density.
6. We routinely freeze MMC-treated MEF cells at three different densities: 6×10^6, 1.2×10^6, and 4×10^5 cells per vial.
7. Starting with mechanically and enzymatically isolated trophoblastic tissue of postimplantation embryos would greatly reduce the chance of contamination by unwanted cells; however, more skill in handling and dissecting embryos is required.
8. Things go slowly, at least for us, with the embryos from mouse strains containing a B6 background (C57BL/6, B6C3F1, and B6D2F1). Eventually, most of them hatch and attach on d 3.
9. We prefer to perform this with P200 Pipetteman set at 95 µL to ensure that the entire contents go through the tip. A drawn Pasteur pipet would also be effective.
10. Again, cells from B6, B6C3F1, and B6D2F1 mice grow more slowly compared with those from CD-1 (ICR). There is also a well-to-well difference even in a single experiment.
11. Uy et al. *(9)* have found that pronase works better in disaggregating TS cell colonies.
12. Other members of the FGF family, FGF1 and FGF2, have been found to successfully replace FGF4 in maintaining TS cells in a proliferative, undifferentiated state *(9,11)*.
13. These are the cell numbers we empirically found to be adequate. Note that we use frozen stocks of MMC-treated MEF cells, and numbers of viable cells are counted before freezing. Numbers may, therefore, be reduced when freshly prepared feeder cells are used. TS cell colonies seem to spread on the surface of the culture dish, pushing MEF cells aside as they expand. Smaller numbers of MEF are used for TS cell culture compared with the case of ES cell culture to leave some space.
14. The general precaution for freezing and thawing (i.e., freeze slowly and thaw quickly) also applies to TS cells.
15. TS cells can be frozen at a lower density. For example, nine vials could be made from an approx 80% confluent 100-mm dish. A vial of cells in such case should be plated on a single 100-mm dish. Keep this 1/9 rule.

References

1. Rossant, J. and Cross, J. C. (2001) Placental development: lessons from mouse mutants. *Nat. Rev. Genet.* **2**, 538–548.
2. Tanaka, S., Kunath, T., Hadjantonakis, A. K., Nagy, A., and Rossant, J. (1998) Promotion of trophoblast stem cell proliferation by FGF4. *Science* **282**, 2072–2075.
3. Yan, J., Tanaka, S., Oda, M., Makino, T., Ohgane, J., and Shiota, K. (2001) Retinoic acid promotes differentiation of trophoblast stem cells to a giant cell fate. *Dev. Biol.* **235**, 422–432.
4. Hughes, M., Dobric, N., Scott, I. C., et al. (2004) The Hand1, Stra13 and Gcm1 transcription factors override FGF signaling to promote terminal differentiation of trophoblast stem cells. *Dev. Biol.* **271**, 26–37.
5. Kunath, T., Strumpf, D., and Rossant, J. (2004) Early trophoblast determination and stem cell maintenance in the mouse—a review. *Placenta* **25**, Suppl. A, S32–S38.
6. Tanaka, T. S., Kunath, T., Kimber, W. L., et al. (2004) Gene expression profiling of embryo-derived stem cells reveals candidate genes associated with pluripotency and lineage specificity. *Genome Res.* **12**, 1921–1928.
7. Sharov, A. A., Piao, Y., Matoba, R., et al. (2003) Transcriptome analysis of mouse stem cells and early embryos. *PLoS Biol.* **1**, 410–419.

8. Shiota, K., Kogo, Y., Ohgane, J., et al. (2004) Epigenetic marks by DNA methylation specific to stem, germ and somatic cells in mice. *Genes Cells* **7,** 961–969.
9. Uy, G. D., Downs, K. M., and Gardner, R. L. (2002) Inhibition of trophoblast stem cell potential in chorionic ectoderm coincides with occlusion of the ectoplacental cavity in the mouse. *Development* **129,** 3913–3924.
10. Nagy, A., Gertsenstein, M., Vintersten, K., and Behringer, R. (2003) *Manipulating the Mouse Embryo,* 3rd ed., Cold Spring Harbor Laboratory Press, New York.
11. Kunath, T., Strumpf, D., Rossant, J., and Tanaka, S. (2001) Trophoblast stem cells, in *Stem Cell Biology* (Marshak, D. R., Gardner, R. L., and Gottlieb, D., eds.), Cold Spring Harbor Laboratory Press, New York, pp. 267–286.

4

Derivation, Maintenance, and Characterization of Rat Embryonic Stem Cells In Vitro

Maren Schulze, Hendrik Ungefroren, Michael Bader, and Fred Fändrich

Summary

The in vitro differentiation of mouse embryonic stem (ES) cells into different somatic cell types such as neurons, endothelial cells, or myocytes is well established, and many mouse ES cell lines have been created so far. The establishment of rat ES cell lines, however, has proven to be difficult. Most attempts to culture rat ES cell lines and maintain them in an undifferentiated state have failed, so researchers were forced to abandon this system and use mouse ES cells. This chapter describes the long-term cultivation of an alkaline phosphatase-positive rat embryonic stem cell-like line (RESC) and their differentiation into neuronal, endothelial, and hepatic lineages. The RESCs can be characterized by typical growth in single cells as well as embryoid bodies when cultivated in the presence of leukemia inhibitory factor. RESC expressed stage-specific-embryonic antigen 1 and the major histocompatibility class 1 molecule. Neuronal differentiation is achieved by standard retinoic acid treatment and endothelial differentiation can be reproducibly induced by growth on or within Matrigel® for 14 d. To induce expression of hepatocyte-specific antigens, RESCs were either grown in hepatocyte-conditioned media or in media containing different combinations of growth factors. The characterization of differentiated cells was done primarily by immunohistochemistry, enzyme-linked immunosorbent assay, and polymerase chain reaction.

Key Words: Differentiation; feeder layer; leukemia inhibitory factor; rat embryonic stem cells; subculturing.

1. Introduction

The culture and maintenance of mouse ES cells are widely established, and initial hazards have been overcome, resulting in much progress in the field. Rat embryonic stem (ES) cells, however, have always been subject to criticism because there have been problems creating stable undifferentiated cell lines, and optimal growing conditions proved difficult to establish, thus making feeder layer-free culture almost impossible. The differentiation of rat ES cells has only been described recently *(1)*.

This chapter describes the culture of rat ES cells, particularly the single cell-cloned line termed C12-WKY that contains approx 70% euploid cells. Special protocols for differentiation into cells of all three germ layers are provided, including a description of hazardous culturing steps and special problems. The characterization of C12-derived somatic cell types by immunohistochemistry and by functional studies is described in a way that it can easily be reproduced by other investigators.

2. Materials

1. Phosphate buffered saline (PBS) (Cambrex, Apen, Germany; cat. no. 17-512F).
2. Dulbecco's modified Eagle's medium (DMEM) (Cambrex, cat. no. Be12-614F).
3. Penicillin/streptomycin (1000 U/mL, 100 mL; Invitrogen, Karlsruhe, Germany; cat. no. 15140144).
4. L-Glutamine (200 mM, 100X, 100 mL; Invitrogen, cat. no. 25030-024).
5. Fetal bovine serum (FBS) (500 mL; Invitrogen, cat. no. 10084-077). Inactivated at 56°C for 30 min (*see* **Note 1**).
6. β-Mercaptoethanol (Merck, Darmstadt, Germany; cat. no. 805740).
7. Trypsin/ethylenediaminetetraacetic acid (Invitrogen, cat. no. 35400-02).
8. Nonessential amino acids (Invitrogen, cat. no. 11140-035).
9. Tissue culture flask T-75 (Greiner, Frickenhausen, Germany; cat. no. 658175).
10. 15-mL tubes (Greiner, cat. no. 188271).
11. 50-mL tubes (Greiner, cat. no. 227261).
12. Cryovials (Nunc, Wiesbaden, Germany).
13. Six-well plates (Sarstedt, Nümbrecht, Germany; cat. no. 83.1839).
14. Cover slips (15 × 15 mm; Microm, Waldorf, Germany; cat. no. 9161015).
15. Slides (Menzel slides; Microm, cat. no. 1000200).
16. Recombinant human insulin (Sigma, Deisenhofen, Germany; cat. no. I9278). Prepare a 10-mg/mL solution in 25 mM HEPES (pH 8.2): dissolve 10 mg in 1 mL PBS with 25 mM HEPES buffer and 0.1% bovine serum albumin (BSA). Aliquot into 50 μL portions, each containing 0.5 mg.
17. Mouse leukemia inhibitory factor (LIF; Chemicon, Hofheim, Germany; cat. no. 2010).
18. RPMI 1640, with out Phenol red and L-glutamine (Cambrex, cat. no. Be12-918F).
19. Carboxy-fluorescein diacetate succinimidyl ester (CFSE; Sigma, cat. no. C5041). Prepare a 10 mM stock solution: dilute 500 μg *CFDA* 8 in 90 μL DMSO. For a 25-μM working solution, dilute 1 μL stock solution in 400 μL PBS.
20. Retinoic acid (Sigma, cat. no. R2625).
21. Fibroblast growth factor (FGF)4 (25 μg; Sigma, cat. no. F8424).
22. Epidermal growth factor (EGF) (100 mg; Calbiochem, Darmstadt, Germany; cat. no. 324851).
23. Hepatocyte growth factor (5 μg; Calbiochem, cat. no. 375228).
24. DiLDL (Biomedical Technologies, Magdeburg, Germany; cat. no. BT-904).
25. Albumin enzyme-linked immunosorbent assay (Bethyl Laboratories, Montgomery, TX; cat. no. BET-E110-125).
26. Red oil (Sigma, cat. no. O-0625).
27. Collagenase, type 1 (Sigma, cat. no. C-0130).
28. Gelatin (Sigma, cat. no. G-7765). For coating tissue culture dishes, prepare a 1% gelatin solution dissolved in sterile aqua dest. To dissolve the gelatin, stir the solution and heat overnight at 30°C. Sterilize, preferably by autoclaving.
29. Matrigel (Becton Dickinson, Heidelberg, Germany; cat. no. 356234).
30. Methanol.

31. peqGOLD RNAPure (peqlab, Erlangen, Germany; cat. no. 301010).
32. SuperScript II reverse transcriptase (Invitrogen, cat. no. 18064-014).
33. Taq-polymerase (Invitrogen, cat. no. 18038-026).
34. Oligo d(T)$_{12-16}$ (Applied Biosystems, Darmstadt, Germany; cat. no. N808-0128).
35. dNTP Set (MBI Fermentas, St-Leon Rod, Germany; cat. no. RO181).
36. Ethidium bromide (Serva Electrophoresis, Heidelberg, Germany; cat. no. 21251).
37. Agarose (Invitrogen, cat. no. 15510-127).
38. Adenosine (Sigma, cat. no. A9251).
39. Guanosine (Sigma, cat. no. A1644).
40. Cytidine (Sigma, cat. no. C122106).
41. Uridine (Sigma, cat. no. A1882).
42. Thymidine (Sigma, cat. no. A4044).
43. Nucleoside solution: 80 mg adenosine, 85 mg guanosine, 73 mg cytidine, 73 mg uridine, 24 mg thymidine dissolved in 100 mL double-distilled water (ddH$_2$O). Freeze 10-mL aliquots at $-20°C$.

2.1. Tissue Culture

2.1.1. Media

1. Medium for C12 rat ES cells: high-glucose DMEM, 10% FBS, 100 U/L penicillin, 100 U/L streptomycin, 1% nonessential amino acids, 5 mL nucleoside solution, insulin (0.09 mg/L), and 1000 U/mL LIF. For 1 L, mix 890 mL DMEM with 90 mL FBS, 10 mL penicillin and streptomycin solution *(2)*, 10 mL nonessential amino acids, 10 µL insulin solution, and 1 mL vial LIF.
2. Media for mouse embryonic feeder layer cells: high-glucose DMEM, 10% FBS, 100 U/L penicillin, 100 U/L streptomycin, 1% nonessential amino acids. For 1 L, mix 890 mL DMEM with 90 mL FBS, 10 mL penicillin and streptomycin solution *(2)*, 10 mL nonessential amino acids.
3. Freezing medium: 9 mL DMEM and 1 mL DMSO. Prepare fresh before use.

2.1.2. Differentiation Media

1. Basal medium without LIF: RPMI 1640 supplemented with 10% FBS (heat inactivated), 1 mM L-glutamine, 100 U/L penicillin, 100 U/L streptomycin. For 1 L, mix 890 mL RPMI with 90 mL FBS, 10 mL penicillin and streptomycin solution *(2)*, 10 mL L-glutamine.
2. Neuronal differentiation: Basal medium without LIF supplemented with $1 \times 10^{-6\text{-}7}$ M retinoic acid. To prepare, dilute 1 M (268.45 g) in 100 mL 10% DMSO (stock) and add 0.5 µL from the stock solution.
3. Endothelial differentiation: place basal medium without LIF on a 5-mm Matrigel® layer. Allow Matrigel to become liquid overnight at 4°C; quickly pipet 200 µL in each well of a 24-well plate, allowing the bottom to be covered equally; place the 24-well plate on ice. Incubate the plate at 37°C for 30 min to allow hardening of Matrigel.
4. Adipocyte differentiation media: supernatant of primary rat adipocytes (isolation is described in **Subheading 3.1.9.1.**) in basal medium without LIF. Use after sterile filtration (0.2 µm pore size; *see* **Subheading 3.1.9.1.**).
5. Hepatocyte-conditioned media (HCM): supernatant of primary rat hepatocytes in basal medium without LIF. Use after sterile filtration (0.2 µm pore size; *see* **Subheading 3.1.10.1.**).
6. HA-hepatocyte differentiation media: basal medium without LIF supplemented with 3 ng/mL FGF4. For 1 L, mix 890 mL RPMI with 90 mL FBS, 10 mL penicillin and streptomycin solution *(2)*, 10 mL L-glutamine, 25 µg of FGF4 are diluted in 1 mL PBS 1% BSA (stock); add 120 µL stock solution.

7. HB-hepatocyte differentiation media: basal medium without LIF supplemented with 3 ng/mL FGF4 and 10 ng/mL EGF. For 1 L, mix 890 mL RPMI with 90 mL FBS, 10 mL penicillin and streptomycin solution (2), 10 mL L-glutamine, 25 µg of FGF4 are diluted in 1 mL PBS 0.1% BSA (stock); add 120 µL stock solution. Dilute 200 µg in 1 mL PBS 0.1% BSA (stock); add 5 µL of stock solution.

2.1.3. Required Equipment for Cell Culture

1. 37°C water bath.
2. Incubator set at 37°C and 5% CO_2.
3. Laminar flow hood (Heraeus, Hanau, Germany).
4. 5, 10, and 25 mL sterile serological pipets.
5. Cell scraper (25 cm; Sarstedt, cat. no. 83.1830).
6. Sterile filters (0.2- and 0.4-mm pores; Schleicher and Schüll, cat. no. 10462200).
7. Pipet boy (Pipetus-Aku, Hirschmann, Eberstadt, Germany).
8. Centrifuge multifuge (Heraeus, model. no. 3s-R).
9. Refrigerator (4°C) and freezers (−20 and −80°C).
10. Five 10-µL, 10 100-µL, 20 200-µL, 200 1000-L pipets (Eppendorf, Hamburg, Germany).
11. Liquid nitrogen storage tank.
12. Inverted microscope.

2.2. Staining

2.2.1. General Reagents

1. Methanol.
2. PBS (*see* **Subheading 2.1., item 1**).
3. BSA (Sigma, cat. no. A4503). 10% blocking solution: mix 10 g BSA powder in 100 mL PBS; stir overnight to dissolve.
4. H_2O_2 (Merck, cat. no. 1.07209.0250). Blocking solution for DAB staining: 3% H_2O_2.
5. Mounting medium: Kaiser's glycerin gelatin (Merck, cat. no. OB 324305).
6. 60% aqueous triethyl phosphate.
7. 0.5% Oil Red O solution: 0.5 g Oil Red O (CI 26125) in 100 mL H_2O. Filter before use.

2.2.2. Primary Antibodies

1. Stage-specific embryonic antigen 1 (DSHB, Iowa City, IA; cat. no. MC480).
2. Alkaline phosphatase (AP) (Dako, Hamburg, Germany; cat. no. 30015).
3. KiS3R (kind gift of Professor Parwaresch, Department of Pathology, Kiel).
4. Microtubule-associated protein (MAP)2 (Chemicon, cat. no. 1284959).
5. S100 (Dako, cat. no. Z0311).
6. Glial fibrillary acidic protein (Sigma, cat. no. G3893).
7. Nestin (abcam, Cambridge, UK; cat. no. ab6023).
8. Vimentin (Dianova, Hamburg, Germany; cat. no. A7295).
9. Synaptophysin (Dianova, cat. no. A6571).
10. CD31 (Dianova, cat. no. 09093).
11. Panendothelial antibody (Serotec, Düsseldorf, Germany; cat. no. MCA639).
12. α-Fetoprotein (αFP) (Dako, cat. no. A0008).
13. Albumin (Dako, cat. no. A0001).
14. Cytokeratin18 (Dianova, cat. no. A7062).
15. α-1-Antitrypsin (Dako, cat. no. A00012).

2.2.3. Secondary Antibodies and Conjugates

1. Rabbit antimouse immunoglobulin (Dako, cat. no. Z259).

2. Mouse antirabbit immunoglobulin (Dako, cat. no. M737).
3. Visualization by APAAP protocol (Dako, cat. no. Do651) *(3)*.
4. Visualization by DAB staining with a biotinylated second antibody rabbit antimouse (Dako, cat. no. E354), followed by a streptavidin complex (Dako, cat. no. Ko377).
5. Hemalum (Merck, cat. no. 1.09249.0500).
6. Alexa fluor 488 nm Fab fragment rat antimouse antibody (Molecular Probes, Eugene, OR; cat. no. A12373).
7. Fluorescein isothiocyanate-coupled secondary goat antimouse (Pharmingen, San Diego, CA).
8. DAPI (Molecular Probes, cat. no. C7590).

2.2.4. DAB Staining

1. DAB substrate kit (Vector, Burlingame, CA; cat. no. SK-4100).
2. 1% DAB (20X) in distilled water: add 0.1 g DAB (included in kit) in 10 mL distilled water. Add three to five drops 10 N HCl; solution turns a light brown color. Shake for 10 min, and DAB should dissolve completely. Aliquot and store at $-20°C$.
3. 1% nickel ammonium sulfate (20X) in distilled water: add 0.1 g nickel ammonium sulfate (included in kit) in 10 mL distilled water. Shake to dissolve. Store at 4°C or aliquot and store at $-20°C$.
4. 1% cobalt chloride (20X) in distilled water: add 0.1 g of cobalt chloride (included in kit) in 10 mL distilled water. Shake to dissolve. Store at 4°C or aliquot and store at $-20°C$.
5. 0.3% H_2O_2 (20X) in distilled water: add 100 µL of 30% H_2O_2 in 10 mL distilled water and mix well. Store at 4°C or aliquot and store at $-20°C$.
6. To prepare DAB working solution:
    a. Add five drops of 1% DAB (one drop = 50 µL) to 5 mL PBS, pH 7.2, and mix well.
    b. Add five drops of 1% nickel ammonium sulfate and mix well.
    c. Add five drops of 1% cobalt chloride and mix well.
    d. Add five drops of 0.3% H_2O_2 and mix well.

2.3. Polymerase Chain Reaction Primers

PCR amplification of αFP, coagulation factor II (CFII), carbamoyl-phosphate synthetase I (CPSI), and transthyretin (TTR) was performed with the following oligonucleotide primers:

1. α-FP: sense 5'-GTCCTTTCTTCCTCCTGGAGAT-3' and antisense 5'-CTGTCACTGCTGATTTCTCTGG-3'.
2. CFII: sense 5'-ACTACATTCACCCCGTGTGCTTGC-3' and antisense 5'-CACAAACCTATCTATGCTGATCAATGAC-3'.
3. CPSI: sense 5'-CTATTCTGAGATGTGAGATGGCTTC-3' and antisense 5'-AGCGCTGTACTGCCTGTAGTGGAA-3'.
4. TTR: sense 5'-CAGCAGTGGTGCTGTAGGAGTA-3' and antisense 5'-GGGTAGAACTGGACACCAAATC-3'.

3. Methods
3.1. Tissue Culture
3.1.1. Freezing Cells (see **Note 2**)

1. Aspirate medium, wash with PBS, add 2 mL trypsin/ethylenediaminetetraacetic acid to each 10-mL dish, and incubate at 37°C until cells float.

2. Carefully transfer cells into a 50 mL tube and add 20 mL medium (DMEM plus 10% FBS) to inactivate the trypsin.
3. Pellet the cells by low-speed centrifugation (i.e., 200g).
4. Remove supernatant and resuspend cells in freezing medium (precooled); immediately transfer cells into freezing vials on ice (1 mL per vial).
5. Transfer vials to a $-20°C$ freezer and after 4 h into $-80°C$.

3.1.2. Thawing Cells (see **Note 3**)

1. Thaw cells and dilute with 10 mL media in a 15 mL tube.
2. Centrifuge the cells at 300g for 5 min.
3. Remove the supernatant and resuspend the cells in fresh medium.
4. Transfer cells to a tissue culture dish and allow them to grow overnight at 37°C in an incubator.
5. Change the media the next day.

3.1.3. Fibroblast Feeder Layers

ES cells from rats and other species require a layer of mitotically inactive fibroblast cells or gelatinized tissue culture dishes and the addition of LIF to remain pluripotent. C12 rat embryonic stem cell-like cells (RESCs) were established on a mouse embryonic fibroblast feeder layer. Rat ES cells are known to differentiate rapidly in culture; despite this fact, we managed to change the culture conditions to gelatin-coated dishes. For initial cultures, we used the feeder layer as described in **ref. 3**. Feeder layers were stored in liquid nitrogen (1 million cells per 1 mL vial).

3.1.4. Rat ES Cells in Culture

3.1.4.1. MYTOMYCIN C TREATMENT OF FEEDER LAYER CELLS

When the feeder layer cells reach confluency they are treated with mitomycin C to induce mitotic arrest. The cells are still capable of conditioning the media.

1. Add 5 mL mitomycin C (1 mg/100 mL) to a 10 mL dish.
2. Incubate the cells for 2–3 h.
3. Remove the mitomycin C solution, rinse the dish five times with PBS and add fresh medium.

3.1.4.2. MAINTENANCE OF RAT ES CELLS

RESC cells were isolated from the inner cell mass of 4- to 5-d-old blastocysts, derived from pregnant WKY-rats as described in Fandrich et al. *(3)*. We used a single cell-cloned line termed C12-WKY, which contains approx 70% euploid cells. C12 RESCs were grown on mitomycin C-treated feeder layer *(3)* or on gelatin-coated dishes (**Fig. 1A**).

1. Culture RESCs at a moderate density and subculture by splitting no more than 1/10 (*see* **Note 4**).
2. Routinely, subculture RESCs every 2–3 d (*see* **Note 5**). To prevent differentiation, the cells should be dissociated into single cells after subculturing.
3. Change the media every day or at least every 2 d (*see* **Note 6**).

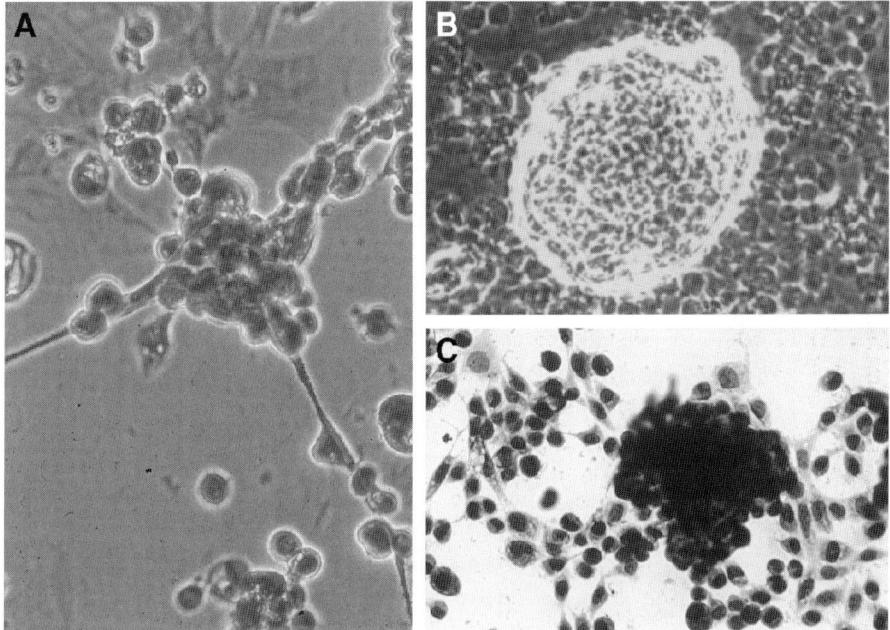

Fig. 1. (**A**) Undifferentiated C12 rat embryonic stem cells grown on gelatin-coated culture dishes; stained for AP. (**B**) Phase contrast microscopy of a differentiating embryoid body from RESCs. (**C**) Signs of differentiation showing flat cells with broad cytoplasm and polygonal shape next to undifferentiated cells. (Used with permission from Alpha Med Press.) (Please *see* companion CD for the color version of this figure.)

3.1.4.3. GELATIN COATING OF TISSUE CULTURE PLASTIC

1. Coat tissue culture dishes with 1% gelatin solution.
2. Incubate for 3–4 h.
3. Wash in three changes of PBS.

3.1.4.4. DIFFERENTIATION INTO EMBRYONIC BODIES

Single-cell differentiation has to be distinguished from differentiation into embryonic bodies. If the cell density increases, characteristic colonies start to grow larger, resulting in rounded bodies (**Fig. 1B**). Within the embryonic bodies, undirected differentiation begins, and these cells should never be subcultured or frozen because even small embryoid bodies are comprised of cells that have already started to differentiate.

3.1.4.5. SIGNS OF DIFFERENTIATION

Cells surrounding the characteristic colonies, with a flattened morphology and a dark and spiky appearance, are typical for different-treated cells. Cells with a clearly visible nucleus and growing within flat colonies are more likely to have undergone differentiation (**Fig. 1C**). For AP staining of ES cells (**Fig. 1A**), use the following protocol:

1. Rinse cells thoroughly with PBS.
2. Fix the cells in 10 mL ice-cold methanol for 10 min.
3. Rinse with aqua dest and incubate in fresh distilled water for 1 min.
4. Freshly prepare AP substrate *(4)*.
5. Incubate for 45 min at room temperature, then rinse with aqua dest.
6. Counterstain nuclei with Hemalum for 5 min.
7. Mount the cells with Kaiser's glycerin gelatin and cover with cover slips.

3.1.5. CSFE Labeling of Cells (see **Note 7**)

CFSE passively diffuses into cells and is converted into a fluorescent carboxyfluorescein succinimidyl ester. During each round of cell division, the CFSE fluorescence is reduced by half, allowing the identification of successive cell generations. CFSE is detected using standard fluorescein filters (492-nm excitation, 517-nm emission).

1. To label cells, add 1 mL prepared cell suspension to a tube containing 100 µL CFSE.
2. Incubate cells for 10 min at 37°C, mixing twice.
3. Transfer cells to a 15-mL conical tube containing 1 mL ice-cold culture medium and mix gently for 1 min.
4. To remove unincorporated CFSE, add 10 mL PBS to the tube.
5. Centrifuge cells for 10 min at 200g and decant supernatant.
6. Wash cells once more as previously described.

3.1.6. Preparation of a Cellular Transplant

1. Harvest 1×10^6 RESCs mechanically by scraping (cell scraper).
2. Centrifuge at 200g for 5 min.
3. Resuspended in 100 µL and inject using a Hamilton syringe (**Note 8**).

3.1.7. Differentiation Into the Neuroepithelial Lineage

3.1.7.1. NEURONAL LINEAGE

For differentiation, RESCs are cultured in six-well plates on cover slips and treated with 10^{-6} to $10^{-7} M$ retinoic acid. After 7 d, neuronal precursors appear (**Fig. 2A**) with a rounded cytoplasm and small bipolar processes. These cells strongly express S100, nestin, and vimentin. Between 14 and 21 d of differentiation in retinoic acid, mature neurons develop, identified by long, slender dendrites and a small cell body (**Fig. 2B,C**). The mature neurons express characteristic antigens MAP2, NF68, and synaptophysin; expression of nestin, vimentin, and S100 diminishes.

1. To maintain newly differentiated neurons in culture, 25 µmol potassium chloride should be added to the culture.
2. At different time points, cover slips are washed in PBS and fixed with methanol for 10 min. Immunohistochemical staining is performed for MAP2, nestin, vimentin, S100, NF68, glial fibrillary acidic protein, and synaptophysin antigens.

3.1.7.2. GLIAL CELLS

In the same culture, flat polygonal cells appear that morphologically resemble glial cells. These cells are growing within the population of neurons (**Fig. 2C**) and express glial fibrillary acidic protein.

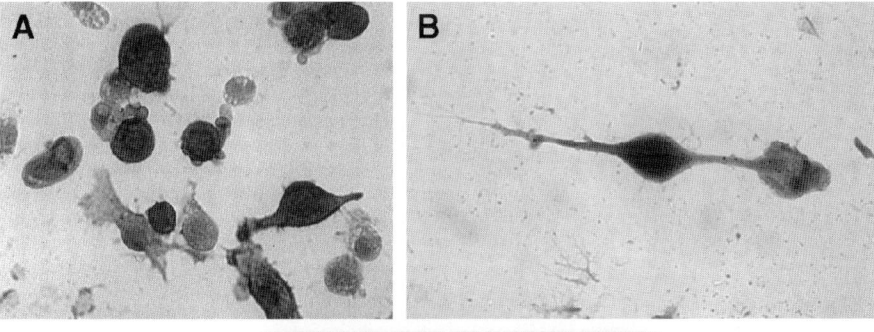
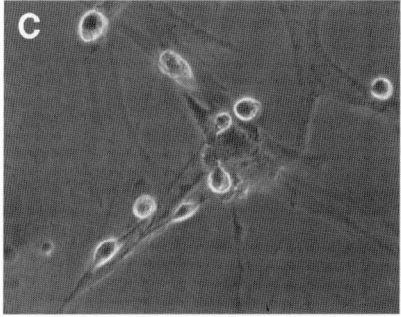

Fig. 2. (**A**) Immature neurons differentiating from C12 cells after 5 d of retinoic acid treatment. Small bipolar cells develop, and here are positive for the precursor antigen nestin. (**B**) Maturing neuron with growing dendrites; Hemalum staining. (**C**) Phase contrast microscopy of a culture after 10 d of retinoic acid treatment. Neurons with long, slender neurites migrate on supporting glial cells and appear as flat and big polygonal-shaped cells underneath the neuronal cells. (Please see companion CD for the color version of this figure.)

3.1.8. Differentiation Into Endothelial Cells (see **Note 9**)

1. Defrost (–20°C) aliquots of Matrigel overnight at 4°C. Precool pipet tips and culture plates to 4°C to prevent premature gelling of the Matrigel.
2. Mix 1 mL Matrigel with 100 mL 10X DMEM in a 1.5-mL Eppendorf tube and start the gelling process by adding 200 mL 7.5% sodium bicarbonate, ensuring complete mixing before placing a drop of approx 80 µL Matrigel mix in the center of each culture well.
3. Incubate the culture plates at 37°C for 10–15 min and then cover the gel in 1 mL SFM; store in a sealed humid box at 4°C until use.

3.1.8.1. RESCs on Matrigel

1. Coat 24-well dishes with a 5-mm Matrigel layer.
2. After the Matrigel has undergone the gelling process, RESCs are plated on top of the Matrigel layer in basal medium (*see* **Subheading 2.2.**). Endothelial differentiation can be observed after 5–6 d (**Fig. 3A**). C12 cells start to build tubelike vascular networks (**Fig. 3B**).
3. Perform immunohistochemistry against CD31 and panendothelial antibody and photodocument tube formation on d 5, 10, and 14.

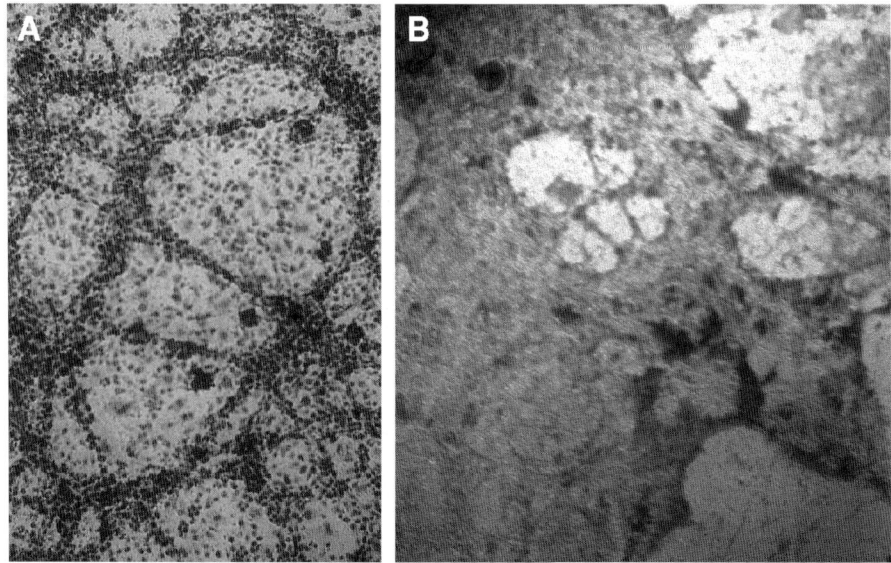

Fig. 3. (**A**) Endothelial networks developing after growth on Matrigel for 5 d, Hemalum staining. (**B**) DAB peroxidase staining against CD31 in a tubular network of C12 cells differentiating into endothelial cells after growth on Matrigel for 14 d; note that confluency is reached in some regions. (Please *see* companion CD for the color version of this figure.)

3.1.9. Differentiation of Adipocytes

3.1.9.1. Generation of Adipocyte-Conditioned Medium

1. Cut 20 g sterile rat omental fat tissue into small pieces (approx 1–2 mm^3).
2. Transfer the pieces to sterile tubes containing PBS with 2.5 mg/mL collagenase and incubate for 15 min.
3. Filter the digest through a nylon screen (500-μm pore size).
4. Wash four times in PBS and place into bacteriological Petri dishes containing RPMI 1640 supplemented with 10% FBS, 5% penicillin/streptomycin, and 4.5 mg insulin.
5. Incubate cultures for 1 h to allow rat primary adipocytes to adhere.
6. Collect the supernatant every 2 d and filter to remove contaminating cells. This is termed fat-cell conditioned media (FCM). FCM is either used immediately or stored at −20°C.

3.1.9.2. Adipocytes

1. Plate C12 cells on cover slips in six-well tissue culture plates or in 75-mL tissue culture flasks in FCM.
2. Culture cells for 14 d; FCM is changed every other day. After 4 d, small intracellular lipid droplets can be observed that closely resemble immature adipocytes (**Fig. 4A**); after 10–14 d, lipid droplets increase in size, and C12 cells mature to become adipocytes (**Fig. 4B**).
3. Stain cells for intracellular fat with Oil Red *(5)*.

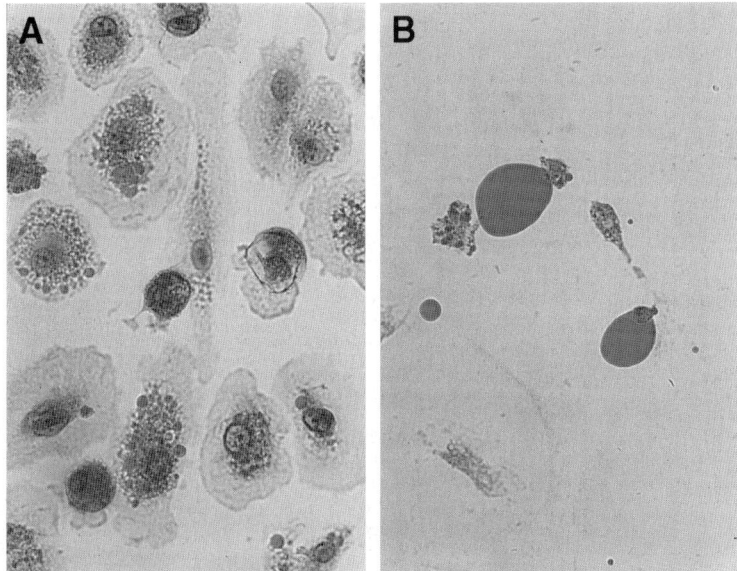

Fig. 4. (**A**) Preadipocyte, developing from C12 cells, with intracellular small lipid droplets stained with Oil Red. (**B**) Mature adipocyte; note that the nucleus has shifted to the periphery of the cytoplasm; Oil Red staining after 10 d FCM treatment. (Please *see* companion CD for the color version of this figure.)

3.1.10. Differentiation of Hepatocytes

3.1.10.1. GENERATION OF HCM *(6)*

1. Perfuse 40 g rat liver tissue with 3 mg/mL collagenase.
2. Incubate for 45 min.
3. Filter the digest through a nylon screen (500 µm) and wash four times in PBS.
4. Plate into 75-mL tissue culture flasks containing RPMI 1640 supplemented with 10% FBS, 5% penicillin/streptomycin, 900 µg/mL insulin, 10 ng/mL EGF, and 40 ng/mL hepatocyte growth factor.
5. Incubate culture dishes at 37°C (5% CO_2, 95% air) in humidified air for 9 d, changing the medium every other day (*see* **Note 10**).
6. Collect medium and filter through a 0.2-µm filter to produce cell- and bacteria-free HCM. HCM is then either directly used for further differentiation protocols or deep frozen at −20°C (*see* **Note 11**).

3.1.10.2. HEPATOCYTES

1. Culture C12 cells on cover slips in six-well tissue culture plates or in 75-mL tissue culture flasks in HCM for 14 d or in hepatocyte differentiation media HA or HB.
2. Change the medium every other day. After 4 d, C12 acquire a phenotype resembling immature hepatocytes (**Fig. 5A**), expressing albumin, CK 18, and αFP. After 10–14 d treatment with HCM and HA, C12 cells differentiated into hepatocytes that secrete albumin (*see* **Note 12**).

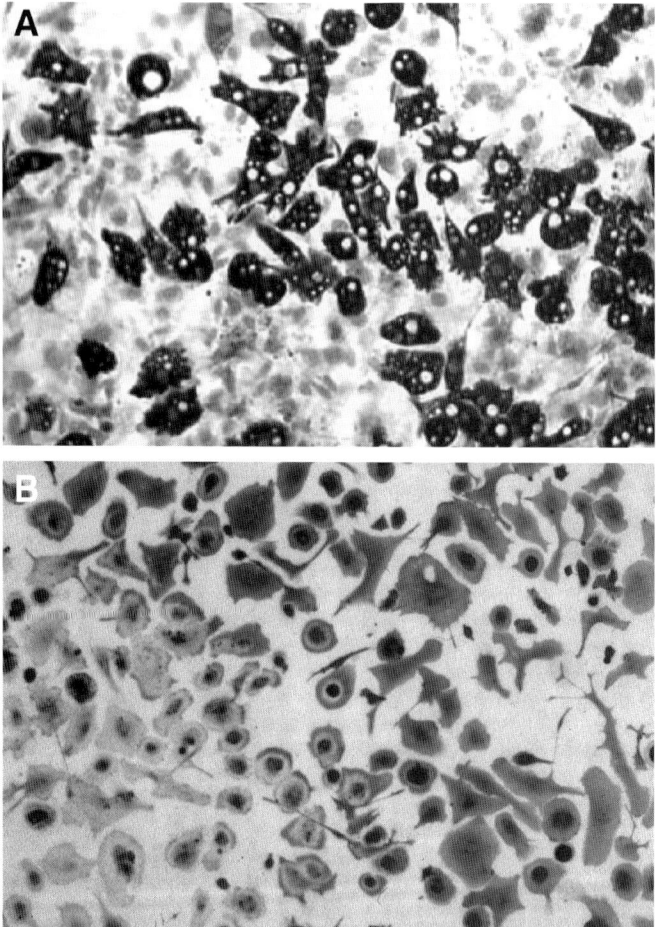

Fig. 5. (**A**) Epithelial polygonal cells developing from C12 cells after 6 d HCM treatment. Note the prominent dark cytoplasm in a subpopulation of the cells. (**B**) Staining for albumin after 10 d HA treatment. The epithelial character of the cells correlates with increasing albumin expression. (Please *see* companion CD for the color version of this figure.)

3. At different time points, fix cells on cover slips with methanol and stain for αFP, albumin, cytokeratin 18, and α1-antitrypsin as described in **Subheading 3.2.** (**Fig. 5B**).

3.2. Analysis of Differentiation

3.2.1. Immunohistochemistry on Cover Slips

1. Fix cells with methanol for 10 min.
2. Rinse cells with PBS for 5 min.
3. Incubate cells with blocking medium (*see* **Subheading 2.2.**, **item 3**).
4. Incubate with primary antibody for 45 min.
5. Rinse cells with PBS for 5 min.

6. Incubate with secondary antibody for 45 min.
7. Visualize with the APAAP protocol (7).

3.2.1.1. DAB Staining

1. Specimens are blocked with 3% H_2O_2 after first blocking step for 20 min.
2. Rinse in PBS for 5 min.
3. Incubate with primary antibody for 45 min.
4. Rinse with PBS for 5 min.
5. Incubate with biotinylated secondary antibody for 45 min.
6. Mix one drop of solution A with one drop of solution B in 500 mL PBS and let it react for 30 min at room temperature, then incubate probes for 30 min. Rinse cells in PBS for 5 min.
7. Stain with DAB staining solution (*see* **Subheading 2.2.3.**, **item 6**) for 5–10 min at room temperature.

3.2.2. Sudan Red Staining for Intracellular Fat Droplets

1. Rinse cells briefly in 60% triethyl phosphate.
2. Stain cells 5–20 min in Oil Red O solution.
3. Rinse cells 1–2 s in 60% triethyl phosphate.
4. Rinse well in distilled water.
5. Counterstain in Hemalum for 15 min.
6. Rinse well in distilled water.
7. Mount in Kaiser's glycerin gelatin.

3.2.3. Low-Density Lipoprotein Uptake Assay for Hepatocytes

To detect the uptake of 1,1′–dioctadecyl–3,3,3′,3′–tetramethylindocarbocyanine–labeled acetylated low-density lipoprotein (DiLDL).

1. Incubate cells with DiLDL (2.4 µg/mL) at 37°C for 1 h.
2. Fix cells with 2% paraformaldehyde for 10 min.
3. Mount with Kaiser's glycerin. Cells that take up the DiLDL fluoresce red.

3.2.4. Reverse Transcriptase PCR Analysis of Hepatocyte-Specific Markers

1. Prepare total RNA from cultures of C12 cells using peqGOLD RNAPure according to the manufacturer's instructions.
2. Synthesize cDNA from 1–5 µg of total RNA from untreated control cultures or from FGF4-treated cells cultured on Matrigel with oligo(dT)$_{12-18}$ and SuperScript II reverse transcriptase as described previously (8).
3. Amplify, by PCR, cDNA corresponding to 100 ng of RNA with Taq-polymerase using oligonucleotide primers (for sequences *see* **Subheading 2.3.**) in a hot-start touch-down program with the following conditions: denaturation at 95°C for 1 min, annealing at 65°C for 1 min, and extension at 72°C for 1 min.
4. Every fourth cycle, reduce the annealing temperature by 2°C until a temperature of 55°C is reached, at which a varying number of cycles (depending on the abundance of the target messenger RNA, αFP, and CFII: 15, CPSI: 25, TTR: 21) is performed (*see* **Note 13**).
5. Separate amplification products on 1.5% agarose gels and visualize by staining with ethidium bromide.

4. Notes

1. Test FBS each time a new batch is chosen. Try to order as many bottles from one batch as possible and use ES cell-tested serum.

2. Cells should always be frozen slowly (0.5°C/min) and kept in liquid nitrogen tank. Cells should be frozen at a density that allows quick recovery *(9)*. For passaging and freezing, it is best when cells are subconfluent.
3. Thawing should be performed quickly in a 37°C water bath.
4. It is important not to let the cells overgrow because otherwise RESCs will rapidly differentiate. Keep a record of how often cells are passaged; do not try to passage them more than 10 times.
5. RESCs should not be grown for too long before starting a differentiation experiment.
6. Do not try to keep the cells in culture for longer than needed.
7. CFSE is provided in 20 single-use tubes. Store tubes at −20°C.
8. After CSFE labeling of the cells and injection into an animal, controls have to be investigated by fluorescent microscopy to visualize autofluorescence and rule out false-positive results.
9. Matrigel should be prepared at least 24 h prior to setting up cultures. Matrigel sets at room temperature; therefore, all preparation should be performed on ice or at 4°C.
10. If the medium has become acidic, discard the cells.
11. When using conditioned media to induce differentiation of a stem cell line, always culture the conditioned media alone to exclude the possibility of cellular contamination, which is responsible for possible cellular fusion events.
12. For the albumin enzyme-linked immunosorbent assay, always use medium as a negative control and preimmune rat serum as a positive control.
13. The number of cycles was chosen to stay within the linear phase of the assay.

References

1. Ruhnke, M., Ungefroren, H., Zehle, G., Bader, M., Kremer, B., and Fandrich, F. (2003) Long-term culture and differentiation of rat embryonic stem cell-like cells into neuronal, glial, endothelial, and hepatic lineages. *Stem Cells* **1,** 428–436.
2. Hill, D. P. and Robertson, K. A. (1998) Differentiation of LA-N-5 neuroblastoma cells into cholinergic neurons: methods for differentiation, immunohistochemistry and reporter gene introduction. *Brain Res. Brain Res. Protoc.* **2,** 183–190.
3. Fandrich, F., Lin, X., Chai, G. X., et al. (2002) Preimplantation-stage stem cells induce long-term allogeneic graft acceptance without supplementary host conditioning. *Nat. Med.* **8,** 171–178.
4. Troy, T. C. and Turksen, K. (2002) Epidermal lineage, in *Methods in Molecular Biology, Vol. 185: Embryonic Stem Cells: Methods and Protocols* (Turksen, K., ed), Humana Press, Totowa, NJ, pp. 229–253.
5. Ramirez-Zacarias, J. L., Castro-Munozledo, F., and Kuri-Harcuch, W. (1992) Quantitation of adipose conversion and triglycerides by staining intracytoplasmic lipids with Oil Red O. *Histochemistry* **97,** 493–497.
6. Hengstler, J. G., Utesch, D., Steinberg, P., et al. (2000) Cryopreserved primary hepatocytes as a constantly available in vitro model for the evaluation of human and animal drug metabolism and enzyme induction. *Drug Metab. Rev.* **32,** 81–118.
7. Cordell, J. L., Falini, B., Erber, W. N., et al. (1984). Immunoenzymatic labeling of monoclonal antibodies using immune complexes of alkaline phosphatase and monoclonal anti-alkaline phosphatase (APAAP complexes). *J. Histochem. Cytochem.* **32,** 219–229
8. Ungefroren, H., Voss, M., Jansen, M., et al. (1998) Human pancreatic adenocarcinomas express Fas and Fas ligand yet are resistant to Fas-mediated apoptosis. *Cancer Res.* **58,** 1741–1749.
9. Matise, M. P., Auerbach, W., and Joyner, A. L. (2000) Production of targeted embryonic stem cell clones, in *Gene Targeting, a Practical Approach*, 2nd ed. (Joyner, A. L., ed.), Practical Approach Series, Oxford University Press, Oxford, UK, pp. 102–132.

5

Derivation, Maintenance, and Induction of the Differentiation In Vitro of Equine Embryonic Stem Cells

Shigeo Saito, Ken Sawai, Arika Minamihashi, Hideyo Ugai, Takehide Murata, and Kazunari K. Yokoyama

Summary

We describe here the isolation and maintenance of pluripotent embryonic stem (ES) cells from equine blastocysts that have been frozen and thawed. Equine ES cells appear to maintain a normal diploid karyotype in culture. These cells express markers that are characteristic of mouse ES cells, namely, alkaline phosphatase, stage-specific-embryonic antigen 1, STAT3, and Oct4. We also describe protocols for the induction of differentiation in vitro to neural precursor cells in the presence of basic fibroblast growth factor (bFGF), epidermal growth factor, and platelet-derived growth factor and to hematopoietic and endothelial cell lineages in the presence of bFGF, stem cell factor, and oncostatin M. Equine ES cells provide a powerful tool for gene targeting and the generation of transgenic clonal offspring.

Key Words: Alkaline phosphatase; bFGF; CD34; CD35; CD45; derivation; EGF; endothelial differentiation; equine blastocysts; equine ES cells; Flk-1; GATA 4; GFAP; hematopoietic differentiation; LIF; maintenance; Nestin; neural differentiation; Oct4; OSM; PDGF; pluripotency; SCF; Sry-1; SSEA-1; STAT3; β-tubulin III.

1. Introduction

Embryonic stem (ES) cells are pluripotent cells that can differentiate into at least three different layers of embryonic germ cells, and the successful differentiation of ES cells in vitro to neurons, hematopoietic cells, and cardiomyocytes has been reported *(1–5)*. ES cells are one of the most useful starting materials for the generation of mutants by homologous recombination for experimental research in mice *(6)*. Moreover, human ES cells have the potential to serve as tools in tissue and cell transplantation therapy because they have the stable developmental capacity to differentiate into any type of cell under specific conditions *(7)*.

Gene targeting in farm animals has been reported, and animals have been cloned by transfer of nuclei from cultured somatic cells *(8–10)*. In view of the probable significant future contribution of ES cells to efforts to manipulate the genome, to generate transgenic clonal offspring, and to replace various tissues and organs, it seems appropriate to adapt ES cell technology to the genetic improvement of farm animals. However, from the economical perspective, the possibilities for application of this technology to farm animals are more limited than they are to laboratory animals, and the gap in basic knowledge between laboratory animals and farm animals is likely to widen.

There are very few reports of the pluripotency of transfected embryo-derived immortalized lines of cells derived from farm animals *(11)*. We reported the first example of the pluripotency of immortalized bovine embryo-derived cell lines. We characterized the marker proteins of the ES cells and produced cloned calves and enhanced green fluorescence protein-transgenic embryos *(11)*.

Methods for the isolation of murine ES cells *(12)* have been modified for application to other species, such as sheep *(13)*, pig *(13)*, rabbit *(14)*, bovine *(15,16)*, mink *(17)*, hamster *(18)*, rat *(19)*, and monkey *(20)*. However, attempts at the primary culture of cells of the inner cell mass (ICM) of various species on a feeder layer of embryo-derived mouse STO cells *(21,22)* or a primary culture of mouse embryonic fibroblasts *(23,24)* in the presence of leukemia inhibitory factor (LIF) have seldom been successful *(25)*. A combination of growth factors might be required for the proliferation of these pluripotent cells, as has been demonstrated in the case of cultures of mouse primordial germ cells *(26)*.

Among studies of farm animals, investigations of the development in vitro of equine embryos at the preimplantation stage have been quite limited *(27–29)*. Isolation of equine ES cells would put horses in the front lines of animal biotechnology and the modification of animal genomes, although cloned mules and horses have been produced by somatic cell cloning, with the offspring a genetic copy of the animal donor of the nuclear material that was transferred into an enucleated oocyte *(30,31)*.

We describe here protocols for the isolation of immortalized equine ES cells from frozen and thawed embryos. We have cultured these cells on a feeder layer derived from bovine umbilical fibroblasts in a highly reproducible manner, and our protocols allow the genetic manipulation of equine ES cells by homologous recombination to generate gene-targeted horses. Furthermore, we describe the conditions under which ES cells undergo neural, endothelial, and hematopoietic differentiation in the presence of various growth factors. The ability to induce and control such differentiation provides a powerful tool for investigation and development of treatments for neurological and hematopoietic disorders.

2. Materials

2.1. Tissue and Cell Culture

1. Ca^{2+}-, Mg^{2+}-free Dulbecco's phosphate-buffered saline (PBS) (Sigma, St. Louis, MO; cat. no. D-5773): 9.6 g of powder is dissolved in 1 L double-distilled water (ddH_2O) and sterilized by filtration through a syringe filter with a 0.22-µm membrane. This solution is used for washing various cells (*see* **Note 1**).
2. Minimum essential medium (MEM)α, 1X (1 L; Gibco BRL, Rockville, MD; cat. no. 12571-048) (*see* **Notes 1** and **2**).
3. Fetal bovine serum (FBS) (100 mL; Gibco BRL, cat. no. 16140-063) (*see* **Note 3**).

Differentiation of Equine ES Cells In Vitro 61

4. Human LIF (10 μg; Sigma, cat. no. L5283). The solution of 1000 ng/mL LIF is prepared by adding 10 mL MEMα to 10 μg LIF.
5. Basic fibroblast growth factor (bFGF) (10 μg; Sigma, cat. no. F-3133) (*see* **Note 4**).
6. Epidermal growth factor (EGF) (100 μg; Sigma, cat. no. E-1257) (*see* **Note 5**).
7. Platelet-derived growth factor (PDGF) (10 μg; Sigma, cat. no. P-3326).
8. Stem cell factor (SCF) (10 μg; Sigma, cat. no. S-7901).
9. Oncostatin M (OSM) (10 μg; Sigma, cat. no. O-9635).
10. β-Mercaptoethanol: 55 mM, 1000X (50 mL; Gibco BRL, cat. no. 21985-023).
11. Penicillin-streptomycin mix, 100X (100 mL; Gibco BRL, cat. no. 15140-122).
12. Mitomycin C (2 mg; Sigma, cat. no. M-4287).
13. Trypsin-EDTA (ethylenediaminetetraacetic acid): 0.25% trypsin, 1 mM EDTA·4Na, 1X (100 mL; Gibco BRL, cat. no. 25200-056).
14. Dimethyl sulfoxide (DMSO) (500-mL; Wako Pure Chemical Inc., Osaka, Japan; cat. no. 045-07215).
15. Falcon 60-mm organ tissue culture dishes with center well (Becton Dickinson, Lincoln Park, NJ; cat. no. 3037).
16. Falcon Primeria 60-mm tissue culture dishes (Becton Dickinson, cat. no. 3803).
17. Nunc 4-well multi dishes (Nunc, Naperville, IL; cat. no. 176740).
18. Falcon 15-mL polystyrene conical tubes (Becton Dickinson, cat. no. 2097).
19. 20-mL syringes (Terumo, Tokyo, Japan; cat. no. SS-20ESZ).
20. 0.22-μm syringe filters (Millipore, Bedford, MA; cat. no. SLGV025LS).
21. Cryotube vials (2.0-mL tubes; Simport, Beloeil, Quebec, Canada; cat. no. T309-2A).
22. Freezing tray plug (for liquid N2 container x C34/18; MVE, Bloomington, MN; cat. no. 97-2084-9).
23. 10-mL serological pipets (Corning, New York, NY; cat. no. 430736).
24. Glycerol (500 mL; Wako Pure Chemical Inc., cat. no. 075-00616).
25. Sucrose (500 g; Wako Pure Chemical Inc., cat. no. 196-00015).
26. Micromanipulator (Narishige Co., Tokyo, Japan).
27. Gelatinized 24-well plate (Iwaki Co., Tokyo, Japan; cat. no. 4820-020).

2.1.1. Media

1. Medium for umbilical cord-derived fibroblasts (UCDFs): UCDFs are established and maintained in MEMα supplemented with 10% heat-inactivated FBS and 1% penicillin-streptomycin. This media is called standard (ST)-MEMα. To prepare 100 mL ST-MEMα, combine 10 mL heat-inactivated FBS, 1 mL penicillin-streptomycin mix, and 89 mL MEMα. The medium is sterilized by filtration through a membrane with 0.22-μm or smaller diameter pores.
2. Medium for ES cells: equine ES cells are isolated and maintained in four-well dishes in ST-MEMα supplemented with 10 ng/mL human LIF. To prepare 100 mL supplemented ST-MEMα, combine 99 mL ST-MEMα and 1 mL 1000 ng/mL human LIF.
3. Medium for initiating differentiation to neural progenitor (NP) cells: NP cells are derived from ES cells in serum-free ST-MEMα that has been supplemented with 10 ng/mL bFGF, EGF, and PDGF, respectively. To prepare 100 mL of medium for NP cells, combine 97 mL ST-MEMα and 1 mL 1000 ng/mL solutions of bFGF, EGF, and PDGF, respectively.
4. Medium for initiating differentiation to hematopoietic and endothelial (HEE) cells: HEE cells are derived from ES cells in ST-MEMα supplemented with 10 ng/mL bFGF, SCF, and OSM, respectively. To prepare 100 mL of medium for HEE cells, combine 97 mL ST-MEMα and 1 mL 1000 ng/mL solutions of bFGF, SCF, and OSM, respectively.

5. Freezing medium: to prepare 10 mL freezing medium, add 2 mL dimethyl sulfoxide to 2 mL FBS and 6 mL MEMα.

2.1.2. General Comments and Required Equipment for Tissue and Cell Culture

All protocols for tissues and cell culture must be followed under sterile conditions with great attention given to the use of clean and detergent-free glassware. All media and solutions must be warmed to 36°C before use *(32)*. The facilities for the culture of ES cells must have the following equipment:

1. Water bath (36°C).
2. Glass pipets designated for tissue culture only (10 and 25 mL).
3. Humidified incubator (38.6°C and an atmosphere of 5% CO_2 in air).
4. Laminar flow hood.
5. Stereomicroscope with magnification from ×10 to ×60.
6. Inverted microscope with a phase contrast objective equipped with fluorescence and photographic systems (×40, ×100, ×200, and ×1000 magnification).
7. Storage tank containing liquid nitrogen.
8. Pipetmen (20, 100, 200, and 1000 µL) designated for tissue culture only.
9. Pasteur pipets with cotton plugs.
10. Refrigerator (4°C) and freezer (−20°C).
11. Tabletop centrifuge.

2.2. Giemsa Chromosome Staining

1. Glass slides (26 × 76 mm, Matsunami, Tokyo, Japan; cat. no. S-226).
2. Methanol (500 mL; Wako Pure Chemical, cat. no. 132-12385).
3. Acetic acid (500 mL; Wako Pure Chemical, cat. no. 014-00266).
4. Fixative: for 100 mL, mix 25 mL methanol and 75 mL acetic acid.
5. Giemsa staining solution (250 mL; Wako Pure Chemical, cat. no. 079-04391). For 100 mL, mix 1 mL Giemsa solution and 99 mL ddH_2O.
6. Hypotonic solution (0.025 *M* KCl): for 500 mL, mix 0.775 g of KCl and 500 mL sterile ddH_2O.
7. Colcemid (20 mL; Wako Pure Chemical, cat. no. 049-16961).

2.3. Alkaline Phosphatase (AP) Staining

1. 4% PBS-buffered formalin (*see* **Subheading 2.4.**, **item 2**).
2. Fast Red TR salt (1 g; Sigma, cat. no. F-8764). For 1-mg/mL solution, combine 0.05 g Fast Red with 50 mL ddH_2O.
3. Naphthol AS-MX phosphate (100 mg; Sigma, cat. no. N-5000). For 100-mg/mL stock solution, dilute 0.1 g naphthol AS-MX phosphate with 1 mL ddH_2O.

2.4. Immunofluorescence Staining

1. Formaldehyde (37% solution; 500 mL; Sigma, cat. no. F-1635).
2. 4% PBS-buffered formalin. For 100 mL, combine 10.8 mL 37% formaldehyde and 89.2 mL PBS and store at 4°C.
3. Mouse STAT3-specific antibody (0.1 mg; Sigma, cat. no. S-5933).
4. Mouse nestin-specific antibody (0.1 mg; Chemicon, Temecula, CA; cat. no. MAB 353).
5. Mouse CD45-specific antibody (50 tests; Sigma, cat. no. F-4149).
6. Mouse CD31-specific antibody (50 tests; Sigma, cat. no. F-8402).

Differentiation of Equine ES Cells In Vitro 63

7. Mouse CD34-specific antibody (50 tests; Sigma, cat. no. C-7714).
8. Mouse Flk-1-specific antibody (0.2 mL; Sigma, cat. no. 129H4861).
9. Mouse GFAP-specific antibody (0.2 mL; Sigma, cat. no. G-9269).
10. Mouse β-tubulin III-specific antibody (0.2 mL; Sigma, cat. no. T-8660).
11. Fluorescein isothiocyanate-conjugated (FITC-conjugated) rabbit secondary antibody against mouse immunoglobulin (Ig) G (5 mL; Sigma, cat. no. F9887).
12. PBS (*see* **Subheading 2.1., item 1**).

2.5. Reverse Transcription Polymerase Chain Reaction (RT-PCR)

1. Thermal cycler (model 2400; Perkin Elmer Applied Biosystem, Foster City, CA; cat. no. 0993-6057).
2. Apparatus for polyacrylamide gel electrophoresis (Nihon Eido Co., Tokyo, Japan; cat. no. NA-1110).
3. Acrylamide (500 g; Bio-Rad, Hercules, CA; cat. no. 161-0101).
4. *bis*-Acrylamide (25 g; Sigma, cat. no. M7279).
5. Tris-HCl (500 g; Sigma, cat. no. T-1503).
6. Boric acid (500 g; Kanto Chemical Co., Tokyo, Japan; cat. no. 04232-00).
7. EDTA (500 g; Sigma, cat. no. 09-1320-5).
8. Acetic acid (500 g; Sigma, cat. no. 01-0280-5).
9. Trizol™ reagents (Invitrogen BV, Groningen, The Netherlands; cat. no. 15596-018).
10. Cell scrapers (Corning, cat. no. 3008).
11. Four-well culture dishes (Nunc, cat. no. 176740).
12. Chloroform (Sigma, cat. no. 06-2320-5).
13. Isopropanol (Sigma, cat. no. 15-2329-5).
14. Ethanol (Sigma, cat. no. 09-0770-5).
15. RNase inhibitor (1000 U; Invitrogen BV, cat. no. 15518-012).
16. DNase (250 U; Invitrogen BV, cat. no. 18162-016).
17. Phenol (Invitrogen BV, cat. no. 15509-237).
18. 50 mM MgSO$_4$ (Invitrogen BV, cat. no. 52044).
19. SuperScript™ one-step RT-PCR with Platinum® Taq (Invitrogen BV, cat. no. 10928-042)
20. 2X reaction mix (Invitrogen BV, cat. no. 51099).
21. 100 mM dNTP set (Invitrogen BV, cat. no. 10298-018).
22. RT/Platinum Taq Mix (Invitrogen BV, cat. no. 53145).
23. Diethylpyrocarbonate (DEPC)-H$_2$O: Add 0.2 mL of DEPC (Sigma, cat. no. D558) to 100 mL water (or the solution to be treated). Shake vigorously to dissolve the DPEC. Autoclave the solution to inactivate the remaining DEPC (*see* **Note 6**).
24. 10X DNase buffer: 200 mM Tris-HCl, pH 8.4, 20 mM MgCl$_2$, 500 mM KCl. For 100 mL, combine 50 mL 1 M KCl, 1.5 mL 1 M MgCl$_2$, 10 µL of a solution of 1 mg/mL gelatin, and 10 mL 2 M Tris-HCl, pH 8.4.
25. A solution of 2.5 mM dNTPs is required for RT: dilute the stock solution of dNTPSs 10-fold with DEPC-H$_2$O.
26. Gateway™ PCR cloning system (Invitrogen BV, cat. no. 11821-014).
27. QIAquick™ PCR purification kit (Qiagen GmbH, Hilden, Germany; cat. no. 28106).
28. Hanks' balanced salt solution Ca^{2+} and Mg^{2+} free, 10X (500 mL; Gibco BRL; cat. no. 14060-057).
29. Digestion buffer: 20 mM Tris-HCl, pH 8.0, 10 mM NaCl, 10 mM EDTA, 0.5% w/v sodium dodecyl sulfate. Store indefinitely at room temperature; add 1 mg/mL proteinase K (500 g; Merck, Whitehouse, NJ; cat. no. EC3.4.21.14) just before use.
30. 10X amplification buffer (KOD Dash buffer: Toyobo).

31. 25 m*M* 4dNTP mix.
32. Sex-determining region Y (SRY) primers (*see* **Subheading 3.2., step 8**). Set 1: sense 5′-GCC ATT CTT CGA GGA GGC ACA GA-3′ (nt 402–425), antisense 5′-TATCGAC-CTCGTCGGAAGGC-3′ (nt 466–486); Set 2: sense 5′-TGG TGT GGT CTC GTG ATC AGG CGC AAG G-3′ (nt 284–314), antisense 5′-TCTGTGCCTCCTCGAAGAATGGC-3′ (nt 402–425).
33. 0.5 U KOD Dash polymerase (250 U; Toyobo, Kyoto, Japan; cat. no. LDP-101).
34. 12% polyacrylamide gel.

3. Methods
3.1. Tissue and Cell Culture
3.1.1. Feeder layers of UCDFs

We established equine ES cell lines (EK1, EY2, and EN3) on a feeder layer of UCDFs *(32)* from a newborn calf with the addition of LIF. This substratum must remain in an undifferentiated state. After expansion of equine ES cells, we froze cells in aliquots as stocks.

3.1.1.1. METHOD FOR THE ISOLATION OF UCDFs

1. Cut a piece from the umbilical cord of a newborn calf with scissors.
2. Wash the sample in PBS supplemented with antibiotic solution (1X penicillin-streptomycin mix) to remove toxic agents and possible contaminants.
3. Transfer the sample to a sterile 60-mm dish and cut into small pieces approx 1–2 mm in diameter using fine sterile scalpel blades without tearing the tissue.
4. Pick up several pieces of tissue with well-cut edges using forceps and place them in a sterile 60-mm culture dish.
5. Add 3 mL ST-MEMα to submerge the small pieces of tissue.
6. Incubate at 38.6°C in a humidified atmosphere of 5% CO_2 in air.
7. Replace the medium with fresh medium every 2 d. When cells have become confluent (4–6 d), trypsinize them with 2 mL 0.25% trypsin-EDTA. Collect all the cells, freeze them in a vial, and store them in liquid nitrogen.

3.1.1.2. MAINTENANCE OF CELLS

1. To thaw UCDF cells, transfer the contents of one vial to three four-well culture dishes.
2. Incubate as described in **Subheading 3.1.1.1.** until the cells become confluent (4–5 d) and then subculture them at dilution ratios of 1:3 to 1:5 for use as feeder layers.
3. To subculture UCDFs, discard medium and rinse cells thoroughly with PBS.
4. Add 1–2 mL 0.25% trypsin-EDTA to each 60-mm dish.
5. Incubate at room temperature for 5–10 min until the cells have detached from the substratum while observing them under a stereomicroscope.
6. Add 5 mL ST-MEMα to neutralize the trypsin and collect cells in a 15-mL Falcon tube.
7. Centrifuge at 700*g* for 5 min to pellet the cells.
8. After centrifugation of cells, remove the medium with a sterile pipet.
9. Add 9 mL ST-MEMα and mix well.
10. Dispense 3 mL of the suspension of cells into each of three 60-mm dishes.
11. Incubate at 38.6°C until UCDFs reach confluence (4–5 d). Change the medium every 2 d. These cells can be maintained until the fifth or sixth passage.

3.1.2. Freezing of Cells (see **Note 7**)

1. Aspirate the medium from a 60-mm dish and wash the surface with 3–5 mL PBS.
2. Flood the dish with 0.8 mL trypsin-EDTA and return the dish to the incubator for 5–6 min at 38.6°C.
3. Remove the dish and swirl it to dislodge the cells. Then, add 2 mL ST-MEMα.
4. Pipet the medium up and down vigorously and check the suspension of cells visually under low-power magnification to ensure that it is relatively free of cell aggregates.
5. Collect the cell suspension in a centrifuge tube and pellet the cells by low-speed centrifugation (700g, 5 min).
6. Ideally, cells should be frozen at a density of $1–2 \times 10^6$/mL. Add 1 mL freezing medium to a 2-mL cryovial. Then, add 1 mL of the suspension of cells in ST-MEMα ($2–4 \times 10^6$/mL), mix gently, and place in the gas phase compartment of a liquid nitrogen storage tank with a freezing tray plug (*see* **Note 8**).
7. Leave the vial for 24 h before transfer to liquid nitrogen.

3.1.3. Thawing of Cells

1. Thaw cells quickly in a water bath at 36°C.
2. Dilute 2 mL of the suspension of cells with 12 mL of medium in a 15-mL tube.
3. Mix gently and pellet cells by low-speed centrifugation at 700g for 5 min.
4. Resuspend cells in fresh ST-MEMα and transfer to a tissue culture dish (four-well dishes).
5. Place the dish in the incubator at 38.6°C and replace the medium every 2 d.

3.1.4. Derivation and Maintenance of ES Cells

3.1.4.1. Preparation of Feeder Layers for the Derivation of ES Cells

Equine ES cells are established on four-well dishes on a feeder layer of mitomycin-treated UCDFs. When UCDFs are nearly confluent, the medium is removed by aspiration, and cells are treated with mitomycin C to stop mitotic expansion while they are still able to feed the other cells.

1. Replace medium by 2 mL ST-MEMα plus 10 µg/mL mitomycin C in four wells using 0.5 mL per well.
2. Incubate at 38.6°C for 2–3 h.
3. Wash each well two or three times with 500 µL PBS and then replace the medium with 500 µL ST-MEMα. The UCDFs are now ready for use as a feeder layer for the derivation of ES cells and these cells can be cultured for up to 3 or 4 d before use.

3.1.4.2. Thawing of Frozen Equine Embryos

Blastocysts are thawed by immersion of straws for 20 s in a water bath at 36°C. Then, they are incubated stepwise as follows:

1. Place in PBS plus 10% FBS that contains 6% glycerol and 0.3 M sucrose for 5 min.
2. Place in PBS plus 10% FBS that contains 3% glycerol and 0.3 M sucrose for 5 min.
3. Place in PBS plus 10% FBS that contains 3% sucrose, with a final wash in PBS plus 10% FBS.

3.1.4.3. Isolation and Culture of ES Cells

ICM cells are isolated microsurgically with a micromanipulator and then cultured as follows.

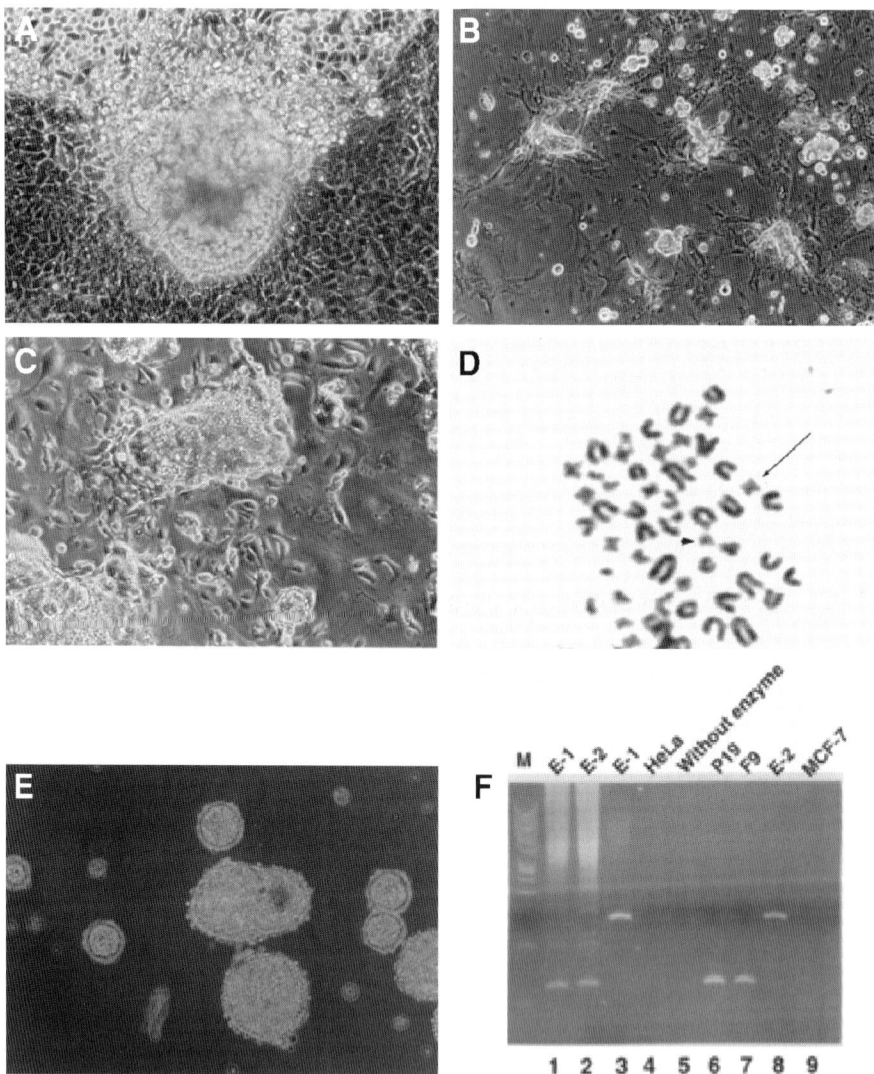

Fig. 1. Characterization of equine embryonic stem (ES) cells. (**A**) Inner cell mass after 6 d in culture. (**B**) Colonies of ES cells. (**C**) ES cells undergoing partial differentiation after culture for 5 d in the absence of a feeder layer of umbilical cord-derived fibroblasts cells but in the presence of leukemia inhibitory factor (LIF) (10 ng/mL). (**D**) A metaphase spread of ES cells that included an X chromosome (arrow) and a Y chromosome (arrowhead). (**E**) Embryoid bodies (EBs) 14 d after the start of suspension culture. Colonies of cells (more than 100 cells per aggregate) were placed in standard minimum essential medium-α without LIF in the absence of a feeder layer. They multiplied rapidly and formed tight aggregates that increased in size with time. These EBs were composed of two layers of cells with heterogeneous cellular particles within each cavity. Endodermlike cells were located on the external surface of the EBs, and ectodermlike cells were

Differentiation of Equine ES Cells In Vitro

1. Plate ICM cells on a feeder layer of UCDFs in supplemented ST-MEMα (*see* **Subheading 2.1.1.**, **item 2**) in four-well culture dishes in a humidified atmosphere of 5% CO_2 in air at 38.6°C (*see* **Notes 9** and **10**).
2. From 3–5 d later, use a Pasteur pipet to pick growing colonies of cells with morphology that resembles that of typical murine ES cells *(12)* and place in small drops (50 µL) of 0.25% trypsin-EDTA (**Fig. 1A**).
3. Incubate each dish that contains a drop in the incubator for 5–6 min.
4. "Pull out" a Pasteur pipet to make an extremely fine capillary tube (*see* **Note 11**). Place in a mouth-controlled holder and, by suction, pick up a small volume (50 µL) of medium.
5. Return the dish to the stage of a binocular stereomicroscope and, using the pipet, break up the growing ICM clump. First, add a small volume (50 µL) of medium into the drop to neutralize the trypsin; then, using mouth control, draw the cell clump in and out of the pipet. It is necessary to repeat this action once or twice.
6. Place the dissociated cells on a fresh feeder layer and culture under the same conditions as previously described.
7. After 10–14 d, cells with ES-like morphology that have become confluent are trypsinized and replated on a fresh feeder layer (passage 1; P1). P1 cells are then subjected to repeated passages at 6- to 7-d intervals. Cultures are examined daily, and the medium is replaced by fresh medium every second day (**Fig. 1B,C**).

3.1.4.4. KARYOTYPE ANALYSIS WITH GIEMSA STAINING

Karyotype analysis must be performed to ascertain the chromosome complement and sex chromosome constitution of cell lines (**Fig. 1D**).

1. Add colcemid at a final concentration of 0.05 µg/mL to ES cells and return plates to the incubator for 3 h.
2. Collect the cells by the standard procedure (*see* **Subheading 3.1.2.**, **steps 3–5**) and pellet the cells in a 15-mL conical tube.
3. Remove the supernatant by aspiration and add 10 mL of a hypotonic 0.56% (w/v) solution of KCl to the pellet. Incubate the mixture at 38.6°C for 30 min to allow hypotonization.
4. Pellet the cells by centrifugation at 700g for 10 min. Discard the supernatant. Add precooled fixative (a mixture of acetic acid and methanol, 1:3 v/v). To avoid agglutination of cells, the fixative should be added slowly, drop by drop, to the cell suspension with gentle shaking of the tube.
5. Incubate for 5 min at room temperature. Change the fixative an additional three times with centrifugation. The volume of the final suspension should be approx 1 mL.
6. Clean a slide by wiping it with 7% ethanol.
7. Draw approx 20 µL of the homogeneous suspension of cells into a micropipet and, holding it vertically, position the end 30 cm above the slide. Then, drop a single drop of suspension onto the slide.

Fig. 1. (*Continued*) located on the internal surface. Magnification: **A**, **C**, and **E**, ×200; **B**, ×100; **D**, ×1800. (**F**) Amplification by polymerase chain reaction of sex-determining region Y (SRY) from equine ES cells. **Lanes 1–3**, ES cells; **lane 4**, HeLa cells; **lane 5**, without KOD Dash polymerase; **lane 6**, P19 cells; **lane 7**, F9 cells; **lane 8**, ES cells; **lane 9**, MCF-7 cells. **Lanes 1**, **2**, **4**, **5–7**, **9**, amplification with primer set 1 for SRY-1; **lanes 3** and **8**, amplification with primer set for SRY-1. Molecular markers (*Hae*III digest of X174). (Please *see* companion CD for the color version of this figure.)

8. Leave the slide at room temperature for a while (10 min) and then rinse it with fresh fixative before the slide is completely dry. This step ensures that the slide is free of debris and any other inappropriate particles.
9. Prepare 5–10 slides this way. Examine the individual slides carefully under the microscope to determine the number and quality of the chromosome spreads.
10. Slides are stained by immersion in a 3% solution of Giemsa in PBS for 5 min. Rinse the slides with two changes of ddH$_2$O and allow to dry before examining (*see* **Note 12**).

3.1.4.5. ALKALINE AP STAINING

1. Remove the medium and wash cells once with PBS.
2. Fix in 4% formaldehyde for 15 min.
3. Wash with ddH$_2$O.
4. Meanwhile, prepare the substrate. For 10 mL, combine 40 µL naphthol AS-MX-PO$_4$ stock solution with 10 µL Fast Red TR salt solution (*see* **Subheading 2.3., item 2**) and 10 mL ddH$_2$O.
5. Pipet 250 µL of the reagent into each well and incubate for 15 min at room temperature in a dark area of the laboratory.
6. Wash with ddH$_2$O and store in PBS (**Fig. 2A**).

3.1.5. Differentiation of ES Cells

ES cells are capable of generating all types of differentiated cells in vitro and in vivo. To develop a powerful system for inducing the differentiation of stem cells, it is important to have proper understanding of the molecular and cellular mechanisms that control the fate of stem cell lineages. When differentiated as suspended multicellular aggregates (embryoid bodies, EBs), ES cells can readily give rise to differentiated lineages of endoderm, mesoderm, and ectoderm. Experiments have shown that treatment with retinoic acid (RA) promotes neural differentiation in EBs. However, it has been suggested that RA, a strong teratogen, suppresses forebrain development in treated embryos. We have developed a system for the differentiation of equine ES cells in vitro into cells with morphological and biochemical characteristics of neural, hematopoietic, and endothelial progenitors *(33)* by culturing cells in media supplemented with a cocktail of cytokines. This system requires neither EBs nor treatment with RA.

3.1.5.1. DIFFERENTIATION OF NP CELLS

1. Culture equine ES cells at a density of 10^4 cells per well in four-well dishes in serum-free ST-MEMα supplemented with bFGF (10 ng/mL), EGF (10 ng/mL), and PDGF (10 ng/mL) for 7 d in the absence of a feeder layer (**Fig. 3**). Cells should develop into bipolar fibrous cells that express nestin, a marker of NP cells (*see* **Subheading 3.2.1.1.**).
2. Remove cytokines from the medium and continue to culture the differentiated cells, allowing them to undergo further differentiation for 7–14 d. Bipolar fibrous cells should develop into astrocytes or neurons that express astrocyte-specific and neuron-specific markers (*see* **Subheading 3.2.1.1.**).

3.1.5.2. DIFFERENTIATION TO HEE CELLS

1. Culture equine ES cells at a density of 10^4 cells per well in four-well dishes in 10% FBS plus ST-MEMα supplemented with bFGF (10 ng/mL), SCF (10 ng/mL), and OSM (10 ng/mL) for 10 d. Two different types of cell colonies will be found in the same well. One type of colony should consist of nonadherent, rounded, cobblestonelike cells, and the other should

Differentiation of Equine ES Cells In Vitro

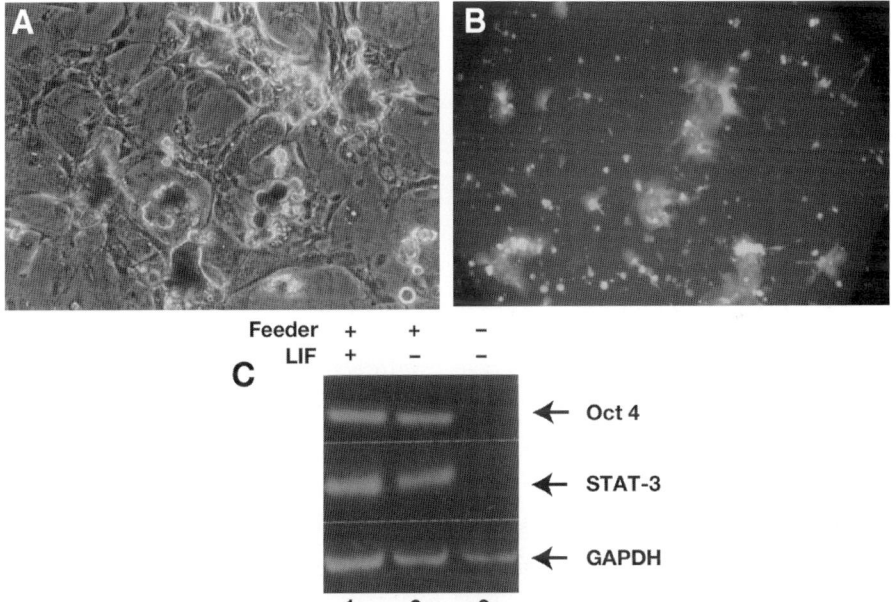

Fig. 2. Expression of molecular markers of undifferentiated equine embryonic stem (ES) cells. (**A**) Alkaline phosphatase (AP) staining of ES cells. ES cells growing on a feeder layer of umbilical cord-derived fibroblasts were stained with AP after 5 d in culture. ES cells are AP positive, indicating that they are in an undifferentiated state. (**B**) Immunostaining of ES cells with STAT3-specific antibody. ES cells were grown in the absence of a feeder layer but in the presence of leukemia inhibitory factor (LIF) (10 ng/mL). Most of the ES cells are STAT3 positive. (**C**) Analysis of the expression of genes for Oct4, STAT3, and GAPDH by reverse transcription polymerase chain reaction. Total RNA was extracted from ES cells that had been cultured for 5 d on a feeder layer with LIF (**lane 1**), on a feeder layer in the absence of LIF (**lane 2**), and in the absence of both LIF and a feeder layer (**lane 3**). STAT3 and Oct4 were detected in the analysis of RNA from undifferentiated ES cells but not from differentiated ES cells. Magnification: **A** and **B**, ×100. (Please *see* companion CD for the color version of this figure.)

    consist of adherent endothelial-like cells (**Fig. 4A**). The clusters of cobblestonelike cells should express a marker of hematopoietic progenitor cells, CD45 (*see* **Subheading 3.2.1.1.**).
2. Remove cytokines from the medium and continue to culture the adherent cells for a further 7–10 d. They should express various markers of endothelial cells, namely Flk-1, CD31, and CD34 (*see* **Subheading 3.2.1.1.**).

3.2. Analysis of Pluripotency and Differentiation

Equine ES cells are pluripotent and can differentiate into various types of cells, which can be analyzed using the techniques described in this section.

3.2.1. Immunocytochemical Staining

Various markers that are used to prove pluripotency and to detect the differentiation of ES cells are indicated in **Table 1**.

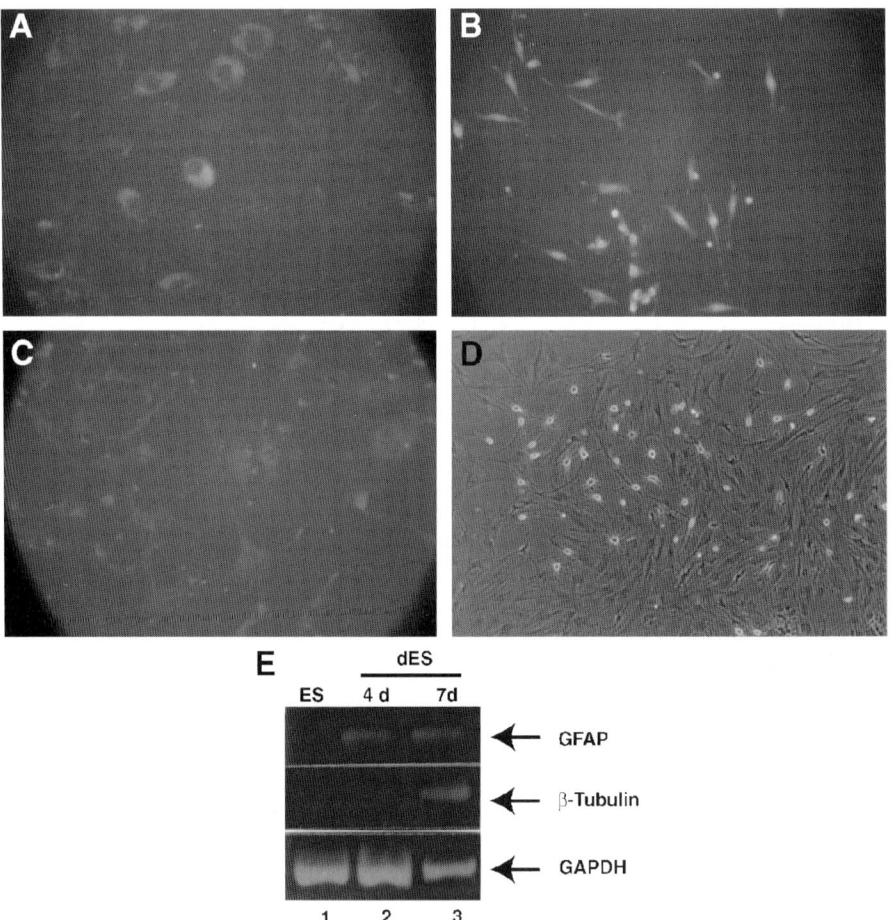

Fig. 3. Morphological differentiation and expression of markers by embryonic stem (ES) cell-derived neural precursors. ES cells were cultured in serum-free standard minimum essential medium-α with basic fibroblast growth factor (bFGF), epidermal growth factor (EGF), and platelet-derived growth factor (PDGF) for 7 d and then analyzed for markers of neural precursor cells. (**A**) Immunostaining with the nestin-specific antibody. (**B**) Immunostaining with the astrocyte-specific GFAP-specific antibody. (**C**) Immunostaining with the neuron-specific β-tubulin III-specific antibody. (**D**) Phase contrast micrograph showing the extensive outgrowth of neural structures from ES cells. Magnification: **A** and **C**, ×200; **B** and **D**, ×100. (**E**) Analysis of the expression of marker genes in equine ES cell-derived neural precursors. Total RNA from ES cells in an undifferentiated state (**lane 1**) and from ES cells that had been treated with bFGF, EGF, and PDGF for 4 d (differentiated ES [dES]; **lane 2**) and 7 d (dES; **lane 3**) was analyzed by reverse transcription polymerase chain reaction for the expression of genes for GFAP, β-tubulin III, and GAPDH. (Please *see* companion CD for the color version of this figure.)

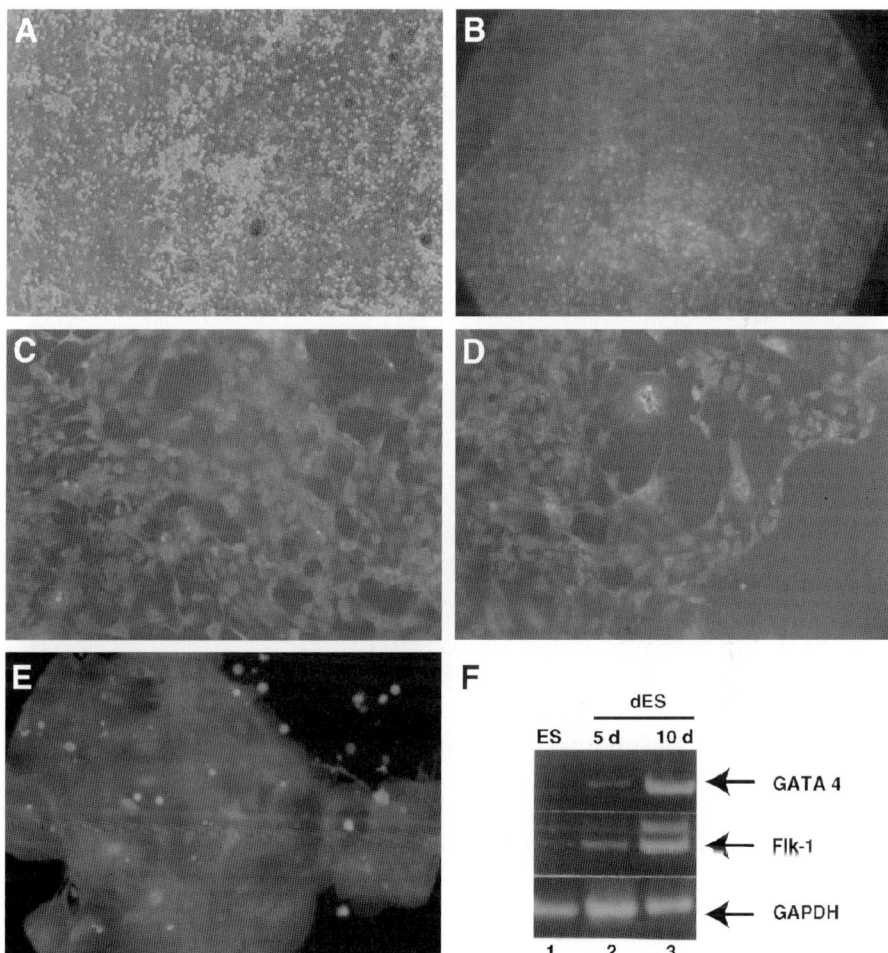

Fig. 4. Morphological differentiation and expression of markers by embryonic stem (ES) cell-derived hematopoietic precursors. ES cells were cultured in standard minimum essential medium-α with basic fibroblast growth factor (bFGF), stem cell factor (SCF), and oncostatin M (OSM) for 10 d. (**A**) Phase contrast micrograph of cobblestonelike cells and adherent endothelial cells. (**B**) Immunostaining of the same culture as in **A** with a CD45-specific antibody. Round cobblestone-like cells were immunopositive. ES cells that had differentiated to endothelial precursors were immunostained with antibodies specific for CD31, CD34, and Flk-1, as shown in panels **C**, **D**, and **E**, respectively. Magnification: **A** and **B**, ×40; **C**–**E**, ×100. (**F**) Expression of genes for markers of hematopoietic and endothelial differentiation in undifferentiated ES cells (ES) and differentiated ES cells (dES). Total RNA from ES cells (**lane 1**) and from ES cells treated with bFGF, SCF, and OSM for 5 d (**lane 2**) and 10 d (**lane 3**) was analyzed by reverse transcription polymerase chain reaction for expression of genes for GATA 4, Flk-1, and GAPDH. (Please *see* companion CD for the color version of this figure.)

Table 1
Markers of Pluripotency and Differentiation

	Primary Antibodies	Secondary Antibodies	
Pluripotency	Mouse anti-STAT3	FITC-conjugated rabbit antimouse IgG	(**Fig. 2B**)
Differentiation	Mouse anti-nestin (neural progenitor cells)	FITC-conjugated rabbit antimouse IgG	(**Fig. 3A**)
	Mouse anti-GFAP (astrocytes)	FITC-conjugated rabbit antimouse IgG	(**Fig. 3B**)
	Mouse anti-β-tubulin III (neurons)	FITC-conjugated rabbit antimouse IgG	(**Fig. 3C**)
	Mouse anti-CD45 (hematopoietic progenitor cells)	FITC-conjugated rabbit antimouse IgG	(**Fig. 4B**)
	Mouse anti-CD31 (endothelial cells)	FITC-conjugated rabbit antimouse IgG	(**Fig. 4C**)
	Mouse anti-CD34 (endothelial cells)	FITC-conjugated rabbit antimouse IgG	(**Fig. 4D**)
	Mouse anti-Flk-1 (endothelial cells)	FITC-conjugated rabbit antimouse IgG	(**Fig. 4E**)

1. Plate frozen and thawed ES cells at 10^4 cells per well in four-well dishes without a feeder layer and culture in the presence of LIF (10 ng/mL) in ST-MEMα for 4 d.
2. When cells are confluent, remove medium and wash cells with PBS.
3. Flood wells with 4% formaldehyde and fix for at least 1 h.
4. Remove fixative and wash three times with PBS, leaving cells in PBS for 1 h after the third wash.
5. Calculate the volume of solution of primary antibody needed given that each well needs 200 µL. Prepare a solution of 1:50 diluted antibody in this volume (*see* **Note 13**).
6. Remove PBS after the final wash and add 200 µL of the diluted solution of antibody to each well. Place wells in the incubator for 1 h.
7. Flood wells with PBS and leave for 10 min at room temperature.
8. Calculate the volume of FITC-conjugated secondary antibody needed. Prepare a solution of 1:200 diluted second antibody (*see* **Note 14**).
9. Remove the solution of primary antibody and wash cells three times with PBS.
10. Add 200 µL of the solution of FITC-conjugated secondary antibody to each well and incubate for 45 min at room temperature (*see* **Note 15**).
11. After incubation, flood each well with PBS and leave for 10 min at room temperature.
12. Remove the PBS and wash cells three times with PBS.
13. Examine the fluorescence caused by FITC, record results, and take pictures.

3.2.2. Amplification by PCR: Sex Determination of ES Cells

The basic protocol involves the enzymatic amplification of DNA that corresponds to the sex-determining region of the Y chromosome by PCR. The oligodeoxynucleotide primers are coprecipitated with the DNA to maximize the efficiency of their annealing to each other (*see* **Notes 16** and **17**).

In general, male ES lines are more useful and are used most commonly because the X chromosome is unstable in female ES lines *(34)*. The following protocol describes a simple screening using PCR for determining the sex of a line of ES cells.

1. Plate ES cells on gelatinized 24-well plates without feeder cells. Culture these cells for 3 d. Dilute cells at least 10-fold at each passage (*see* **Note 18**). ES cells can be grown in standard ES cell medium without antibiotics at this stage. Some differentiation will occur in the absence of a feeder layer.
2. Remove medium and rinse cells well with Hanks' balanced salt solution.
3. Add 300 µL digestion buffer, transfer the mixture to a 1.5-mL microcentrifuge tube, and incubate overnight at 55°C.
4. Add 150 µL of a saturated solution of NaCl to the microcentrifuge tube and mix vigorously on a vortex mixer (the solution will turn milky white). Add two volumes of 95% ethanol (the solution will turn clear, and DNA will form a precipitate).
5. After centrifugation at 13,000g for 10 min, wash the precipitates with 70% ethanol and resuspend the pellet DNA in 50 µL ddH$_2$O. Determine the concentration of DNA by measuring the absorbance at 260 nm.
6. For PCR, mix the following reagents in a 0.5-mL thin-wall PCR tube for each set of primers: 2.5 µL 10X PCR buffer (1.5 mM final concentration of MgCl$_2$); 0.2 µL 25 mM 4dNTP mix; 0.5 µM sense primer; 0.5 µM antisense primer; 0.5 µg DNA template; 0.5 U KOD polymerase; make up to 25 µL with water.
7. Using the following parameters, perform PCR for 35 cycles of denaturation at 94°C for 30 s, with annealing at 55°C for 30 s and extension at 72°C for 120 s.
8. Fractionate the PCR products on a 12% polyacrylamide gel by electrophoresis with the appropriate molecular weight markers (**Fig. 1E**). The SRY primers are derived from the sex-determining region of the Y chromosome. Male cells will yield a product of 83 bp with primer set 1 and of 140 bp with primer set 2; female cells will yield no products.

3.2.3. Expression of Marker Proteins During the Differentiation of ES Cells

This section describes the methods for enzymatic amplification of RNA by PCR. The reaction consists of three steps: annealing, reverse transcription, and amplification. These steps are performed under optimal considerations, maximizing efficiency and recovery.

3.2.3.1. EXTRACTION OF RNA WITH TRIZOL

The following procedure was derived from the manufacturer's protocol (*see* **Note 19**).

1. Remove the medium from cell cultures and rinse the dishes three times with PBS. Lyse cells by adding the appropriate amount of Trizol reagent (i.e., for a 35-mm dish, use 1 mL; for a 100-mm dish, use 6 mL).
2. Detach the cells from the substratum with a scraper, pipet the suspension several times to collect all the cells, and transfer the suspensions to a sterile culture tube. Incubate the sample for 5 min at room temperature.
3. Add 0.2 mL chloroform (RNase free) per 1 mL Trizol used (i.e., for a 35-mm dish, use 200 µL; for a 100-mm dish, use 1.2 mL). Shake vigorously for 30 s and then incubate at room temperature for 3 min.

4. Centrifuge at 13,000g for 15 min at 4°C. The mixture will separate into a lower phase (phenol and chloroform), an interface, and a colorless upper aqueous phase. The RNA is in the aqueous phase, and the volume is approx 60% of the original volume of Trizol used.
5. Transfer the aqueous phase into a fresh tube. Precipitate the RNA by adding 0.5 mL RNase-free isopropanol per milliliter of Trizol originally added (i.e., for a 35-mm dish, use 0.5 mL; for a 100-mm dish, use 3 mL). Incubate the mixture for 10 min at room temperature.
6. Centrifuge at 13,000g for 10 min at 4°C. The RNA should form a pellet.
7. Remove the supernatant and wash the pellet with RNase-free 75% ethanol. Use 1 mL ethanol for each milliliter of Trizol used originally (i.e., for a 35-mm dish, use 1 mL; for a 100-mm dish, use 6 mL). Mix the sample on a vortex mixer and centrifuge at 750g for 5 min at 4°C.
8. Dry the pellet and resuspend it in DEPC-treated water. If it does not dissolve, incubate for 10 min at 55°C.
9. Measure the A260 and A280. The ratio A260:A280 should be between 1.6 and 2.0 (*see* **Note 20**). Determine the concentration of the sample (*see* **Note 21**).

3.2.3.2. Discrimination Between RNA and DNA

Given the proclivity of enzymatic amplification to yield rare molecules, one must consider whether the amplified products of PCR arise from RNA or DNA templates. In some cases, it is possible to choose primers so that specific amplification products cannot arise from DNA. For example, primers from different exons will yield products of different sizes if DNA or mRNA is used as a template. In other cases, it will not be possible to select such discrimination primers. Enzymatic treatment of samples with RNase-free DNase, followed by phenol extraction and ethanol precipitation, may be helpful in such cases.

1. Digest 30 µg of RNA so that the volume of RNA plus DEPC-H_2O equals 87 µL. Add 10 µL 10X DNase buffer; 2 µL RNase inhibitor; and 1 µL DNase I.
2. Mix and incubate for 15 min at room temperature.
3. Prepare a mixture of chloroform and isoamyl alcohol (24:1 v/v) in a 20-mL tube using 12 mL RNase-free chloroform plus 0.5 mL RNase-free isoamyl alcohol.
4. Add an equal volume of phenol plus chloroform to the RNA sample (i.e., for a 100-µL sample, add 50 µL phenol and 50 µL chloroform plus isoamyl alcohol, 24:1).
5. Mix vigorously and centrifuge at 13,000g at room temperature for 5 min.
6. Transfer the aqueous phase (the top phase) to a fresh RNase-free microtube.
7. Add 1/10 vol of RNase-free 3 *M* sodium acetate (i.e., for a 100-µL sample, add 10 µL 3 *M* sodium acetate).
8. Add two volumes of 100% ethanol (RNase free, (−20°C) (i.e., for a 100-µL sample, add 200 µL 100% ethanol).
9. Mix and incubate tube at −80°C for 1 h.
10. Centrifuge immediately at 13,000g for 20 min at 4°C.
11. Wash the pellet with 200 µL 70% ethanol (RNase-free, −20°C).
12. Centrifuge again for 10 min at 13,000g at 4°C.
13. Dry the pellet and resuspend in DEPC-H_2O.
14. Quantify the DNased RNA (*see* **Note 22**).
15. Dilute samples to 1 µg/µL with DEPC-H_2O for RT-PCR.

3.2.3.3. Denaturation of RNA

Before the RT-PCR reaction, the RNA is denatured by heating to facilitate the binding of primers to the template RNA.

1. Mix 1 µL of the solution of RNA with 2 µL DEPC-H_2O in a 100-µL PCR tube.
2. Place the tubes in a thermal cycler and incubate at 65°C for 5 min. Then, lower temperature to 4°C for at least 5 min before proceeding to the SuperScript one-step RT-PCR reaction.

3.2.3.4. SuperScript One-Step RT-PCR With Platinum Taq

The SuperScript one-step RT-PCR with Platinum Taq system is designed for the convenient, sensitive, and reproducible detection and analysis of RNA by RT-PCR. Both complementary DNA (cDNA) synthesis and PCR are performed in a single tube using gene-specific primers and target RNAs from either total RNA or messenger RNA. The RT/Platinum Taq mix is a mixture of recombinant Taq DNA polymerase complexed with proprietary antibody that inhibits polymerase activity and SuperScript II reverse transcriptase. The activity of Taq DNA polymerase is blocked at ambient temperatures but is regained after the denaturation step in PCR cycling at 94°C. This reaction provides an automatic "hot start" for Taq DNA polymerase in PCR. Hot starts are typically used in PCR to increase sensitivity, specificity, and yield.

The system consists of two major components: SuperScript II RT Platinum Taq Mix (RT/Platinum Taq mix) and 2X reaction mix. The RT/Platinum Taq mix contains a mixture of SuperScript II RT and Taq DNA polymerase for optimal cDNA synthesis and PCR amplification. The 2X reaction mix consists of a proprietary buffer system optimized for RT and amplification by PCR, Mg^{2+} ion optimized for universal use, deoxyribonucleotide triphosphates, and stabilizers.

1. Mix 25 µL of 2X reaction mix containing 0.4 mM of each dNTP and 2.4 mM $MgSO_4$ (*see* **Note 22**); 1.0 µL RNase inhibitor; 1.0 µL of solution of template RNA (0.1 µg); 1.0 µL sense primer (10 µM); 1.0 µL antisense primer (10 µM); 1.0 µL RT/Platinum Taq mix; and ddH_2O to 50 µL.
2. Program the thermal cycle so that cDNA synthesis is followed automatically and immediately by PCR.
    a. cDNA synthesis and predenaturation: perform one cycle of incubation at 45°C for 30 min and at 94°C for 2 min (*see* **Note 23**).
    b. PCR amplification: perform 40 cycles of denaturation at 94°C for 15 s, annealing at 55°C for 30 s, and extension at 72°C for 2 min (*see* **Note 24**).
    c. Final extension: one cycle at 72°C for 2 min (*see* **Note 25**).
3. Gently mix and make sure that all the components are at the bottom of each amplification tube. Centrifuge briefly if needed. Depending on the thermal cycler used, overlay the mixture with silicone oil if necessary.
4. Analyze the product of amplification by polyacrylamide gel (10%) electrophoresis with the appropriate molecular markers (**Figs. 2C, 3E,** and **4F**).
5. Purify the DNA of a QIAquick™ PCR spin column.
6. Subclone the product of PCR into Gateway™ vector and sequence (**Table 2**).

Table 2
Primer Sequences

Genes	Primers (upper, sense; lower, antisense)	Positions
Oct4	5'-TCCCAGGACATCAAAGCTCTGCAGA-3'	(nt 848–873)
	5'-TCTGGGCTCTCCCATGCATTCAAACTGA-3'	(nt 1495–1524)
STAT3	5'-TCTGGCTAGACAATATCATCGACCTTG-3'	(nt 1908–1935)
	5'-TTATTTCCAAACTGCATCAATGAATCT-3'	(nt 2413–2439)
GFAP	5'-GCCCTGGACATCGAGATCGCCACCTACAGG-3'	(nt 1091–1120)
	5'-ATGTTCCTCTTGAGGTGGCCTTCTGACAC-3'	(nt 1214–1244)
β-*Tubulin III*	5'-CAGAGCAAGAACAGCAGCTACTT-3'	(nt 999–1022)
	5'-GTGAACTCCATCTCGTCCATGCCCTC-3'	(nt 1201–1227)
GATA-4	5'-CTCTGGAGGCGAGATGGGACGGG-3'	(nt 971–994)
	5'-GAGCGGTCATGTAGAGGCCGGCAGGCATT-3'	(nt 1452–1481)
Flk-1	5'-CTGCCTACCTCACCTGTTTCCTGTATGG-3'	(nt 3833–3861)
	5'-GGATATCTTGAAATGTTTTTACACTCAC-3'	(nt 4003–4031)
GAPDH	5'-GGGCTTGGCTTCGGTGACAACACCAAGGCGG-3'	(nt 646–678)
	5'-CGAGCAAAGGCCTCTGCCACCTTGCGGTT-3'	(nt 808–837)

4. Notes

1. Use deionized, distilled water in all solutions and procedures. Prepare all solutions for cell culture from tissue culture-grade reagents. Use tissue culture-grade water (high resistance and endotoxin free). Sterilize all final solutions by filtration or prepare from sterile stocks. Use disposable sterile plasticware to prevent microbial and detergent contaminations. Storage of liquid media at freezer temperatures is not recommended because insoluble complexes can develop during freezing and thawing processes.
2. This media contains L-glutamine, ribonucleosides, and deoxyribonucleosides.
3. Because this serum has already been heat inactivated, it can be stored at −20°C for 1–2 yr.
4. Addition of bFGF and PDGF has been shown to promote the proliferation of glial precursor cells.
5. The media containing EGF and bFGF induces the differentiation of ES cells to astrocytes and glial cells.
6. Wear gloves and use a fume hood when using DEPC because it is a suspected carcinogen.
7. For indefinite storage, cells must be kept at −196°C.
8. When cells are confluent, one cryovial can usually be prepared from each four-well or 60-mm dish.
9. We have found that Nunc four-well dishes (made in Denmark) are the best dishes for the isolation and maintenance of animal (equine and bovine) ES cells. Nunc four-well dishes are superior to other plastic dishes in terms of cell attachment affinity.
10. We recommend that equine embryos should be cultured singly in individual wells. This procedure eliminates the requirement for recloning the derived stem cells. A cell line derived from a single embryo is effectively a clonal population in terms of genotype and sex chromosome complement *(35)*.

11. The diameter of the end of the tube should be approximately a quarter of the diameter of the clump of ICM cells to be dislodged.
12. The diluted solution of Giemsa tends to form a "metallic scum" on the surface. It is necessary to remove this scum by drawing a clean slide glass across the surface of solution before staining because the scum sticks to preparations.
13. For example, for 1000 µL, dilute 20 µL of the preparation of antibody in 980 µL PBS.
14. For example, for 1000 µL, dilute 5 µL of the solution of FITC-conjugated second antibody in 995 µL PBS.
15. It is essential to protect fluorescent conjugates and labeled dishes from light. Incubate samples in the dark and cover whenever possible.
16. Primers should not be self-complementary or complementary to each other at the 3′ ends.
17. Investigators may wish to consider the use of commercially available kits for performing PCR. Several kits are available that offer advantages of convenience and speed, albeit with some increase in terms of cost and a decrease in the ability to customize reaction conditions for specific target DNAs.
18. The cells must be diluted out to prevent contamination of the PCR by any residual feeder cells.
19. During the extraction of RNA, all equipment and reagents should be free of RNase. Moreover, gloves are changed frequently to minimize the contamination of RNase.
20. Higher-quality intact RNA is essential for successful synthesis of full-length cDNA.
21. Dilute 1 µL in 100 µL DEPC-H_2O. (X) µg/µL = (A260 × 40 × Dilution factor) / 1000.
22. Keep all reaction mixtures and samples on ice and prepare reaction mixture on ice. The 2X reaction buffer provided with the system was optimized to provide higher specific amplification by RT-PCR. The final concentration of Mg^{2+} ion is 1.2 mM.
23. SuperScript II RT is inactivated, Taq DNA polymerase is reactivated, and the RNA/DNA hybrid is denatured during the 2-min incubation at 94°C.
24. The extension time varies with the size of the amplicon (approx 1 min per 1 kb amplicon).
25. The reaction mixture should be on ice before starting the reaction. A reaction cocktail can be made when multiple reactions are assembled.

References

1. Evans, M. J. and Kaufmann, H. M. (1981) Establishment in culture of pluripotent cells from mouse embryos. *Nature* **292,** 154–156.
2. Brüstle, O., Jones, K. N., Lealish, R. D., et al. (1999) Embryonic stem cell-derived glial precursors: a source of myelinating transplants. *Science* **285,** 754–756.
3. Choi, K., Kennedy, M., Kazarov, A., Papadimitriou, J. C., and Keller, G. (1998) A common precursor for hematopoietic and endothelial cells. *Development* **125,** 725–732.
4. Klug, M. G., Soompao, M. M., Koh, G. Y., and Field, L. J. (1996) Genetically selected cardiomyocytes from differentiating embryonic stem cells form stable intracardiac grafts. *J. Clin. Invest.* **98,** 216–224.
5. Bain, G., Kitchens, D., Yao, M., Hunnetner, J. E., and Gottlieb, D. I. (1995) Embryonic stem cells express neuronal properties in vitro. *Dev. Biol.* **168,** 342–357.
6. Thompson, S., Clarke, A. R., Pow, A. M., Hooper, M. L., and Melton, D. W. (1989) Germ line transmission and expression of a corrected HPRT gene produced by gene targeting in embryonic stem cells. *Cell* **56,** 313–321.
7. Smith, A. G. (2001) Embryonic-derived stem cells of mice and men. *Annu. Rev. Dev. Biol.* **17,** 435–462.

8. McCreath, K. J., Howcroft, J., Campbell, K. H. S., Colman, A., Schnieke, A. E., and Kind, A. J. (2000) Production of gene-targeted sheep by nuclear transfer from cultured somatic cells. *Nature* **405,** 1066–1069.
9. Lai, L., Kolber-Simmons, D., Park, K.-W., et al. (2002) Production of α-1,3-galactsyltransferase knockout pigs by nuclear transfer cloning. *Science* **295,** 1089–1092.
10. Denning, C., Burl, S., Ainslie, A., et al. (2001) Deletion of the α-(1,3)-galactsyltransferase (GCTA1) gene and the prion protein (PrP) gene in sheep. *Nat. Biotechnol.* **19,** 559–562.
11. Saito, S., Sawai, K., Ugai, H., et al. (2003) Generation of cloned calves and transgenic chimeric embryos from bovine embryonic stem-like cells. *Biochem. Biophys. Res. Commun.* **309,** 104–113.
12. Robertson, E. J. (1978) Embryo-derived stem cell lines, in *Teratocarcinomas and Embryonic Stem Cells: A Practical Approach* (Robertson, E. J., eds.), IRL Press, Oxford, pp. 19–49.
13. Notarianni, E., Galli, C., Laurie, S., Moor, R. M. and Evans, M. J. (1991) Derivation of pluripotent embryonic cell lines from pig and sheep. *J. Reprod. Fert. Suppl.* **43,** 255–260.
14. Graves, K. H., and Moreadith, R. W. (1993) Derivation and characterization of putative pluripotential embryonic stem cells from preimplantation rabbit embryos. *Mol. Reprod. Dev.* **36,** 424–433.
15. Saito, S., Strelchenko, N., and Niemann, H. (1992) Bovine embryonic stem cell-like cell lines cultured over several passages. *Roux's Arch. Dev. Biol.* **201,** 134–141.
16. Strelchenko, N. (1996) Bovine pluripotent stem cells. *Theriogenology* **45,** 131–140.
17. Sukoyan, M. A., Vatolin, S. Y., Golobitsa, A. N., Zhalezova, A. I., Semenova, L. A., and Serov, O. L. (1993) Embryonic stem cells derived from morulae, inner cell mass, and blastocysts of mink: comparisons of their pluripotencies. *Mol. Reprod. Dev.* **36,** 148–258.
18. Doetchman, T., Williams, P., and Maeda, N. (1988) Establishment of hamster blastocysts-derived embryonic stem cells. *Dev. Biol.* **127,** 224–227.
19. Iannacone, P. M., Taborn, A. U., Garton, R. L., Caplice, M. D., and Brenin, D. R. (1994) Pluripotent embryonic stem cells from the rat are capable of producing chimeras. *Dev. Biol.* **163,** 288–292.
20. Thomson, J. A., Kalishman, J., Golos, T. G., et al. (1995) Isolation of a primate embryonic stem cell line. *Proc. Natl. Acad. Sci. USA* **92,** 7844–7848.
21. Hogan, B., Beddington, R., Costantini, F., and Lacy, E. (1994) *Manipulating the Mouse Embryo: A Laboratory Manual*, 2nd ed., Cold Spring Harbor Laboratory Press, Cold Spring Harbor, NY.
22. Martin, G. R. and Evans, M. J. (1975) Differentiation of clonal lines of teratocarcinoma cells: formation of embryoid bodies in vitro. *Proc. Natl. Acad. Sci. USA* **72,** 1441–1445.
23. Evans, M. J. and Kaufman, M. H. (1981) Establishment in culture of pluripotential cells from mouse embryos. *Nature* **292,** 154–156.
24. Martin, G. R. (1981) Isolation of a pluripotent cell line from early mouse embryos cultured in medium conditioned by teratocarcinoma stem cells. *Proc. Natl. Acad. Sci. USA* **78,** 7634–7636.
25. Smith, A. G. and Hooper, M. L. (1987) Buffalo rat liver cells produce a diffusible activity which inhibits the differentiation of mouse embryonal carcinoma and embryonic stem cells. *Dev. Biol.* **121,** 1–9.
26. Matsui, Y., Zseki, K., and Hogan, B. L. (1992) Derivation of pluripotential embryonic stem cells from murine primordial germ cells in culture. *Cell* **70,** 841–847.
27. Freeman, D. A., Buth, J. E., Weber, J. A., Geary, R. T., and Woods, G. L. (1991) Co-culture of day-5 and day-7 equine embryos in medium with oviductal tissue. *Theriogenology* **36,** 815–822.

28. Clarke, K. E., Squires, E. L., Mckinnon, A. O., and Seidel, G. E., Jr. (1987) Viability of stored equine embryos. *J. Animal Sci.* **65,** 534–542.
29. Hinricks, K., Schmidt, A. L., Memon, M. A., Selgrath, J. P., and Evert, K. M. (1990) Culture of 5-day horse embryos in microdroplets for 10–20 days. *Theriogenology* **34,** 643–653.
30. Woods, G. L., White, K. L., Vanderwall, D. L., et al. (2003) A mule cloned from fetal cells by nuclear transfer. *Science* **301,** 1063.
31. Galli, C., Lagutina, I., Crotti, G., et al. (2003) A clone horse born to its dam twin. *Nature* **424,** 635.
32. Troy, T. C. and Turksen, K. (2001) Epidermal lineage, in *Embryonic Stem Cells: Methods and Protocols* (Turksen, K., ed.), Humana Press, Totowa, NJ, pp. 229–253.
33. Saito, S., Ugai, H., Sawai, K., et al. (2002) Isolation of embryonic stem-like cells from equine blastocysts and their differentiation in vitro. *FEBS Lett.* **531,** 389–396.
34. Robertson, E. J., Evans, M. J., and Kaufman, M. H. (1983) X-chromosome instability in pluripotential stem cell lines derived from parthenogenetic embryos. *J. Embryol. Exp. Morphol.* **74,** 297–309.
35. Saito, S., Liu, B., and Yokoyama, K. K. (2004) Animal embryonic stem (ES) cells: Self-renewal, pluripotency, transgenesis, and nuclear transfer. *Human Cell* **17,** 107–115.

6

Generation and Characterization of Monkey Embryonic Stem Cells

Hirofumi Suemori and Norio Nakatsuji

Summary

Nonhuman primate embryonic stem (ES) cells are very important for preclinical research of the medical application of human ES cells. Because primate ES cells show significant differences from mouse ES cells, we have been optimizing protocols for the establishment and maintenance of monkey ES cells. The latest methods for derivation and culture of cynomolgus monkey ES cells are described in detail.

Key Words: Cynomolgus monkey; differentiation; embryonic stem cells; nonhuman primate; pluripotency.

1. Introduction

For many years, mouse embryonic stem (ES) cell lines have been important tools for basic biology to investigate cell differentiation and development in mammals and to carry out gene targeting for analysis of gene function in gene-disrupted "knockout" mice. Establishment of human ES cell lines in 1998, from surplus human embryos produced at infertility clinics, indicated the great potential of ES cells in medical research and applications such as cell therapy and drug discovery *(1,2)*. ES cells have two important abilities: unlimited rapid proliferation to enable the production of a large number of human cells and pluripotency to differentiate into almost any type of cells. Such abilities also enable genetical alteration of ES cells, including insertion of various gene constructs designed for particular purposes and gene targeting to modify endogenous genes by homologous recombination.

To advance such medical applications of human ES cells, it is very important to devise reliable methods to establish and maintain ES cells in defined conditions and to find out various methods of inducing their differentiation into specific cell types. There have been reports of the production of various types of neurons, glia, cardiac muscle, hematopoietic cells, and endothelial cells, all of which have important medical applications.

From: *Methods in Molecular Biology, vol. 329: Embryonic Stem Cell Protocols: Second Edition: Volume 1*
Edited by: K. Turksen © Humana Press Inc., Totowa, NJ

1.1. Establishment of Monkey ES Cell Lines

ES cell lines have been established from nonhuman primates, including rhesus monkey (*Macaca mulatta*), common marmoset (*Callithrix jacchus*), and cynomolgus monkey (*Macaca fascicularis*) *(3–5)*. These primate ES cell lines have very similar characteristics to each other as well as to human lines. They express alkaline phosphatase activity, and stage-specific embryonic antigen (SSEA) 4, and, in most cases, SSEA-3. Their pluripotency is confirmed by the formation of embryoid bodies and differentiation into various cell types in culture and by the formation of teratomas that contain many types of differentiated tissues, including derivatives of three germ layers after transplantation into severe combined immunodeficient (SCID) mice.

When compared to mouse ES cells, the noneffectiveness of the leukemia inhibitory factor (LIF) in maintenance of stem cells makes culture of primate and human ES cell lines difficult and prone to undergo spontaneous differentiation. Also, these ES cells are more susceptible to various stresses, causing difficulty with subculturing using enzymatic treatment, cloning from single cells, and gene transfection. However, with various improvements in culture methods, it is now possible to maintain stable colonies of monkey ES cells using a serum-free medium and subculturing with trypsin treatment. Under such conditions, we can maintain cynomolgus monkey ES cell lines in an undifferentiated state with a normal karyotype and pluripotency even after prolonged periods of culture over 1 yr *(5)*.

1.2. Significance of Monkey ES Cells

Nonhuman primate ES cell lines provide important research tools for basic and applicative research. First, they provide wider aspects of investigation of the regulative mechanisms of stem cells and cell differentiation among primate species. Second, their usage can be valuable in preparation for research using human ES cells, which is under strict ethical regulation in many countries. Last and most important, they are indispensable for animal models of cell therapy to test effectiveness, safety, and immunological reaction of the allogenic transplantation in a setting similar to the treatment of human diseases.

The safety of cell transplantation must be tested using animal models before clinical application of ES cells. ES cells are not malignant tumor cells, but they show unlimited proliferation and form benign tumors in immunorepressed animals. Therefore, removal of stem cells from differentiated cell populations is necessary before cell transplantation. For even a higher degree of safety, implementation of a suicidal or cell ablation gene in ES cells may be necessary to trigger cell death if something goes wrong after transplantation.

Another important aspect is the immunological response after transplantation of allogenic ES-derived cells or, possibly, after genetic alteration of ES cell lines to reduce their antigenicity. The major histocompatibility complex (MHC) genes of ES cells could be altered by gene targeting and subsequent introduction of desired MHC-type genes. Complete matching of the many MHC types would be impossible between the modified ES cells and recipients, but matching of the major types might reduce immunological rejection to clinically manageable levels. For such immunological tests, nonhuman primates provide the best allogenic combination of ES cell lines and disease model animals.

Generation and Characterization of Monkey ES Cells

For preclinical research of actual cell transplantation, it is also important that the sizes and structures of various organs and tissues in animal models are similar to those of humans. This is especially important for surgical procedures; the absolute size and depth of structures in the brain, for example, determine the cell number and transplantation protocol that needs to be evaluated.

Among the monkeys, macaques such as the rhesus and cynomolgus monkeys are most suitable as model nonhuman primates. They are bred as experimental animals and widely used for medical research. Also, various disease models in macaques are available for research purposes. For such reasons, we established cynomolgus monkey ES cell lines and devised improved methods for culture and manipulation of monkey ES cells.

2. Materials

2.1. Cell Culture

1. Dulbecco's modified Eagle's medium (DMEM) (Sigma, St. Louis, MO; cat. no. D5796).
2. DMEM nutrient mixture F-12 Ham (CMEM/F12) (Sigma; cat. no. D6421).
3. Knockout™ serum replacement (KSR) (Invitrogen, Carlsbad, CA; cat. no. 10828).
4. Fetal bovine serum (FBS).
5. Minimum essential medium nonessential amino acids solution (100X) (Sigma; cat. no. M7145).
6. 200 mM L-glutamine (Sigma; cat. no. G7513).
7. β-Mercaptoethanol (Sigma; cat. no. M-7522).
8. Recombinant human LIF (Chemicon, Temecula, CA; cat. no. LIF1010).
9. Recombinant human basic fibroblast growth factor (Upstate, Charlottesville, VA; cat. no. 01-106).
10. Mytomycin C (Wako, Osaka, Japan; cat. no. 134-07911) (see **Note 1**).
11. Phosphate-buffered saline (PBS) (Ca^{2+} Mg^{2+} free) (Sigma; cat. no. D-5652).
12. Trypsin, 2.5% (Invitrogen; cat. no. 15090-046).
13. 0.25% trypsin, 1 mM ethylenediaminetetraacetic acid (Invitrogen; cat. no. 25200-056).
14. Collagenase type IV (Invitrogen; cat. no. 17104-019).
15. Pronase E (Sigma; cat. no. P8811).
16. $CaCl_2$ (0.1 M). For 100 mL, dissolve 1.11 g calcium chloride in 100 mL H_2O. Filter-sterilize and store at 4°C.
17. Mouse fibroblast medium: DMEM supplemented with 10% FBS.
18. CMK medium: mix aseptically 400 mL DMEM/F12, 4 mL 100X nonessential amino acids solution, 5 mL 200 mM L-glutamine, 4 µL β-mercaptoethanol, and 100 mL KSR. After mixing, the culture medium can be stored at 4°C for 2 wk.
19. Establishing medium: during establishment of ES cell lines, 0.4 µL recombinant human basic fibroblast growth factor and 100 µL LIF are added to 100 mL CMK medium. After mixing, the medium can be stored for 2 d at 4°C.
20. Dissociation solution: mix aseptically 10 mL 2.5% trypsin solution, 10 mg/mL collagenase type IV solution, 20 mL KSR, 1 mL 0.1 M $CaCl_2$, and 59 mL PBS(−). Make aliquots and store at −20°C. They can be thawed once, and refreezing should be avoided.

2.2. Immunosurgery

1. 0.5% Pronase E solution: dissolve 0.5 g Pronase E in 100 mL DMEM. Filter-sterilize, make aliquots, and store at −20°C. This can be thawed once.

2. Antimonkey antiserum: immunize rabbits with monkey spleen cells or lymphocytes. Antiserum is heat inactivated at 56°C for 30 min and filter-sterilized. Make aliquots and store at −20°C.
3. Guinea pig complement (Sigma; cat. no. S1639). After reconstitution, make aliquots and store at −80°C. They can be thawed once.
4. M16 medium (Sigma; cat. no. M7292).
5. Mineral oil (Sigma; cat. no. M5310).
6. Glass capillaries of approx 300-µm inner diameter for blastocysts handling and of approx 100-µm inner diameter for inner cell mass (ICM) isolation.

3. Methods

3.1. Establishment and Maintenance of Cynomolgus ES Cell Lines

The methods to establish monkey ES cell lines are essentially the same as used to generate mouse ES cell lines *(5,6)*.

1. ICMs isolated from blastocysts are cultured on a feeder layer of mouse embryonic fibroblasts.
2. Stem cell colonies are selected from differentiated derivatives and propagated until undifferentiated ES cells can be maintained stably in culture.

3.2. Preparation of Feeder Cell Layer

A primary culture of mouse embryonic fibroblasts is used as the feeder layer (*see* **Note 2**).

1. Prepare a confluent culture of embryonic fibroblasts in a 100-mm culture dish.
2. Add mitomycin C solution at the final concentration of 10 µg/mL.
3. Incubate the cells for 2 h.
4. Wash with PBS(−) twice and add culture medium.
5. Incubate the cells for at least 6 h, preferably overnight.
6. Aspirate the culture medium and wash the cells with PBS(−) twice.
7. Add 1 mL trypsin/ethylenediaminetetraacetic acid and incubate for approx 5 min at room temperature.
8. Add 9 mL culture medium and dissociate cells.
9. Pellet the cells by centrifugation at approx 170*g* for 5 min.
10. Resuspend the cells and seed the cells onto gelatinized dishes at a density of 1.5–1.75 × 10^4/cm². Feeder cells should be used within 4 d after preparation.

3.3. ICM Isolation From Blastocysts

1. Prepare a feeder layer in a four-well plate 1 d before immunosurgery (**Fig. 1**).
2. Prepare expanded monkey blastocysts of about 7 d in vitro culture after in vitro fertilization (IVF) or intracytoplasmic sperm injection (ICSI) *(7)*.
3. Make drops of solutions as shown in **Fig. 1** and cover them with mineral oil.
4. Remove zona pellucida by incubating in pronase solution for 10–20 min at room temperature.
5. Wash three times with M16 medium.
6. Incubate denuded blastocysts in 1:10 dilution of antiserum in M16 for 1 h in a CO_2 incubator.
7. Wash three times with M16.
8. Incubate blastocysts in 1:10 dilution of complement for 30 min in a CO_2 incubator.
9. Trophoblast cell debris is removed to isolate ICM by pipetting with a glass capillary. Isolated ICMs are immediately transferred onto a feeder layer in ES cell medium (*see* **Note 3**).

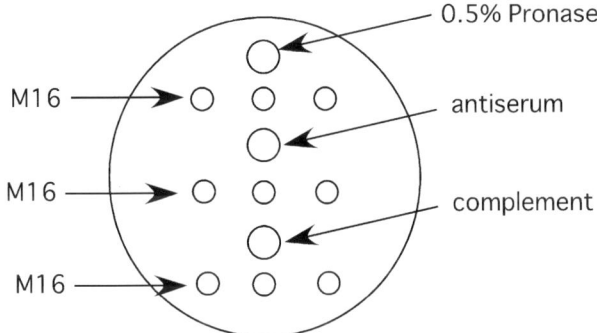

Fig. 1. Preparation of reagent drops for immunosurgery. Overlay mineral oil to prevent evaporation.

3.4. Culture of ICM

Stem cell colonies, typically one colony from one ICM, surrounded by differentiated cells can be observed in approx 10 d. Stem cell colonies are distinguished by their morphology, such as a small cytoplasm/nucleus ratio, and prominent nucleoli. Stem cell colonies can be transferred to a new feeder layer when they reach approx 500 μm in diameter.

1. Aspirate the medium and rinse once with PBS.
2. Add 0.5 mL dissociation solution.
3. Cut the stem cell colony into several pieces using a 27-gage hypodermic needle.
4. Incubate the cells at room temperature until the edges of the colony begin to lift from the plate.
5. Pick up the clusters of stem cells with a glass capillary.
6. Wash the cells twice with the culture medium in culture dishes.
7. Transfer the cell clumps onto a new feeder layer in a four-well plate.
8. Change the culture medium daily until the stem cells grow enough to be subcultured (see **Note 4**).

3.5. Expansion and Maintenance of Monkey ES Cells

Mechanical dissociation as previously described should be performed in the first three to five passages, until the stem cells proliferate to the confluency on a 35-mm culture dish. After that, cells can be subcultured by enzyme treatment and pipetting and can usually be maintained stably. Differentiated cell colonies are sometimes observed. They are scraped and aspirated out using a Pipetman tip. ES cells should be passaged when the cells cover approx 50–60% of the surface of the feeder layer (**Fig. 2A,B**).

To subculture ES cells confluent in a 60-mm culture dish:

1. Aspirate medium and rinse the cells with PBS.
2. Add 1 mL dissociation solution.
3. Incubate for 5–6 min at 37°C. Examine the cells under a microscope and confirm that the feeder cells show a rounded shape and that most colonies of ES cells are partially detached from the dish (**Fig. 2C,D**).
4. Add 4 mL of CMK medium and dissociate ES cell colonies by gentle pipetting into clusters consisting of approx 100 cells (**Fig. 2E**; see **Notes 5** and **6**).

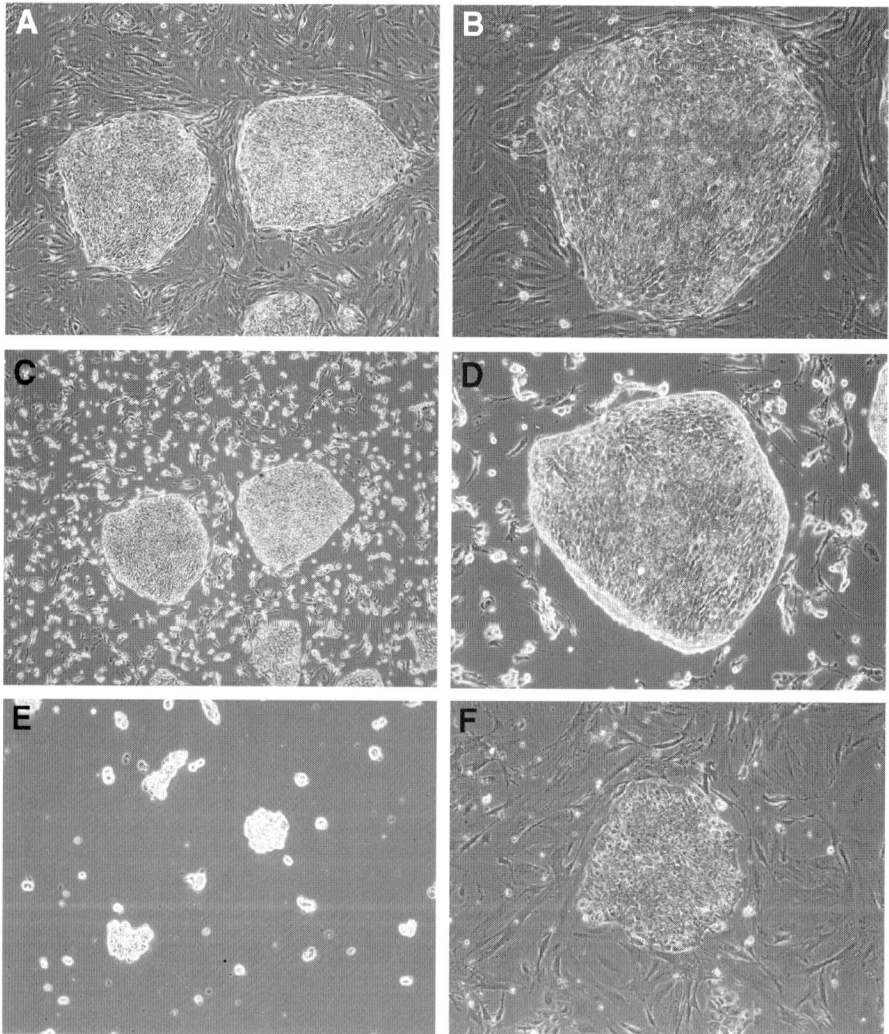

Fig. 2. (**A,B**) A confluent culture of cynomolgus monkey embryonic stem (ES) cells. (**C,D**) Appearance of ES cell colonies after enzyme treatment. (**E**) ES cell clusters after dissociation by pipetting. Cell clusters must be kept as clusters of 50–100 cells. (**F**) A view of an ES cell colony 1 d after subculturing. (Please *see* the companion CD for the color version of this figure.)

5. Transfer cell suspension to a 15-mL centrifuge tube.
6. Centrifuge the cells at approx 170*g* for 5 min.
7. Aspirate supernatant and resuspend the cell pellet in 15 mL CMK medium.
8. Dispense 5 mL cell suspension to a new 60-mm culture dish with a fresh feeder layer.
9. Change the medium daily. ES cells will become confluent in 3–4 d, at which point cells will be subcultured (*see* **Note 7**).

Table 1
Expression of Markers in ES Cells

	Human ES	Cynomolgus ES	Mouse ES
SSEA-1	−	−	+
SSEA-3	+	−[a]	−
SSEA-4	+	+	−
TRA-1-60, TRA-1-81	+	+	[b]
Oct3, Nanog	+	+	+

[a]Rhesus monkey ES cells express SSEA-3 antigen.
[b]Antibodies do not recognize mouse antigens.

3.6. Characterization of ES Cell Lines

3.6.1. Expression of Markers

Several cell surface markers are known to be expressed in undifferentiated money ES cells. Primate ES cells express alkaline phosphatase activity as well as SSEA-4, TRA-1-60, and TRA-1-81 antigens but not SSEA-1 antigen. Whereas human and rhesus ES cells have been reported to express SSEA-3 antigen at variable levels, cynomolgus ES cells are negative for SSEA-3. The expression in undifferentiated ES cells of genetic markers such as Oct3/4, Rex-1, and Nanog have been detected by reverse transcriptase polymerase chain reaction. Expression of markers in cynomolgus monkey ES cells is summarized by comparing with mouse and human ES cells in **Table 1**.

3.6.2. Karyotype

ES cell lines retain normal karyotype even after a long period in culture. Both male and female ES cell lines are obtained in primates, whereas male mouse ES cell lines are preferentially established. Periodic examination of the chromosome number of each cell line is recommended for routine qualification. For more detailed karyotype analysis, chromosome band staining should be performed. Standard protocols for chromosome spread preparation from cultured cells, chromosome staining, and banding are applicable to monkey ES cells.

3.6.3. Differentiation Potency

ES cells spontaneously differentiate when they are cultured in the absence of the feeder layer. However, only a few cell types, mostly flattened, extraembryonic, endodermlike cells, are formed by such a method. To induce differentiation into more various cell types, formation of embryoid bodies (EBs) is an effective method. EBs are formed by culturing ES cell aggregates floating in Petri dishes.

Teratoma formation from ES cells is a simple and reliable method to analyze differentiation potency of ES cells. To produce teratomas, monkey ES cells are transplanted into SCID mice. About 10^7 cells, which corresponds to a confluent cell layer in a 100-mm dish, are injected subcutaneously or intraperitoneally into SCID mice.

Teratoma formation becomes apparent in 2–3 mo. Teratomas can be examined by standard histological analyses and should contain tissues derived from three embryonic germ layers.

3.6.4. EB Formation

1. Prepare a confluent culture of monkey ES cells. Colonies of about 500–1000 μm in diameter are required (*see* **Note 8**).
2. Aspirate the medium and rinse the cells with PBS.
3. Cover the cells with the collagenase solution.
4. Incubate at 37°C for 20 min. Add CMK medium and detach colonies from the feeder layer by tilting the dish to make colonies float in the medium.
5. Transfer the suspension of colonies to a 15-mL conical tube and allow the ES cell colonies to sink to the bottom of the tube.
6. Collect ES cell aggregates in medium and transfer them to a new Petri dish.
7. ES cell aggregates will form simple EBs in a few days. They can be cultured in suspension until apparent cell differentiation is observed (*see* **Note 9**).

4. Notes

1. Dissolve 10 mg mitomycin C to 5 mL tissue culture-grade water, filter-sterilize, make aliquots, and store at −80°C. Use immediately after thawing in a 37°C water bath.
2. Quality of the feeder layer is very important for establishment and maintenance of ES cell lines. Early passage, up to five passages, of embryonic fibroblast cells should be used to prepare feeder cells. The feeder layer prepared from the STO cell line, a mouse embryonic fibroblast cell line, can also be used.
3. It is important to confirm that each ICM is in close contact with the feeder cells. Cell debris remaining after immunosurgery sometimes inhibits ICM attachment to feeder cells. The next day, ICM adhesion to feeder cells should be seen. If not, the ICM will differentiate or die very soon. The ICM does not appear to grow in the following several days but should spread out a little bit. Occasionally, the ICM will differentiate and grow rapidly, in which case undifferentiated stem cells cannot be recovered.
4. Many dead cells are surrounding the stem cell colonies on the next day. This is normal.
5. Excess dissociation will damage cells and result in loss of ES cells.
6. Feeder cells secrete a viscous matrix resistant to trypsin digestion in CMK medium, and sheets of feeder cells remain after pipetting. Do not try to dissociate these sheets into small clumps. It will cause overdissociation of ES cell clusters.
7. Monkey ES cell lines can be maintained in an undifferentiated state for more than 1 yr.
8. Because ES cells start differentiation immediately after detaching from the feeder layer, ES cells do not proliferate enough to form EBs if the starting cell aggregates are too small. Thus, it is important to start from larger ES cell aggregates to obtain good EBs.
9. In 2–3 wk or longer, beating heart muscle and hematopoietic cells may be observed in EBs. EBs can also be plated to a tissue culture dish from the suspension culture at any time. EBs will attach to the dish and undergo cell differentiation into various tissues, such as neurons and cardiac muscles.

Acknowledgment

Our original study included in this chapter was supported in part by a Research for the Future (RFTF) program of the Japan Society for the Promotion of Science.

References

1. Thomson, J. A., Itskovitz-Eldor, J., Shapiro, S. S., et al. (1998) Embryonic stem cell lines derived from human blastocysts. *Science* **282,** 1145–1147.
2. Reubinoff, B. E., Pera, M. F., Fong, C.-Y., Trounson, A., and Bongso, A. (2000) Embryonic stem cell lines from human blastocysts: somatic differentiation in vitro. *Nat. Biotechnol.* **18,** 399–404.
3. Thomson, J. A., Kalishmanm, J., Golos, T. G., Durning, M., Harris, C. P., and Hearn, J. P. (1996) Pluripotent cell lines derived from common marmoset (*Callithrix jacchus*) blastocysts. *Biol. Reprod.* **55,** 254–259.
4. Thomson, J. A., Kalishman, J., Golos, T. G., et al. (1995) Isolation of a primate embryonic stem cell line. *Proc. Natl. Acad. Sci. USA* **92,** 7844–7848.
5. Suemori, H., Tada, T., Torii, R., et al. (2001) Establishment of embryonic stem cell lines from cynomolgus monkey blastocysts produced by IVF or ICSI. *Dev. Dynamics* **222,** 273–279.
6. Nagy, A., Gertsenstein, M., Vinterstein, K., et al. Isolation and culture of blastocyst-derived stem cell lines, in *Manipulating the Mouse Embryo,* 3rd ed. Cold Spring Harbor Laboratory Press, Cold Spring Harbor, NY, pp. 359–397.
7. Torii, R., Hosoi, Y., Masuda, Y., Iritani, A., and Nigi, H. (2000) Birth of the Japanese monkey (*Macaca fuscata*) infant following in-vitro fertilization and embryo transfer. *Primates* **41,** 39–47.

7

Derivation and Propagation of Embryonic Stem Cells in Serum- and Feeder-Free Culture

Jennifer Nichols and Qi-Long Ying

Summary

The availability of murine embryonic stem (ES) cells has revolutionized the study of mammalian development and disease. We recently developed a culture medium that has enabled us to identify the essential signaling pathways required for maintenance of pluripotency in vitro. Addition of leukemia inhibitory factor and bone morphogenetic protein4 to this medium is sufficient to activate the signal transducer and activator of transcription3 and mammalian homolog of *Drosophila* mothers against decapentaplegic pathways, respectively. We have successfully derived and propagated ES cells in the absence of feeder cells and serum. This chapter describes a simple protocol for efficient derivation and maintenance of ES cells from embryos of the 129 and C57Bl/6 strains of mice.

Key Words: BMP; diapause/delayed implantation; ES cell derivation; gp130; Id; LIF; pluripotency; SMAD; STAT3.

1. Introduction

Embryonic stem (ES) cells are invaluable as an in vitro model for the epiblast and as a means to create genetic modifications in mice for the study of gene function and disease. They are derived from the inner cell mass of mammalian blastocysts by allowing the embryo to attach to the substrate for a few days, then disrupting the cell contacts by disaggregation and replating. ES cells were first isolated in 1981 using a feeder layer of mitotically inactivated fibroblasts and medium containing serum *(1,2)*. Subsequently, it became possible to replace the feeder cells with addition of leukemia inhibitory factor (LIF) to the culture medium, both for routine culture *(3–5)* and for *de novo* generation of ES cells *(6,7)*.

LIF functions by activation of the signal transducer and activator of transcription, (STAT3), via a membrane receptor complex that includes glycoprotein (gp) 130. Interleukin 6 and related cytokines utilize receptors incorporating gp130 that also signal via STAT3. Receptors for some of these cytokines are not expressed in embryos. However, it is possible to derive germline-competent ES cells in the absence of feeders using medium supplemented with interleukin 6 and a soluble form of its receptor instead of LIF *(8)*. The procedure for deriving ES cells using a mitotically inactivated feeder layer has been clearly described previously *(9)*. It is modified only very slightly when the feeder cells are replaced with recombinant cytokines *(6)*.

Since ES cells were first created, it has become clear that different strains of mice vary considerably in their facility for derivation *(10)*. The inbred 129 strain is the most permissive; approx 30% of 129 embryos can be expected to give rise to ES cell lines. Embryos of the C57Bl/6 inbred strain will also produce ES cells but with a lower frequency than 129, and most other strains, such as CBA, rarely produce ES cells. The reason for this intriguing variability is not yet known. Several modifications have been introduced to the protocol that increased the success rates and have allowed derivation of ES cells from recalcitrant strains, but the discrepant efficiency between strains still persists. For example, subjecting the embryos to delayed implantation or diapause can improve the efficiency *(9)*, and dissecting away the extraembryonic tissues from delayed blastocysts can increase the yield to 100% for 129 and to greater than 50% for CBA embryos *(11)*. However, this technique is not widely adopted because dissecting epiblasts from delayed blastocysts is far from trivial. In addition to removing the differentiative signals from the surrounding tissues, it is possible to encourage the self-renewal aspect of stem cell proliferation by preferentially inhibiting signaling pathways that promote differentiation. For example, commercially available inhibitors of the extracellular signal-regulated kinase (Erk)-activating mitogen-activated extracellular-regulated kinase (MEK) pathway such as PD98059 have been used to promote ES cell self-renewal and improve the efficiency of ES cell derivation *(12–14)*.

The composition of serum is ill defined and variable between batches. Therefore, reliance on its inclusion in media for derivation and propagation of ES cells does not facilitate study of the minimal requirements and molecular responses involved in self-renewal and differentiation. We established culture conditions that allow efficient propagation and derivation of germline-competent ES cell lines in the absence of serum *(19)*. Using this medium (N2B27), we identified the critical signaling pathways involved in maintenance of pluripotency, namely, the gp130/STAT3 and mammalian homolog of *Drosophila* mothers against decapentaplegic (SMAD) pathways, which can be stimulated by addition of LIF and bone morphogenetic protein (BMP4), respectively. Subsequently, we have been able to derive ES cells from C57Bl/6 embryos at an improved frequency of 30% using serum-free culture conditions.

2. Materials

1. Insulin (Sigma, St. Louis, MO; cat. no. I-1882). For a 25-mg/mL stock solution, dissolve 100 mg in 4 mL sterile 0.01 M HCl overnight at 4°C. Store at -20°C for up to 1 yr.
2. Apo-transferrin (Sigma, cat. no. T-1147). For a 100-mg/mL stock solution, dissolve 500 mg in 5 mL distilled water (dH$_2$O) overnight at 4°C. Store at -20°C for up to 1 yr.
3. Progesterone (Sigma, cat. no. P8783). For a 0.6-mg/mL stock solution, dissolve 6 mg in 10 mL ethanol. Filter through a 0.22-µm syringe filter and store at -20°C for up to 1 yr.
4. Ethanol (BDH Laboratory Supplies, Poole, UK; cat. no. UN1170).

5. Putrescine (Sigma, cat. no. P5780). For a 160-mg/mL stock solution, dissolve 1.6 g in 10 mL dH$_2$O. Filter through a 0.22-μm syringe filter and store at −20°C for up to 1 yr.
6. Sodium selenite (Sigma, cat. no. S5261). For a 3 mM stock solution, dissolve 2.59 mg in 5 mL dH$_2$O. Filter through a 0.22-μm syringe filter and store at −20°C for up to 1 yr.
7. Bovine serum albumin (BSA) (Invitrogen, Carlsbad, CA; cat. no. 15260-037). For a 75-mg/mL stock solution, dissolve 750 mg in 10 mL of phosphate-buffered saline (PBS). Store at −20°C for up to 1 yr.
8. Dulbecco's modified Eagle's medium nutrient mixture F-12 Ham (DMEM/F12) (Invitrogen, cat. no. 42400-010).
9. Neurolbasal™ medium (Invitrogen, cat. no. 21103-049).
10. B27 (Invitrogen, cat. no. 17504-044).
11. L-Glutamine (Sigma, cat. no. G5763).
12. β-Mercaptoethanol (BDH, cat. no. 441433A). For a 0.1 M solution, mix 200 μL of the solution supplied (14.3 M) with 28.2 mL tissue culture-grade water and store aliquots at 4°C for up to 1 mo.
13. LIF (Sigma, cat. no. L5158).
14. BMP4 (R + D Systems, Minneapolis, MN; cat. no. 314-BP-010).
15. Tamoxifen (Sigma, cat. no. T5648). For a 10-mg/mL stock solution, dissolve 100 mg in 10 mL ethanol. Store at 4°C for up to 1 mo. For the working solution, dilute 100 μL of this in 9.9 mL propylene glycol (100 μg/mL) and store at 4°C for up to 1 mo. Inject 0.1 mL per mouse.
16. Depo-Provera (medroxyprogesterone 17-acetate) (Sigma, cat. no. M1629). Dilute with PBS to a concentration of 30 mg/mL and inject 0.1 mL/mouse. Store at 4°C for up to 1 yr.
17. Gelatin (Sigma, cat. no. G1890). Make a 0.1% solution in PBS and store at 4°C for up to 1 mo.
18. Trypsin (2.5% solution) (Invitrogen, cat. no. 15090-046).
19. Trypsin-versine PBS (TVP): to 500 mL PBS, add 0.186 g ethylenediaminetetraacetic acid, 5 mL chicken serum, and 5 mL 2.5% trypsin. Filter, aliquot, and freeze at −20°C; store frozen until required. Once thawed, do not refreeze, but store at 4°C for up to 1 mo.
20. Ethylenediaminetetraacetic acid (Sigma, cat. no. E6758).
21. Chicken serum (Sigma, cat. no. C5405).
22. Dimethyl sulfoxide (Sigma, cat. no. D8799).
23. Single-use filters, 0.22 μL (Sartorius AG, Goettingen, Germany; cat. no. 16534).
24. Tissue culture plastics.
25. Pasteur pipets.

2.1. Media

1. N2 100X stock solution. For 10 mL, mix 1 mL insulin (2.5 mg/mL final concentration) with 1 mL apo-transferrin (10 mg/mL final concentration), 0.67 mL BSA (5 mg/mL final concentration), 33 μL progesterone (2 μg/mL final concentration), 100 μL putrescine (1.6 mg/mL final concentration), 10 μL sodium selenite (3 mM final concentration), and 7.187 mL DMEM/F12. Store at 4°C and use within 1 mo.
2. DMEM/F12-N2 medium: to 100 mL DMEM/F12, add 1 mL N2 100X stock solution. The final concentration of each component of N2 in the DMEM/F12 medium is 25 mg/mL insulin; 100 mg/mL apo-transferrin; 6 ng/mL progesterone; 16 mg/mL putrescine; 3 nM sodium selenite; 50 mg/mL BSA. Store at 4°C and use within 1 mo.
3. Neurolbasal/B27 medium: to 100 mL Neurolbasal medium, add 2 mL B27 and 0.5–1 mL 200 mM L-glutamine. Store at 4°C and use within 1 mo.

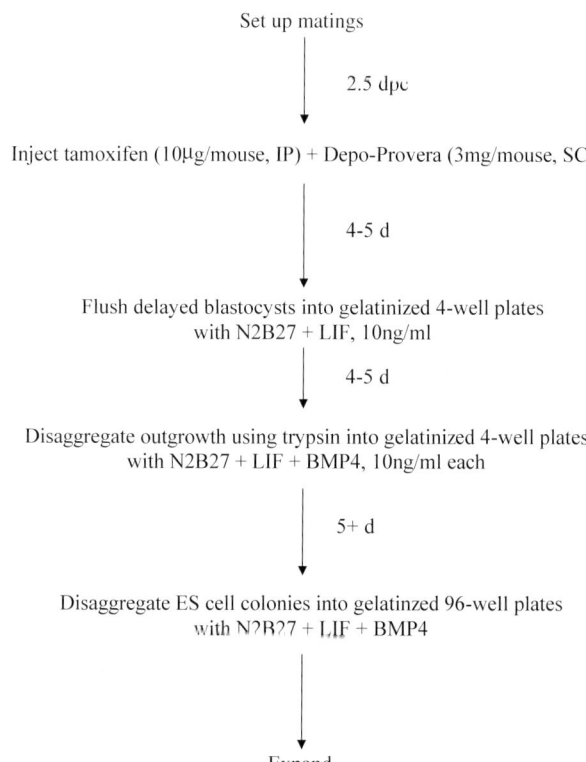

Fig. 1. Summary of procedure for embryonic stem cell derivation in serum-free medium.

4. N2B27 medium: mix DMEM/F12-N2 medium with Neurolbasal/B27 medium in the ratio of 1:1. Add β-mercaptoethanol to a final concentration of 0.1 mM from the 0.1 M stock. Store at 4°C and use within 1 mo.

3. Methods

This protocol is summarized in **Fig. 1**.

3.1. Recovery of Embryos

1. Induce diapause (delayed implantation) in mice either by ovariectomy *(16)* or by intraperitoneal injection of tamoxifen (10 μg per mouse) at 2.5 d postcoitum (*see* **Note 1**) *(17)*.
2. Inject Depo-Provera (1–3 mg/mouse) subcutaneously in either case as a source of progesterone (*see* **Notes 2** and **3**).
3. Flush the embryos from the uterus 4 or 5 d after ovariectomy or administration of tamoxifen (**Fig. 2A**) and plate into gelatinized four-well plates containing 0.5 mL/well of N2B27 with 10 ng/mL LIF (*see* **Note 4**). Preequilibrate the plates to 7% CO_2 and 37°C and place up to 10 embryos in each well (*see* **Note 5**).

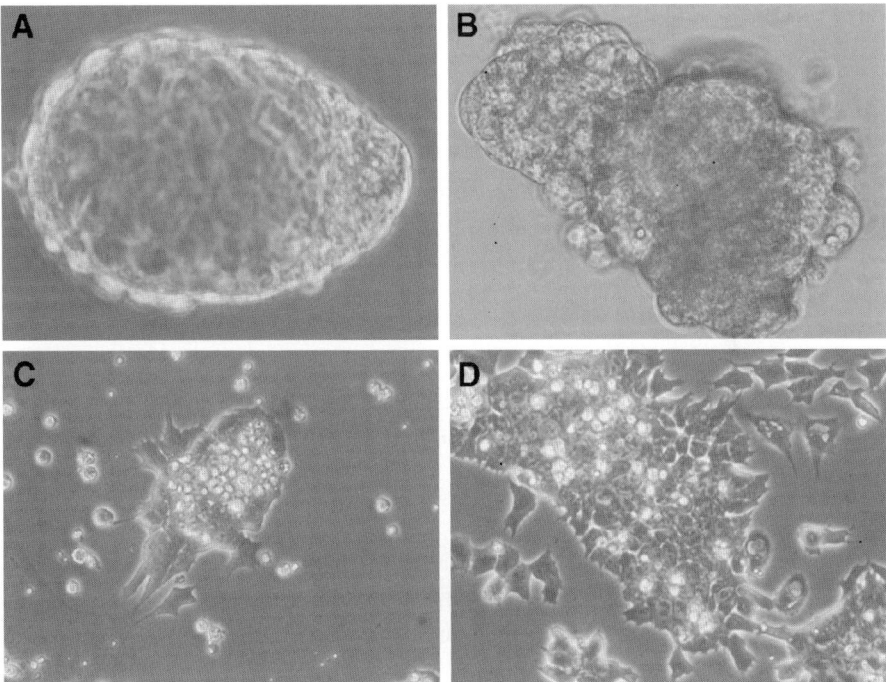

Fig. 2. (**A**) Delayed blastocyst flushed from the uterus 4 d after tamoxifen injection. (**B**) Embryo outgrowth in serum-free medium after 4 d in culture. (**C**) Primary colony from a disaggregated outgrowth from a C57Bl/6 embryo. Many dead cells are visible on the surface and surrounding the colony. (**D**) Colony of established C57Bl/6 embryonic stem cells (passage 5) prior to trypsinization and injection into host blastocysts.

3.2. Disaggregation and Expansion

1. Disaggregate cultures after 3–5 d (*see* **Note 6**) by gently detaching each cell clump using a mouth-controlled, finely drawn, plugged Pasteur pipet with a tip diameter that is just bigger than the outgrowth. Each outgrowth is handled separately at this point and in all subsequent manipulations.
2. Wash twice in drops of PBS in a sterile bacteriological plastic Petri dish before transferring to a small (approx 5 µL) drop of TVP.
3. Incubate the cell clumps in trypsin at 37°C for a few min until they begin to dissociate (*see* **Note 7**).
4. Aspirate each outgrowth into a finely drawn, plugged Pasteur pipet containing a little culture medium, ensuring minimal uptake of the surrounding TVP.
5. Transfer to a new well of a four-well plate containing N2B27 with 10 ng/mL each of LIF and BMP4 and triturate into small clumps of approx 1–5 cells (*see* **Note 8**). ES cell colonies should become identifiable after approx 5 d (**Fig. 2C**).
6. Expand any ES cell colonies by trypsinization of the whole well, centrifugation for 5 min at 200*g*, and resuspension in 50 µL N2B27 with LIF and BMP4 (*see* **Note 9**) into a 96-well plate (*see* **Notes 10** and **11**).

3.3. Routine Passaging

1. Plate $1\text{–}2 \times 10^6$ cells into a gelatinized small flask (T25) in 8 mL N2B27 with LIF and BMP4, pre-equilibrated to 7% CO_2 and 37°C. Under these conditions, ES cells should be passaged every 2–4 d.
2. Change the medium every 2 d if the cells are not ready for passaging. ES cells can be plated at clonal density.
3. To passage cells, rinse once with PBS, add TVP to cover the cells (about 2 mL), and incubate at 37°C for a few minutes.
4. Dilute the trypsin with 8 mL N2B27 without cytokines, centrifuge, aspirate off the medium, resuspend in 0.5 mL N2B27 with LIF and BMP4, and replate into a pre-equilibrated gelatinized flask.

3.4. Freezing Serum-Free ES Cells

1. Trypsinize ES cells as described in **Subheading 3.3.** for routine passaging but resuspend in 1.5 mL N2B27 medium with 10% dimethyl sulfoxide if freezing cells from a T25 flask (*see* **Note 12**).
2. Place 0.5 mL in each of three cryovials.
3. Store at −80°C for a few days.
4. Move to liquid nitrogen cell bank for indefinite storage.

3.5. Injection of ES Cells Into Blastocysts

1. Trypsinize the cells and resuspend in a small volume (0.5 mL for a slightly subconfluent T25 flask) of serum-containing medium (*see* **Note 13**).
2. Inject 10–20 cells into each host blastocyst (*see* **Note 14**).

4. Notes

1. Tamoxifen is a nonsteroidal antiestrogen that prevents the embryos from implanting in the uterus by blocking the estrogen secreted from the corpora lutea when the embryos are at the morula stage in the oviduct. The amount injected is critical; too little will fail to prevent implantation, and too much is toxic to the embryos.
2. The role of progesterone is to maintain the pregnancy. Depo-Provera does not dissolve in aqueous media, so it is important to ensure that it is properly dispersed prior to injection.
3. The embryos will continue to progress along the reproductive tract into the uterus and will hatch from their zonae pelucidae. They can remain in diapause for many days, but the viability begins to decrease after about 2 wk *(18)*.
4. Gelatinize all tissue culture plastics by incubating in 0.1% gelatin in PBS at room temperature for about 30 min. Aspirate off the solution and allow to dry if possible.
5. Over the next few days, the embryos will attach to the plastic. Very little outgrowth of extraembryonic cell types is observed in serum-free conditions (**Fig. 2B**), and each outgrowth remains distinct.
6. This protocol has been applied so far only to embryos of the 129 and C57Bl/6 inbred strains of mice. It may be applicable to other strains; it is possible to maintain established BALBC and CBA ES cell lines under these conditions. ES cells have been established from 129s by disaggregating after 3 or 4 d and from C57Bl/6 after 4 or 5 d.
7. Dissociation can be recognized morphologically; the outlines of the individual cells become more distinct, and the clumps assume a "blackberry-like" appearance.

Serum- and Feeder-Free ES Cell Culture

Fig. 3. Chimeras generated by injection of C57Bl/6 embryonic stem cells into MF1 blastocysts. The mouse on the left shows a high coat color contribution, whereas the mouse on the right exhibits only a small amount of black fur.

8. The 96-well plates can be used at this stage, but it is more difficult to insert the pipet for disaggregation, and the optical qualities of small wells are inferior for identifying any ES cell colonies.
9. In subsequent passages, the size of the wells can be gradually increased.
10. To inhibit the trypsin, N2B27 is used without added cytokines for reasons of economy. It is important to dilute the trypsinized cells in a large volume because no serum is included in the medium. Also, cells must be centrifuged and resuspended in medium to ensure that no trypsin is carried over into the culture.
11. At this early expansion stage, the cells are very susceptible to death or differentiation (*see* **Fig. 2C**). They seem to fare better in a small volume of medium, which may facilitate autocrine effects. As they become established, cell death is reduced (*see* **Fig. 2D**).
12. N2B27 is used without added cytokines in the freezing mixture. There is no need to add serum.
13. The addition of serum to the medium for cells for blastocyst injection reduces stickiness.
14. The choice of host blastocyst will depend on the strain of mice from which the ES cells were derived. For 129 ES cells, C57Bl/6 blastocysts are appropriate, but for ES cells derived from C57Bl/6 embryos the selection of host blastocyst may be more complicated. The blastocysts of albino outbred mice are regarded as poor hosts for ES cell injection. This may be because the comparatively vigorous outbred embryos tend to overgrow the injected ES cells. However, although the overall frequency of obtaining chimeras using blastocysts from MF1 outbred mice is disappointing, we have managed to obtain some high-level chimeras (**Fig. 3**). An alternative for generating chimeras from C57Bl/6 ES cells may be the use of an albino derivative of C57Bl/6 as a source of recipient blastocysts *(19)*.

Acknowledgments

We would like to thank Austin Smith and Tilo Kunath for comments on the manuscript. This work is funded by the Biotechnology and Biological Sciences Research Council of the United Kingdom, the Medical Research Council, the Juvenile Diabetes Research Foundation, and EuroStemCell.

References

1. Williams, R. L., Hilton, D. J., Pease, S., et al. (1988) Myeloid leukaemia inhibitory factor maintains the developmental potential of embryonic stem cells. *Nature* **336,** 684–687.
2. Robertson, E. J. (1987) *Teratocarcinoma and Embryo-Derived Stem Cells: A Practical Approach,* IRL Press, Oxford, UK.
3. Kawase, E., Suemori, H., Takahashi, N., Okazaki, K., Hashimoto, K., and Nakatsuji, N. (1994) Strain difference in establishment of mouse embryonic stem (ES) cell lines. *Int. J. Dev. Biol.* **38,** 385–390.
4. Hunter, S. and Evans, M. (1999) Non-surgical method for the induction of delayed implantation and recovery of viable blastocysts in rats and mice by the use of tamoxifen and Depo-Provera. *Mol. Reprod. Dev.* **52,** 29–32.
5. Nichols, J., Chambers, I., Taga, T., and Smith, A. (2001) Physiological rationale for responsiveness of mouse embryonic stem cells to gp130 cytokines. *Development* **128,** 2333–2339.
6. Buehr, M. and Smith, A. (2003) Genesis of embryonic stem cells. *Philos. Trans. R. Soc. Lond. B Biol. Sci.* **358,** 1397–402, discussion 1402.
7. Burdon, T., Stracey, C., Chambers, I., Nichols, J., and Smith, A. (1999) Suppression of SHP-2 and ERK signalling promotes self-renewal of mouse embryonic stem cells. *Dev. Biol.* **210,** 30–43.
8. Brook, F. A. and Gardner, R. L. (1997) The origin and efficient derivation of embryonic stem cells in the mouse. *Proc. Natl. Acad. Sci. USA* **94,** 5709–5712.
9. Ying, Q.-L., Nichols, J., Chambers, I., and Smith, A. (2003) BMP induction of Id proteins suppresses differentiation and sustains embryonic stem cell self-renewal in collaboration with STAT3. *Cell* **115,** 281–292.
10. Nichols, J., Chambers, I., and Smith, A. (1994) Derivation of germline competent embryonic stem cells with combination of interleukin-6 and soluble interleukin-6 receptor. *Exp. Cell Res.* **215,** 237–239.
11. Evans, M. J. and Kaufman, M. (1981) Establishment in culture of pluripotential cells from mouse embryos. *Nature* **292,** 154–156.
12. Martin, G. R. (1981) Isolation of a pluripotent cell line from early mouse embryos cultured in medium conditioned by teratocarcinoma stem cells. *Proc. Natl. Acad. Sci. USA* **78,** 7634–7638.
13. Moreau, J. F., Donaldson, D. D., Bennett, F., Witek-Giannotti, J., Clark, S. C., and Wong, G. G. (1988) Leukaemia inhibitory factor is identical to the myeloid growth factor human interleukin for DA cells. *Nature* **336,** 690–692.
14. Smith, A. G., Heath, J. K., Donaldson, D. D., et al. (1988) Inhibition of pluripotential embryonic stem cell differentiation by purified polypeptides. *Nature* **336,** 688–690.
15. Seong, E., Saunders, T. L., Stewart, C. L., and Burmeister, M. (2004) To knockout in 129 or in C57BL/6: that is the question. *Trends Genet.* **20,** 59–62.
16. Nichols, J., Evans, E. P., and Smith, A. G. (1990) Establishment of germ-line competent embryonic stem (ES) cells using differentiation inhibiting activity. *Development* **110,** 1341–1348.
17. Pease, S., Braghetta, P., Gearing, D., Grail, D., and Williams, R. L. (1990) Isolation of embryonic stem (ES) cells in media supplemented with recombinant leukemia inhibitory factor (LIF). *Dev. Biol.* **141,** 344–352.
18. Buehr, M., Nichols, J., Stenhouse, F., et al. (2003) Rapid loss of Oct-4 and pluripotency in cultured rodent blastocysts and derivative cell lines. *Biol. Reprod.* **68,** 222–229.
19. Hogan, B., Beddington, R., Costantini, F., and Lacy, E. (1994) *Manipulating the Mouse Embryo: A Laboratory Manual,* Cold Spring Harbor Laboratory Press, Cold Spring Harbor, NY, pp. 497.

II

SIGNALING IN EMBRYONIC STEM CELL DIFFERENTIATION

8

Internal Standards in Differentiating Embryonic Stem Cells In Vitro

Christopher L. Murphy

Summary

Embryonic stem (ES) cell lines are important for use in developmental biology studies, and because these cells are totipotent, they may provide a much-needed source of differentiated cells for certain therapeutic applications. The phenotype of the ES cell in culture is often assessed by (semi)quantitative RNA analyses. In such cases, it is critical to use appropriate internal standards to correct for experimentally induced sources of error. This is particularly true for ES cell differentiation because it is heterogeneous in nature. We describe protocols for determining the suitability of housekeeping genes to act as internal controls in differentiating ES cell cultures. Such assessment is needed for every experimental condition under investigation. The protocol focuses on polymerase chain reaction; however, the principle and experimental design are applicable to any (semi)quantitative RNA assay.

Key Words: Embryoid body; embryonic stem cells; endogenous control; GAPDH; gene expression; housekeeping gene; internal standard; PCR.

1. Introduction

The inner cell mass of the preimplantation blastocyst consists of transient populations that differentiate into all the somatic cells present in the adult (*1*). Embryonic stem (ES) cells are the in vitro counterpart of these cells; therefore, ES cell lines represent a powerful model for developmental studies (*2*), as well as a potential source of differentiated cells for therapeutic applications (*3–5*).

In vitro, ES cells can be induced to differentiate by forming cell aggregates termed embryoid bodies (EBs) that form cells of the three germ layers: endoderm, mesoderm, and ectoderm (*6–8*). The presence of particular mediators (e.g., growth factors, cytokines) can enhance the differentiation of specific lineages (*9–13*). The differentiated status of the cell is often assessed by (semi)quantitative RNA analyses such as polymerase chain reaction (PCR), Northern blots, microarrays, and ribonuclease protection assays. However, care must be exercised to use appropriate internal standards, particularly because ES cell differentiation is heterogeneous. These internal standards are therefore necessary to correct for experimentally induced sources of error *(14)*. However, to meet this requirement, housekeeping gene expression should be stable under all experimental conditions, and because no one gene is expressed unvaryingly under all conditions *(15)*, it is important to test the ability of different housekeeping genes to act as suitable internal standards under the particular conditions investigated.

Hypoxanthine phosphoribosyltransferase (HPRT) *(11,13,16–19)* β-tubulin *(11,16)* and glyceraldehyde-3-phosphate dehydrogenase (GAPDH) *(10,12)* are among the more commonly used housekeeping genes as internal RNA standards in ES cell/EB studies examining gene transcription effects. However, only one of these housekeeping genes has been suitable as an internal standard in experiments investigating the differentiation of ES cells through EBs and in the presence of transforming growth factor-β growth factors *(20)*.

We describe protocols for assessing the suitability of housekeeping genes to act as true internal controls in differentiating ES cell cultures. Such assessment is needed for each new experimental condition. The protocol focuses on PCR; however, the principle and experimental design are applicable to any (semi)quantitative RNA analysis.

2. Materials

2.1. Tissue Culture

1. 1X phosphate-buffered saline (PBS) made from 10X stock, without Ca^{2+} or Mg^{2+} (Cambrex Bio Science, Wokingham, UK; cat. no. BE17-517Q), with distilled water (dH_2O).
2. ES cell screened fetal bovine serum (FBS) (500 mL) (Hyclone, South Logan, UT; cat. no. SH30071.03E). Store at −20°C in 50-mL aliquots (*see* **Note 1**).
3. Dulbecco's modified Eagle's medium (DMEM) with high glucose (Life Technologies, Paisley, UK; cat. no. 11960-044), without Na pyruvate, and without L-glutamine.
4. Penicillin and streptomycin (100X; 100-mL bottle) (Life Technologies, cat. no. 15070-014). Store at −20°C in 5-mL aliquots.
5. L-Glutamine (100X; Life Technologies, cat. no. 25030-016), stored at −20°C in 5-mL aliquots.
6. β-Mercaptoethanol (14.4 *M*, Sigma-Aldrich, Poole, UK; cat. no. M6250), diluted in 1X PBS to 0.1 *M* (1000X); aliquot and store at −20°C.
7. ESGRO® leukemia inhibitory factor (LIF; 1×10^6 U/mL) (Chemicon International Inc., Temecula, CA; cat. no. ESG1106). Store at 4°C.
8. Gelatin, 2% (Sigma, cat. no. G1393). Make a 0.1% working solution with 1X PBS. Autoclave to sterilize and store at 4°C.
9. Mitomycin C (2-mg vial; Sigma, cat. no. M4287). Dissolve vial contents in 2 mL 1X PBS and transfer to a tube containing 18 mL of media. This is a 10X stock solution and can be stored at −20°C (1-mL aliquots) for up to 4–6 mo (*see* **Note 2**).
10. Trypsin-EDTA (ethylenediaminetetraacetic acid; 0.05%) (500 mL; Life Technologies, cat. no. 25300-062). Aliquot and store at −20°C.
11. Falcon T-75 tissue culture flasks (Becton Dickinson, Franklin Lakes, NJ; cat. no. 353024).
12. 100-mm Petri dishes (Fisher Scientific, Loughborough, UK; cat. no. 08-757-12).
13. 15-mL polypropylene tubes (Becton Dickinson, cat. no. 352097).
14. 70-μm nylon cell strainer (Becton Dickinson, cat. no. 352350).

Internal Standards for ES Cells

15. Dimethyl sulfoxide (DMSO) (100 mL; Sigma, cat. no. D2650).
16. Nalgene Nunc cryovials (1-mL tubes; Fisher, cat. no. 375353).
17. "Mr. Frosty" Nalgene freezing containers (Fisher, cat. no. 15-350-50).
18. Media for SNLs-a murine fibroblast cell line DMEM supplemented with 10% FBS, 1% penicillin (50 U/mL) and streptomycin (50 µg/mL), 1% L-glutamine (2 m*M*), and 0.1% β-mercaptoethanol (0.1 m*M*) (*see* **Note 3**). For 500 mL, take 60.5 mL from a 500-mL bottle of DMEM and replace with 50 mL FBS, 5 mL penicillin-streptomycin, 5 mL L-glutamine, and 0.5 mL β-mercaptoethanol.
19. ES cell media: CCE ES cells *(20,21)* are cultured in media as in **step 1** for SNLs but with the addition of LIF (500 U/mL final concentration). For 100 mL of media, add 50 µL LIF to 100 mL SNL media.

2.2. Reverse Transcriptase Polymerase Chain Reaction

2.2.1. RNA Isolation

1. RNeasy® minikit (Qiagen, Crawley, UK; cat. no. 74104).
2. Active RLT buffer: for 1 mL, add 10 µL β-mercaptoethanol to 1 mL RLT buffer. Store at room temperature and discard after 4 wk.
3. QIAshredders (Qiagen, cat. no. 79654).
4. RNase-free DNase set (Qiagen, cat. no. 79254).
5. 14.3 *M* β-mercaptoethanol.
6. Sterile RNase-free pipet tips.
7. Ethanol.

2.2.2. RNA Quantitation and Purity

1. Spectrophotometer (A_{260}/A_{280}).
2. Ultraviolet grade, DNA, RNase, and protein free, disposable cuvets (Eppendorf Uvettes; Fisher, cat. no. 0030 106.300).
3. DNA-, RNase-free water.

2.2.3. Reverse Transcription

1. Heat block.
2. AMV Reverse transcriptase system (Promega, Southhampton, UK; cat. no. A3500).

2.2.4. PCR Amplification

1. Thermal cycler.
2. 0.2-mL Eppendorf PCR reaction tubes (Eppendorf UK, Cambridge, UK; cat. no. 0030 124.332).
3. AmpliTaq Gold PCR master mix (2X) (Applied Biosystems, Warrington, UK; cat. no. 4318739) containing AmpliTaq Gold DNA polymerase, dNTPs, Gold Buffer, and $MgCl_2$. Store at 4°C.
4. PCR primers: dilute in TE buffer to 10 µ*M*, aliquot, and store at −20°C.
5. 1 *M* Tris-Cl, pH 8.0. For 1 L, 121.1 g Tris-base, 800 mL dH_2O. Adjust pH to 8.0 by adding 42 mL HCl. Adjust volume to 1 L with distilled water and sterilize by autoclaving. Store at room temperature and discard solution when it turns yellow.

2.2.5. Gel Electrophoresis

1. Agarose and gel apparatus.
2. 10 mg/mL aqueous solution ethidium bromide (EtBr) (Sigma, cat. no. E1510).

3. TAE buffer: For 1 L of 50X TAE buffer, add 242 g Tris-base, 57.1 mL glacial acetic acid, 100 mL 0.5 M EDTA (pH 8.0). Adjust volume to 1 L with distilled water. Store at room temperature. Dilute with distilled water to generate a 1X working solution (40 mM Tris-acetate, 1 mM EDTA).
4. 5X loading dye for PCR products: 3.75 g ficol 400, 63 mg bromophenol blue. Bring to 50 mL with water. Aliquot and freeze until needed.
5. Formaldehyde gel buffer (10X): 200 mM MOPS, 50 mM sodium acetate, 10 mM EDTA; bring pH to 7.0 with NaOH.
6. 5X loading buffer for RNA samples: 16 µL saturated aqueous bromophenol blue solution, 80 µL 500 mM EDTA, pH 8.0, 720 µL 37% (12.3 M) formaldehyde, 2 mL 100% glycerol, 3084 µL formamide, 4 mL 10X formaldehyde gel buffer. RNase-free water to 10 mL (*see* **Note 4**).

2.2.6. Quantitation of PCR-Amplified Products

1. DNA quantitative standards (Bio-Rad, Hemel Hempstead, UK; cat. no. 170-8207).
2. Fluor-S MultiImager (Bio-Rad; model no. 170-7700).
3. *Quantity One* software for image analysis (Bio-Rad).

3. Methods
3.1. Tissue Culture
3.1.1. Freezing Cells (see **Note 5**)

1. Aspirate the culture media and wash the cells (twice) with sterile 1X PBS.
2. Add enough trypsin to cover the surface of the culture flask and place in the incubator for 5–10 min until cells start to detach and float off. Gentle agitation helps the process.
3. Add an equal volume of FBS-containing media, aspirate the cell suspension, and spin at 500g for 5 min.
4. Aspirate the supernatant and resuspend the cell pellet in a volume of cold (i.e., from the refrigerator) FBS.
5. Add dropwise an equal volume of cold 20% DMSO in FBS to obtain the desired concentration of cells in a final concentration of 10% DMSO (*see* **Note 6**).
6. Aliquot 1 mL into prelabeled cryovials placed in a Mr. Frosty container (*see* **Note 7**).
7. Transfer cells immediately to a $-80°C$ freezer and leave overnight.
8. Transfer the cells to liquid nitrogen for long-term storage.

3.1.2. Thawing Cells (see **Note 8**)

1. Take a frozen vial of cells and submerge in a 37°C water bath so that the cells contained in the vial are beneath the water level.
2. Keep checking the vial and remove before the last ice crystal has disappeared, which is usually approx 60–90 s. Once thawed, quickly sterilize the outside of the vial with 70% ethanol, wipe, and transfer to the flow cabinet.
3. Transfer the cells to a 15-mL centrifuge tube containing nine times the volume of cold (4–8°C) medium (e.g., 9 mL for an initial 1 mL thawed cell suspension). Briefly resuspend the cells by pipetting.
4. Centrifuge the cells at 500g for 5 min.
5. Aspirate the medium, resuspend in the required amount of medium, and plate as desired.
6. Change the medium after 24 h to remove any nonadherent/dead cells.

3.1.3. Gelatinizing Tissue Culture Flasks

1. Prior to culture of SNLs, flasks are treated with 0.1% gelatin for 10 min at room temperature. Sufficient volume is added to flasks to cover the whole surface area.
2. The gelatin is aspirated after 10 min, and cells are seeded.

3.1.4. Culture of SNLs

1. SNLs are grown in standard medium on pregelatinized plates.
2. When confluent, cultures are passaged at 1:8.
3. Such cultures take approx 3 d to reach confluence and are either subcultured or inactivated using mitomycin C to prepare feeder layers for ES cell culture (*see* **Note 9**).

3.1.5. Preparation of Feeder Layers

1. Aspirate the medium from the flask of SNLs and replace with fresh medium containing 1X mitomycin C (10 µg/mL). Leave in the incubator for 2–3 h (*see* **Note 10**).
2. Remove the mitomycin C-containing medium and wash the cells (three times) with sterile PBS.
3. Add sufficient trypsin to cover the surface area of the flask and incubate for 10 min in the incubator.
4. Once trypsinized, add twice the volume of medium and resuspend to obtain a single-cell suspension.
5. Centrifuge the cells for 5 min at 500g, aspirate the supernatant, and resuspend the cell pellet in fresh medium.
6. Seed the inactivated fibroblasts at a confluent level (1.3×10^5 cells/cm^2).
7. Allow the inactivated SNLs (iSNLs) to attach overnight (or at least 2 h) before seeding ES cells on the feeder layer.

3.1.6. Culture of ES Cells

1. ES cells (from frozen stocks) are seeded at a density of 2×10^4/cm^2 on a feeder layer of mitomycin C-iSNLs in media containing 500 U/mL LIF.
2. Medium is changed every day (*see* **Note 11**).

3.1.7. Subculture of ES Cells

ES cells are subcultured while colonies are just starting to touch each other but with distinct, clearly defined edges (**Fig. 1**). Subculturing is typically required every 2–3 d, and cultures should not be split more than 1:5 to help prevent the cells from differentiating. Mitomycin C-iSNLs must be ready prior to subculturing ES cells.

1. Aspirate medium and rinse twice with 1X PBS.
2. Add trypsin to the culture flask.
3. Return to incubator for 5–10 min.
4. Add an equal volume of medium, aspirate off the cell suspension, and pellet at 500g for 5 min.
5. Resuspend the pellet in fresh medium and reseed the cells on freshly inactivated feeder layers.
6. To help prevent their differentiation, ES cells should not be split more than 1:5.
7. Medium is replaced every day (*see* **Note 12**).

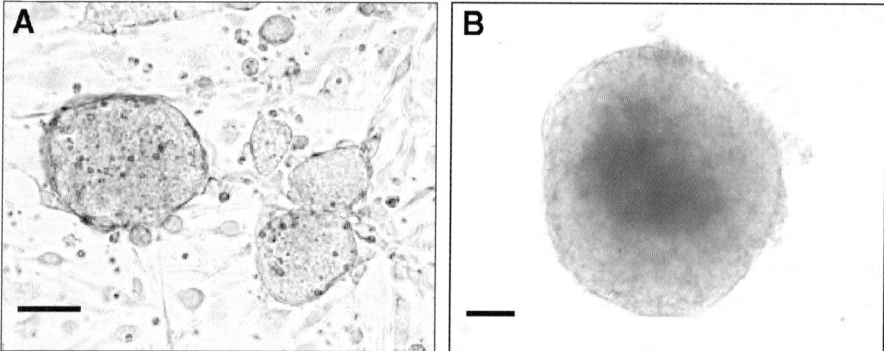

Fig. 1. Murine CCE embryonic stem cell colonies on (**A**) an SNL feeder layer, and (**B**) a d-5 embryoid body. Bar = 50 μm.

3.1.8. EB Formation

1. To initiate ES cell differentiation, cultures are briefly trypsinized until the ES cell colonies just lift off from the surface of the culture flask.
2. Medium is added to neutralize the trypsin, and the aggregates are spun down at 500g for 2 min.
3. The cell aggregates are gently resuspended in fresh medium and placed in bacteriological-grade Petri dishes in the absence of both feeder layer and LIF to enable spontaneous differentiation in the form of EB over a period of 5 d (*see* **Note 13**).
4. Medium is replaced after 24 h. This is done by gently aspirating the medium (containing the EBs) into a 25-mL pipet. They quickly settle to the bottom of the pipet and can be placed back in the dish; the remaining medium is discarded and replaced with fresh medium.
5. After 3 d, EBs are placed on a 70-μm nylon cell strainer to remove smaller aggregates and single cells, which pass through and are discarded. The EBs collected on the membrane are then placed in fresh medium. If they adhere to the Petri dish during culture, change to a new dish.

3.1.9. Preparation of ES Cells for RNA Isolation

1. ES cells are removed from feeder cultures by differential adhesion.
2. After trypsinization, plate ES and feeder layer cell suspension on a nongelatinized plate for 20–30 min at 37°C. This is sufficient time for SNLs to attach; the supernatant predominantly contains ES cells.
3. These can be aspirated off and spun down at 500g for 5 min.

3.2. Reverse Transcriptase Polymerase Chain Reaction

3.2.1. RNA Isolation (see **Note 14**)

The following procedure is derived from the manufacturer's protocol:

1. Wash the ES cells (or EBs) in ice-cold (1X) PBS and pellet by centrifugation (500g for 5 min).
2. Lyse the cell pellet in RLT buffer (350 μL for up to 5×10^6 cells).
3. Resuspend with a 1-mL pipet and apply the sample to a QIAshredder spin column for homogenization. This is achieved by centrifugation at maximum speed for 2 min (*see* **Note 15**).

4. Add 1 volume (350 µL) of 70% ethanol and mix by pipetting.
5. Apply to an RNeasy minicolumn placed in a 2-mL collection tube and centrifuge for 20 s at >8000g to bind the RNA to the silica gel membrane of the column. Discard the flow-through.
6. Add 350 µL buffer RW1 to the column and centrifuge again at 8000g for 20 s. Discard the flow-through.
7. Add 10 µL DNase I stock solution to 70 µL buffer RDD. Mix and pipet this solution directly onto the column and leave for 15 min (see **Note 16**).
8. Add another 350 µL buffer RW1 to the column and centrifuge again at 8000g for 20 s. Discard the flow-through.
9. Transfer the column to a new 2-mL collection tube and add 500 µL buffer RPE. Centrifuge at >8000g for 20 s and discard flow-through.
10. To dry the membrane, add another 500 µL buffer RPE and centrifuge for 2 min at >8000g.
11. Transfer the column to a new 2-mL collection tube and carefully add 50 µL RNase-free water to the center of the membrane. Centrifuge at >8000g for 1 min to elute the RNA.

3.2.2. RNA Quantitation and Purity

Following RNA isolation, measurement of RNA concentration is performed spectrophotometrically at 260 nm. An A_{280} reading is also taken, with an A_{260}/A_{280} ratio between 1.8 and 2.0 indicating purity of isolated RNA. In addition, the integrity of the RNA can be further checked by denaturing agarose gel electrophoresis with EtBr staining. Distinct, sharp bands should appear for the 18S and 28S ribosomal species (**Fig. 2**). If the bands are smeared, with smaller size RNA showing, it is likely that the RNA has undergone significant degradation. The procedure for formaldehyde agarose gel electrophoresis is given next.

1. Make a 1.2% gel by mixing 1.2 g agarose, 10 mL 10X gel buffer, and RNase-free water to a final volume of 100 mL.
2. Heat to melt the agarose and then cool to 65°C (in a water bath).
3. Add 1.8 mL 37% (12.3 M) formaldehyde and 1 µL EtBr (10-mg/mL stock solution).
4. Mix and pour onto gel support.
5. After the gel has set, equilibrate in 1X formaldehyde gel running buffer for 45 min.
6. Add 1 vol 5X loading buffer per 4 vol of RNA sample and mix before incubating at 65°C for 5 min.
7. Chill on ice before loading samples onto the gel.
8. Run at 5–7 V/cm in 1X formaldehyde gel running buffer.

3.2.3. Reverse Transcription

Reverse transcribe 1 µg RNA using random hexamer primers and the Promega reverse transcription kit; 4% of the reverse-transcribed solution is used in all subsequent PCR reactions (see **Note 17**).

1. 1 µg RNA is made up to 10-µL volume with RNase-free water in 0.2-mL (PCR-grade) tubes.
2. 1 µL primer solution is added, and the tube is heated at 65°C for 10 min on a heat block.
3. 10 mL master mix is added to the tube, which is left at room temperature for 10 min prior to heating to 42°C for 30 min. Master mix contains 4 µL $MgCl_2$, 2 µL 10X buffer, 2 µL DNTP, 0.5 µL RNase inhibitor, and 0.5 µL AMV transcriptase.
4. Followed at 99°C for 5 min; finally, the samples are placed on ice for 5 min.

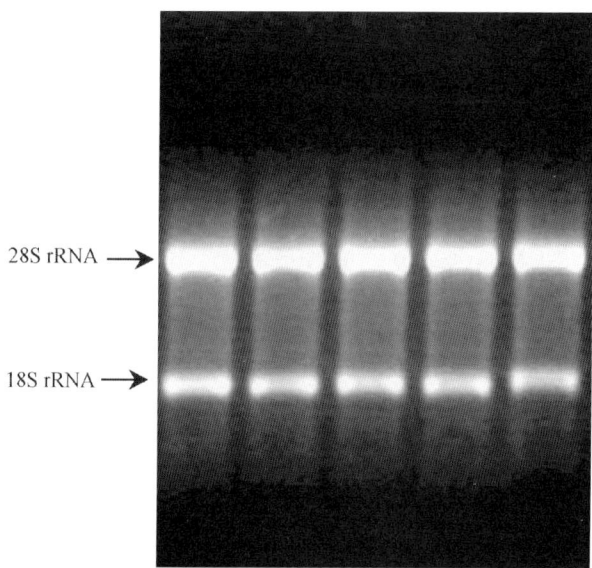

Fig. 2. Integrity of total RNA purified from five ES cell cultures as assessed by denaturing gel electrophoresis and ethidium bromide staining. Sharp, distinct bands corresponding to 28S and 18S ribosomal RNA indicate intact RNA. In addition, the 28S band should be approximately twice as intense as that of the 18S ribosomal RNA band.

5. The tubes are briefly spun down to settle the condensed liquid off the sides and top of the inside of the tube. The 20-μL reaction volume is brought to 100 μL with RNase-free water and, when not used immediately, is aliquoted prior to storage at −20°C.

3.2.4. PCR Amplification

To ensure measurements are made before the plateau phase of amplification is reached, the PCR reaction is run for each gene for increasing cycle numbers, and the amplified products are measured. The plateau phase is detected by a marked decrease in the rate of product accumulation. Therefore, following preliminary experiments, cycle numbers are chosen only in the exponential phase of amplification.

1. For PCR reactions, 4% (4 μL) of the 1 μg of RNA that was reverse transcribed is used. These are performed using AmpliTaq Gold DNA polymerase.
2. Samples are cycled as follows: 94°C (30 s), 59°C (30 s), and 72°C (30 s) in an Eppendorf Mastercycler thermal cycler.
3. Master mix consists of 31.7 μL RNase-/DNase-free water, 5 μL 10X buffer, 3 μL $MgCl_2$ (1.5 mM final concentration), 1 μL DNTP (200 μM final concentration), 2.5 μL primers (both forward and reverse, 0.5 μM final concentration; **Table 1**), and 0.3 μL AmpliTaq Gold DNA polymerase (1.5 U). To this, 4 μL reverse-transcribed RNA gives a final reaction volume of 50 μL (*see* **Note 18**).

Table 1
Primmer Sequences for Commonly Used Housekeeping Genes: GAPDH (32 cycles, 500-bp product), HPRT (32 cycles, 507-bp product), and β-Tubulin (29 cycles, 317-bp product)

Gene	Antisense (5′ to 3′)	Sense (5′ to 3′)
GAPDH	CATCACCATCTTCCAGGAGC	ATGCCAGTGAGCTTCCCGTC
HPRT	GCCTGTATCCAACACTTCG	GCGTCGTGATTAGCGATG
β-Tubulin	GGAACATAGCCGTAAACTGC	TCACTGTGCCTGAACTTACC

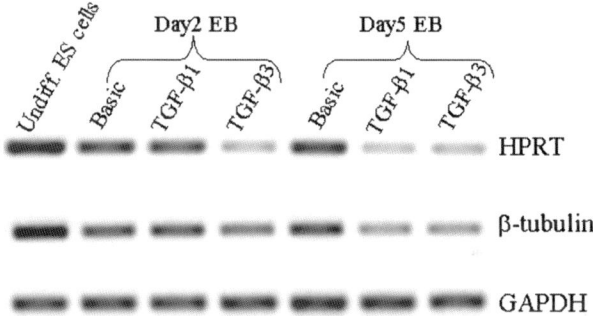

Fig. 3. EtBr-stained gels showing PCR-amplified products from undifferentiated ES cells and growth factor-treated embryoid body cultures for HPRT, β-tubulin, and GAPDH. Undiff., undifferentiated; basic, cultured in basic media.

3.2.5. Gel Electrophoresis of PCR-Amplified Products

Amplified products are separated by electrophoresis on a 1.2% agarose gel containing 0.5 μg/mL EtBr in both the agarose gel and running (TAE) buffer (*see* **Note 19**). Samples from each PCR reaction are run in triplicate to correct for pipetting/loading errors.

1. Mix 10 μL amplified product with 2.5 μL 5X loading buffer before loading onto the gel.
2. To one lane, load quantitative DNA standards; 2.5 μL is mixed with 7.5 μL 1X running buffer and 2.5 μL loading buffer. This contains five products consisting of 100 ng of 1000 bp, 70 ng of 700 bp, 50 ng of 500 bp, 20 ng of 20 bp, and 10 ng of 100 bp.
3. Run at 5–7 V/cm in 1X running buffer containing EtBr.
4. Stop when the bromide front has reached the bottom of the gel.
5. Image the gel.

3.2.6. Detection and Quantitation of Amplified PCR Products

1. Digital images of EtBr-stained gels (**Fig. 3**) are captured using the Bio-Rad Fluor-S MultiImager system, which consists of an enclosed flatbed ultraviolet light scanner and charge-coupled device camera connected to a computer.
2. Images are analyzed using Bio-Rad *Quantity One* quantitation software. This system allows detection of the individual bands and subtraction of background "noise," yielding intensity values caused solely by the gene-specific amplified products.

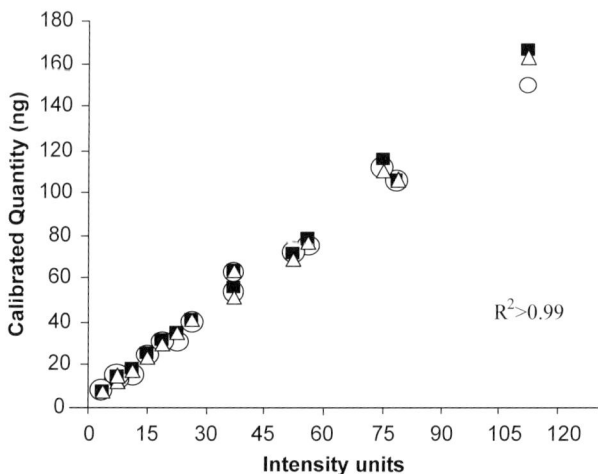

Fig. 4. Calibration curve obtained using DNA standards of known size and quantity. Unknowns from PCR-amplified products must lie within this linear range of the system if accurate quantitation is to be achieved.

3. For each gel, a solution of known DNA standards (Bio-Rad) is loaded, which allows quantitation of the amount of gene-specific amplified product and normalization for staining and imaging conditions.
4. The DNA standard calibration curve is linear in the range of 5 to 150 ng DNA (**Fig. 4**). PCR-amplified product quantities must always lie within this linear detection range of the system for reliable quantitation.

3.3. Experimental Design and Statistical Analysis

1. To adequately assess the stability of housekeeping gene expression under a given set of experimental conditions, RNA must be isolated separately from (at least) six independent cultures for *each* condition tested.
2. After reverse transcription of these samples, PCR reactions are run in triplicate, and the average values are taken to help correct for any pipetting errors.
3. For ease of comparison, calibrated quantities can be expressed relative to undifferentiated ES cell culture values.
4. For each gene under investigation, statistical comparisons can be made between experimental treatments by a one-way analysis of variance with Tukey's multiple comparison test.
5. An appropriate internal control would not show any statistical differences ($p > 0.05$) between any of the experimental treatments.

4. Notes

1. This FBS is prescreened, which involves plating efficiency, colony morphology, and toxicity tests. In addition, it is filtered through sequential 0.04-µm pore filters.
2. Mitomycin C is toxic; wear gloves and prepare in tissue culture hood.
3. SNL STO is a fibroblast cell line that produces LIF.
4. Solution is stable for 3 mo at 4°C.

5. For long-term storage, liquid nitrogen must be used. Prepare and label all cryovials prior to addition of DMSO to cells. Cells should be frozen at a concentration of $1-5 \times 10^6$ cells/mL.
6. We have found slightly better viability of cells retrieved from frozen stocks if they were frozen with chilled DMSO/FBS (i.e., at refrigerator temperature) rather than at room temperature.
7. Fill inner container with isopropyl alcohol as directed on the Mr. Frosty container. This gives a controlled cooling rate of 1°C per min. Replace solution after five uses.
8. When thawing cells from liquid nitrogen, work quickly and efficiently, that is, have everything ready at hand before you start to thaw the cells.
9. Frozen stocks should be made of both SNLs and iSNLs. Frozen iSNLs should be thawed the day before they are needed for ES cell culture.
10. When handling mitomycin C, keep in foil because the product is light sensitive. A 1X solution (10 µg/mL) can be stored for up to 2 wk at 2–8°C in the dark. Prepare a fresh solution if a precipitate forms because this is known to be toxic to cells.
11. LIF is stable for 7 d at 37°C, 5% CO_2, during ES cell culture.
12. Fresh medium is preequilibrated in an incubator (37°C, 5% CO_2) for 2 h before it is added to cell cultures.
13. Bacteriological-grade Petri dishes are used to prevent cell attachment.
14. During RNA isolation, it is important to wipe down surfaces and keep them RNase free. In addition, gloves should be frequently changed. The procedure should be carried out quickly at room temperature. If storing, keep the RNA at $-70°C$. Always check concentration and purity of RNA prior to further processing.
15. An alternative method of homogenization is to pass the cell lysate five times through a 20-gage needle using a 1-mL syringe.
16. Generally, DNase treatment is not essential, and this step may be omitted. However, it is recommended for certain RNA assays that are sensitive to even very small levels of DNA, such as TaqMan PCR.
17. Kit components should be aliquoted to reduce freeze/thaw cycles. Aliquots are thawed and kept on ice during the procedure.
18. We routinely reconstitute our PCR primers in TE buffer because we have found that this gives better results than distilled water. Samples are run in triplicate (i.e., three separate tubes) to correct for pipetting errors.
19. EtBr is mutagenic; wear appropriate clothing and gloves. Dispose of used EtBr to labeled glass containers. Sunlight will break down EtBr and may be used to decontaminate stocks. For use, keep EtBr stocks out of the light.

References

1. Smith, A. G. (2001) Embryo-derived stem cells: of mice and men. *Annu. Rev. Cell Dev. Biol.* **17,** 435–462.
2. Bradley, A., Evans, M., Kaufman, M. H., and Robertson, E. (1984) Formation of germ-line chimeras from embryo-derived teratocarcinoma cell lines. *Nature* **309,** 255–256.
3. Boheler, K. R. and Fiszman, M. Y. (1999) Can exogenous stem cells be used in transplantation? *Cells Tissues Organs* **165,** 237–245.
4. D'Amour, K. and Gage, F. H. (2001) New tools for human developmental biology. *Nat. Biotechnol.* **18,** 381–382.
5. Odorico, J. S., Kaufman, D. S., and Thomson, J. A. (2001) Multilineage differentiation from human embryonic stem cell lines. *Stem Cells* **19,** 193–204.
6. Abe, K., Niwa, H., Iwase, K., et al. (1996) Endoderm-specific gene expression in embryonic stem cells differentiated to embryoid bodies. *Exp. Cell Res.* **229,** 27–34.

7. Leahy, A., Xiong, J. W., Kuhnert, F., and Stuhlmann, H. (1999) Use of developmental marker genes to define temporal and spatial patterns of differentiation during embryoid body formation. *J. Exp. Zool.* **284**, 67–81.
8. Itskovitz-Eldor, J., Schuldiner, M., Karsenti, D., et al. (2000) Differentiation of human embryonic stem cells into embryoid bodies compromising the three embryonic germ layers. *Mol. Med.* **6**, 88–95.
9. Buttery, L. D., Bourne, S., Xynos, J. D., et al. (2001) M. Differentiation of osteoblasts and in vitro bone formation from murine embryonic stem cells. *Tissue Eng.* **7**, 89–99.
10. Gualandris, A., Annes, J. P., Arese, M., Noguera, I., Jurukovski, V., and Rifkin, D. B. (2000) The latent transforming growth factor-beta-binding protein-1 promotes in vitro differentiation of embryonic stem cells into endothelium. *Mol. Biol. Cell.* **11**, 4295–4308.
11. Kramer, J., Hegert, C., Guan, K., Wobus, A. M., Muller, P. K., and Rohwedel, J. (2000) Embryonic stem cell-derived chondrogenic differentiation in vitro: activation by BMP-2 and BMP-4. *Mech. Dev.* **92**, 193–205.
12. Schuldiner, M., Yanuka, O., Itskovitz-Eldor, J., Melton, D. A., and Benvenisty, N. (2000) Effects of eight growth factors on the differentiation of cells derived from human embryonic stem cells. *Proc. Natl. Acad. Sci. USA.* **97**, 11,307–11,312.
13. Wobus, A. M., Kaomei, G., Shan, J., et al. (1997) Retinoic acid accelerates embryonic stem cell-derived cardiac differentiation and enhances development of ventricular cardiomyocytes. *J. Mol. Cell Cardiol.* **29**, 1525–1539.
14. Thellin, O., Zorzi, W., Lakaye, B., et al. (1999) Housekeeping genes as internal standards: use and limits. *J. Biotechnol.* **75**, 291–295.
15. Spanakis, E. (1993) Problems related to the interpretation of autoradiographic data on gene expression using common constitutive transcripts as controls. *Nucleic Acids Res.* **21**, 3809–3819.
16. Vassilieva, S., Guan, K., Pich, U., and Wobus, A. M. (2000) Establishment of SSEA-1- and Oct-4-expressing rat embryonic stem-like cell lines and effects of cytokines of the IL-6 family on clonal growth. *Exp. Cell Res.* **258**, 361–373.
17. Phillips, B. W., Belmonte, N., Vernochet, C., Ailhaud, G., and Dani, C. (2001) Compactin enhances osteogenesis in murine embryonic stem cells. Biochem. *Biophys. Res. Commun.* **284**, 478–484.
18. Weinhold, B., Schratt, G., Arsenian, S., et al. (2000) Srf(−/−)ES cells display non-cell-autonomous impairment in mesodermal differentiation. *EMBO J.* **19**, 5835–5844.
19. Wiles, M. V. and Johansson, B. M. (1999) Embryonic stem cell development in a chemically defined medium. *Exp. Cell Res.* **247**, 241–248.
20. Murphy, C. L. and Polak, J. M. (2002) Differentiating embryonic stem cells: GAPDH, but neither HPRT nor β-tubulin is suitable as an internal standard for measuring RNA levels. *Tissue Eng.* **8**, 551–559.
21. Robertson, E., Bradley, A., Kuehn, M., and Evans, M. (1986) Germ-line transmission of genes introduced into cultured pluripotential cells by retroviral vector. *Nature* **323**, 445–448.

9

Matrix Assembly, Cell Polarization, and Cell Survival

Analysis of Peri-Implantation Development With Cultured Embryonic Stem Cells

Shaohua Li and Peter D. Yurchenco

Summary

A variety of mutations, including those affecting laminin expression and basement membrane, cause early embryonic lethality in the peri-implantation period. However, low cell numbers and inaccessibility of these small embryos make it difficult to study the molecular mechanisms that underlie these defects. Embryoid bodies cultured as suspended spherical cell aggregates derived from normal and defective embryonic stem cells provide a tractable experimental system with which the early developmental processes can be recapitulated under defined conditions. Thus, endoderm formation and maturation, basement membrane assembly and its signaling consequences, epiblast polarization, apoptosis, and cavitation can be studied using a combination of genetic, biochemical, cell, and molecular biology approaches.

Key Words: Anoikis; basement membrane; basal lamina; ectoderm; embryoid body; embryonic development; endoderm; epiblast; epithelial; laminin; type IV collagen.

1. Introduction

A distinguishing attribute of a tissue is the presence of an extracellular matrix (ECM). Basement membranes, among the most ancient of ECMs, are required for the development and functional maintenance of tissues, exerting their organizing influence early in embryonic development. Composed of laminins, type IV collagens, nidogens, perlecan, and agrin, basement membranes are layered supramolecular assemblies that adhere to cells. Members of the laminin family, key constituents of these ECMs, have been found to be both necessary and sufficient to form an ECM with the ability to induce cell differentiation *(1,2)*. Cell adhesion appears to critically depend on the binding of the LG4 module located at the end of the long arm of laminin 1 with cell surface molecules, possibly acidic glycolipids *(3)*, with signaling mediated by β1-integrins (e.g., α6β1), α-dystroglycan, and other receptors. In epithelia, basement membranes are deposited

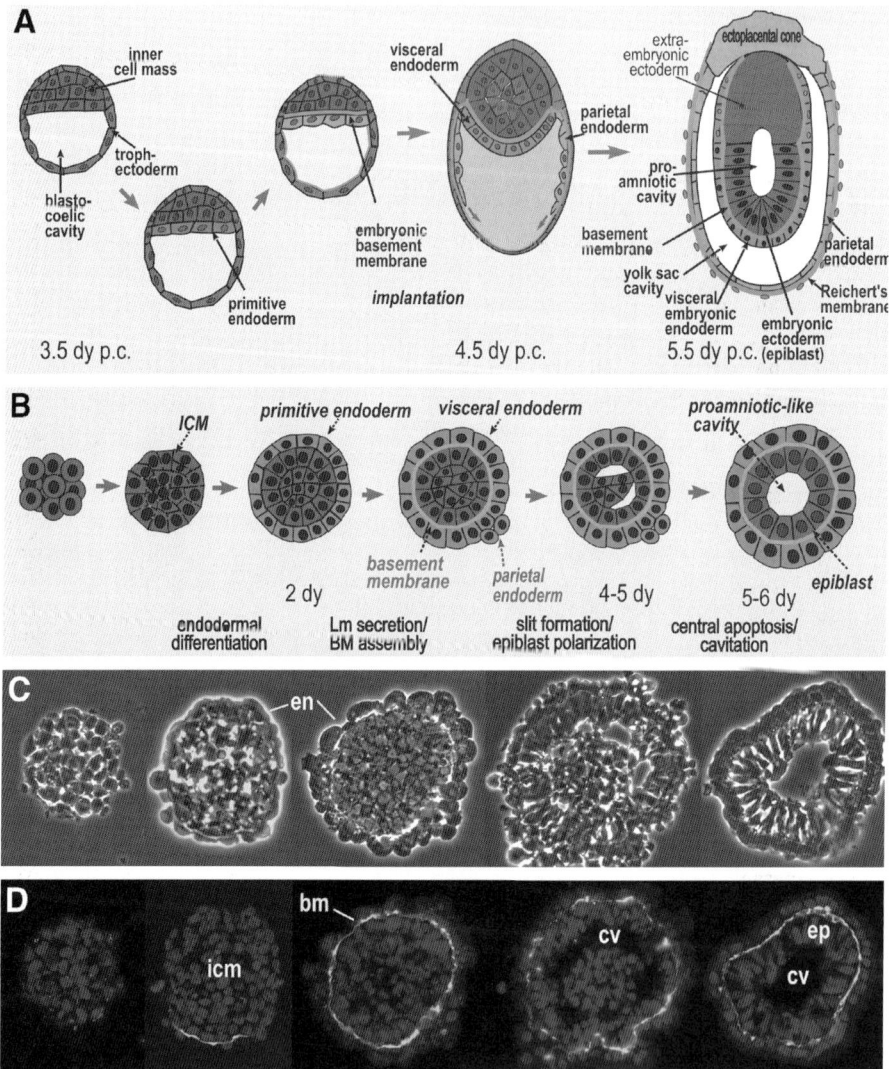

Fig. 1. Embryonic and embryoid body development. (**A**) The compacted morula gives rise to the blastocyst consisting of inner cell mass (ICM) and trophectoderm by 3.5 d postcoitum. The primitive endoderm develops from the ICM and in turn gives rise to visceral and parietal endoderm. Each secretes laminin(s) and type IV collagen. These components accumulate as basement membranes between visceral endoderm and ICM (embryonic basement membrane [BM]) and parietal endoderm and trophectoderm (Reichert's membrane). Following assembly, ICM undergoes central cavitation with polarization of the surviving cells (epiblast). (**B–D**) embryonic stem cell aggregates develop into spherical embryoid bodies when cultured in suspension. As the cells proliferate, they undergo steps of primitive endoderm formation, conversion to actively secretory endoderm (mostly visceral), basement membrane assembly, slit formation accompanied by

on what becomes the basal side of the cell layer and are found to participate in the induction of cell polarization (3).

During mouse embryogenesis, basement membranes first assemble in the preimplantation blastocyst between visceral endoderm and inner cell mass (ICM; embryonic basement membrane) and between parietal endoderm and trophectoderm (Reichert's membrane) (4). Assembly requires the synthesis and secretion of laminins bearing the α1/α5-, β1-, and γ1-subunits, that is, laminin 1 (α1β1γ1) and laminin 10 (α5β1γ1), with both laminins found in the embryonic basement membrane and mostly laminin 1 in Reichert's membrane (5,6). In the absence of laminins, the epiblast layer, the source of the three definitive germ layers, fails to differentiate, and development is arrested right after implantation. The question of elucidating the mechanisms that underlie this critical laminin function demanded the use of new experimental approaches, a challenge partly met by the approach of culturing embryonic stem (ES) cells bearing different relevant mutations under conditions that allow them to form differentiated embryoid bodies (EBs).

EBs are spherical cell aggregates that compact and differentiate to form a tissue consisting of an outer secretory endodermal layer, basement membrane, polarized epiblast (primitive ectoderm layer), and central cavity (**Fig. 1**). The sequence of events recapitulates the embryonic transition from undifferentiated ICM to a two germ-layer structure corresponding to the early egg cylinder stage embryo as it begins to enter gastrulation (i.e., from about embryonic days [E] 3–6). Although parietal endodermal cells will also differentiate from the primitive endoderm, they lack the trophectoderm normally adjacent to them and instead tend to form small peripheral aggregates of cells. This lack of trophectoderm prevents analysis of the events that occur during extraembryonic parietal/trophectodermal differentiation and formation of Reichert's membrane.

EB differentiation in vitro has widespread applications beyond the analysis of basement membrane assembly and its immediate consequences, allowing one to study peri-implantation lethal mutations in general, and a system with which to characterize the function of lineage-specific precursor cells, to examine the cellular interaction during endodermal and epiblast differentiation, to study mechanisms of apoptosis and cavitation, and to analyze the processes of cell polarization. In this chapter, we focus on the role of basement membrane and on the methodological approaches that can be used to study them. In part, because basement membrane components are secreted macromolecules that act on target cells, the effects of basement membrane components and their domains can be evaluated in EBs null for those components using exogenous proteins, providing a nontransfection tool to supplement genetic ones. The two approaches, used together, are particularly advantageous.

Fig. 1. *(Continued)* early ICM polarization, and then completion of polarization, central apoptosis, and cavitation. Phase contrast and DAPI/Lm8$_1$ immunostained images of frozen sections of EBs are shown in rows **C** and **D**, respectively. (Rows **C** and **D** reprinted from **ref. 3** with permission from Cell Press.) (Please *see* the companion CD for the color version of this figure.)

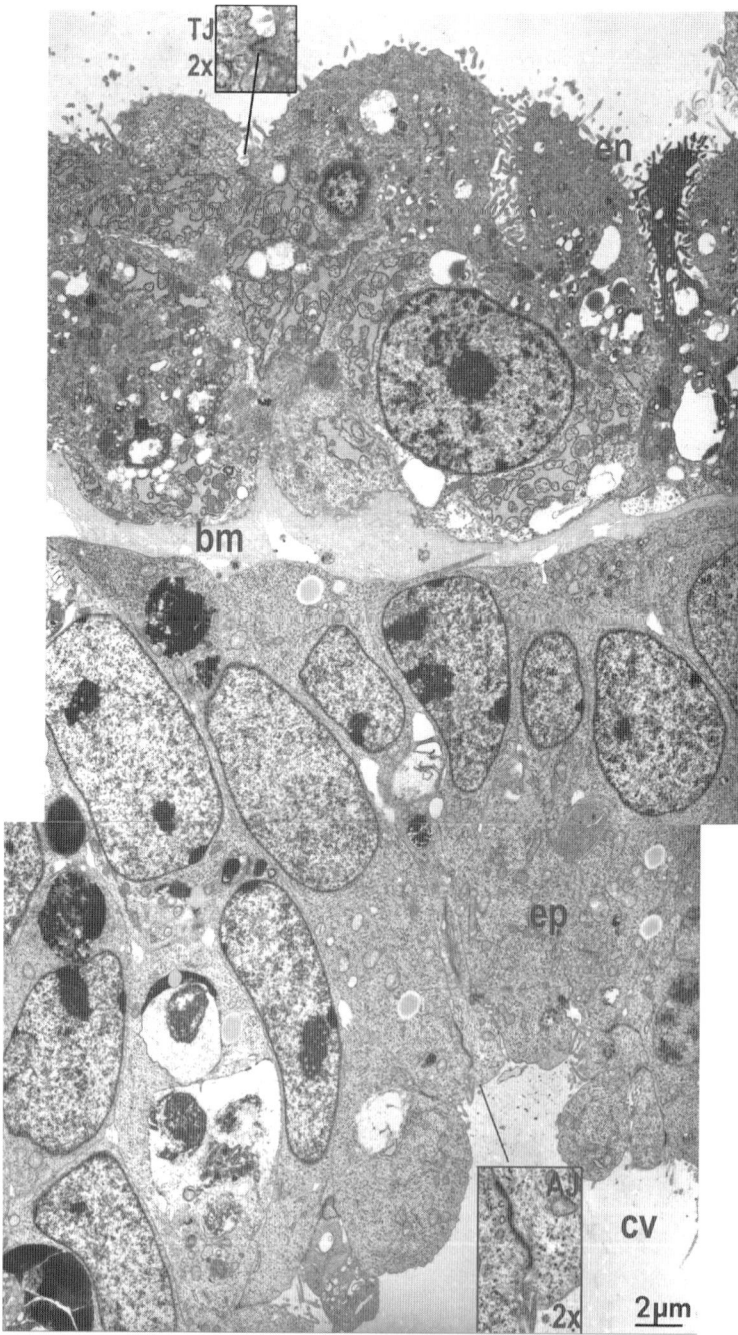

Fig. 2.

1.1. Sequence of EB Differentiation

Several morphological methods can be used to follow and analyze EB differentiation. Phase contrast microscopy allows one to follow formation of the endoderm, epiblast, and cavity at low resolution during culturing without sacrificing the EBs. Immunofluorescence microscopy of EBs treated with fluorescent proteins (e.g., laminin) or transfected with expression constructs that employ green fluorescent protein fusion proteins can also be used to study living EBs. Following harvesting of EBs at the end of an experiment, we use wide-field and confocal immunofluorescence microscopy of frozen (5- to 20-μm) sections of EBs stained with fluorescent-tagged antibodies to analyze the tissue distribution of differentiation, polarization, and other antigenic markers backed by analysis of methylene blue-stained thick (1-μm) sections and transmission electron microscopy (**Fig. 2**) of gluteraldehyde-fixed thin sections as tools of higher-resolution analysis *(1,3,7)*.

The conversion of small clusters of wild-type ES cells to cystic EBs takes about 5 d of culturing in suspension and up to 7 d in ES cells that cannot express laminin but are rescued with exogenous laminin. Initially, the ES cell clusters are loosely adherent but become compacted within 2 d to form spheroid entities with a smooth outer surface. It is possible that this step is analogous to the process of compaction that occurs with the developing morula at the 8- to 16-cell stage.

A single outer layer of primitive endoderm, with a flat squamous or cuboidal morphology, becomes apparent by 3 d. This is a critical step required for subsequent differentiation and is observed in as many as 90% of EBs cultured from pluripotent ES cells under optimal conditions. At the ultrastructural level, the flat primitive endodermal cells have very little endoplasmic reticulum. The endoderm becomes positive for α-fetoprotein (which can be detected by immunofluorescence microscopy), marking the conversion of primitive endoderm to visceral endoderm. At the ultrastructural level, the initially flat layer of outer cells become rounded to cuboidal and develops a prominent rough endoplasmic reticulum, mostly between the nucleus and basal surface, with cisternae that become dilated with proteins (notably laminins) destined for secretion. The endodermal layer becomes the exclusive "factory" for laminins 1 and 10, secreting also most of the type IV collagen into the potential space that lies between the endoderm and ICM. In addition, scattered individual and groups of larger and more rounded parietal endodermal cells, also secretory cells, develop (generally at the periphery of the EBs). The secreted basement membrane components also accumulate in the tissue culture medium but at concentrations below that required for polymer self-assembly *(8,9)*.

Formation of a thick (0.5 to several microns) basement membrane residing between endoderm and adjacent polygonal cells of the ICM occurs following endoderm entry into the secretory phase, with the ECM visualized at 3–4 d in wild-type EBs. Initially,

---

Fig. 2. Embryoid body (EB) ultrastructure. Composite electron micrograph of a wild-type EB cultured for 7 d. Note presence of secretory endoderm (en), thick basement membrane (bm), polarized epiblast (ep), and central cavity (cv). Tight junctions (TJ) are found at endodermal cell junctions; adherens-type junctions (AJ) are found at epiblast junctions adjacent to the cavity.

the polygonal ICM cells, internal to the border of the basement membrane, look unchanged. However, within 1–2 d of its formation the ICM undergoes a profound transformation characterized by the accumulation of a basal distribution of β1-integrin and dystroglycan epitopes; cell and nuclear elongation along a radial axis; pericentral slit cavities; apical accumulation of F-actin, β-catenin, and E-cadherin; central apoptosis; and extension of the slit cavities to form a large central cavity (cavitation).

1.2. Mutations That Affect Early Differentiation

Homozygous null ES cells for genes affecting early embryogenesis are increasingly available or can be generated for analysis during EB differentiation. Mutations that affect laminin subunit synthesis and heterotrimer assembly prevent basement membrane assembly, with downstream failures of epiblast polarization and cavitation. These include null mutations for the laminin β1- and γ1-subunits and for the β1-integrin receptor subunit *(1,5,6,10,11)*. Overexpression of a dominant-negative tailless version of the fibroblast growth factor (FGF) receptor (fgfr)-2 also prevents endoderm development and laminin expression and hence results in a lack of basement membrane assembly and epiblast development *(12)*. Other mutations appear to primarily affect steps downstream of basement membrane assembly, preventing epiblast polarization or apoptosis/cavitation to varying degrees. Such defects are seen in a null mutation for integrin-linked kinase and for apoptosis-inducing factor *(7,13)*. Culturing of these ES cells or their EB derivatives can require modifications of the standard protocol. For example, β1-integrin null ES cells do not adhere well to feeder cell layers and require culturing on plastic dishes. ES cells that are null for dystroglycan (*Dag1*−/−), on the other hand, will spontaneously assemble a basement membrane, form epiblast, and cavitate when cultured as EBs. The basement membranes are unusually thick, and the epiblast that forms subsequently undergoes an accelerated apoptosis.

1.3. Analytical Applications

The ability to manipulate normal and genetically modified ES cells so that they recapitulate steps in mammalian embryogenesis is, we think, a promising approach with which to facilitate the elucidation of fundamental mechanisms of developmental biology. Although care is required to properly maintain both stem cells and EBs, we have found that it is possible for the ES cell aggregates to undergo a stereotyped sequence of transitions of endoderm formation, endoderm conversion to a secretory state, basement membrane assembly, epiblast polarization, and cavitation. Furthermore, cells within long-term cultured EBs can differentiate to a variety of cell lineages that appear in more advanced embryonic stages, such as endothelial cells, hematopoietic progenitors, cardiomyocytes, neuronal cells, and adipocytes. For example, EBs have been successfully employed to define the earliest lineage commitment, leading to the discovery of the hemoangioblast, a common ancestor of endothelial cells and hematopoietic precursors *(14,15)*. With the derivation of human ES cell lines and their potential applications in replacement therapy, EB differentiation will also be a powerful model system to address questions regarding regulatory mechanisms of lineage commitment, cell cycle progression, and tumor suppression, which will provide guidance for safe and efficient clinical usage of human ES cells.

2. Materials

2.1. Tissue Culture Reagents and Media

1. ES cell medium: minimum essential medium (MEM)-α medium (Life Technologies, Grand Island, NY; cat. no. 1257-063) containing 15% ES-grade fetal bovine serum (FBS) (Gemini, Woodland, CA; cat. no. 100-500), 1000 U/mL leukemia inhibition factor (LIF) (Chemicon, Temecula, CA; cat. no. ESG1107), 0.1 mM nonessential amino acids (Life Technologies, cat. no. 11140-050), 1 mM sodium pyruvate (Life Technologies, cat. no. 11360-070), 100 U/mL penicillin/100 µg/mL streptomycin (Life Technologies, cat. no. 15140-122), and 0.1 mM β-mercaptoethanol. For 1 L, use 820 mL MEM-α medium, 150 mL FBS, 10 mL 10X nonessential amino acids, 10 mL 10X sodium pyruvate, 10 mL 10X penicillin/streptomycin, 100 µL LIF (10^7 U/mL), and 100 µL β-mercaptoethanol (1 mol/L).
2. EB differentiation medium: EBs are allowed to differentiate in ES medium without LIF. For 1 L, use 820 mL MEM-α medium, 150 mL ES-grade FBS, 10 mL 10X nonessential amino acids, 10 mL 10X sodium pyruvate, 10 mL 10X penicillin/streptomycin, and 100 µL β-mercaptoethanol (1 mol/L).

2.2. Basement Membrane Components

1. Laminin 1 (2-[diethylamino]ethylamine [DEAE]-unbound fraction) and laminin fragments E1' (short-arm complex), E3 (α1-LG modules 4–5), E4 (β1 domains VI and V), E8 (lower coiled-coil with LG1–3), and C1–4 (polymerizing α1β1γ1 short-arm complex) are prepared from the mouse Englebreth-Holm-Swarm (EHS) tumor as described in **ref. 16**.
2. Nonpolymerizing laminin 1: treat laminin 1 with 5 mM aminoethyl benzene sulfonyl fluoride (AEBSF) in 50 mM Tris-HCl, 90 mM NaCl, pH 7.4 at 4°C overnight, followed by dialysis to remove AEBSF *(17)*.
3. AEBSF-E1' (a control laminin fragment with the ability to inhibit polymerization that has been blocked by the AEBSF treatment): incubate E1' under the same conditions described in **Subheading 2.2., item 2**.
4. Laminin 2 (α2β1γ1) and laminin 4 (α2β2γ1): digest human placenta with bacterial collagenase and separate the laminins from the digested materials by DEAE ion exchange, heparin affinity, and gel filtration chromatography as described in **ref. 18**.

2.3. Antibody Reagents

1. Rat monoclonal antimouse laminin-γ1 (clone A5; Upstate Biotechnology, Lake Placid, NY; cat. no. 05-206).
2. Rabbit anti-type IV collagen antibody (Rockland Immunochemicals, Gilgertville, PA; cat. no. 600-401-106-0.5).
3. Rat antimouse perlecan monoclonal antibody (Clone A7L6, Chemicon, cat. no. MAB1948).
4. Rabbit antimouse type I collagen antibody (Chemicon, cat. no. AB765P).
5. Rabbit polyclonal antibodies specific for laminin 1, E4 (β1-subunit), mouse laminin 1 RG50 (α1 LG 4–5) fractionated from recombinant G-domain were prepared by immunization of New Zealand white rabbits with purified fragment E4 or recombinant baculovirus laminin-α1-G domain, and affinity purification of the antisera on columns of immobilized E4 and E3 fragments, respectively.
6. Rabbit polyclonal nidogen-specific antibody was generated with purified EHS nidogen 1, affinity purified with immobilized nidogen, and cross-absorbed with laminin.

7. Mouse monoclonal immunoglobulin (Ig) M antibody IIH6 hybridoma medium specific for α-dystroglycan *(19)*, a gift of Kevin Campbell (University of Iowa), is used as conditioned hybridoma medium.
8. Mouse monoclonal antibody specific for β-dystroglycan (Novocastra, Newcastle upon Tynecity, UK; cat. no. NCL-b-DG).
9. Rat antimouse integrin β1-chain monoclonal antibody (Clone 9EG7, Pharmingen, San Diego, CA; cat. no. 553715).
10. Hamster antimouse integrin β3-chain monoclonal antibody (Clone 2C9.G2, Pharmingen, cat. no. 550541).
11. Fluorescein isothiocyanate- and Cy5-conjugated antibodies (Jackson Immunochemicals, West Grove, PA) specific for mouse IgG (cat. no. 715-095-151 and cat. no. 715-175-151, respectively), mouse IgM (cat. no. 715-095-140 and cat. no. 715-175-140, respectively), and rabbit IgG (cat. no. 711-095-152 and cat. no. 711-175-152, respectively).
12. Horseradish peroxidase-linked antibodies (Amersham-Pharmacia, Piscataway, NJ) specific for mouse IgG (cat. no. NA931), rat IgG (cat. no. NA935), and rabbit IgG (cat. no. NA934).

3. Methods

3.1. Dissociation of ES and Feeder Cells

1. Add 3 mL 0.25% trypsin-0.53 mM ethylenediaminetetraacetic acid (EDTA) to culture dishes and incubate for 2–5 min at room temperature until cells begin to detach from the dish.
2. Neutralize trypsin by adding 6 mL feeder cell medium (Dulbecco's modified Eagle's medium, 10% FBS, and penicillin/streptomycin) and then transfer the cells to a 15-mL conical tube (*see* **Note 1**).
3. Allow ES cell clumps to pellet by gravity for 10 min. Decant the supernatant that contains unsettled STO feeder cells.
4. Add 2 mL ES cell medium and gently pipet the cells up and down three times in a fine-tip Pasteur pipet to break up cell clumps.
5. Add 0.2 mL cell suspension to a 100-mm dish with confluent STO feeder cells and 10 mL ES medium.
6. Change the medium every day and split ES cells every other day at 1:10 ratios.

3.2. Maintenance of ES Cells

3.2.1. Generation of STO Cells as a Feeder Layer

1. STO cell (thioguanine/oubain-resistant cell line of SIM mouse fibroblasts) cultures are grown in feeder cell medium to 90–95% confluence (*see* **Note 2**).
2. Treat for 2 h with 10 µg/mL mitomycin C to block mitosis.
3. Wash the cells twice with 10 mL PBS and incubate in 3 mL trypsin/EDTA for 3 min at room temperature.
4. Add 6 mL feeder cell medium to dislodge the cells from the dish. Pellet the cells by centrifugation at 100g for 5 min.
5. Resuspend the cells in 10 mL ES cell medium and plate onto a fresh 100-mm tissue culture dish. Allow STO cells to attach for at least 6 h and to form a monolayer before seeding ES cells on the STO cell layer.

3.2.2. Maintenance of R1 ES Cells

Mouse ES cells (R1) are routinely maintained on the STO feeder layer in ES medium (*see* **Note 3**).

Peri-Implantation Development With ES Cells

1. Thaw one vial of frozen ES cells quickly at 37°C. Add the cells immediately to a 15-mL conical tube containing 10 mL feeder cell medium. Centrifuge at 100g for 5 min and decant the supernatant.
2. Resuspend the ES cells in 5 mL prewarmed ES cell medium and add to a 100-mm dish containing STO feeders and 5 mL ES cell medium.
3. Check the growing cells twice a day and change the medium every day.

3.2.3. Maintenance of β1-Integrin-Null ES Cells

β1-Integrin-null ES cells are cultured directly on plastic surface of tissue culture dishes (*see* **Note 4**) as follows:

1. Thaw β1-integrin-null ES cells as described in **Subheading 3.2.2.**
2. Resuspend the cells in 10 mL ES cell medium and directly add to a 100-mm tissue culture dish without feeder cells or gelatin coating.
3. Change the medium every day and pass the cells every other day at 1:5–6 ratios.

3.3. EB Differentiation

3.3.1. Preparation of Undifferentiated ES Clusters

1. ES cultures are usually ready for conversion to EBs after culturing on STO cell layers for 3–4 d (*see* **Note 5**).
2. Incubate ES cells with trypsin/EDTA, neutralize trypsin with two volumes of feeder cell medium, and allow the cells to pellet by gravity (*see* **Subheading 3.1., steps 1–3**).
3. Resuspend the pellet in a fine-tip pipet (*see* **Note 6**) to disperse any differentiated ES cells but maintain the balls of undifferentiated ES cells.
4. Let the ES cell clumps settle by gravity a second time. The supernatant (containing feeder cells) is discarded (*see* **Note 7**). Resuspend the ES cells in 2 mL EB medium and disperse with a fine-tip Pasteur pipet into clusters consisting of 3–7 cells (*see* **Note 8**).
5. Seed cells into tissue culture dishes and culture for 1–3 h to selectively remove the adherent feeder cells through panning (*see* **Note 9** and **Fig. 3**).

3.3.2. Initiation of EB Differentiation in Suspension Culture

1. Transfer ES cell clusters to bacteriological Petri dishes at a density of 10^3 cell clusters/dish, serving as the initial small EBs.
2. Culture under 5% CO_2 at 37°C for 7 d (*see* **Note 10**). Change the medium on d 3 for the first time and every other day thereafter (*see* **Notes 11** and **12**).

3.3.3. Preparation of EBs for Immunofluorescence and Electron Microscopy

Settle EBs in a 15-mL conical tube by gravity and then wash with PBS by resuspension and settling prior to preparation for light or electron microscopy.

3.4. Frozen Sectioning and Immunostaining for Light Microscopy

1. Collect EBs from **Subheading 3.3.2., step 2**, into a 15-mL conical tube and allow to settle by gravity.
2. Wash once in PBS containing 0.5% bovine serum albumin.
3. Fix EBs with 3% paraformaldehyde in PBS for 30 min at room temperature.
4. Incubate in 7.5% sucrose-PBS for 3 h at room temperature.
5. Incubate in 15% sucrose-PBS at 4°C overnight.
6. Embed the EBs in tissue-Tek OCT compound.

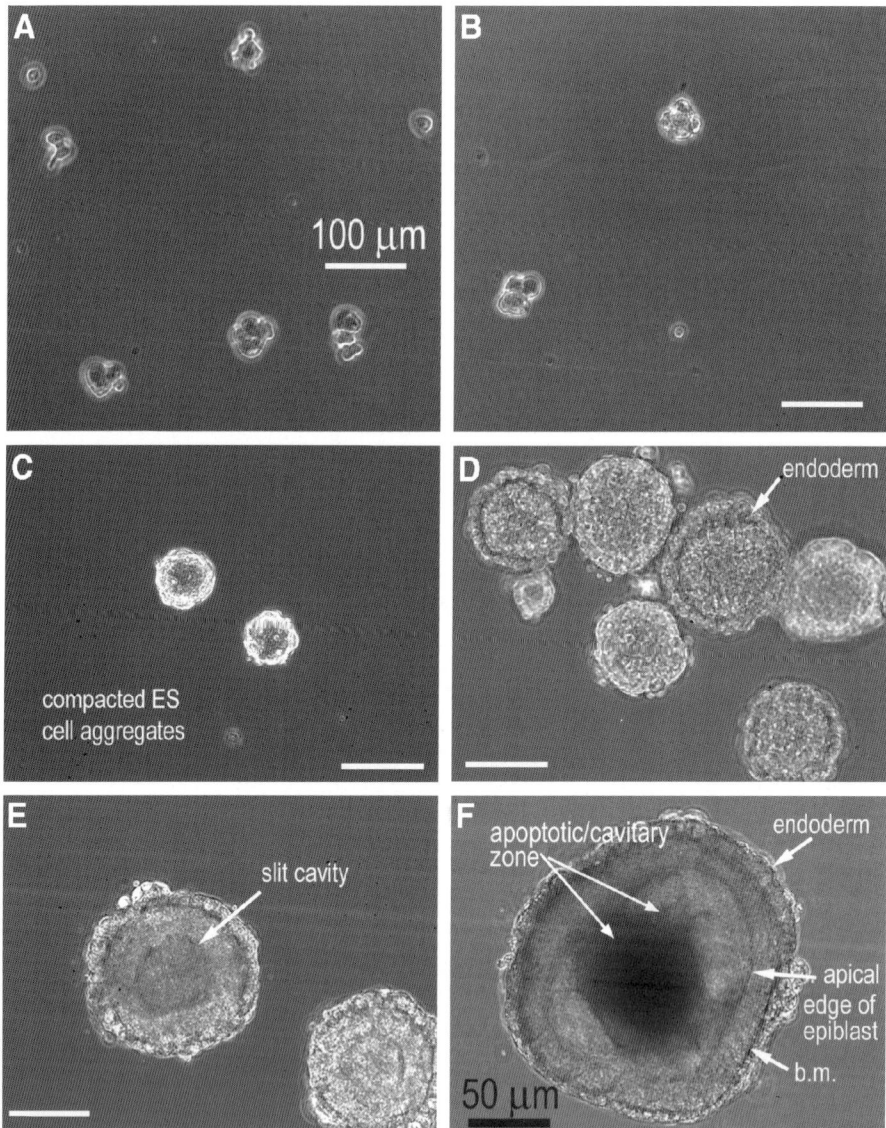

Fig. 3. Embryoid body (EB) culturing. Phase contrast images of (**A**) and (**B**) live embryonic stem (ES) cell aggregates and (**C–F**) EBs. The endoderm, epiblast, and cavity can be distinguished as they develop. Images of developing EBs were recorded at (**C**) d 2, (**D**) d 3, (**E**) d 4, and (**F**) d 7.

7. Cut frozen sections to 4–20 μm in thickness on a cryostat.
8. Block nonspecific binding sites for 20 min with 5% goat or donkey serum.
9. Incubate EB sections with primary antibodies for 1 h.
10. Detect specific binding with fluorescein isothiocyanate-, Cy3-, or Cy5-conjugated secondary antibodies.

11. Counterstain nuclei by incubation of EB sections with 1 µg/mL DAPI in PBS for 10 min. Wash slides three times in PBS and mount with 50 mM Tris-HCl buffer (pH 8.5) containing 6% DABCO and 50% glycerol.
12. View slides using an inverted fluorescent microscope (*see* **Note 13**).

3.5. Electron Microscopy

1. Fix EB pellet from **Subheading 3.3.2.**, **step 2**, in 0.5% gluteraldehyde and 0.2% tannic acid in PBS for 1 h at room temperature.
2. Wash three times with 1 mL 0.1 M sodium cacodylate buffer for 5 min each.
3. Transfer EBs to modified Karnovsky's fixative for 1 h.
4. Postfix in 1% osmium tetroxide for 1 h.
5. Dehydrate and embed in Epon/SPURR resin.
6. Cut thick (1-µm) and thin (approx 90-nm) sections with a diamond knife in an ultramicrotome.
7. Stain thick sections with 1% methylene blue in 1% sodium borate for light microscopy by overlaying the slide with dye for several seconds or longer, followed by washing with water. To stain thin sections, grids are floated (sample facing down) onto a drop of saturated uranyl acetate resting on Parafilm for 20 min followed by a brief water wash (transfer to a drop of water). The grid is then transferred to a drop of 0.2% lead citrate for 2–5 min followed by an additional water wash.

4. Notes

1. Trypsinization and resuspension of the ES/STO culture is a critical step. Trypsinization should be monitored under a microscope and stopped when the caps and balls of ES cells start to dissociate (approx 3–4 min).
2. The treated STO cells may be trypsinized and replated if overgrown. Early-passage STO cells that contain few senescent cells should be used.
3. β1-integrin-null ES cells adhere poorly to mouse embryonic fibroblast or STO feeder layers, and most of the cells grow in suspension to form small balls.
4. β1-integrin-null cells adhere poorly to the feeder cell layer
5. An ideal culture contains colonies of ES cells that appear to grow as "balls" on top of the STO layer. The balls have tightly packed cells and a refractile edge. An overgrown culture has differentiated epithelioid cells spreading at the edges. The undifferentiated cells in the latter type of culture can often still be rescued by selective trypsinization and resuspension.
6. The pipet tip is drawn in a flame from the long variety of Pasteur pipet such that liquid flows as a fine stream on mild pressure from the bulb—if the liquid exits as a series of droplets, the tip is too small. The outer diameter will be 0.2 mm with a drawn length of 4–8 mm.
7. The remaining small cell aggregates are the desired EB precursors. Single ES cells (which make poor EBs) along with some of the feeder cells will remain in the supernatant that is discarded.
8. The ES cells should be resuspended with a fine-drawn pipet and checked (in the 15-mL conical tube) that aggregates of appropriate size have formed. These aggregates produce differentiated EBs. In contrast, we have found that large cell aggregates differentiate in a more disorganized fashion, and single cells grow too slowly in suspension and usually fail to differentiate beyond endoderm formation over the usual time-course (7 d) of suspension culturing.
9. Panning allows the more adherent feeder cells to attach to the plastic, enriching the population of small ES cell aggregates in suspension.

10. It is possible to extend EB cultures for as much as 14 d.
11. During this time, the cell clusters (EBs) increase substantially in number and size. We think that it is important to shake the cell cultures as little as possible as this increases the fraction of EBs that adhere to the plastic (or, in 35-mm dishes, to each other). Agitation tends to remove some of the developing endoderm, leaving behind exposed ICMs that are more likely to adhere. If the density of EBs is high, a suboptimal condition, one will need to change the medium *every day* after d 3.
12. Manipulation of EBs bearing mutations that affect differentiation: EBs that are null (−/−) for the γ1-subunit of laminin or the β1-integrin subunit develop a secretory-type endoderm similar to that seen in wild-type EBs. However, they fail to secrete laminins, fail to undergo formation of slits and central cavity, and fail to develop a polarized epiblast layer *(1,10)*. (1) Culture EBs that are null (−/−) for the γ1-subunit of laminin or the β1-integrin subunit with the addition of purified EHS-type laminin 1 to the culture medium at a final concentration of 25–50 µg/mL, concentrations of laminin below its critical concentration (approx 100 µg/mL) of polymerization *(1)*. The laminin can be added to the medium as early as the initial small aggregate stage or after the endoderm has formed. If added before endoderm formation, basement membrane assembly is delayed as much as a day. If added after endoderm formation, basement membrane assembly is delayed up to several days. This basement membrane forms in the correct location between endoderm and adjacent ICM and is followed by epiblast differentiation and cavitation. If laminin is added at (and presumably also above) its critical concentration of polymerization prior to formation of the endoderm, then an outer ECM surrounds an EB that fails to develop endoderm, yet goes on to convert ICM to polarized epiblast with cavitation *(12)*. (2) β1-Integrin-null ES cells, cultured on uncoated dishes in LIF-containing ES medium, tend to differentiate to epithelioid cells as their colonies become larger. They should be dispersed into single cells before plating.
13. We have acquired digital images with a MicroMax 5-mHz charge-coupled device camera (Princeton Instruments) controlled by IP Lab 3.0 (Scanalytics).

Acknowledgments

We thank Karen McKee, PhD, and David Harrison for their help in assembling the protocols. This work was supported by a grant from the National Institutes of Health (R37-DK36425).

References

1. Li, S., Harrison, D., Carbonetto, S., et al. (2002) Matrix assembly, regulation, and survival functions of laminin and its receptors in embryonic stem cell differentiation. *J. Cell Biol.* **157,** 1279–1290.
2. Pöschl, E., Schlotzer-Schrehardt, U., Brachvogel, B., Saito, K., Ninomiya, Y., and Mayer, U. (2004) Collagen IV is essential for basement membrane stability but dispensable for initiation of its assembly during early development. *Development* **131,** 1619–1628.
3. Li, S., Edgar, D., Fässler, R., Wadsworth, W., and Yurchenco, P. D. (2003) The role of laminin in embryonic cell polarization and tissue organization. *Dev. Cell* **4,** 613–624.
4. Leivo, I., Vaheri, A., Timpl, R., and Wartiovaara, J. (1980) Appearance and distribution of collagens and laminin in the early mouse embryo. *Dev. Biol.* **76,** 100–114.
5. Smyth, N., Vatansever, H. S., Murray, P., et al. (1999) Absence of basement membranes after targeting the LAMC1 gene results in embryonic lethality due to failure of endoderm differentiation. *J. Cell Biol.* **144,** 151–160.

6. Miner, J. H., Li, C., Mudd, J. L., Go, G., and Sutherland, A. E. (2004) Compositional and structural requirements for laminin and basement membranes during mouse embryo implantation and gastrulation. *Development* **131**, 2247–2256.
7. Sakai, T., Li, S., Docheva, D., et al. (2003) Integrin-linked kinase (ILK) is required for polarizing the epiblast, cell adhesion, and controlling actin accumulation. *Genes Dev.* **17**, 926–940.
8. Yurchenco, P. D. and Furthmayr, H. (1984) Self-assembly of basement membrane collagen. *Biochemistry* **23**, 1839–1850.
9. Yurchenco, P. D., Cheng, Y. S., and Colognato, H. (1992) Laminin forms an independent network in basement membranes. *J. Cell. Biol.* **117**, 1119–1133.
10. Murray, P. and Edgar, D. (2000) Regulation of programmed cell death by basement membranes in embryonic development. *J. Cell Biol.* **150**, 1215–1221.
11. Aumailley, M., Pesch, M., Tunggal, L., Gaill, F., and Fässler, R. (2000) Altered synthesis of laminin 1 and absence of basement membrane component deposition in β-1 integrin-deficient embryoid bodies. *J. Cell Sci.* **113**, 259–268.
12. Li, X., Chen, Y., Scheele, S., et al. (2001) Fibroblast growth factor signaling and basement membrane assembly are connected during epithelial morphogenesis of the embryoid body. *J. Cell Biol.* **153**, 811–822.
13. Joza, N., Susin, S. A., Daugas, E., et al. (2001) Essential role of the mitochondrial apoptosis-inducing factor in programmed cell death. *Nature* **410**, 549–554.
14. Choi, K., Kennedy, M., Kazarov, A., et al. (1998) A common precursor for hematopoietic and endothelial cells. *Development* **125**, 725–732.
15. Kennedy, M., Firpo, M., Choi, K., et al. (1997) A common precursor for primitive erythropoiesis and definitive haematopoiesis. *Nature* **386**, 488–493.
16. Yurchenco, P. D. and Cheng, Y. S. (1993) Self-assembly and calcium-binding sites in laminin. A three-arm interaction model. *J. Biol. Chem.* **268**, 17,286–17,299.
17. Colognato, H., Winkelmann, D. A., and Yurchenco, P. D. (1999) Laminin polymerization induces a receptor-cytoskeleton network. *J. Cell Biol.* **145**, 619–631.
18. Cheng, Y. S., Champliaud, M. F., Burgeson, R. E., Marinkovich, M. P., and Yurchenco, P. D. (1997) Self-assembly of laminin isoforms. *J. Biol. Chem.* **272**, 31,525–31,532.
19. Ervasti, J. M. and Campbell, K. P. (1991) Membrane organization of the dystrophin-glycoprotein complex. *Cell* **66**, 1121–1131.

10

Phosphoinositides, Inositol Phosphates, and Phospholipase C in Embryonic Stem Cells

Leo R. Quinlan

Summary

The stimulation of inositol phospholipid metabolism via phospholipase C (PLC) is an important signal transduction pathway in a wide variety of cell types. Activation of the pathway is associated with many aspects of cellular activity, including cell growth and differentiation. Activation of hormone-sensitive PLC results in the rapid breakdown of polyphosphoinositides to generate two second messengers: inositol trisphosphate and diacylglycerol. The water-soluble inositol trisphosphate is involved in the release of intracellular calcium from internal stores, whereas the lipophilic diacylglycerol is involved in protein kinase C activation. Inositol supplementation is essential for the in vitro growth of rabbit blastocysts, and studies have shown that the components of the signaling system are present in mouse and cattle embryos and in mouse embryonic stem (ES) cells. In ES cells, the signaling system appears to be constitutively active and essential for normal ES cell proliferation. Here, we describe in detail the materials required and some of techniques involved in studying the phosphoinositide signaling system in mouse ES cells. Furthermore, we describe methods of analyzing the effects of modulating the PtdIns signaling system on ES cell proliferation and the induction of apoptosis.

Key Words: Apoptosis; autoradiography; cell division; chromatography; Dowex chromatography; drug effects; high-pressure liquid chromatography; inositol phosphates; metabolism; phosphatidylinositols; phospholipase C; protein kinase C; stem cells; thin-layer chromatography.

1. Introduction

Inositol is a cyclohexanol with the empirical formula $C_6H_{12}O_6$. Of the nine possible steroisomers, only myo-inositol occurs naturally. This optically inactive form of inositol (possessing full biological activity) is characterized by a single axial hydroxyl group-attached carbon at position 2 on the backbone. All carbons have a hydroxyl group attached, all of which can be substituted with phosphate groups; thus, there are 63 possible myo-inositol phosphates *(1)*.

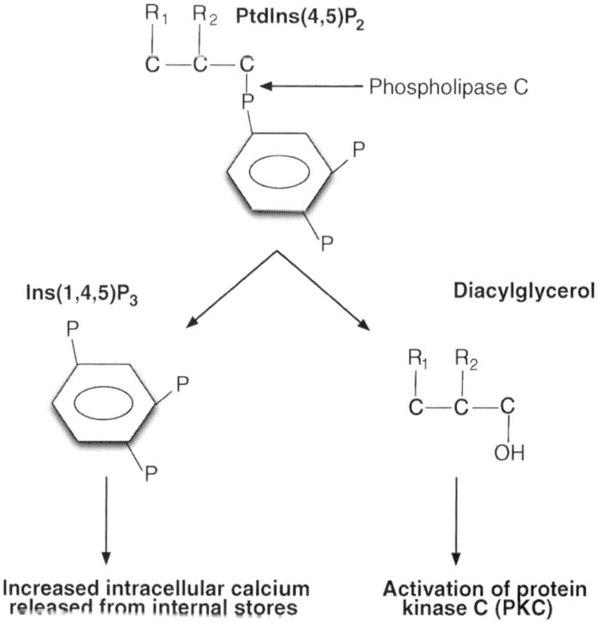

Fig. 1. Overview of the PtdIns(4,5)P$_2$ signaling system. PLC activity generates two second messengers: Ins(1,4,5)P$_3$ and DAG.

The three major inositol-containing lipids, phosphatidylinositol (PtdIns), phosphatidylinositol-phosphate (PtdIns P), and phosphatidylinositol(4,5)bisphosphate [PtdIns(4,5)P$_2$], comprise only approx 2–10% of the total cell membrane phospholipids. Despite this limited occurrence, they form key elements in one of the major transmembrane signaling pathways *(2)*. The PtdIns signal transduction pathway is involved in many and varied aspects of cellular life, from cell growth and differentiation to neuromodulation and sensory perception *(3)*. Inositol is also a component of glycosylphosphatidylinositol (GPI) protein anchors as well as a group of related compounds of similar structure (inositol phosphoglycans), which it is proposed plays a role as second messengers *(4)*.

In developmental terms, inositol itself has been shown to be an essential nutrient for the in vitro growth of rabbit blastocysts *(5)* and stimulates the hatching of hamster blastocysts in vitro *(5,6)*. Studies have also shown that the components of the PtdIns signaling system are present in mouse *(7)*, rabbit *(8)*, and cattle *(9)* embryos. The components of the system have also been demonstrated in mouse embryonic stem (ES) cells *(10)*, and the signaling system appears to be constitutively active and essential for normal ES cell proliferation *(11)*.

The basis of the PtdIns signaling system (**Fig. 1**) is the agonist-activated enzymatic (phospholipase C, PLC) cleavage of a parent phospholipid [PtdIns(4,5)P$_2$], yielding two second messengers: inositol-1,4,5-trisphosphate [Ins(1,4,5)P$_3$] and diacylglycerol (DAG). PLC activity has been associated with a number of cell responses, such as cell

Phosphoinositides, Inositol Phosphates, and Phospholipase C

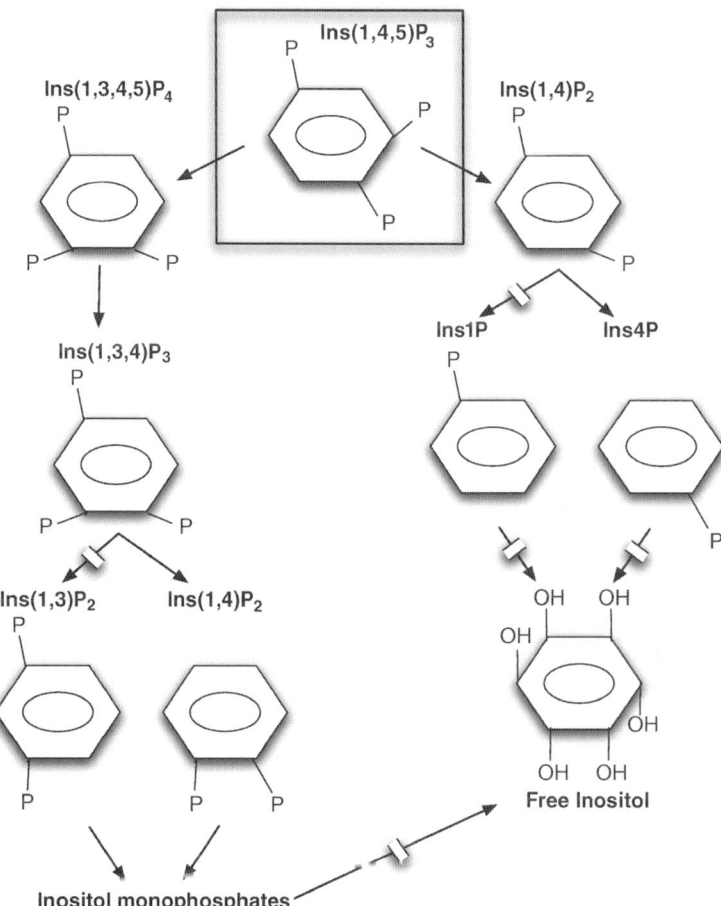

Fig. 2. Inositol recycling. The sketch illustrates the pathways of InsP metabolism and inositol recycling. Reactions inhibited by lithium are indicated by ‖ lines.

proliferation and differentiation *(12)*. Increased PLC activity has also been found in transformed cell types (reviewed in **ref. 12**). One of these messengers [Ins(1,4,5)P$_3$] is key in regulating the release of intracellular calcium from internal stores, while the other (DAG) is involved in the activation of protein kinase C, a central protein in the regulation of many cellular activities *(2,13)*.

This chapter describes the methods associated with the analysis of phosphoinositides and the water-soluble inositol phosphates (InsPs), (**Fig. 2**) as well as the effects of modulating the PtdIns signaling system on cell proliferation and apoptosis. To demonstrate how these measurements can be applied to ES cells, we base our description of the methods around a simple experiment to characterize the effects of the PLC inhibitor U-73122 on InsP production and phosphoinositide turnover. Furthermore, we detail methods involved in assaying the effect of U-73122 and bisindolylmaleimide II (Bis II) on ES cell proliferation, cell viability, and apoptosis.

The most common and convenient method of analyzing InsP production and the classical phosphoinositides is through the use of tritiated labels, by [^{3}H]-inositol incorporation into the phosphoinositides or InsPs of ES cells. The components of the PtdIns signaling system are extracted using a bifurcating extraction pathway, allowing the isolation of both lipid and aqueous InsP components concurrently *(14)*. A modification of this method is described here; the identity and quantity of each of the components can then be determined by different chromatographic methods as detailed in the relevant sections.

Cell proliferation can be assessed in many ways; in this chapter, we describe the use of a simple, fast and reliable assay based on tetrazolium salts. Tetrazolium salts, such as 3-[4,5-dimethylthioizol-2-yl]-2,5, diphenyl-tetrazolium bromide (MTT), 3'-[(phenylamino)-carbonyl]-3,4-tetrazolium-bis (4-methoxy-6-nitro) benzenesulfonic acid hydrate (XTT), and 2-(4-Iodophenyl)-3-(4-nitrophenyl)-5-(2,4, disulfophenyl)-2H-tetrazolium (WST-1), are now in common use as methods of assessing cell proliferation. We confine our discussion to MTT; however, further information on XTT and WST-1 can be found on the Roche Web site (http://www.roche-applied-science.com/index.jsp). The assay principle is based on the fact that living cells can reduce tetrazolium salts to colored formazan compounds. The MTT assay is based on the reduction of the tetrazolium salt by microsomal enzymes that require pyridine nucleotides (i.e., Nicotinamide adenine dinucleotide (NADH) and to a lesser extent nicotinamide adenine dinucleotide phosphate), leading to the formation of an insoluble formazan product *(15,16)*. In the MTT assay, a water-insoluble formazan compound is produced, which must be dissolved for measurement. It is worth noting that using the XTT assay, a soluble dye is produced; thus, a solubilization step is not required. Use of MTT, and to a greater extent XTT, provides a simple procedure for measuring ES cell proliferation/viability. In both cases, dye absorbance is proportional to the number of viable cells present. To provide a check on the use of the MTT assay, the simple procedure of counting cells microscopically using a hemocytometer slide is also described.

For assessment of cell viability, a simple and convenient method using live/dead counterstaining is described. For determination of apoptosis, we use a Roche kit (*In Situ Cell Death Detection Kit, Fluorescein*). The basis of this kit is the labeling and detection of fragmented DNA strands in individual cells by fluorescence microscopy. The free 3′ OH ends of the DNA strand are labeled with fluorescein-dUTP using terminal transferase (TdT).

2. Materials

2.1. Tissue Culture *(see Note 1)*

1. Phosphate-buffered saline (PBS): to prepare a 1X PBS solution, to ultrapure water add 8.0 g sodium chloride (NaCl), 0.2 g potassium chloride (KCl), 2.89 g disodium orthophosphate 12-hydrate (Na$_2$HPO$_4$·12H$_2$O), and 0.2 g potassium dihydrogen phosphate (KH$_2$PO$_4$). Make up to 1 L and filter-sterilize through Startolab-P filter.
2. Startolab-P sterile pressure filtration filter, 0.2-µm pore size (Sartorius, Wicklow, Ireland; cat. no. 18053).
3. Ministart single-use sterile syringe filter (Sartorius, cat. no. 16534) 0.2 µm pore size.

4. Tissue culture plastics (24-well plates and tissue culture flasks, all Nunc brand, supplied by Biosciences, Dublin, Ireland; cat. no. 146485 and cat. no. 147589, respectively).
5. Dulbecco's modified Eagle's medium (DMEM) 1X (Gibco, Paisley, Scotland; cat. no. 52100039). To prepare 1 L of DMEM, add 13.3 g of powered media, 3.7 g of NaHCO$_3$, and 0.11 g of sodium pyruvate (Sigma-Aldrich-Aldrich, Dublin, Ireland; cat. no. P5280) to ultrapure water and make up to 1 L. Adjust pH to 7.4 with 1 N HCl. Filter-sterilize through Startolab-P filter and store at 4°C for no longer than 2 wk. This basic serum-free medium is referred to as DMEM.
6. Penicillin G (Sigma-Aldrich, cat. no. P3032)/streptomycin sulfate (Sigma-Aldrich, cat. no. S9137): to prepare a 100X solution of penicillin/streptomycin (pen/strep), to 50 mL PBS add 0.436 g penicillin G and 0.724 g streptomycin sulfate. Filter-sterilize through a Ministart filter; store in 5-mL aliquots below 0°C for no longer than 1 mo.
7. Inositol-free DMEM (Gibco, cat. no. 11963-022).
8. Fetal bovine serum (FBS) (Gibco, cat. no. 10270-106).
9. Bovine serum albumin (essentially fatty acid free) (Sigma-Aldrich, cat. no. A6003).
10. Mouse leukemia inhibitory factor (LIF; 1×10^6 U, ESGRO™) (Gibco, cat. no. 13275-011): to prepare a working solution of LIF, add 1×10^6 U to 9 mL DMEM, aliquot (1 mL per tube) into 1.8-mL cryovials (Nalge Nunc International, cat. no. 363401) and store below 0°C for no longer than 2 mo.
11. β-Mercaptoethanol (Sigma-Aldrich, cat. no. M7522): to prepare a working solution, add 7 µL to 10 mL PBS in a 15-mL test tube. Filter-sterilize through a Ministart filter and store at 4°C for no longer than 1 mo.
12. Trypsin/ethylenediaminetetraacetic acid (EDTA) solution (Gibco, cat. no. 45300-027).
13. Gelatin (type A, from porcine skin) (Sigma-Aldrich, cat. no. G2500): to prepare a gelatin solution, dissolve 0.1 g gelatin in 100 mL ultrapure water. Sterilize by autoclaving at 10 psi for 15 min and store at 4°C for no longer than 1 mo.
14. L-Glutamine (Sigma-Aldrich, cat. no. G5792): to prepare 100X (200 mM) L-glutamine, add 1.461 g to 50 mL PBS, sterilize through a Ministart filter, and store in 10-mL aliquots below 0°C for no longer than 1 mo.
15. U-73122: the PLC inhibitor U73122 and its inactive analogue U-73433 are prepared as 10 mM stock solutions in dimethyl sulfoxide (DMSO) and stored at −20°C in 200-µL aliquots for no longer than 1 mo (*see* **Note 2**).
16. Centrifuge (Beckman; CS-15R).

2.1.1. Tissue Culture Media

1. ES cells are routinely maintained in DMEM containing a number of supplements. To prepare 100 mL of complete culture medium, add 92 mL DMEM, 5 mL FBS, 1 mL pen/strep (100X), 1 mL β-mercaptoethanol (100X), and 200 µL LIF solution. Store at 4°C for no longer than 1 wk.
2. For studies involving inositol labeling, an inositol-free DMEM (Gibco; cat. no. 11963-022) is used, which greatly increases incorporation of the label. The final medium is serum free but contains 1% bovine serum albumin and is supplemented with 1X of the following (Gibco; all obtained as 100X stocks): L-glutamine (cat. no. 25033-010), L-cysteine (cat. no. 41035-015), L-leucine (cat. no. 11079-019), L-methionine (cat. no. 11088-010), L-arginine (cat. no. 21007-018), and D-glucose (cat. no. 19004-019).

2.2. Labeling ES Cells With [³H]-Inositol and Stimulation of PLC Activity

1. Myo-[2-H³]-inositol (Amersham Biosciences, Buckinghamshire, UK; cat. no. TRK317).
2. Analar grade lithium chloride (Sigma-Aldrich, cat. no. L4408).

3. Phytic acid hydrolysate *(17)* (*see* **Note 3**): to prepare an 11.1-mg/mL solution:
    a. Dissolve 1 g sodium phytate (Sigma-Aldrich, cat. no. P8810) in 5 mL 0.2 M sodium acetate/concentrated acetic acid (pH 4.0) in a glass-stoppered, round-bottom test tube and heat in a boiling water bath for 8 h.
    b. Prepare a 4-mL column of Amberlite IR-120 [H^+]-cation exchange resin (BDH chemicals, Poole, UK; cat. no. 1151310500) by washing the column with 3 bed volumes of 1 M HCl, followed by a subsequent wash with 12 bed volumes of ultrapure water.
    c. Desalt the hydrolysate by passing it through the column (gravity feed) and wash the hydrolysate through the column with 100 mL ultrapure water; collect in a 250-mL glass bottle. Determine the inorganic phosphorous content using one of the numerous methods available (e.g., EnzChek® Phosphate Assay Kit, Molecular Probes, Leiden, The Netherlands; cat. no. E6646).
4. On determination of the inorganic phosphorous content, the entire volume of hydrolysate is freeze-dried and reconstituted with ultrapure water to give a final concentration of 11.1 mg phosphorous/mL solution. This is aliquoted and stored indefinitely below 0°C.

2.3. Extraction of InsPs and Phosphoinositides

1. Conical-bottomed glass tube (15-mL borosilicate glass).
2. Perchloric acid (PCA) (70% solution, analar grade) (BDH Chemicals, cat. no. 101761V).
3. 50:50 v/v freon:tri-*n*-octylamine: freon (Sigma-Aldrich, cat. no. 254991) and tri-*n*-octylamine (Sigma-Aldrich, cat. no. T8631).
4. 5-mL polypropylene plastic test tubes.
5. Glass transfer pipets (3-mL capacity, borosilicate glass).
6. EDTA (Sigma-Aldrich, cat. no. E5134).
7. Chloroform (analar grade).
8. Methanol (analar grade).
9. HCl (analar grade).

2.4. InsP and Phosphoinositide Analysis

2.4.1. InsP Analysis

1. Dowex 1X 8-400 resin (Sigma-Aldrich, cat. no. 44340). The resin is supplied in the chloride form crosslinked with 8% divinylbenzene with an effective pH range of 1.0–14.0 pH units. The resin needs to be converted to the formate form before use (i.e., bound chloride is replaced with formate); this is achieved as follows:
    a. Prepare resin slurry with an equal volume of ultrapure water and stir the slurry using a magnetic stirrer for approx 30 min.
    b. Allow the slurry to settle by gravity and remove the excess water, leaving approx 20% by volume of water in excess.
    c. Mix the slurry again until uniform and aliquot at a volume of 1 mL per polyprep chromatographic column (Bio-Rad Laboratories, Hertfordshire, UK; cat. no. 731-1550).
    d. Conversion to the formate form is completed by washing through the column with 15 mL 5% (w/v) sodium formate (H.COONa, analar grade) in three 5-mL steps.
    e. After the final step, wash the column with 30 mL ultrapure water in three 10-mL steps. In all cases, the columns should be allowed to drip-dry before the next step begins (*see* **Note 4**).
2. Ammonium formate (H.COONH$_4$, analar grade).
3. Disodium tetraborate 10-hydrate (Na$_2$B$_4$O$_7$·10H$_2$O, analar grade).

4. InsP standards: all tritiated InsP standards are purchased from PerkinElmer (Bucks, UK) and should be of the form myo-[2-H^3]. Inositol-4-phosphate [Ins(4)P] (cat. no. NET963); inositol-1,4-phosphate [Ins(1,4)P$_2$] (cat. no. NET912); Ins(1,4,5)P$_3$ (cat. no. NET911); and inositol-1,3,4,5-phosphate [Ins(1,3,4,5)P$_4$] (cat. no. NET941).
5. 20-mL scintillation vials (Sarstedt Ltd., Wexford, Ireland; cat. no. 73.662).
6. ReadySafe Beckman scintillation cocktail (Labplan Ltd., Kildare, Ireland; cat. no. 141349).
7. Liquid scintillation counter (e.g., Wallac, model 1409).
8. MonoQ 5/50 GL strong anion exchange column (Amersham Biosciences, cat. no. 17-5166-01): two eluents are required for efficient separation. Eluent A is composed of 0.1 mM ZnSO$_4$·7H$_2$O (Sigma-Aldrich, cat. no. Z0501), 0.1 mM EDTA (Sigma-Aldrich, cat. no. E9884), and 10 mM HEPES (Sigma-Aldrich, cat. no. H7006). Eluent B is composed of 0.5 M Na$_2$SO$_4$ (Sigma-Aldrich, cat. no. S6264), 0.1 mM ZnSO$_4$·7H$_2$O, 0.1 mM EDTA, and 10 mM HEPES. To prepare 1 L of eluent A, add 28.75 mg ZnSO$_4$·7H$_2$O, 29.22 mg EDTA, and 2.6 g HEPES to ultrapure water. To prepare 1 L of eluent B, add 71 g Na$_2$SO$_4$, 28.75 mg ZnSO$_4$·7H$_2$O, 29.22 mg EDTA, and 2.6 g HEPES. The pH of both eluents should be adjusted to 7.4 using 1 N NaOH.
9. MonoQ internal standards (required to help in peak identification): 10 nmol of each (purchased from Sigma-Aldrich) adenosine and guanosine monophosphate (cat. no. A1752 and cat. no. G5662, respectively), diphosphate (cat. no. A5285 and cat. no. G7127, respectively), triphosphate (cat. no. A7699 and cat. no. G9002, respectively); and guanosine tetraphosphate (cat. no. G8378).

2.4.2. Phosphoinositide Analysis

1. Pressurized 100% N$_2$ gas.
2. Precoated, silica gel 60 F-254 thin-layer chromatography (TLC) plates (Merck Biosciences Ltd., Nottingham, UK; cat. no. 1.05729.0001).
3. 1% (w/v) solution of potassium oxalate (Sigma-Aldrich, cat. no. O0501).
4. Shandon tank (8.75 × 3 × 9 in.) (*see* **Note 5**).
5. Filter paper (Bio-Rad filter paper, cat. no. 165-0962).
6. Phosphoinositide standards (lyophilized powder, Sigma-Aldrich): PtdIns (L-α-phosphatidylinositol) (cat. no. P8443); PtdIns(4)P (L-α-phosphatidylinositol 4-monophosphate) (cat. no. P9638); PtdIns(4,5)P$_2$ (L-α-phosphatidylinositol 4,5-bisphosphate) (cat. no. P9763). The use of standards serves two roles: the identification of lipids based on comparing their mobilities to that of the standards and the presence of the standard in the lane aids the running (separation) of the unknown lipids. To prepare standards (e.g., 1 mg lipid supplied in a dark glass bottle), reconstitute in 1 mL of chloroform, dry the lipid onto the glass under a stream of nitrogen, and store below 0°C in the dark.
7. Hamilton syringe (50-μL capacity).
8. Hyperfilm-^{3}H X-ray film (Amersham Biosciences UK Ltd., cat. no. 301 PRN 535).

2.5. ES Cell Proliferation, Viability, and Apoptosis

2.5.1. Hemocytometer Cell Counting

1. Improved Neubauer hemocytometer slides (AGB Scientific Ltd., Dublin, Ireland; cat. no. MNK-540-010X).
2. Special cover slips (20 × 26 mm) (AGB Scientific Ltd., cat. no. MNK-550-230Q).

2.5.2. MTT Cell Proliferation Assay

1. MTT salt (Sigma-Aldrich, cat. no. M2128): prepare a stock solution of MTT (5 mg/mL) fresh each time just prior to use. Add 50 mg MTT salt to 10 mL PBS and warm at 37°C

until the salt is fully dissolved. Filter-sterilize through a Ministart filter and return to 37°C for immediate use; otherwise, store for no more than 2 d at 4°C in the dark.
2. 2-propanol.
3. 1X liquid phenol red-free DMEM (Sigma-Aldrich, cat. no. D1145).

2.5.3. Cell Viability Staining

1. Ethidium bromide (Sigma-Aldrich, cat. no. E8751).
2. Acridine orange (Sigma-Aldrich, cat. no. A6014).
3. Prepare a 100X stock solution of ethidium bromide/acridine orange as follows:
   a. Dissolve 50 mg ethidium bromide and 15 mg acridine orange in 1 mL absolute ethanol.
   b. Dilute 1:50 with ultrapure water.
   c. Store the solution in 1-mL aliquots below 0°C in the dark.
   d. To prepare a 1X working solution, dilute the stock solution 1:100 with PBS, mix well, and store in the dark at 4°C for up to 1 mo.

2.5.4. TUNEL Staining for Apoptosis

1. TUNEL In Vitro Cell Death Detection Kit (Roche Diagnostics Ltd., East Sussex, UK; cat. no. 1684795).
2. Vectashield kit (Vector Laboratories Ltd., Peterborough, UK; cat. no. H-1300).
3. Paraformaldehyde (analar grade).
4. Bis II (Merck Biosciences Ltd., cat. no. 203292).
5. Clear nail polish.

3. Methods

3.1. Tissue Culture

All cell culture procedures should be carried out in a laminar flow hood, taking standard sterile precautions, and cells should be grown in an incubator at 37°C in an atmosphere of 5% CO_2 in air. Exponentially growing cells at approx 60–75% confluency are detached using trypsin/EDTA solution. The enzyme neutralized by the addition of serum (serum contains trypsin inhibitors). The cell suspension is centrifuged, and the pellet is resuspended in 1 mL complete DMEM. The cells are now ready to begin the experimental procedure. The labeling protocol is detailed next, taking as an example a standard experiment examining the effect of the PLC inhibitor U-73122 on FBS-stimulated InsP formation. In this experiment, levels of U-73122 for 0–10 μM are employed, and 10% FBS is used to stimulate InsP turnover. Typical results for such an experiment are shown in **Figs. 3** and **4**.

3.2. Labeling ES Cells With [³H]-Inositol and Stimulation of PLC Activity in the Presence of Inhibitor

1. Count cells by whatever method available; the simplest method is hemocytometer counting (see **Subheading 3.4.1.**, steps 1–9). Seed cells at a density of 20×10^4 cells/mL in 24-well plates, 1 mL per well, in complete DMEM. Allow the cells to attach for 24 h.
2. At the end of the 24-h attachment period, aspirate the medium and replace with inositol-free DMEM containing 2.5 μCi/mL [³H]-inositol. The labeling period continues for 24 h (see **Note 6**).
3. At 30 min from the end of the labeling period, U-73122 is added over a range of concentrations, such as 0, 0.5, 1, 5, and 10 μM; lithium chloride is added 25 min from the end of

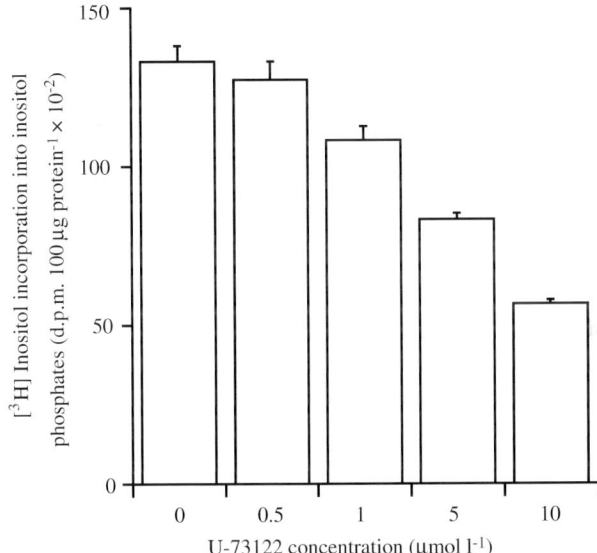

Fig. 3. Effect of the phospholipase C inhibitor U-73122 on FBS-stimulated InsP turnover. Increasing levels of U-73122 decrease the production of total InsPs.

the incubation period to a final concentration of 10 mM (*see* **Note 7**), and 15 min from the end of the labeling period, FBS is added to a final concentration of 10%.
4. At the end of the 24-h labeling period, the reaction is stopped by removing the medium and adding 0.4 mL 10% PCA, 0.4 mL PBS, and 3 µL phytic acid hydrolysate.
5. Place the entire plate in the refrigerator and incubate for 15 min.

3.2.1. Extraction of InsPs and Phosphoinositides

1. After **Subheading 3.2.**, **step 5**, remove the cells by scraping and transfer to a conical-bottom glass test tube and centrifuge at 800g for 5 min.
2. Aspirate the resulting supernatant as completely as possible and place in a 5-mL plastic tube; this supernatant contains the InsPs, which can be extracted later.
3. Wash the remaining pellet with 2 mL ice-cold 2% PCA (a 50-µL aliquot can be taken at this point for future protein determination) and store at –20°C. This solution is the starting point for extraction of the phospholipids.

3.2.1.1. Extraction of InsPs

1. Transfer the supernatant from **Subheading 3.2.1.**, **step 1** above to a 5-mL plastic tube and add 1 mL 50:50 (v/v) freon:tri-*n*-octylamine and 0.8 mL ultrapure water (*see* **Note 8**).
2. Vortex the solution once for 40 s and then leave on ice until the interphase (middle phase) hardens (**Fig. 5**); this can take approx 10–12 min. The molecules of interest are in the uppermost aqueous phase.
3. Centrifuge at 800g for 5 min (*see* **Note 9**) and return to ice for 8–10 min; after this time, the middle phase should be hardened sufficiently to allow the rapid and efficient removal of the upper phase.
4. Transfer the upper phase using a glass transfer pipet to a glass borosilicate test tube and bring the volume up to 3 mL using ultrapure water; cover top with Parafilm and store below 0°C.

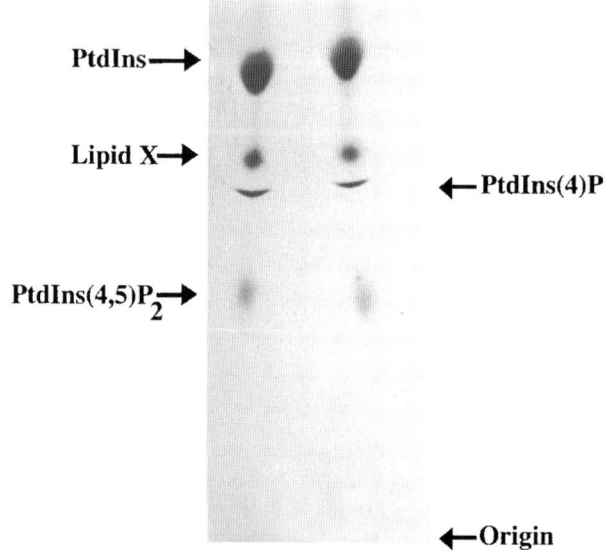

Fig. 4. Representative TLC separation. Autoradiogram showing the TLC separation of phospholipids extracted from mouse ES cells labeled with [^{3}H]-inositol for 24 h. Lipid identification of PtdIns, PtdIns(4)P, and PtdIns(4,5)P2, is on the basis of co-migration with unlabeled lipid standards visualized by iodine staining. Lipid X is an inositol-containing glycophospholipid *(4)*. Using the autoradiogram as the template, stained regions are marked on the silica plate, and the silica is scraped off so that the radioactivity may quantified.

3.2.1.2. EXTRACTION OF PHOSPHOINOSITIDES

1. Wash the pellet resulting from **Subheading 3.2.1.**, **step 1** above in 2 mL ice-cold 2% PCA. Centrifuge at 800*g* for 5 min and discard supernatant.
2. Resuspend the pellet in 3.75 mL chloroform/methanol/HCl (500:1000:6), leave on ice for 10 min, then centrifuge at 800*g* for 5 min.
3. Transfer the resulting supernatant to a conical-bottom glass test tube (15-mL capacity) and add the following: 1 mL ultrapure water, 0.25 mL EDTA (100 m*M*), 1.25 mL NaCl (0.9%), and 1.25 mL chloroform. Vortex this mixture for 30 s (two clear phases should be clearly visible), then centrifuge at 800*g* for 5 min.
4. Discard the upper phase and add 1.25 mL of chloroform/methanol/1 *N* HCl (3:48:47). Vortex the mixture for 30 s and centrifuge at 800*g* for 5 min to wash the lower phase.
5. Discard the upper phase as completely as possible and dry the lipid-containing lower phase onto the glass tube under a stream of N_2 (this can take up to 45 min). Once fully dried, cover the tube with Parafilm, trying to retain some N_2 in the tube, and store below 0°C (*see* **Note 10**).

3.3. InsP and Phosphoinositide Analysis

3.3.1. InsP Analysis

The extracted InsPs can be separated by a number of methods; the most effective and simple method is anion exchange chromatography. The two most common are Dowex resin and high-performance liquid chromatography (HPLC); the former is

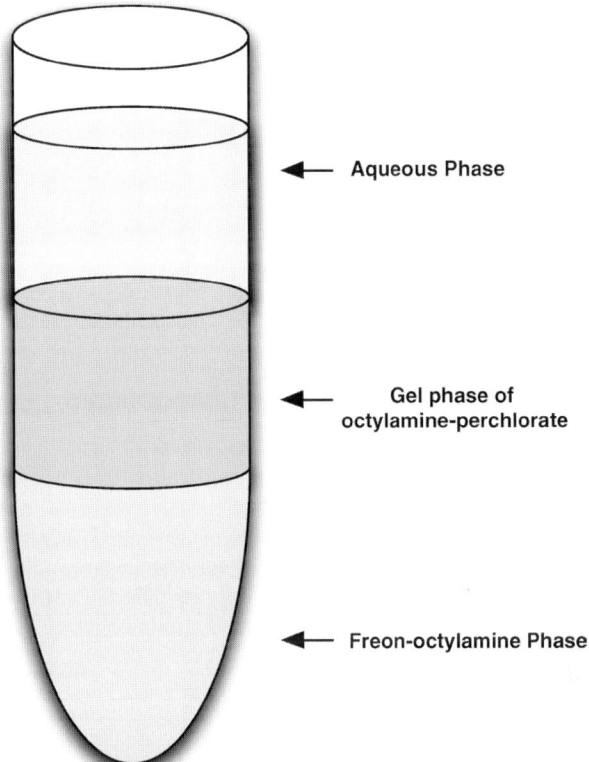

Fig. 5. InsP extraction. The sketch illustrates the appearance of phase separation during InsP extraction. The water-soluble InsPs are found in the upper phase.

technically simpler, very cost-effective, and higher throughput is allowed. Both methods are detailed. Regardless of the method employed, columns need to be standardized using HPLC-grade tritiated InsPs Ins(4)P_1, Ins(4,5)P_2, Ins(3,4,5)P_3, and Ins(1,3,4,5)P_4; these radiolabeled standards should be run through the columns singly and as a cocktail; the output from the columns is collected, and the radioactivity is determined. Representative traces for both methods are shown in **Figs. 6** and **7**, respectively.

3.3.1.1. INSP SEPARATION USING DOWEX RESIN

1. Dowex resin and columns are prepared as described in **Subheading 2.4.1.**, **steps 1–5**.
2. The sample is prepared by adding 1 mL sample to 9 mL loading buffer (i.e., 5 m*M* disodium tetraborate/0.5 m*M* EDTA), and the entire 10 mL are applied directly to the column; the column is allowed to drip-dry (*see* **Note 11**).
3. Buffers of increasing formate concentration (10 mL total of each) are then passed through the column (*see* **Table 1**). Buffers 1 and 2 are passed through the column in numerical order in two 5-mL fractions for each, allowing the column to drip-dry between additions. Buffers 3–6 are passed through the column in five 2-mL fractions and collected into scintillation vials.

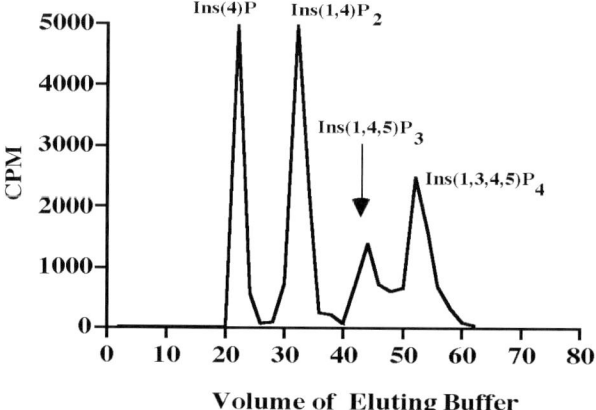

Fig. 6. Dowex column standardization. Standardization is performed once in one column. The standardization indicates the relative position of eluting InsPs as the salt concentration is increased. The tracing shows the separation achieved on a Dowex column when loaded with radiolabeled InsP standards. All standards were loaded concurrently on one column and individually on separate columns, 20 nCi of each standard was added to 10 mL loading buffer and applied directly to the column. The standards were eluted and radioactivity determined as described in **Subheading 3.3.1.1., steps 1–4**. The position of the various InsP isomers is indicated on the trace.

4. To each vial, add 8 mL scintillation cocktail, vortex for 30 s, and incubate overnight. Vortex again for 30 s and count the radioactivity on a liquid scintillation counter (*see* **Note 12**).

3.3.1.2. INSP SEPARATION USING HPLC

Efficient and reproducible separation of InsPs can be carried out on strong anion exchange HPLC columns using a HPLC workstation, which can use preprogrammed methods to control eluent flow rate and gradients. The procedure here is based on that of Meek (*18*). Even with a good setup and employing best practice, variations occur during separations; thus, to aid peak identification nonradioactive internal standards should be applied to the column along with the test sample. These internal standards (10 nmol of each) are adenosine mono-, di-, and triphosphate and their guanosine equivalents and guanosine tetraphosphate. These standards can be picked up by monitoring the absorbance of the eluent at 254 nm, thus, giving an indication of the relative position of different species as they are washed from the column. HPLC separation allows a cleaner (i.e., better resolution) separation of InsPs. It also allows the experimenter to detect InsPs above $InsP_4$, something that the Dowex-based system does not. The method described here is based on a MonoQ column, but other columns are also in common use (e.g., a Partisil SAX column, normally equilibrated with ammonium formate or ammonium phosphate) (*19*).

1. Using a 1-mL syringe, apply 0.1 mL of the sample of interest (to which has been added the internal standards) to the column.
2. Most modern HPLC systems are computer controlled, so from within the software set a flow rate of 1 mL/min (i.e., 1 mL of buffer is passing through the column per minute).

Phosphoinositides, Inositol Phosphates, and Phospholipase C

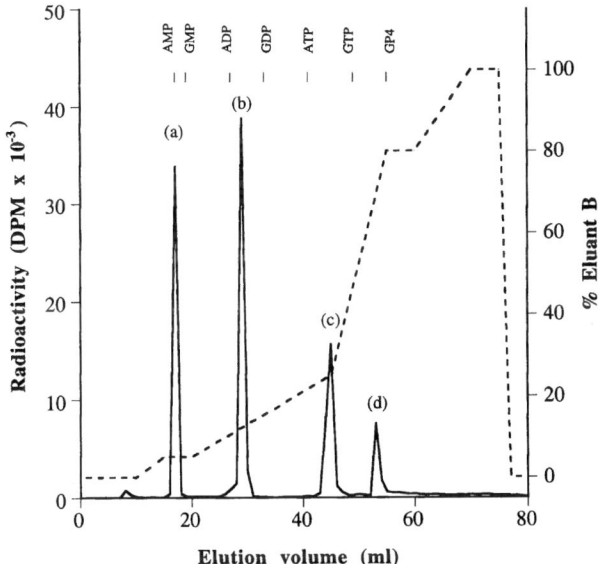

Fig. 7. MonoQ column standardization. The tracing shows the separation achieved on a MonoQ column when loaded with radiolabeled InsP standards. All standards were loaded concurrently, with 2 µCi/mL of each standard applied in a 1-mL volume directly to the column. The standards were eluted and radioactivity was determined as described in **Subheading 3.3.1.2., steps 1–5**. Shown are the elution profiles of water-based sample containing 2 µCi/mL tritiated standards (**A**) Ins4P, (**B**) Ins(1,4)P_2, (**C**) Ins(1,4,5)P_3, and (**D**) Ins(1,3,4,5)P_4. Also shown are the positions (indicated by |) of the added internal nucleotide standards (10 nmol).

Table 1
Dowex Column Eluent Buffers

Buffer	Constituent	Compound eluted
1	H_2O	Free inositol
2	5 mM disodium tetraborate/0.06 M ammonium formate	Glycerophosphoinositol
3	0.2 M ammonium formate	Ins(4)P
4	0.4 M ammonium formate	Ins(1,4)P_2
5	0.8 M ammonium formate	Ins(1,4,5)P_3
6	1.0 M ammonium formate	Ins(1,3,4,5)P_4
7	2.0 M ammonium formate	

Buffers 1 and 2 should be prepared in ultrapure water. Buffers 3–7 should be made in 0.1 M formic acid.

3. After loading the sample onto the column, wash column through with 40 mL eluent A. This initial wash is to remove any free [^{3}H]-inositol, thus, the eluent can be allowed to run to waste (*see* **Note 13**).
4. A salt elution gradient as per **Table 2** should then be applied to the column, and the eluent should be collected in 1-mL fractions into 5-mL scintillation tubes for later assessment of radioactive content.

Table 2
HPLC Salt Gradient

Elution volume (mL)	Eluent B (%)
0.0	0.0
10.0	0.0
15.0	5.0
20.0	5.0
45.0	25.0
55.0	80.0
60.0	80.0
70.0	100
75.0	100
77.0	0.0

The salt elution gradient refers to the mixing of eluents A and B; that is, 0% eluent B equals 100% eluent A.

5. At the end of the separation, add 4 mL scintillation cocktail to each tube, vortex for 30 s, and count in a liquid scintillation counter.

3.3.2. Phosphoinositide Analysis

3.3.2.1. Phosphoinositide Separation (see Note 14)

1. Extracted phospholipids are separated by TLC using precoated silica gel 60 F-254 plates; however, these plates need to be activated before use: spray the entire plate evenly from about a 10-cm distance with 1% (w/v) potassium oxalate and "activate" by heating at 110°C for 1 h.
2. While the plate is baking, the chromatography tank (see **Note 15**) can be prepared and allowed to equilibrate as follows:
   a. Ensure the tank is clean by washing with detergent (soapy water is sufficient) followed by methanol. It is important that the tank be fully clean; the final methanol wash ensures final cleaning and rapid drying of the solvent.
   b. Line the tank with thick blotting/filter paper. Using the recommended filter paper, lining the tank will require one sheet cut in two pieces; ensure at least a 2-cm overlap at joints and cut any filter paper off the edge of the tank.
   c. Prepare 250 mL of the solvent methanol/chloroform/ultrapure water/concentrated ammonia solution (100:70:25:15) (see **Note 16**) and pour the solution into the tank, filling the bottom of the tank to a height of approx 2 cm; the solvent will then slowly rise up the filter paper, reaching the top. Having reached the top of the filter paper, there should be approx 1 cm solvent standing at the bottom of the tank (see **Note 17**).
3. Place the tank in a fume hood and air seal the solvent into the tank by placing a clean glass plate on top, sealing the edges with silicone grease or simply vaseline. The solvent will rise up the filter paper, and the tank should be equilibrated in approx 2 h (i.e., 20 min after the solvent reaches the top of the filter paper) (see **Note 18**).
4. When the plate is activated, allow it to cool; then, using a soft pencil, draw a horizontal line 1.5 cm from the edge of the plate at both ends. Choose one end as the start line and mark with the pencil points where lanes will begin. It is normal to run five lanes per plate, one

Phosphoinositides, Inositol Phosphates, and Phospholipase C 141

  lane for standards and four lanes for the samples of interest. Ensure the lanes are at least 1.5 cm apart on the horizontal axis to ensure that the lanes do not cross as they run.
5. To apply the samples, they first need to be redissolved. Redissolve the dried lipids from **Subheading 3.2.1.2.**, **step 5**, in 20 µL chloroform. Apply the sample to the chosen point using a glass 50-µL Hamilton syringe in 0.5-µL steps, allowing the spot to air-dry between steps; ideally, the lipid spot should not exceed 4 mm in diameter.
6. To each sample lane, add 5 µg cold (i.e., unlabeled) standard for PtdIns, PtdIns4P and PtdIns(4,5)P in the same way as the sample. The addition of the unlabeled standards aids the running of the lanes. A standards-only lane is also run and should contain 10 µg of each standard.
7. When loading is complete, place the plate into the equilibrated tank, replace the lid on the tank, seal, and allow the solvent to run up to the top line (this can take up to 2.5 h), then remove the plate from the tank. Place the plate in the fume hood and allow it to air-dry for 1.5 h. Wrap the plate in aluminum foil and store until you are ready to identify the lipid spots.

3.3.2.2. Phosphoinositide Identification and Quantification

1. The location of lipids is based on co-migration with unlabeled standards. The relative positions on the TLC plate can be identified by one of two methods or both combined. The methods are iodine vapor staining and autoradiography (*see* **Note 19**).
2. The simplest method to visualize the lipid spots is to place the plate into a sealed tank in which iodine crystals have been allowed to sublimate; the vapor will stain the lipid spots a yellow color. Once the spots are clearly visible, remove the plate and then ring the spot with a soft pencil because the staining will fade quickly.
3. The spots can be visualized by autoradiography. This is performed under dark room conditions by placing the plate face down on the emulsion layer of a high-performance autoradiography film in an autoradiography cassette. The cassette is left in a light-sealed container (a simple method is to double wrap in aluminum foil and place inside a heavy duty black refuse sack) for 1–4 wk at room temperature.
4. The film is developed in the dark room by immersing in Kodak LX 24 developer for 4–5 min, then fixing in Kodak FX-40 for 1–2 min, followed by washing in ultrapure water. Alternatively, an automatic developer may be used. This film is then used as a template to mark points of interest on the TLC plate. To quantitate the phosphoinositides, the marked regions based on the template are scraped, and the radioactivity is counted.
5. If you are only interested in specific regions, then before scraping the region of interest on the plates is initially marked using a soft pencil. Then, using a scalpel blade scrape the silica from the glass and place the fragments into a scintillation vial.
6. Add 1 mL methanol to each vial, vortex for 30 s, and incubate overnight at room temperature. Add 8 mL of scintillation cocktail, vortex for 30 s, and count in a scintillation counter.

3.4. ES Cell Proliferation, Viability, and Apoptosis

Again, here we take the example of our simple experiment looking at the effect of the PLC inhibitor U-73122 on ES cell number measured by cell counting as well as by an MTT assay. Also, for the apoptosis example we use an agent we have shown to cause apoptosis in ES cells: the protein kinase C inhibitor Bis II.

3.4.1. Hemocytometer Cell Counting (see **Note 20**)

1. Harvest cells using trypsin/EDTA as described (*see* **Subheading 3.1.**) and pellet at 800*g* for 5 min.
2. Resuspend pelleted cells gently in an appropriate volume of medium to ensure a unicellular

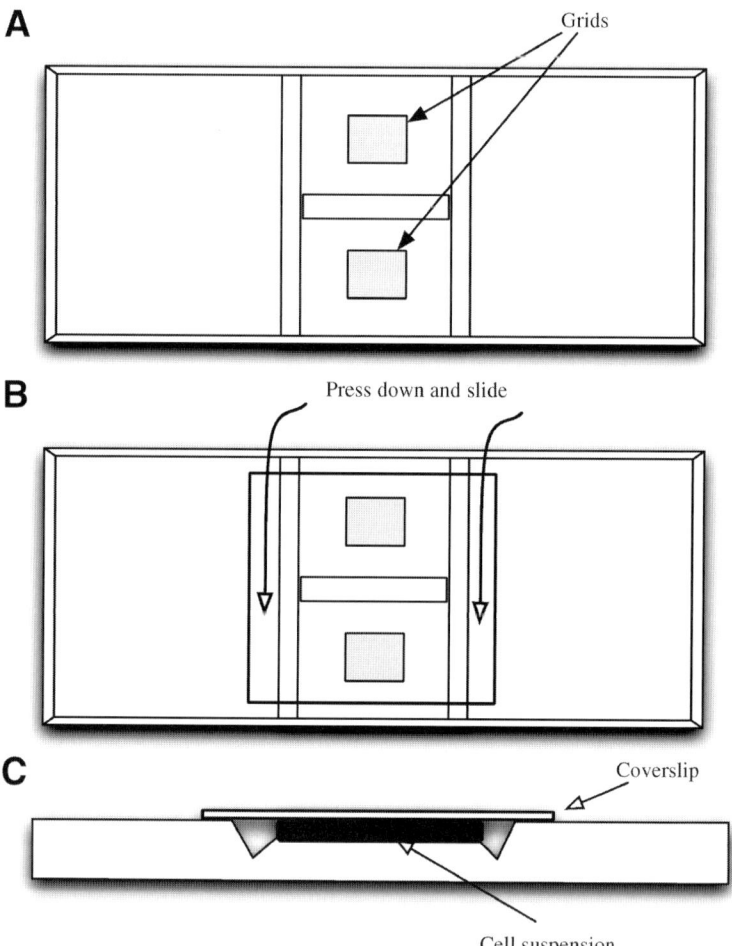

Fig. 8. Views of an improved Neubauer hemocytometer slide. (**A**) Top view of the slide. Indicated by arrows is the location of the grids used as the counting area. (**B**) Application of glass cover slip to slide. Gentle pressure is required to fix the cover slip to the glass slide; too much pressure will break the cover slip. (**C**) Side view of slide with cell suspension applied. Notice that there are no cells or liquid in the troughs on either side.

   suspension is achieved; the actual volume is not important, but from experience when harvesting cells from a 25-mm^3 flask at 70% confluency, an appropriate volume is 1 mL.
3. Clean a glass cover slip (special glass cover slips are available; *see* **Subheading 2.5.1., step 2**) and the hemocytometer slide with 70% ethanol and allow to air-dry.
4. Moisten the edges of the cover slip and place over the counting chamber using gentle downward pressure to fix the cover slip to the slide (**Fig. 8B**).
5. Dilute a 50-µL aliquot of dissociated cells 1:2 with PBS; using a micropipeter, transfer the cell suspension to the chamber by placing the pipet tip at the edge of the chamber and allow it to fill completely via capillary action (*see* **Note 21**).

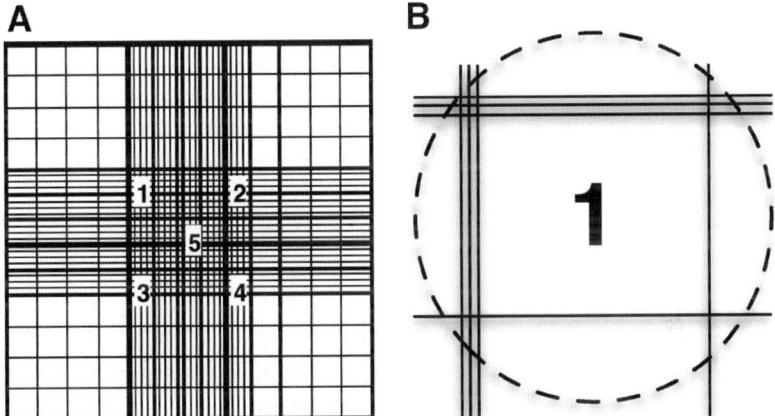

Fig. 9. View of improved Neubauer hemocytometer slide down microscope. (**A**) View at approx ×10. Indicated are count areas labeled 1 to 5, which are designated for counting. (**B**) On the right is a high-power view (×40) of box 1, which contains 16 small squares.

6. Repeat **step 5** using another aliquot sample for the second chamber.
7. Place the hemocytometer slide on the microscope stage and, using a ×10 objective, focus on the counting chamber grid lines (**Fig. 9**) (*see* **Note 22**).
8. Count the total number of cells present in regions marked 1–5 in **Fig. 9**; in doing this, you will count 80 (5 × 16) of the small squares. To avoid double counting, include cells that are on the top and left borders of each small square and exclude those on the left and bottom borders.
9. Repeat the count for the second chamber. If no second chamber exists, then the slide should be cleaned and the process repeated (*see* **Note 23** for sample calculation)

3.4.2. MTT Cell Proliferation Assay

1. Immediately prior to use, dilute the MTT stock solution to a working concentration of 1 mg/mL with culture medium (*see* **Note 24**).
2. At the end of the cell culture treatment period, remove the medium and replace it with the prewarmed MTT working solution prepared in **step 1**. Incubate for 3 h at 37°C (this time varies depending on cell density and cell type).
3. Remove the MTT solution. The cells at this stage should be stained with the blue formazan product. Incubate in 2-propanol using a plate shaker for approx 3 min to ensure that the product goes fully into solution. Read absorbance at 550 nm. (Results presented here show the effect of U-73122 on ES cell proliferation. Also included is the effect of U-73433, an inert analogue of U-73122. **Figs. 10** and **11**).

3.4.3. Cell Viability Staining

1. Cells can be grown in two ways:
    a. Seed cells at 20×10^4 cells/mL onto gelatinized 20 × 20-mm glass cover slips. Place the cover slips individually into large Petri dishes and place in the incubator for 24 h at 37°C in 5% CO_2. After the 24-h attachment period, remove the medium and add the desired treatment medium; for our example experiment, this is complete DMEM with

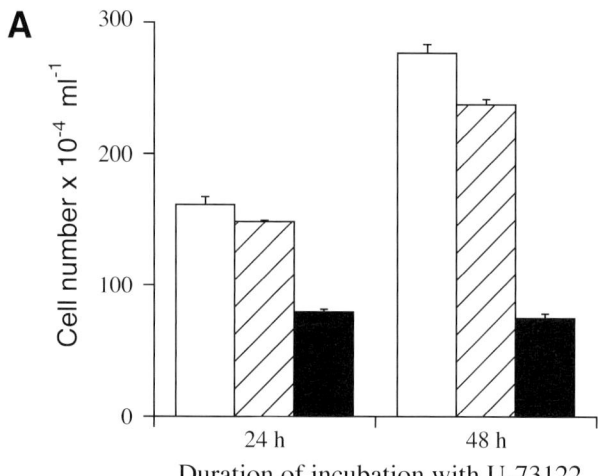
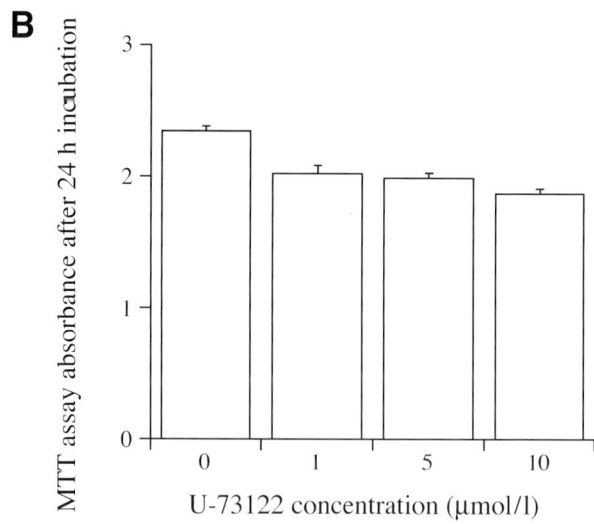

Fig. 10. The effects of the PLC inhibitor U-73122 on ES cell numbers in culture. **(A)** Effect of U-73122 (0, □; 1, ▨; and 10 μmol L^{-1}, ■) for 24 and 48 h on ES cell numbers as determined by hemocytometric counting. **(B)** Effect of U-73122 on ES cells over 24 h as measured by the MTT cell proliferation assay. Absorbance values at 560 λ–690 λ are means ± SEM for eight replicates.

      the PLC inhibitor added to the required concentration. Then, return the dish to the incubator for a further 24 h.
- b. Alternatively, you can seed cells as above onto gelatinized, round, 10-mm diameter glass cover slips. Place the cover slips individually into individual wells of a 24-well plate and treat as previously described.
2. At the end of the treatment period, remove the medium and replace with sufficient trypsin/EDTA solution to cover the cells; return to incubator for 2–3 min. Once the cells are

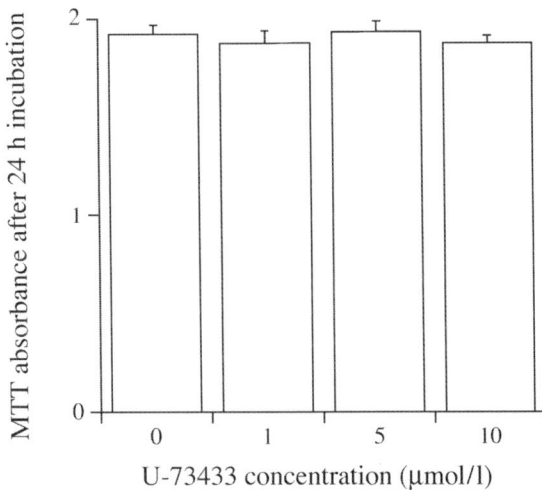

Fig. 11. Effect of U-73433 for 24 h on ES cells as measured by the MTT proliferation assay. Absorbance values at 560λ–690λ are means ± SEM for eight replicates.

    detached, add double the volume of complete medium, aspirate the cells, and centrifuge at 500g for 4 min in a 3-mL Eppendorf and resuspend the pellet in 1 mL PBS.
3. Take 25 µL cell suspension and add an equal volume of ethidium bromide/acridine orange working solution (*see* **Subheading 2.5.3.**, **item 3**, **part d**) and mix gently.
4. Place a small volume (just enough to fill the chamber) underneath the cover slip on a hemocytometer slide.
5. Observe cells under the microscope first with visible light, then switch to fluorescent light (*see* **Note 25**).
6. Count cells and calculate viability. Live cells will fluoresce green (with acridine orange), and dead cells will fluoresce orange (with ethidium bromide).

3.4.4. TUNEL Staining for Apoptosis

1. Set up the cells as per the second method described in **Subheading 3.4.3.**, **step 1**, **part b**. However, in this case we also look at the effect of the protein kinase C inhibitor Bis II.
2. At the end of the treatment period, remove the cover slip, place it into a fresh Petri dish, and fix the cells with 2% (w/v) paraformaldehyde. The cover slips can at this point be stored under PBS for analysis later; store at 4°C for no longer than 2 d.
3. Stain the cells using one of the commercially available TUNEL labeling kits as per manufacturer's instructions (*see* **Note 26**).
4. TUNEL-stained cells can then be counterstained with propidium iodide (PI) using Vectashield® mounting medium with PI. In this way, the Vectashield serves two roles: to stain nuclei with PI and to help preserve the fluorescence.
   a. Apply one drop (dropper supplied) of Vectashield to the slide (more is not necessarily better).
   b. Place a glass cover slip over the drop and press lightly on the cover slip, allowing the drop to spread beneath the cover slip.
   c. Seal the edges of the cover slip using clear nail polish. Store the slide wrapped in foil at 4°C for up to 1 mo.

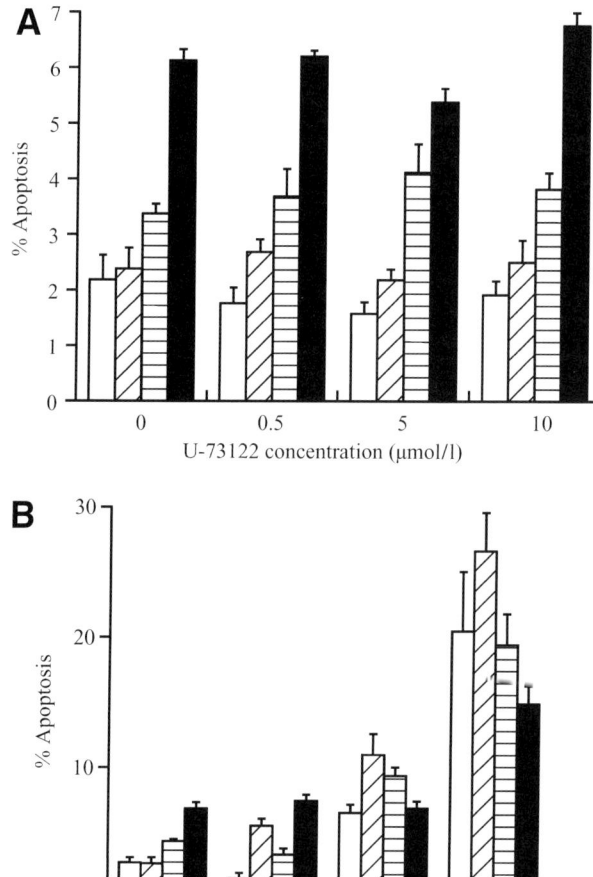

Fig. 12. Effect of PLC and protein kinase C inhibitors on the level of apoptosis in ES cells over various time periods (4 h, ☐; 8 h, ▨; 12 h, ☰; and 24 h, ■). (**A**) Effect of U-73122 on percentage apoptosis. U-73122 had a slight effect on apoptosis levels in ES cells. The level found is small and acceptable for cells in culture. Thus, U-73122 does not significantly induce apoptosis in mouse ES cells. (**B**) Effect of Bis II on percentage apoptosis. Bis II at the highest level tested (10 µM) significantly affected the level of apoptosis in ES cells.

5. The stained cells can then be visualized using standard epifluorescence or on a confocal microscope. Count apoptotic nuclei to measure apoptosis. **Figure 12** shows the results of examining the effects of U-73122 and Bis II, respectively, in inducing apoptosis in ES cells over time.

4. Notes

1. All media and solutions should be made up with ultrapure water (>18 MΩ resistance) from a Milli Q water system (Elix 3 system from Millipore) or equivalent.

2. The supplier suggests that U-73122 is soluble in DMSO; however, in our hands it was extremely difficult to dissolve U-73122 in DMSO alone. Thus, to prepare a solution of U-73122, we recommend adding chloroform stepwise in 25-µL drops until all the salt is dissolved. Aliquot the U-73122 solution into 1.8-mL cryovials (100 µL per tube) and dry onto the plastic under a stream of N_2 gas. To resuspend the U-73122, add 20 µL chloroform and swirl the tube around to ensure the drug is fully resuspended. Then, add sufficient DMSO to make the solution up to the correct concentration. It is important to remember that, for proper controls, the correct volume of the chloroform/DMSO mixture alone be added to control treatments.
3. This is employed to avoid loss of labeled InsPs caused by nonspecific binding during the extraction procedure. The theory is that the hydrolysate occupies potential InsP-binding sites *(20)*.
4. In all cases, the columns should be allowed to drip-dry (i.e., until the buffer stops dripping from column, before the next step begins). This is important to ensure efficient separation of InsP classes. When the columns stop dripping, they are considered ready for the application of the next buffer in the series.
5. A Shandon tank (8.75 × 3 × 9 in.) is the ideal heavy-duty glass container for this kind of work. If unavailable, then any glass container that can be sealed and is large enough to accept the plate is sufficient.
6. It is safer and more reliable to add medium containing the [^{3}H]-inositol to wells rather than first adding unlabeled medium and then adding 2.5 µCi as single additions to each well. Thus, for a standard 24-well plate, add 62.5 µCi to 25 mL and then aliquot 1 mL per well. Using a 25-mL volume allows for some minor inaccuracies in pipetting, which are always present and unavoidable.
7. Lithium is added to prevent breakdown of InsPs and thus amplify agonist-induced inositol formation *(21,22)* (**Fig. 2**).
8. The freon:tri-*n*-octylamine mixture is required to neutralize the PCA in the solution.
9. This step sediments any floating particles firmly in the interphase.
10. It is essential to use N_2 gas and important to try to seal some N_2 into the tube. Lipids oxidize easily and quickly; using N_2 keeps this oxidizing effect to a minimum.
11. It would be standard to run 24 columns simultaneously. This is achieved by building a simple stand for the columns so that they are arranged in two parallel rows of 12 columns per row. Also, the bottom tip of the column should have a clearance of at least 5 cm. This clearance is required to place a scintillation vial under each of the columns. For most experiments, the eluents resulting from buffers 1 and 2 are of no major importance; thus, the stand containing the 24 columns can be placed over a large plastic basin, and the eluents can be allowed to run into the basin. Remember that the eluent is radioactive and needs to be disposed of correctly. For buffers 3–6, remove the stand from the basin and place a scintillation vial under each of the columns to collect the 2-mL fractions.
12. The columns need to be regenerated before reuse; wash the columns with 20 mL buffer 7 followed by 40 mL ultrapure water The resin works consistently well through five cycles of regeneration. Between cycles, the columns can be stored in approx 2 mL of ultrapure water.
13. Take care to remember that this eluent is radioactive and should be dealt with appropriately.
14. Wear gloves at all times.
15. Shandon tank (8.75 × 3 × 9 in.) or similar.
16. When preparing the running solvent, add the chloroform last; otherwise, the solution may have a cloudy appearance that can affect your separation. Do not use the solvent if it becomes cloudy.

17. The actual initial volume of solvent prepared is not important but try to match the volume of solvent prepared to the tank you are using so the tank is fully equilibrated while at the same time leaving approx 1 cm solvent standing at the bottom. Try not to splash the sides as you introduce the solvent to the tank. You want to create a uniform atmosphere within the tank; allowing the solvent to rise evenly from the bottom achieves this most efficiently.
18. The temperature is important for efficient separation; thus, the fume hood should be on a low setting, and the tank should not be disturbed for the duration of the separation.
19. If you plan to perform autoradiography, then it is best not to stain with iodine vapor until after the autoradiography is completed.
20. The hemocytometer slide is a specialized glass slide on which two grids have been etched in a central region that is 0.1 mm lower than the rest of the slide (**Fig. 8A**). Each grid is comprised of 25 large squares, each containing 16 smaller squares with a 1/400 mm^2 area. When a glass cover slip is placed over the central region, this creates a region of known volume (0.1 mm^3) between the cover slip and the slide. Approximately 10 µL cell suspension solution is then pipeted under the cover slip, and cells are counted in a proportion of the grid squares.
21. Some slight downward pressure on the micropipeter piston may be required to fill the chamber; however, do not flood the troughs in the slide.
22. Adjust the contrast as needed to clearly see both grid and cells.
23. Sample calculation: area of each small square = 1/400 mm^2. Volume of each small square = 1/400 mm^2 × 0.1 mm. Volume of 80 squares = (1/400 mm^2 × 0.1 mm) × 80 = 1/50 mm^3. For example, if chamber 1 total cells counted (80 squares) = 482 cells and chamber 2 total cells counted (80 squares) = 574 cells, then the average count = 528 cells. If n is the number of cells counted, then the concentration of cells in the sample = $50 \times n$ per 1 mm^3. Using the previous example with an average count of 528 cells, the concentration of cells = 26,400 cells/mm^3 = 26,400 cells/µL (because 1 mm^3 = 1 µL). If the original sample of cells was diluted 1 in 2, then the cell density of original sample was 26,400 × 2 cells/µL = 5.28×10^4 cells/µL or 5.28×10^7 cells/mL.
24. Phenol red-free medium is used because phenol red is known to interfere with assay efficiency; also, this medium is prepared without protein or mercaptoethanol because these are known to interfere with the assay.
25. It is best to use an upright microscope for this procedure; although you can use an inverted microscope by simply placing the hemocytometer slide cover slip side down, it is not a good solution because count errors and leakage of a potential mutagen (ethidium bromide) are likely.
26. Staining the cells is a simple process using the TUNEL *In Situ* Cell Death Detection Kit from Roche Diagnostics as per manufacturer's instructions.

References

1. Majerus, P. W., Connolly, T. M., Bansal, V. S., Inhorn, R. C., Ross, T. S., and Lips, D.L. (1988) Inositol phosphates: synthesis and degradation. *J. Biol. Chem.* **263,** 3051–3054.
2. Berridge, M. J. (1992) Phosphoinositides and cell signaling. *Fidia Research Foundation Neuroscience Award Lectures* **6,** 5–45.
3. Michell, R. H. (1988) Inositol lipids and phosphates in growing, stimulated and differentiating cells. *Biochem. Soc. Trans.* **17,** 1–3.
4. Quinlan, L. R. and Kane, M. T. (2001) Nature of glycosylphosphatidylinositols produced by mouse embryonic stem cells. *Reproduction* **122,** 785–791.
5. Kane, M. T. (1988) The effects of water soluble vitamins on the expansion of rabbit blastocysts. *J. Exp. Zool.* **245,** 220–223.

6. Kane, M. T. and Bavister, B. D. (1988) Protein-free culture medium containing polyvinylalcohol, vitamins and amino acids supports development of eight-cell hamster embryos to hatching blastocysts. *J. Exp. Zool.* **247,** 183–186.
7. Kane, M. T., Norris, M., and Harrison, R. A. P. (1992) Uptake and incorporation of inositol by preimplantation mouse embryos. *J. Reprod. Fertil.* **96,** 617–625.
8. Fahy, M. M. and Kane, M. T. (1993) Incorporation of [^{3}H]inositol into phosphoinositides and inositol phosphates by rabbit blastocysts. *Mol. Reprod. Dev.* **34,** 391–395.
9. Hynes, A. C., Sreenan, J. M., and Kane, M. T. (2000) Uptake and incorporation of myo-inositol by bovine preimplantation embryos from two-cell to early blastocyst stages. *Mol. Reprod. Dev.* **55,** 265–269.
10. Duffy, C. and Kane, M. T. (1996) Investigation of the role of inositol and the phosphatidylinositol signal transduction system in mouse embryonic stem cells. *J. Reprod. Fertil.* **108,** 87–93.
11. Quinlan, L. R., Faherty, S., and Kane, M. T. (2003) Phospholipase C and protein kinase C involvement in mouse embryonic stem-cell proliferation and apoptosis. *Reproduction* **126,** 121–131.
12. Berridge, M. J. (1987) Inositol trisphosphate and diacylglycerol: two interacting second messengers. *Annu. Rev. Biochem.* **56,** 159–193.
13. Bootman, M. D., Berridge, M. J., and Roderick, H. L. (2002) Activating calcium release through inositol 1,4,5-trisphosphate receptors without inositol 1,4,5-trisphosphate. *Proc. Natl. Acad. Sci. USA* **99,** 7320–7322.
14. Roldan, E. R. and Harrison, R. A. (1989) Polyphosphoinositide breakdown and subsequent exocytosis in the Ca^{2+}/ionophore-induced acrosome reaction of mammalian spermatozoa. *Biochem. J.* **259,** 397–406.
15. Denizot, F. and Laing, R. (1986) Rapid colorimetric assay for cell growth and survival: modifications to the tetrazolium dye procedure giving improved sensitivity and reliability. *J. Immunol. Methods* **89,** 271–277.
16. Mosmann, T. (1983) Rapid colorimetric assay for cellular growth and survival: application to proliferation and cytotoxicity assays. *J. Immunol. Methods* **65,** 55–63.
17. Desjobert, A. and Petek, F. (1956) Chromatographie sur papier des esters phosphoriques de l'Inositol. Application a l'etude de la degradation hydrolytique de l'inositol hexaphosphate. *Bull. Soc. Chim. Biol.* **38,** 871–883.
18. Meek, J. L. (1986) Inositol bis-, tris-, and tetrakis(phosphate)s: analysis in tissues by HPLC. *Proc. Natl. Acad. Sci. USA* **83,** 4162–4166.
19. Woodcock, E. A. (1997) Analysis of inositol phosphates in heart tissue using anion-exchange high-performance liquid chromatography. *Mol. Cell. Biochem.* **172,** 121–127.
20. Wreggett, K. A. and Irvine, R. F. (1987) A rapid separation method for inositol phosphates and their isomers. *Biochem. J.* **245,** 655–660.
21. Berridge, M. J., Downes, C. P., and Hanley, M. R. (1982) Lithium amplifies agonist-dependent phosphatidylinositol responses in brain and salivary glands. *Biochem. J.* **206,** 587–595.
22. Sherman, W. R. (1989) Inositol homeostasis, lithium and diabetes, in *Inositol Lipids in Cell Signalling* (Michell, R. H., Drummond, A. H., and Downes, C. P., eds.), Academic Press, London. pp. 39–79.

11

Cripto Signaling in Differentiating Embryonic Stem Cells

Gabriella Minchiotti, Silvia Parisi, and M. Graziella Persico

Summary

Embryonic stem (ES) cells have been suggested as candidate therapeutic tools for regenerative medicine approaches. In this scenario, great efforts are made to define protocols to preferentially direct ES cells toward a defined cell type. To this end, it becomes crucial to characterize the molecular mechanisms as well as the signaling pathways implicated in ES cell differentiation. Findings highlight a key role of *cripto*, the founding member of a new class of extracellular factors, called EGF-CFC. Indeed, Cripto signaling is strictly required in an early acting window to negatively regulate neural differentiation and to permit differentiation of ES cells to cardiac fate. The protocols defined in this chapter allow preferential direction of ES cell differentiation as embryoid bodies toward either cardiomyocytes or neurons. Although referred to as modulation of Cripto signaling, these methods build the basis for the use of other classes of secreted molecules to control ES cell differentiation.

Key Words: Cardiomyocytes; Cripto; differentiation; EGF-CFC; embryoid bodies; ES cells; hanging drops; neurons.

1. Introduction

Breakthroughs in stem cell research have opened new possibilities for cell replacement therapy. In this regard, embryonic stem (ES) cells provide access to the earliest stages of development and may serve as a source of specialized cells for regenerative medicine. However, the successful therapeutic use of stem cells is still limited. Indeed, our knowledge of the molecular mechanisms implicated in the determination of cell fate in ES cells is incomplete; thus, their differentiation is poorly controlled.

To this end, the introduction of developmental control genes into stem cells may be a useful strategy to direct their differentiation. Actually, the fate of a cell can be determined by insertion or deletion of genes encoding factors that control the several different stages of development. Although this approach may represent a powerful strategy, it suffers from several adverse side effects caused by clonal variance, promoter dependence, and the apparent ability of some ES cells to silence the expression of transgenes *(1)*.

From: *Methods in Molecular Biology, vol. 329: Embryonic Stem Cell Protocols: Second Edition: Volume 1*
Edited by: K. Turksen © Humana Press Inc., Totowa, NJ

To overcome these problems, great effort is made to identify secreted molecules capable of promoting controlled differentiation of stem cells. For example, a good proportion of neural differentiation is achieved by treatment of multicellular aggregates derived from ES cells (embryoid bodies, EBs) with retinoic acid (RA), in the presence of serum *(2)* or by co-culture of ES cells with a particular stromal cell line, PA6 *(3)*. Although these methodologies can improve the efficiency of neural cell generation, they both show some limits. The action of RA is pleiotropic and of indeterminate physiological relevance; it is therefore preferable to avoid RA treatment. On the other hand, the effect of PA6 cells is attributed to an undefined neural inducing activity.

In this scenario, the mouse *cripto* gene turned out to have a crucial role. Cripto is the founding member of a new class of extracellular factors, called epidermal growth factor (EGF)-cripto-FRL1-cryptic (CFC), strictly required in the early phases of vertebrate development *(4)*. Indeed, using EBs derived from Cripto$^{-/-}$ ES cells, *cripto* has been shown to be essential for cardiomyocyte differentiation *(5)*. Data obtained in our laboratory indicated that stimulation *in trans* with soluble Cripto protein is sufficient to promote cardiomyocyte induction and differentiation in Cripto$^{-/-}$ ES cells, and that impaired Cripto signaling results in enhanced neural differentiation ability of ES cells *(4,6)*. Taken together, our data suggest that Cripto could represent a key molecule required for both induction of cardiomyocyte differentiation and repression of neural differentiation in ES cells.

Here, we describe a simple method based on the EBs system: how to obtain efficient cardiomyocyte differentiation of Cripto$^{-/-}$ ES cells using recombinant Cripto protein. Furthermore, the same protocol is described to induce neural differentiation of Cripto$^{-/-}$ ES cells in the absence of any inducing factor or defined culture conditions.

2. Materials

2.1. Tissue Culture

1. 1X phosphate-buffered saline (PBS) (*see* **Subheading 2.1.1.**).
2. Gelatin solution (*see* **Subheading 2.1.1.**).
3. Dulbecco's modified Eagle's medium (DMEM) without sodium pyruvate 1X (500 mL; Gibco BRL, Carlsbad, CA; cat. no. 41965-062).
4. 100 mM 100X sodium pyruvate solution (100 mL; Gibco BRL, cat. no. 11360-039).
5. 100X minimum essential medium (MEM) nonessential amino acid solution, 100X (100 mL; Gibco BRL, cat. no. 11140-035).
6. 100X penicillin/streptomycin solution (100 mL; Gibco BRL, cat. no. 15070-063).
7. 200 mM 100X L-glutamine (100 mL; Gibco BRL, cat. no. 25030-024).
8. Mitomicin C (2 mg/vial; Sigma, St. Louis, MO; cat. no. M-0503) (*see* **Subheading 2.1.1.**).
9. 2-Mercaptoethanol (Sigma, cat. no. M-7522, *see* **Subheading 2.1.1.**).
10. Trypsin solution (Gibco BRL, cat. no. 25090-028; *see* **Subheading 2.1.1.**).
11. Falcon 100-mm tissue culture dishes (Falcon, San Jose, CA; cat. no. 353003).
12. Falcon 48-well tissue culture dishes (Falcon, cat. no. 353047).
13. Falcon 100-mm Petri dishes (Falcon, cat. no. 351029).
14. Cryotube vials, 1.8-mL tubes (Nalge Nunc International, Rochester, NY; cat. no. 368632).
15. Pipetman (2-, 10-, 100-, 200-, 1000-µL) designated for tissue culture use only.
16. Falcon 60-mm Petri dishes (Falcon, cat. no. 351016).

Cripto Role in ES Cell Differentiation 153

17. Hematocytometer.
18. Trypan blue solution (Sigma, cat. no. T8154).
19. Leukemia inhibitory factor (LIF), ESGRO 10^7 U/mL (Chemicon International, Temecula, CA; cat. no. ESG1107).
20. 8-well permanox Lab-Tek Chamber slides (Nalge Nunc International, cat. no. 177445).
21. Disposable pipets (2-, 5-, 10-, 25-mL).
22. Pipet-aid.
23. Multichannel pipeter.
24. Fetal bovine serum (FBS) defined and screened for ES cell growth (500 mL; Hyclone, Logan, UT; cat. no. SH30070.03) (see **Note 1**).
25. FBS characterized (500 mL; Hyclone, cat. no. SH30071.03) (see **Notes 1** and **2**).
26. FBS (500 mL; Euroclone, Milano, Italy; ECS0180L).
27. High-glucose DMEM (500 mL; Euroclone, cat. no. ECS0050D).
28. Gelatin (Sigma, cat. no. G9391).
29. Chicken serum (100 mL; Euroclone, cat. no. ECS0050D).
30. Syringe filter (0.22-µm), low protein binding (Millipore, Billerica, MA; cat. no. GVSLGV033RB).
31. 15-mL polypropylene conical tube (Falcon, cat. no. 352096).
32. 50-mL polypropylene conical tube (Falcon, cat. no. 352070).
33. Nickel-NTA-agarose (100 mL; Quiagen, Milano, Italy; cat. no. 30230).
34. Amicon concentrator (Millipore, cat. no. 13142).

2.1.1. Tissue Culture Solutions (see **Note 3**)

1. 1X PBS: dissolve 8 g sodium chloride (NaCl), 0.2 g potassium chloride (KCl), 1.15 g sodium monohydrogen phosphate (Na_2HPO_4), 0.2 g potassium dihydrogen phosphate (KH_2PO_4) in 800 mL pure water (Milli-Q or equivalent) in a glass beaker. Adjust the volume to 1 L. Dispense the solution into approx 500-mL aliquots and sterilize by autoclaving. Store at room temperature.
2. Gelatin solution (0.1%): dissolve 2 g gelatin in 2 L pure water (Milli-Q or equivalent) in a glass beaker. Heat to assist dissolution on a magnetic stirrer. Dispense into 500-mL aliquots and sterilize by autoclaving. Store at 4°C.
3. Trypsin solution (0.1%): to 235 mL of 1X PBS, add 2.5 mL chicken serum, 2.5 mL 0.1 M Na-EDTA (ethylenediaminetetraacetic acid) (pH 7.4), and 10 mL 2.5% trypsin in a 500-mL sterile glass bottle. Dispense into aliquots and store at $-20°C$.
4. Mitomycin C: to a vial containing 2 mg mitomycin C, add 2 mL 1X PBS and shake vigorously to assist dissolution. Transfer mitomycin C solution into a 2-mL cryotube vial wrapped in aluminum foil to prevent damage by light. Mitomycin C solution can be stored at 4°C up to 1 mo.
5. 500X 2-mercaptoethanol: to 25 mL sterile PBS, add 88 µL 2-mercaptoethanol (14.4 M solution). Store at 4°C.
6. Media for mouse embryonic fibroblasts (MEFs) cells (MEF media): MEF cells are maintained in high-glucose DMEM supplemented with 10% FBS, 100 U/mL penicillin/streptomycin, and 2 mM glutamine. To 500 mL of DMEM (Euroclone), add 50 mL heat-inactivated FBS (Euroclone), 5 mL glutamine, and 5 mL penicillin/streptomycin. Store complete media at 4°C.
7. Media for ES cells (ES media): both RI (wild type) and DE7 (Cripto$^{-/-}$) ES cells are maintained in high-glucose DMEM medium supplemented with 15% FBS, 0.1 mM 2-mercaptoethanol, 1 mM sodium pyruvate, 1X nonessential amino acids, 2 mM glutamine,

100 U/mL penicillin/streptomycin, and 10^3 U/mL LIF. To 500 mL DMEM (Gibco), add 75 mL heat-inactivated defined, ES-screened FBS (Hyclone), 6 mL glutamine, 6 mL penicillin/streptomycin, 6 mL nonessential amino acids, 6 mL sodium pyruvate, 1 mL 2-mercaptoethanol 500X stock solution, and 60 µL LIF. Store complete media at 4°C.
8. ES cell differentiation media: for in vitro differentiation, ES cells are cultured in high-glucose DMEM media supplemented with: 15% FBS, 0.1 mM 2-mercaptoethanol, 1 mM sodium pyruvate, 1X nonessential amino acids, 2 mM glutamine, and 100 U/mL penicillin/streptomycin. To 500 mL of high-glucose DMEM (Gibco), add 75 mL heat-inactivated FBS characterized (Hyclone), 6 mL glutamine, 6 mL penicillin/streptomycin, 6 mL nonessential amino acids, 6 mL sodium pyruvate, and 1 mL 2-mercaptoethanol (500X stock solution). Store at 4°C.
9. Freezing medium (10 mL): add 1 mL dimethyl sulfoxide to 9 mL appropriate FBS.

2.1.2. General Comments and Required Equipment for Tissue Culturing (see **Note 4**)

The tissue culture facility requires the following:

1. Humidified incubator at 37°C and 6% CO_2 (see **Note 5**).
2. Water bath.
3. Glass bottles, beakers, and Pipetman designated for tissue culture use only.
4. Laminar flow cabinet.
5. Inverted microscope with a range of phase contrast objectives (×10 to ×25).
6. Refrigerator (+4°C) and freezer (−20°C).
7. Tabletop centrifuge.
8. Liquid nitrogen storage tank.

2.2. Cell Lysates and Western Blotting

1. 1X PBS (see **Subheading 2.1.1.**).
2. Lysis buffer (see **Subheading 2.2.1.**).
3. Protease inhibitor cocktail (tablets; Roche, Basel, Switzerland; cat. no. 1697498).
4. Cell scraper (Costar, Corning Inc., New York, NY; cat. no. 3008).
5. Bio-Rad protein assay (500 mL; Bio-Rad, Hercules, CA; cat. no. 500-0006).
6. 1X sodium dodecyl sulfate (SDS) gel-loading buffer (see **Subheading 2.2.1.**).
7. 10X Tris-glycine electrophoresis buffer (5 L; Bio-Rad, cat. no. 171 0771).
8. Acrylamide/*bis* solution 29:1 (Bio-Rad, cat. no. 161-0156).
9. TEMED (Bio-Rad, cat. no. 161-0801).
10. Ammonium persulfate (Bio-Rad, cat. no. 161-0700).
11. Rainbow marker (Amersham Pharmacia Bioscience, Uppsala, Sweden; cat. no. RPN755).
12. Mini Protean 3 apparatus (Bio-Rad).
13. Trans-Blot Semi-Dry Electrophoretic Transfer Cell (Bio-Rad, cat. no. 170-3940).
14. Immobilon-P Transfer membrane (Millipore, cat. no. IPVH00010).
15. Methyl alcohol (Carlo Erba Reagenti, Milano, Italy; cat. no. 412533).
16. Low-fat milk powder.
17. Horseradish peroxidase-conjugated goat antirabbit immunoglobulin G (Bio-Rad, cat. no. 170-6515).
18. ECL Western blotting analysis system (Amersham Pharmacia Biosciences, cat. no. RPN2109).

19. Rabbit polyclonal anti-Cripto antibodies (*see* **Note 6**).
20. 25X protease inhibitor cocktail solution: dissolve 1 tablet in 2 mL 1X PBS. Dispense into aliquots wrapped in aluminum foil to prevent damage by light and store at $-20°C$.
21. Lysis buffer: 10 mM Tris-HCl (pH 8.0), 140 mM NaCl, 2 mM Na-EDTA (pH 8), and 1% NP-40 (*see* **Note 7**). To 80 mL pure water, add 1 mL 1 M Tris-HCl, pH 8.0, 2.8 mL 5 M NaCl, 0.4 mL 0.5 M Na-EDTA, pH 8.0, and 1 mL NP-40; adjust the volume to 100 mL and sterilize by filtration. Dispense into 10-mL aliquots and store at 4°C up to 1 mo. Immediately before use, to 1 mL of lysis buffer add 40 µL 25X protease inhibitor cocktail solution.
22. 4X SDS gel-loading buffer: 62.5 mM Tris-HCl, pH 6.8, 2% SDS, 10% glycerol, 2.5% 2-mercaptoethanol, and 0.1% bromophenol blue. To 5 mL pure water, add 0.625 mL Tris-HCl, pH 6.8, 1 mL 20% SDS, 1 mL of glycerol, 0.25 mL 2-mercaptoethanol, and 0.2 mL 5% of bromophenol blue. Adjust the volume to 10 mL with pure water and dispense into aliquots wrapped in aluminum foil to prevent damage by light; store at 4°C.
23. Tris-glycine electrophoresis buffer: to 900 mL pure water, add 100 mL 10X electrophoresis buffer and 5 mL 20% SDS. Mix on a magnetic stirrer.
24. Transfer buffer: 48 mM Tris-HCl, 39 mM glycine, 0.04% SDS, and 10% methanol. To 800 mL pure water, add 5.82 g Tris-HCl, 2.93 g glycine, 1.85 mL 20% SDS, and 100 mL methanol and mix on a magnetic stirrer. Adjust the volume to 1 L with pure water. Prepare fresh before use.
25. 1X Tris-buffered saline (TBS): 25 mM Tris and 150 mM NaCl. To prepare 10X stock solution, dissolve 30 g Tris-base, 80 g NaCl, and 2 g KCl in 800 mL deionized water. Adjust the volume to 1 L with pure water and store at room temperature.
26. Tween-20 in 1X TBS (TBS-T): to 495 mL 1X TBS, add 5 mL 10% Tween-20. Store at 4°C.
27. Blocking solution: dissolve 2.5 g low-fat milk powder into 30 mL 1X PBS. Add 0.5 mL 10% Tween-20 and adjust the volume to 50 mL with 1X PBS.
28. SDS-polyacrylamide gel:
    a. Resolving gel (12%): to prepare 5 mL solution, mix 1.7 mL deionized water, 2 mL 30% acrylamide/*bis* acrylamide solution, 1.3 mL 1.5 M Tris-HCl, pH 8.8, and 50 µL 20% SDS. Immediately before casting the gel, add 50 µL 10% ammonium persulfate and 2 µL TEMED.
    b. Staking gel (5%): to prepare 2 mL solution, mix 1.4 mL deionized water, 330 µL 30% acrylamide/*bis*-acrylamide solution, 250 µL 1.5 M Tris-HCl, pH 6.8, 20 µL 20% SDS. Immediately before casting the gel, add 20 µL 10% ammonium persulfate and 2 µL TEMED.

2.3. Immunofluorescence

1. Methanol (Carlo Erba Reagenti, cat. no. 412533).
2. Acetone (Sigma, A-4206).
3. Disposable pipets.
4. PBS (*see* **Subheading 2.1.1.**).
5. Paraformaldehyde (PFA; Merck, Whitehouse Station, NJ; cat. no. 1.04005.1000).
6. Triton X-100 (Fluka, St. Gallen, Switzerland; cat. no. 93420).
7. Tween-20 (Sigma, cat. no. P-2287).
8. Normal rabbit serum (Dako, Glostrup, Denmark; cat. no. X0902).
9. Vecta Shield medium (Vector Laboratories, Burlingame, CA; cat. no. H-1000).
10. Rabbit antimouse fluorescein isothiocyanate (Dako, cat. no. F0261).

11. Bovine serum albumin (BSA; Sigma, cat. no. A4503).
12. Goat antimouse Cy3 (Jackson ImmunoResearch Laboratories, West Grove, PA; cat. no. 115-165-003).
13. Antisarcomeric myosin (MF-20; monoclonal supernatant obtained from the Developmental Studies Hybridoma Bank, Iowa City, IA).
14. Monoclonal anti-β-tubulin isotype III (Sigma, cat. no. T-8660).
15. DAPI (4′, 6-diamidino-2-phenylindole; Sigma, cat. no. D 8417).
16. 0.22-μm syringe-driven filter units (Millipore, cat. no. SLGV033RB).
17. Cover slips (Nalge Nunc International, cat. no. 174952).
18. Methanol:acetone solution (7:3): mix 140 mL methanol and 60 mL acetone in a 250-mL sterile glass bottle; store at −20°C.
19. 4% PFA: to 500 mL 1X PBS, add 20 g PFA. Heat to approx 60–70°C on a magnetic stirrer with constant mixing until PFA is completely dissolved. Take care not to boil PFA solution. Dispense the chilled solution into 10-mL aliquots using 15-mL polypropylene disposable tubes. Store at −20°C. Once thawed, store PFA 4% at 4°C up to 24 h.
20. Triton X-100 (10% solution): dissolve 1 mL Triton X-100 in 9 mL sterile pure water.
21. Tween-20 (10%): dissolve 1 mL Tween-20 in 9 mL sterile pure water.
22. Tween-20 in PBS (PBS-T): to 495 mL 1X TBS, add 5 mL of 10% Tween-20.
23. BSA in PBS: dissolve 2.5 g BSA in 50 mL 1X PBS; heat to 37°C in a water bath. Sterilize by filtration using a 0.22-μm filter. Dispense into 10-mL aliquots and store at −20°C.
24. DAPI: dissolve 0.5 mg DAPI powder in 1 mL pure water to make 0.5-mg/mL stock solution. Dispense into aliquots and store protected from light at 4°C.

3. Methods
3.1. Tissue Culture
3.1.1. Freezing Cells

1. Trypsinize cells in the exponential phase of growth from a 100-mm dish.
2. Pellet by centrifigation (800g, 3–4 min) and gently resuspend in appropriate amount of freezing medium.
3. Dispense into 1-mL aliquots using freezing vials. Make two vials of frozen cells from one 100-mm confluent plate.
4. Store the vials at −80°C for 24 h.
5. Transfer the tubes to liquid nitrogen for long-term storage.

3.1.2. Thawing Cells

1. Thaw vials of frozen cells at 37°C in a water bath.
2. Immediately after the cells are thawed, transfer the vial content into 10 mL appropriate growth media in a 15-mL Falcon tube.
3. Collect cell pellet by centrifugation (800g, 3–4 min).
4. Remove the media and gently resuspend the cells with fresh growth media.
5. Transfer to 100-mm tissue culture dish and incubate at 37°C overnight to allow the cells to adhere.
6. Change media the next day.

3.1.3. ES Cells in Culture

ES cells are grown on a feeder layer of mitotically inactivated MEF cells to remain pluripotent in culture.

3.1.3.1. GELATIN-COATING PROCEDURE

To prepare gelatin-coated tissue culture dishes, distribute a thin layer of 0.1% gelatin solution to cover the bottom of the plate. Immediately remove gelatin solution and keep the plate at room temperature for at least 15 min. You can prepare several gelatin-coated plates to be stored at 37°C up to 2 wk before use.

3.1.3.2. MAINTENANCE OF MEF CELLS

MEF cells are maintained in complete MEF media. When thawing MEF cells, the content of a cryovial is transferred to one 100-mm tissue culture dish. At confluence, subculture MEF cells 1:3 using the following procedure:

1. Remove MEF media and rinse with 1X PBS; repeat three times.
2. Add 1 mL of trypsin solution to each plate, incubate in the humidified incubator at 37°C for 5 min.
3. While incubating the cells, fill three gelatin-coated tissue culture dishes with MEF media. Add 7 mL MEF media for each 100-mm dish.
4. Remove the plate from the incubator and swirl to dislodge the cells from the bottom of the plate.
5. Add 8 mL MEF media to floating cells and gently pipet up and down several times to obtain a single-cell suspension.
6. Plate 3 mL cell suspension to each of the three 100-mm dishes. Gently agitate the plate back and forth and side to side to distribute the cells. Incubate in the humidified incubator at 37°C. Track the passage number of MEF cells (*see* **Note 8**). MEF cells can be subcultured twice a week (every Monday and Friday; *see* **Table 1**).

3.1.3.3. MITOMYCIN C TREATMENT OF MEF FEEDER LAYERS FOR ES CELLS

At confluence, MEF cells are treated with mitomycin C to halt the division of the cells when they are still able to condition the media. Prepare plates of mitomycin C-treated MEF cells once a week to provide a fresh substrate for undifferentiated ES cells using the following procedure:

1. To one plate of confluent MEF cells containing 10 mL MEF media, add 100 µL mitomycin C solution.
2. Incubate at 37°C for at least 3 h. Alternatively, add 10 µL mitomycin C solution and incubate overnight.

3.1.3.3.1. Subculture

1. Remove mitomycin C-containing media and rinse with 1X PBS. Repeat three times.
2. Add 1 mL 0.1% trypsin solution and return the plate to the incubator for 5 min.
3. Remove the plate from the incubator and swirl to dislodge the cells from the bottom of the plate. Add 9 mL MEF media and collect all the cells.
4. Prepare five gelatin-coated plates; add 8 mL MEF media to each plate. Dispense 2 mL cell suspension to each plate. Gently move the plate back and forth and side to side to distribute the cells. Incubate at 37°C. The cells are now ready to be used as feeder layers. Mitomycin C-treated MEF cells can be stored at 37°C up to 1 wk.

Table 1
Tissue Culture Timetable

Week	Monday	Tuesday	Wednesday	Thursday	Friday	Saturday	Sunday
1	Thaw one vial MEF cells	Change medium to MEF cells	Change medium to MEF cells	Subculture MEF cells 1:3	Change medium to MEF cells	Mitomycin C treatment of 2X MEF cell plates Thaw one vial of ES cells Freeze one plate of MEF cells	Change medium to ES cells
2	Change medium to ES cells	Subculture ES cells 1:3 2X m/c MEF cells plates 1X gelatin-coated dish	Start hanging drop culture from ES cells	Freeze two plates of ES cells	Culture of EBs in suspension		
3	Culture of EBs in adhesion	Change medium to adherent EBs	Change medium to adherent EBs Beating heart observation	Change medium to adherent EBs Beating heart observation	Change medium to adherent EBs Beating heart observation	Change medium to adherent EBs Beating heart observation	
4	Fix EBs for MF-20 staining	Fix EBs for β-III-tubulin staining					

3.1.3.4. MAINTENANCE OF ES CELLS

Thaw ES cells according to **Subheading 3.1.2.** The content of one vial is thawed onto one 100-mm plate of mitomycin C-treated MEF cells.

1. Remove one feeder layer plate from the incubator. Change MEF media with ES media.
2. Plate the content of one vial of ES cells. Gently move the plate back and forth and side to side and incubate at 37°C. The ES media must be changed every day and the ES cells subcultured every 2–3 d.

3.1.3.4.1. Subculture of ES Cells

1. Remove one plate of ES cells from the incubator and wash the cells three times with 5 mL 1X PBS.
2. Add 1 mL 0.1% trypsin solution.
3. Return the plate to the incubator for 4–5 min.
4. Remove the plate from the incubator and swirl to dislodge the cells from the bottom of the plate.
5. Add 2 mL ES media to the plate and pipet up and down at least 20 times with a 1-mL Pipetman to obtain a single-cell suspension.
6. Collect cell suspension into a 15-mL Falcon centrifuge tube and pellet the cells by low-speed centrifugation (180g for 5 min).
7. While spinning, remove four dishes of mitomycin C-treated MEF cells from the incubator. Remove MEF media and add 8 mL ES media to each plate.
8. After the spin has completed, aspirate the media with a Pasteur pipet. Resuspend the pellet in 8 mL ES media.
9. Distribute 2 mL ES cell suspension (~10^6 cells) to each of the four dishes. Gently agitate the plate back and forth and side to side to distribute the cells evenly and incubate at 37°C. Track passage number of ES cells (*see* **Notes 9** and **10**).

3.2. Differentiation of ES Cells Into Cardiomyocytes: EB Formation

ES cell differentiation can be induced in vitro by removing ES cells from feeder cell layers and LIF and allowing them to form multicellular three-dimensional EB structures, which resemble early postimplantation embryos. Following the induction of mesodermal cells, a portion of the cells develop into beating cardiomyocytes; however, ES cell differentiation yields a heterogeneous population of cells because of numerous cell lineages undergoing dynamic changes in cell type and cell number. We have adopted and modified the hanging drop culture assay *(7,8)* to enrich for differentiated cardiomyocytes. Hanging drops provide a method for obtaining aggregates of uniform size.

We use the following weekly schedule for ES cell differentiation (*see* **Table 1**): d 1 (Wednesday), hanging drop culture; d 3 (Friday), culture of EBs in suspension; d 6 (Monday), culture of EBs in adhesion; d 9 (Wednesday), morphological analysis (appearance of beating cardiomyocytes).

From d 9 to 13 of the protocol, RI-derived EBs show rhythmically contracting cardiomyocytes. Beating activity of cultured EBs containing cardiomyocytes is measured using videomicroscopy. Adopting the following described protocol, you can obtain approx 100% of single EBs showing beating areas.

3.2.1. Day 1: Hanging Drop Culture

1. Trypsinize one plate of undifferentiated ES cells; harvest the cells and remove the trypsin by centrifugation (800g, 3–4 min). Use 10 mL ES differentiation media to resuspend the pellet.
2. Collect cell suspension, plate on one gelatin-coated dish, and incubate at 37°C for 1 h (*see* **Note 11**). This step will remove most of the feeder cells, which quickly attach to the plate.
3. Remove the plate from the incubator and carefully recover cell suspension. Collect the cells into a 15-mL Falcon tube and count (*see* **Note 12**).
4. Adjust the volume using ES differentiation media to make a suspension of 2×10^4 cells/mL.
5. Using a multichannel pipeter, dispense 100 drops of cell suspension (20 µL/drop, 400 cells/drop) concentrically onto the lid of one 100-mm tissue culture dish.
6. Invert the lid and place over the bottom of a tissue culture dish filled with 5 mL sterile water; incubate at 37°C. Once the lid is inverted, the cells fall to the bottom of the drop and aggregate into single clumps within 2 d.

3.2.2. Day 3: Culture of the EBs in Suspension

1. Remove the plates containing 2-d-old hanging drops from the incubator. Carefully invert the lid and flush with 5 mL ES differentiation media to collect EBs (*see* **Note 13**).
2. Transfer EBs to 100-mm Petri dishes and adjust the volume to 10 mL with ES differentiation media. Note that it is extremely important to use Petri dishes in this step to avoid attachment of EBs to the plate (*see* **Note 14**).
3. Incubate at 37°C for an additional 3 d.

3.2.3. Day 6: Culture of the EBs in Adhesion

1. Prepare gelatin-coated tissue culture dishes. Use 48 multiwell plates (for morphological analysis), 8-well chamber slides (for immunofluorescence analysis) or 100-mm plates (for Western blotting analysis).
2. Fill the plates with ES differentiation media.
3. Transfer single EBs from suspension to each well using a 1-mL Pipetman and incubate at 37°C. For Western blotting analysis, *see* **Subheading 3.3**.
4. The ES differentiation media must be changed every day. Spontaneous contractile cardiomyocytes start to appear after 3 d and can be easily identified by microscopy. Alternatively, EBs are fixed and processed for immunofluorescence analysis (*see* **Subheading 3.6.**).

3.3. Cripto Expression in ES Cell Differentiation

3.3.1. Protein Extraction (see **Fig. 1**)

3.3.1.1. Day 1

Prepare 72 plates of wild-type hanging drops (*see* **Note 15**) following the standard hanging drop protocol (*see* **Subheading 3.2.**).

3.3.1.2. Day 3

1. Carefully remove 18 plates of hanging drops from the incubator. Flush with 1X PBS (5 mL/plate) to recover the EBs from the lid and transfer the EB suspension to a 15-mL tube. Allow the EBs to settle to the bottom of the tube; *do not centrifuge*. This step takes only 3–4 min.

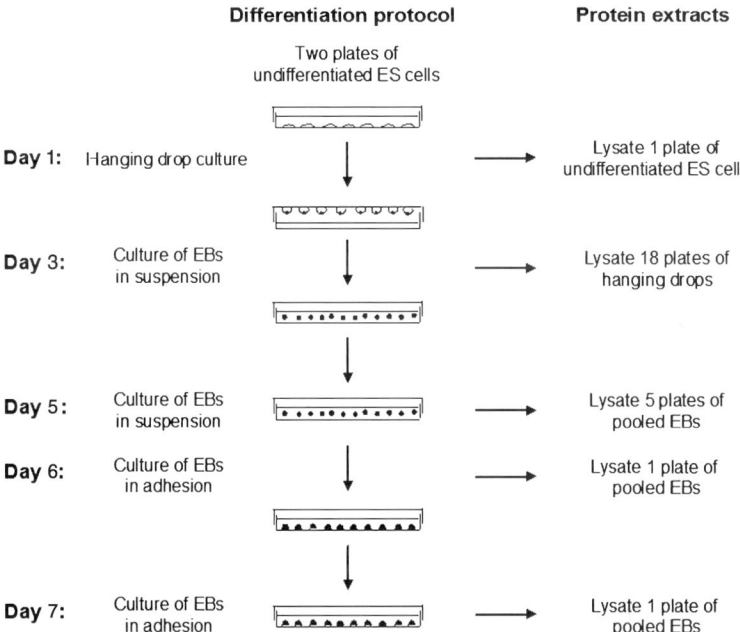

Fig. 1. Schematic representation of the protein extraction timing during embryonic stem cell differentiation.

2. While waiting, carefully remove the remaining 54 plates of drops from the incubator. Flush with 1X PBS to collect EBs. Pool the EBs collected from three plates of drops and plate into one 100-mm Petri dish. Incubate at 37°C in a humidified incubator for 2 d (*see* **Subheading 3.2.**).
3. Carefully remove PBS from EB suspension (*see* **step 1**). Add 200 µL prechilled lysis buffer and vortex to dissolve the EBs completely (*see* **Note 16**). Transfer the lysates to a 1.5-mL Eppendorf tube and place on ice for 1 h.
4. Centrifuge at top speed for 30 min at 4°C using a refrigerated centrifuge. Transfer the supernatant to a fresh 1.5-mL Eppendorf tube; store at −80°C.

3.3.1.3. Day 5

1. Remove five plates of pooled EBs growing in suspension (*see* **Subheading 3.2.**), transfer the EBs to a 50-mL polypropylene tube, and allow the EBs to settle to the bottom. *Do not centrifuge*.
2. Carefully remove the media and wash the EBs with 10 mL 1X PBS. Proceed as described in **Subheading 3.3.1.2., steps 3–4**.

3.3.1.4. Day 6

1. Remove five plates of EBs growing in suspension (*see* **Subheading 3.2.**), transfer the EBs to a 50-mL polypropylene tube, and allow the EBs to settle to the bottom. *Do not centrifuge*.

2. Remove the media and wash EBs with 10 mL 1X PBS. Proceed as described in **Subheading 3.3.1.2., steps 3–4**.
3. Remove eight plates of EBs and two gelatin-coated dishes from the incubator. Combine EBs collected from four plates and transfer to a 50-mL polypropylene tube. Allow the EBs to settle to the bottom. Remove the media, resuspend EBs in fresh differentiation media (12 mL), and plate each pool into one gelatin-coated dish. Return the plates to the incubator and proceed with the standard protocol (*see* **Subheading 3.2.**).

3.3.1.5. Day 7

1. Remove one plate of EBs growing in adhesion from the incubator; wash extensively with 1X PBS.
2. Add 200 µL lysis buffer and scrape adherent EBs with a sterile cell scraper.
3. Transfer the cell lysates to a 1.5-mL Eppendorf tube, vortex, and incubate on ice for 1 h. Proceed as described in **Subheading 3.3.1.2., steps 3–4**.

3.3.1.6. Day 8

Proceed as described in **Subheading 3.3.1.5., steps 1–3**.

3.3.1.7. Preparation of Cell Lysates From Undifferentiated ES Cells

1. Subculture one plate of confluent ES cells 1:3 directly onto gelatin-coated dishes.
2. Incubate at 37°C for 1 or 2 d to reach confluence; change media every day.
3. Recover the plate from the incubator and wash the cells with 10 mL 1X PBS.
4. Add 350 µL prechilled lysis buffer and scrape adherent cells using a sterile cell scraper.
5. Transfer cell lysates to a 1.5-mL Eppendorf tube, vortex, and incubate on ice for 1 h.
6. Centrifuge at top speed for 30 min at 4°C in a refrigerated microfuge.
7. Transfer the supernatant to a fresh Eppendorf tube and store at 80°C. To quantify the proteins in the cell lysates, use the Bio-Rad protein assay (according to the manufacturer's instructions).

3.3.2. Gel Electrophoresis and Western Blotting

1. On a 12% polyacrylamide gel, load 7 µL rainbow marker and 50 µg total lysates from each sample in a predetermined order.
2. Run in 1X Tris-glycine electrophoresis buffer at 100 V using Mini Protean 3 apparatus. Stop running when the red-color band of rainbow marker (14.3 kDa) reaches the gel bottom. Crypto protein moves as approx 21 kDa.
3. Blot the gel onto an Immobilon-P membrane, presoaked in transfer buffer, at 15 V for 90 min using the Trans-Blot Semi-Dry Electrophoretic Transfer Cell.
4. Recover the membrane and rinse in methanol. Dry onto a filter paper.
5. Incubate the blotted membrane at room temperature for 1 h in blocking solution.
6. Remove the blocking solution and add anti-Crypto antibodies (*see* **Note 6**) 1:2000 to 10 mL blocking buffer. Incubate at room temperature for 1 h.
7. Wash three times for 5 min with TBS-T.
8. Incubate the membrane with antirabbit horseradish peroxidase-conjugated secondary antibodies 1:10,000 in 10 mL blocking buffer at room temperature for 1 h.
9. Wash three times for 5 min with TBS-T.
10. Use the ECL Western blotting analysis system for antibodies detection. Crypto is detectable already in undifferentiated ES cells, and its expression peaks at d 5 in wild-type (wt) EBs.

3.4. Differentiation of Cripto$^{-/-}$ ES Cells Into Cardiomyocytes by Addition of Recombinant Cripto Protein

The ability of Cripto$^{-/-}$ ES cells to differentiate into beating cardiomyocytes can be rescued by adding recombinant Cripto directly to the culture media *(6)*. To achieve full rescue of the Cripto$^{-/-}$ ES cells, recombinant Cripto protein is added at d 1, 2, or 3 of the differentiation protocol (*see* the weekly schedule described in **Subheading 3.2.**) using the following procedure.

3.4.1. Cripto-His Soluble Protein Purification

Cripto-His soluble protein is purified by nickel affinity chromatography from conditioned media of human embryonic kidney cells (293) transfected with pCDNA3 Cripto-His vector as previously described *(9)*. Briefly,

1. Grow cells in DMEM supplemented with 10% FCS to 80% confluence.
2. Carefully wash the cells twice with 10 mL 1X PBS.
3. Add 10 mL serum-free DMEM per plate. Incubate the cells at 37°C for 16–24 h.
4. Harvest conditioned media and spin at 1600*g* for 10 min to eliminate cell debris.
5. Start the purification from 200 to 300 mL conditioned media.
6. Dialyze and concentrate (at least 50 mL) Cripto-conditioned media (*see* **Note 17**) with 50 mM NaPO$_4$, pH 8.0, 300 mM NaCl, and 10 mM imidazole (binding buffer). Perform this step at 4°C.
7. To 5 mL concentrated conditioned media, add 0.8 mL of preequilibrated Nickel-NTA-agarose slurry (1:1). Place overnight at 4°C on a rotating platform with gentle oscillation.
8. Transfer the resin to a Bio-Rad column.
9. Wash the column with at least 10 vol binding buffer.
10. Elute with 50 mM NaPO$_4$ buffer at pH 8.0, 300 mM NaCl, and 300 mM imidazole. Use 1 vol of the column. Repeat three times.
11. Dialyze in 50 mM NaPO$_4$ buffer at pH 8.0 (overnight at 4°C).

3.4.2. Addition of Recombinant Cripto at Day 1

1. On Wednesday, prepare 5 mL ES cell differentiation media containing 10 µg/mL Cripto protein and sterilize the solution by filtering with a 0.22-µm syringe filter. Use the Cripto-containing media to prepare a suspension of approx 2×10^4 cells/mL to dispense cell drops (*see* **Subheading 3.2.1.** and **Note 18**) onto the lid of two plates (100-mm tissue culture dishes; approx 100 drops/plate).
2. From this point, follow the standard protocol (*see* **Subheading 3.2.**).

3.4.3. Addition of Recombinant Cripto at Day 2

1. On Wednesday, prepare two plates (100 mm) of drops following the protocol previously described (*see* **Subheading 3.2.** and **Note 19**) and incubate the plates at 37°C for 1 d.
2. The following day (Thursday), to 2 mL of ES cell differentiation media add recombinant Cripto protein to a final concentration of 30 µg/mL. Sterilize the solution by filtering using a 0.22-µm syringe filter.
3. Carefully remove the plates containing the hanging drops from the incubator. Using a multichannel pipeter, add 10 µL Cripto-containing media to each drop to achieve a final concentration of 10 µg/mL Cripto in each drop (30 µL; *see* **Note 20**).
4. Incubate for an additional day to allow the cells in the drop to aggregate correctly (*see* **Note 18**). From this point, follow the standard protocol (*see* **Subheading 3.2.**).

3.4.4. Addition of Recombinant Cripto at Day 3

1. On Wednesday, make four plates (100 mm each) of hanging drops following the standard protocol (*see* **Subheading 3.2.**).
2. On the following Friday, to 4.5 mL of ES differentiation media add recombinant Cripto protein to a final concentration of 10 µg/mL. Sterilize the solution using a 0.22-µm syringe filter. Carefully remove the plates of hanging drops from the incubator and collect EBs in ES differentiation media using a 5-mL pipet. Transfer the EBs to a 15-mL polypropylene tube and allow the EBs to settle to the bottom of the tube. This step takes only 3–4 min. *Do not centrifuge*.
3. Remove the supernatant and gently resuspend the EBs in the Cripto-containing media. Plate the EBs suspension into 60-mm Petri dishes. Incubate at 37°C and proceed with the standard protocol (*see* **Subheading 3.2.** and **Note 18**).

3.5. Neural Differentiation Potential of Cripto$^{-/-}$ ES Cells

Several in vitro systems allowing potential neuronal differentiation of ES cells are described *(3,10–12)*. We have found that Cripto$^{-/-}$ ES cells spontaneously differentiate into neurons using the hanging drop protocol *(6)*. Following the standard protocol (*see* **Subheading 3.2.**), plate Cripto$^{-/-}$-derived EBs at d 6 on gelatin-coated surfaces. At d 11 of the in vitro differentiation protocol, a population of cells with a neuronlike morphology appears, producing a network surrounding the EBs. Neuritic outgrowths become more complex over the following 3 d. At this time (d 14), EBs are processed for subsequent immunofluorescence analysis (*see* **Subheading 3.6.2.**).

3.6. Immunofluorescence Analysis

3.6.1. Cardiomyocyte Detection

1. Recover 9- to 13-d-old adherent EBs from the incubator and wash three times with 1X PBS for 5 min (*see* **Note 21**). Use 0.3 mL/well of 1X PBS.
2. Fix EBs with methanol:acetone (7:3) on ice for 20 min.
3. Remove fixing solution and dry EBs at room temperature. At this step, you can either store fixed EBs (*see* **Note 22**) or proceed with the immunofluorescence protocol.
4. Incubate with 5% BSA in 1X PBS for 15 min at room temperature to block. Use 200 µL/well.
5. Remove blocking solution; incubate with 5% BSA in 1X PBS for additional 30 min at room temperature (*see* **Note 23**).
6. Incubate with antisarcomeric myosin MF-20 antibodies in 5% BSA, 1X PBS, for 2 h at room temperature. Use MF-20 antibodies at a 1:50 dilution.
7. Wash three times with PBS-T (*see* **Subheading 3.2.1.**) for 5 min at room temperature; use 0.3 mL/well.
8. Incubate with antimouse Cy3 secondary antibodies in 5% BSA, 1X PBS, at room temperature for 30 min. Use the antibodies at a 1:400 dilution (*see* **Note 24**).
9. Wash five times with PBS-T for 5 min at room temperature.
10. Incubate with DAPI (250 ng/mL) in 1X PBS for 5 min at room temperature.
11. Wash with 1X PBS for 5 min.
12. Remove the gasket of the chamber slides and dry the liquid drops surrounding the EBs very carefully using a piece of soft paper. Mount with Vecta Shield medium, adding a drop of mount medium on each EB, and lay the cover slip gently on one edge onto mount medium, avoiding introducing bubbles under cover slip.
13. Observe and photograph (*see* **Fig. 2**).

Cripto Role in ES Cell Differentiation

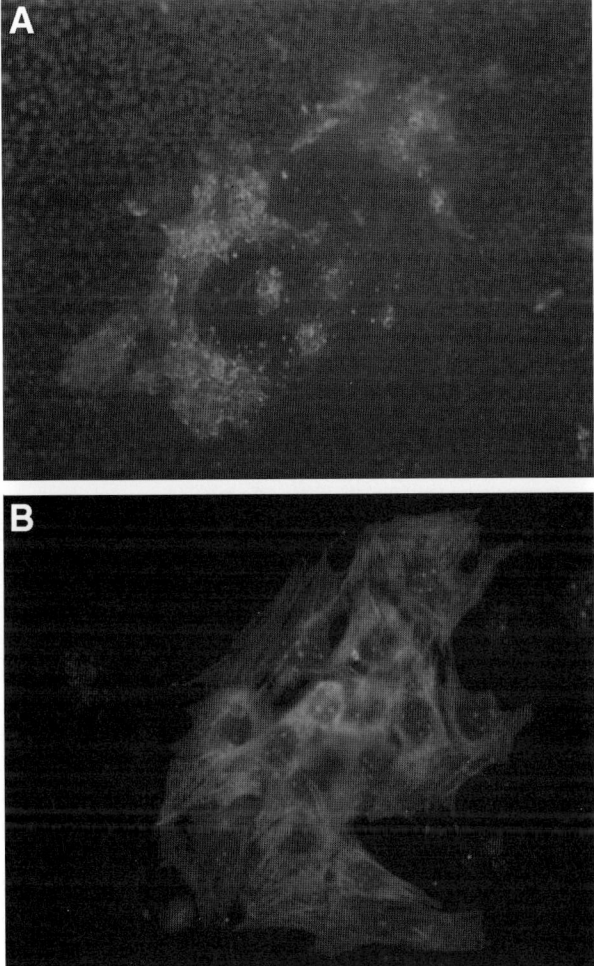

Fig. 2. Immunofluorescence analysis of differentiated wt (RI) embryonic stem cells. On d 13 of in vitro differentiation, the presence of cardiomyocytes is revealed by indirect immunofluorescence using antisarcomeric myosin (MF-20) antibodies (see **Subheading 3.6.1.**). (**A**) ×10 and (**B**) ×40 magnification.

3.6.2. Neuron Detection

1. Remove from the incubator 14-d-old adherent EBs plated on eight-well chamber slides, wash three times with 1X PBS for 5 min, and fix EBs with 4% PFA in PBS 1X at room temperature for 30 min.
2. Wash three times with 1X PBS for 5 min (see **Note 21**); store at −20°C (see **Note 25**) or proceed with the immunofluorescence.
3. Incubate EBs with 10% normal rabbit serum in 1X PBS for 15 min at room temperature to block and permeabilize. Use 200 µL/well.

4. Incubate with anti-β-tubulin isotype III antibodies in 10% normal rabbit serum, 1X PBS, for 2 h at room temperature (120 µL/well). Use antibodies at a 1:400 dilution.
5. Wash three times with 1X PBS for 5 min at room temperature.
6. Incubate with antimouse fluorescein isothiocyanate-conjugated antibodies at 1:40 working dilution in 10% normal rabbit serum, 1X PBS, for 30 min at room temperature (*see* **Note 24**).
7. Wash five times in 1X PBS for 5 min at room temperature.
8. Incubate with DAPI (250 ng/mL) in 1X PBS for 5 min at room temperature.
9. Wash with 1X PBS for 5 min.
10. Remove the gasket of the chamber slides and dry the liquid drops surrounding the EBs very carefully using a piece of soft paper. Mount with Vecta Shield medium, adding a drop of mount medium on each EB, and lay the cover slip gently on one edge onto mount medium, avoiding introduction of bubbles.
11. Observe and photograph (*see* **Fig. 3**).

4. Notes

1. The FBS is heat inactivated before use. Thaw the bottle of serum either overnight at 4°C or at 37°C in a water bath. Incubate at 54°C for 30 min and mix extensively. The heat-inactivated serum is aliquoted into sterile 100-mL bottles and stored at −20°C.
2. ES cells are very sensitive to variations of serum compositions. We have found that the yield of EBs showing beating cardiomyocytes varies significantly with different batches of serum analyzed. Based on our experience, we strongly advise testing different batches of FBS before starting a series of experiments. Use the following criteria: choose a serum that (1) supports EB aggregation in the first 2 d of differentiation and (2) yields a high percentage of wild-type-derived EBs showing areas of beating cardiomyocytes (at least 70–80%).
3. **Caution:** prepare all solutions for tissue culture under a tissue culture hood (*see* **Subheading 2.1.1.**).
4. All tissue culture protocols must be performed using sterile techniques with great attention using clean and detergent-free glassware. Before use, prewarm all media and solutions in a 37°C water bath.
5. We have found that EB aggregation is improved at 6% CO_2.
6. Rabbit polyclonal anti-Cripto antibodies are raised against the mouse recombinant Cripto protein produced in bacteria.
7. It is important to use NP-40 as the detergent in the lysis buffer. Cripto is a GPI-anchored protein *(13)*; GPI-anchored proteins are associated with large vesicular membrane structure (rafts) containing cholesterol, phospholipids, and sphingolipids that are insoluble in Triton X-100.
8. MEF cells are maintained in culture up to the 25th passage before new cells are thawed.
9. It is important not to keep ES cells in culture for long periods to maintain pluripotency. Extensive culturing can result in inconsistent differentiation. However, we have noted that in vitro differentiation is often more efficient in ES cell lines that have many passages and thus are inefficient in germline chimera formation. Finally, we generally keep ES cells in culture up to the 25th passage before new cells are thawed.
10. To keep the ES cells in an undifferentiated state, use the following criteria:
    a. Do not plate ES cells too sparsely.
    b. Use the proper quantity of MEF feeder layer.
    c. Do not let ES cells overgrow.

Cripto Role in ES Cell Differentiation

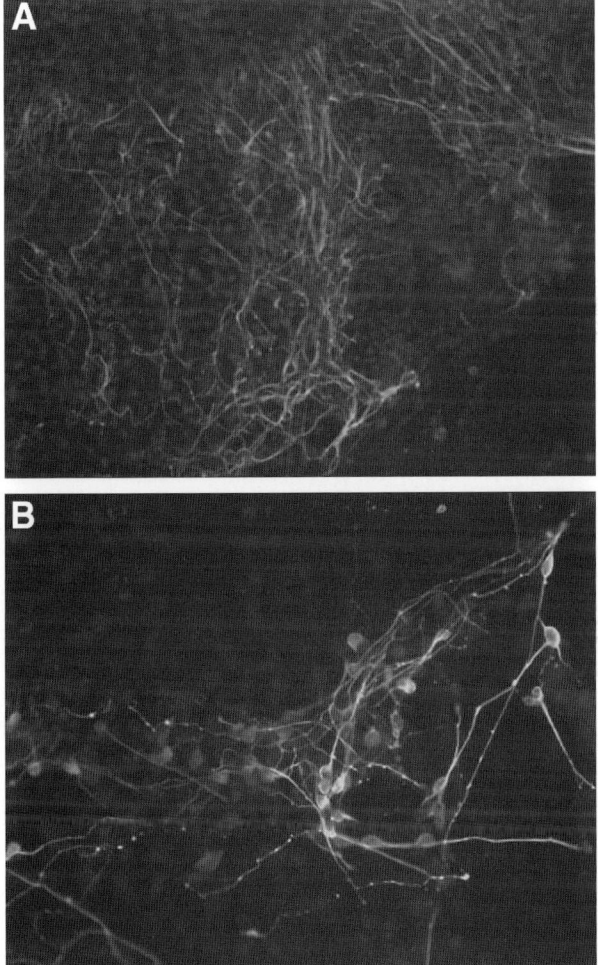

Fig. 3. Immunofluorescence analysis of differentiated Cripto$^{-/-}$ (DE7) embryonic stem (ES) cells. On d 14 of in vitro differentiation, the presence of neural cells in differentiated Cripto$^{-/-}$ ES cell culture is revealed by indirect immunofluorescence using anti-β-III-tubulin antibodies (*see* **Subheading 3.6.2.**). (**A**) ×10 and (**B**) ×20 magnifications.

11. Alternatively, ES cells can be subcultured directly on gelatin-coated dishes 1 or 2 d before starting the hanging drop culture. This step will remove most of the feeder cells.
12. Take 50 µL cell suspension and mix with an equal volume of Trypan blue in a 1.5-mL Eppendorf tube. Count the cells using a hematocytometer. ES cells can be easily distinguished from MEF cells because of their smaller size and round shape.
13. To avoid contamination, do not touch the outer edge of the plate.
14. Note that some EBs can attach to the plate.
15. Follow the same protocol for DE7 Cripto$^{-/-}$ ES cell line.

16. EBs are usually hard to dissolve. You can freeze and thaw to improve dissolution.
17. This step is crucial. We use the Amicon concentrator with ultrafiltration membranes.
18. We have noted that, at this step of the protocol, the FBS composition is extremely critical. In our hands, several batches of FBS did not allow efficient aggregation of cells in the hanging drops in the presence of recombinant Cripto. We thus advise adding recombinant Cripto protein at d 3 (*see* **Subheading 3.4.4.**).
19. We suggest dispensing 50 drops/lid of each plate so the drops are more dispersed.
20. Be careful to avoid breaking drops when inverting the lid of the dish.
21. To remove the solution, turn the chamber slides upside-down on a piece of blotting paper to avoid touching EBs with tips.
22. You can store dried EBs at $-20°C$ up to 6 mo. Remove the frozen EBs from the $-20°C$ and rehydrate by incubation with 1X PBS at room temperature for 30 min before proceeding with immunofluorescence.
23. We have noted that this additional step significantly reduced the background of the MF-20 antibodies.
24. From this point, protect EBs from light to avoid quenching of fluorescence.
25. Add 1 mL 50% glycerol in 1X PBS to each eight-well chamber slide and store at $-20°C$. To proceed with the immunofluorescence protocol, remove the glycerol solution and wash three times with 1X PBS.

References

1. Boheler, K. R., Czyz, J., Tweedie, D., Yang, H. T., Anisimov, S.V., and Wobus, A. M. (2002) Differentiation of pluripotent embryonic stem cells into cardiomyocytes. *Circ. Res.* **91,** 189–201.
2. Bain, G., Kitchens, D., Yao, M., Huettner, J. E., and Gottlieb, D. I. (1995) Embryonic stem cells express neuronal properties in vitro. *Dev. Biol.* **168,** 342–357.
3. Kawasaki, H., Mizuseki, K., Nishikawa, S., et al. (2000) Induction of midbrain dopaminergic neurons from ES cells by stromal cell-derived inducing activity. *Neuron* **28,** 31–40.
4. Adamson, E. D., Minchiotti, G., and Salomon, D. S. (2002) Cripto: a tumor growth factor and more. *J. Cell Physiol.* **190,** 267–278.
5. Xu, C., Liguori, G., Adamson, E. D., and Persico, M. G. (1998) Specific arrest of cardiogenesis in cultured embryonic stem cells lacking Cripto-1. *Dev. Biol.* **196,** 237–247.
6. Parisi, S., D'Andrea, D., Lago, C. T., Adamson, E. D., Persico, M. G., and Minchiotti, G. (2003) Nodal-dependent Cripto signaling promotes cardiomyogenesis and redirects the neural fate of embryonic stem cells. *J. Cell Biol.* **163,** 303–314.
7. Wobus, A. M., Wallukat, G., and Hescheler, J. (1991) Pluripotent mouse embryonic stem cells are able to differentiate into cardiomyocytes expressing chronotropic responses to adrenergic and cholinergic agents and Ca^{2+} channel blockers. *Differentiation* **48,** 173–182.
8. Maltsev, V. A., Rohwedel, J., Hescheler, J., and Wobus, A. M. (1993) Embryonic stem cells differentiate in vitro into cardiomyocytes representing sinusnodal, atrial and ventricular cell types. *Mech. Dev.* **44,** 41–50.
9. Minchiotti, G. Manco, G., Parisi, S., Lago, C. T., Rosa, F., and Persico, M. G. (2001) Structure-function analysis of the EGF-CFC family member Cripto identifies residues essential for nodal signalling. *Development* **128,** 4501–4510.
10. Lee, S. H., Lumelsky, N., Studer, L., Auerbach, J. M., and McKay, R. D. (2000) Efficient generation of midbrain and hindbrain neurons from mouse embryonic stem cells. *Nat. Biotechnol.* **18,** 675–679.

11. Tropepe, V., Hitoshi, S., Sirard, C., Mak, T. W., Rossant, J., and van der Kooy, D. (2001) Direct neural fate specification from embryonic stem cells: a primitive mammalian neural stem cell stage acquired through a default mechanism. *Neuron* **30,** 65–78.
12. Ying, Q. L., Stavridis, M., Griffiths, D., Li, M., and Smith, A. (2003) Conversion of embryonic stem cells into neuroectodermal precursors in adherent monoculture. *Nat. Biotechnol.* **21,** 183–186.
13. Minchiotti, G., Parisi, S., Liguori, G., et al. (2000) Membrane-anchorage of Cripto protein by glycosylphosphatidylinositol and its distribution during early mouse development. *Mech. Dev.* **90,** 133–142.

12

The Use of Embryonic Stem Cells to Study Hedgehog Signaling

Sandy Becker and Laura Grabel

Summary

Because they are capable of differentiating into cell types of all three primary germ layers, embryonic stem cells provide an ideal in vitro system in which to study the signals that regulate differentiation toward a specific cell type. Here, we describe methods for using embryonic stem cells to study the signals that control differentiation into neurectoderm and into vascular cell types, focusing on the Hedgehog signaling pathway.

Key Words: Embryonic stem cells; Hedgehog; neurectoderm; neurogenesis; vasculogenesis.

1. Introduction

Embryonic stem (ES) cells are pluripotent, capable of differentiating into derivatives of all three of the primary germ layers and even into eggs and sperm *(1,2)*. Since their isolation *(3,4)*, this property has made them an ideal starting material with which to work out the in vitro conditions that direct differentiation toward a specific cell type. These studies provide valuable information not only for the developmental biologist interested in identifying the cues that regulate cell fate decisions but also for scientists and physicians interested in designing ES-cell-based transplantation therapies to treat a wide range of degenerative diseases. ES cells can also be used to evaluate the role a specific signal transduction cascade may play in promoting differentiation down a particular pathway. We focus here on the Hedgehog (Hh) signaling pathway.

Hh is a secreted signaling molecule that acts by binding to its receptor, the 12-membrane pass protein Patched1 (Ptc1) *(5,6)*. This binding relieves Ptc1-mediated repression of Smoothened (Smo) action (**Fig. 1**). Although the precise manner in which Hh binding facilitates activation of Smo is unknown, downstream events focus on the transcription factor Cubitus interuptus (Ci) in *Drosophila* and the Gli family in vertebrates *(7)*. By binding to Ptc1, Hh inhibits the cleavage of Ci/Gli to its repressor form, permitting the full-length protein to promote expression of mediators of the Hh response *(7)*. Regulation of Ci/Gli processing includes the action of Costal2, Fused, and Suppressor

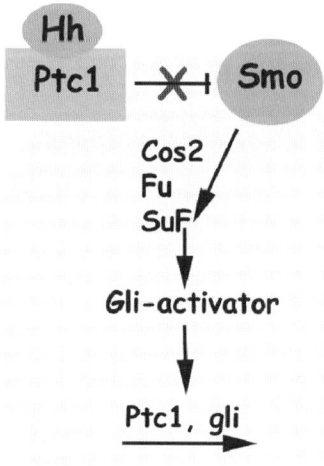

Fig. 1. Hedgehog signaling pathway. Hh binding to Ptc1 relieves Ptc1-mediated suppression of Smo action by inhibiting cleavage of the Gli transcription factor to a repressor form. Full-length Gli promotes the transcription of downstream targets, including Ptc1. Cos2, Fu, and SuF facilitate formation of Gli-containing complex that binds to microtubules. (Please *see* the companion CD for the color version of this figure.)

of Fused, which form a scaffold that links Ci/Gli to microtubules *(8)*. In the absence of the Hh signal, the kinases protein kinase A, glycogen synthase kinase (GSK3), and casein kinase 1 phosphorylate Ci/Gli and mediate its degradation to the repressor form *(9)*. Interestingly, one of the downstream targets of the Hh cascade is the transcription of *Ptc1* itself *(5)*.

There are three *Hh* genes in the mouse: *Sonic hedgehog (Shh), Indian hedgehog (Ihh),* and *Desert hedgehog (Dhh)* *(6)*. The consensus is that the signals encoded by these genes all activate the same downstream signaling cascade, and that the presence of three genes controlled by separate regulatory elements facilitates the expression of the signal at multiple sites and times during embryogenesis. Hh signaling is used throughout embryogenesis in many differentiating tissues to establish cell fate, promote cell proliferation, and mediate programmed cell death. Specific roles include establishing ventral cell identity in the neural tube *(10)* and posterior identity in the developing limb bud *(11)* and promoting proliferation of adult neural progenitors *(12)*. Genes of the Hh cascade have also been implicated in a variety of cancers *(13)*. For example, mutations in the *Ptc1* gene are associated with basal cell carcinomas and with medulloblastomas *(14,15)*, and the *Gli* genes are named for their role in glioma formation *(16)*.

When ES cells are plated in suspension culture, they form embryoid bodies (EBs), which consist of an outer layer of extraembryonic endoderm and an inner core of undifferentiated stem cells. EBs are frequently used as an intermediate in the generation of a variety of differentiated cell types. They provide a model that mimics the interactions

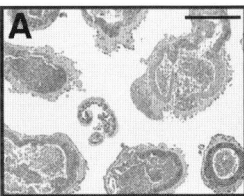 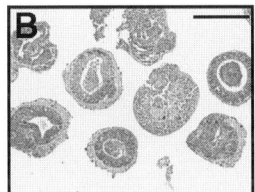 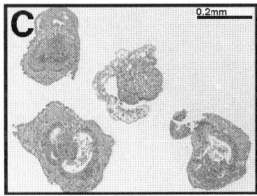

Fig. 2. Expression of *Ihh*, *Ptc1*, and *Ptc1-LacZ* in embryonic stem cell embryoid bodies (EBs). *In situ* hybridization analysis for (**A**) *Ihh* (**A**) or (**B**) *Ptc1* mRNA and β-gal activity in the +/− Ptc1-LacZ cell line. Note *Ihh* is expressed in the outer endoderm layer, and *Ptc1* is expressed in the EB core. (Please *see* the companion CD for the color version of this figure.)

between the visceral endoderm and the epiblast cells that occur in the embryo. During EB culture, the core differentiates first into primitive ectoderm and subsequently into a combination of neurectoderm and mesoderm derivatives, including neurons and blood islands *(17–19)*.

ES cell-derived EBs establish a Hh signaling system as they differentiate in vitro *(20)*. The outer visceral endoderm layer begins to express *Ihh* at approx 4 d in suspension, and the core responds by upregulating expression of *Ptc1* and *gli1* (**Fig. 2**) *(20)*. At approx d 10, *Shh* is also expressed, but in the now-complex EB core *(20)*.

The availability of a variety of cell lines and reagents makes it possible to examine the role of Hh signaling in the differentiation of ES-derived cell types. There are heterozygous and homozygous mutant cell lines for the cascade members *Ihh*, *Smo*, and *Ptc1* *(19,21)*, which can be used to examine loss-of-function and gain-of-function phenotypes. It should be noted that the *Ihh*-deficient cultures still express Shh at later times in culture, but the *Smo*-deficient cell line does provide a system in which all Hh response is absent. In contrast, mutation of the *Ptc1* gene removes repression of *Smo* action and allows constitutive activation of downstream events in Smo-expressing responsive cells. Although these cell lines provide a useful means of establishing a role for Hh signaling in a specific differentiation event, it may prove difficult to define this role fully using mutational analysis alone, particularly if Hh is involved in multiple steps in the pathway from ES cell to the differentiated cell type.

To overcome this limitation, a variety of reagents is available that either inhibit or promote Hh signaling, and these can be added at specific points during in vitro differentiation experiments. Compounds that promote the action of protein kinase A, such as cyclic adenosine monophosphate (cAMP) and forskolin, inhibit the Hh cascade by promoting phosphorylation and therefore cleavage of Ci/Gli to its repressor form *(20,22)*. Natural product plant alkyloids, such as cyclopamine, are antagonists of Hh signaling and act at the level of Smo *(23)*. Another antagonist of Hh signaling, Cur61414, also appears to act at the level of Smo *(24)*. Recombinant Shh peptide is frequently used to promote Hh signaling, and derivatives of the leiosamine family have been identified as potent Hh agonists, promoting the action of Hh in both in vitro and in vivo assays *(24,25)*.

To determine if antagonist or agonist treatment has the desired effect on Hh signaling, expression levels of downstream target genes can be assayed. In addition to using

174   *Becker and Grabel*

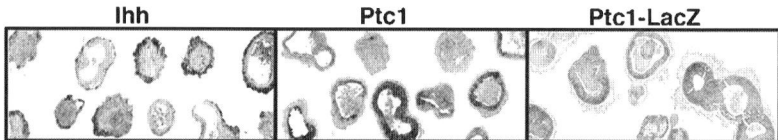

Fig. 3. Cyclopamine, but not the control compound veratramine, inhibits Hh response, decreasing Ptc1-LacZ expression. β-Gal activity in d-10 embryoid bodies (**A**) untreated, treated (**B**) with 2 μ*M* veratramine, or (**C**) with 2 μ*M* cyclopamine. (Please *see* the companion CD for the color version of this figure.)

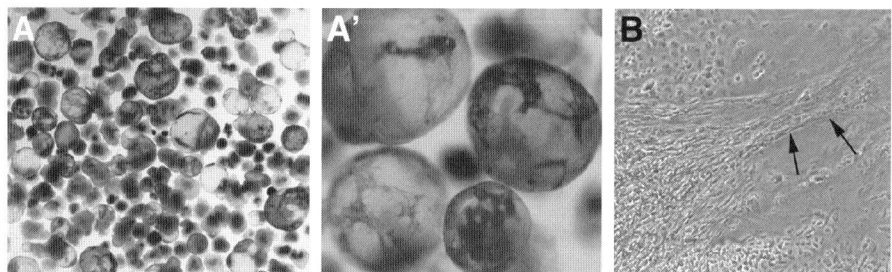

Fig. 4. Differentiation of blood island and neurons. Expression of PECAM1 and benzidine staining for blood cells in d-14 embryoid bodies (**A,A'**), appearance of regions associated with expression of neuronal and glial markers after attachment culture using the retinoic acid protocol (*see* arrows in **B**). (Please *see* the companion CD for the color version of this figure.)

conventional methods for protein and messenger RNA detection, such as Western blot and reverse transcriptase polymerase chain reaction analysis, levels of β-galactosidase (β-gal) can be readily assayed using an ES cell line that carries a *Ptc1-LacZ* allele, driving expression of β-gal from the *Ptc1* promotor *(18,21)*. Cyclopamine, but not the control compound veratramine, inhibits Hh signaling in ES EBs based on decreased expression of Ptc1-LacZ (**Fig. 3**).

Using these cell lines and reagents, we have established a role for Hh signaling in vasculogenesis/angiogenesis and neurectoderm differentiation in ES-derived cultures *(18,19)*. Blood island-containing EBs are shown in **Fig. 4A,A'**, and a culture containing neurons and glia is shown in **Fig. 4B**.

2. Materials (see Note 1)

2.1. ES Cell Maintenance

1. Medium for STO cells and feeder layers: high-glucose Dulbecco's modified Eagle's medium (DMEM; Sigma, St. Louis, MO; cat. no. D8900), 2 m*M* L-glutamine (Gibco, Carlsbad, CA; cat. no. 25030-081), 50 μg/mL penicillin/streptomycin (pen/strep; Gibco, cat. no. 15140-122), 10% newborn calf serum (Hyclone, Logan, UT; cat. no. SH30072.03), 5% fetal calf serum (either ES cell or EB batch; *see* **Note 2**). For 500 mL, in sterile hood, add 50 mL calf serum, 25 mL fetal calf serum, 5 mL pen/strep, and 5 mL glutamine to 450 mL DMEM in a 500-mL media bottle. Store complete medium at 4°C for up to 1 mo; after 1 mo, add more glutamine at the original dilution.

2. Medium for R1 ES cells and their derivatives: DMEM, L-glutamine, and pen/strep as in **step 1**; add 15% fetal calf serum (*see* **Note 2**), 0.1 mM nonessential amino acids solution (Gibco, cat. no. 11140-050), 1 mM sodium pyruvate (mix using Sigma, cat. no. P-2256). For 100 mL, add 11.1 g sodium pyruvate to 100 mL distilled water, sterile filter, aliquot, and store at 4°C), 0.1 mM monothioglycerol (MTG; Sigma, cat. no. M-6145), and leukemia inhibitory factor (LIF; Chemicon, Temecula, CA; cat. no. 2010). For 500 mL, add 75 mL fetal calf serum, 5 mL pen/strep, 5 mL L-glutamine, 5 mL sodium pyruvate solution, 5 mL nonessential amino acids, 7 µL MTG, and 500 µL LIF to 450 mL DMEM in a 500-mL media bottle. For *Smo+/−* and *Smo−/−* ES cells (**item 6**), omit the sodium pyruvate and include 100X nucleotides (**item 3**). Store complete medium at 4°C for up to 1 mo, after which time do not use for ES cells, although it is still fine for STO cells.
3. 100X nucleotides: for 100 mL, add 80 mg adenosine (Sigma, cat. no. A-4036), 85 mg guanosine (Sigma, cat. no. G-6264), 73 mg cytidine (Sigma, cat. no. C-4654), 73 mg uridine (Sigma, cat. no. U-3003), and 24 mg thymidine (Sigma, cat. no. T-1895) to 100 mL distilled water. Dissolve by heating to 37°C. Filter-sterilize and aliquot while still warm. Store at 4°C. Warm to 37°C before use to redissolve the precipitate. Use 1 mL of this stock per 100 mL ES medium.
4. Medium for EB differentiation: same as for stem cells (**item 2**) except that LIF and MTG are not used, and a different batch of fetal calf serum may be needed (*see* **Note 2**). Store complete medium at 4°C for up to 1 mo, after which time do not use for EBs, although it is still fine for STO cells.
5. Mitomycin C (Sigma, cat. no. M-0503): for 2 mL, with 3-mL syringe and needle, add 2 mL phosphate-buffered saline (PBS) to 2-mg vial. Withdraw solution with syringe and filter-sterilize. Mix in small batches because the solution is only good for 1 wk or so at 4°C (*see* **Note 3**).
6. Cell lines: STO cells can be purchased from ATCC (Rockville, MD; cat. no. ATCC CRL 1503). The *Ihh+/−* and *Ihh−/−* cell lines were derived in this laboratory via electroporation of R1 ES cells with a construct from Andrew McMahon's laboratory *(19)*. *Smoothened +/−* and *−/−* cell lines were derived in McMahon's laboratory *(26)*. *Ptc+/−/LacZ* R1 cells were derived in Matthew Scott's laboratory *(21)* (*see* **Note 4**).
7. Gelatin for coating tissue culture dishes (Sigma, cat. no. G-2500). For 1 L, dissolve 1 g gelatin in 1 L distilled water (it is not necessary to boil), aliquot into 100-mL bottles and autoclave. Store at room temperature.
8. Trypsin (Gibco, cat. no. 25300-062): thaw and aliquot 500-mL bottle into approx 75-mL aliquots, freeze at −20°C. Thaw and use as needed, storing small bottles at 4°C. Warm to 37°C before using but avoid long incubations at this temperature as potency will decline.
9. Freezing media for STO and ES cells: make a 2X solution containing 50% FCS, 20% dimethyl sulfoxide (DMSO; Baker, Phillipsbrug, NJ; cat. no. 9444-01), and 30% DMEM (no additives). This can be aliquoted and frozen at −20°C.
10. Freezing media for mitomycin-treated STO cells: make a 1X solution containing 50% fetal calf serum, 10% DMSO, and 40% DMEM (no additives). This can be aliquoted and frozen at −20°C.
11. PBS: 8 g/L sodium chloride, 0.2 g/L potassium chloride, 1.15 g/L sodium phosphate (Na_2HPO_4), 0.2 g/L potassium phosphate (KH_2PO_4), 0.1 g/L calcium chloride, and 0.1 g/L magnesium chloride. Adjust pH to 7.2. Use this for any manipulations involving live cells. It must be sterilized by filtration, not autoclaving. For other purposes, use CMF, which is the same but omits the calcium and magnesium. CMF can be autoclaved.

2.2. In Vitro Differentiation Into Neural Cells

1. Laminin: 1-mg/mL solution (Sigma, cat. no. L-2020). Aliquot 50 µL into 0.5-mL Eppendorf tubes and store at −70°C. Once an aliquot is thawed, store at −20°C. Prepare 10-µg/mL working solution just before use by diluting 1/100 in PBS.
2. EB medium as in **Subheading 2.1.**, **item 4.**
3. Retinoic acid (RA; Sigma, R-2625): make a 10^{-2} M stock solution by adding 6 mg RA to 100% EtOH; store at −70°C. For a 10^{-4} M working solution, dilute stock 1/100 in 100% ethanol; store at −70°C. Use working solution at 1/1000 in EB medium to obtain a final concentration of 10^{-7} M.
4. Four-well Permanox chamber slides (Fisher Scientific, Pittsburgh, PA; cat. no. 12-565-21) (*see* **Note 5**).

2.3. Characterization of ES-Derived Neural Cells

1. CMF (*see* **Subheading 2.1.**, **item 11**).
2. Formaldehyde: dilute formalin (Sigma, cat. no. 1635) 1/10 in PBS or CMF. Diluted solution can be stored at 4°C for a few days.
3. X-gal solution: 1 mg/mL X-gal (Stratagene, La Jolla, CA; cat. no. 300201-61), 5 mM potassium ferricyanate, 5 mM potassium ferrocyanate, and 2 mM magnesium chloride in CMF. Make this fresh before each use from the following stock solutions: for 5 mL 20 mg/mL X-gal, add 100 mg X-gal to 5 mL DMSO and store wrapped in foil to protect from light at −20°C; for 10 mL 200 mM potassium ferricyanate, add 6.59 g to 10 mL distilled water, aliquot, and store at −20°C; for 10 mL 200 mM potassium ferrocyanate, add 7.37 g to 10 mL distilled water, aliquot, and store at −20°C; for 1 M magnesium chloride, store at room temperature.
4. Triton X-100 (Sigma, cat. no. X-100). For stock solution, use 10% in water (*see* **Note 6**).
5. Tween-20 (Sigma, cat. no. P-5927). For stock solution, use 10% in water (*see* **Note 6**).
6. PBT: CMF plus 0.1% Tween-20.
7. PBTr: CMF plus 0.1% Triton X-100.
8. Hoechst staining solution: stock 10-mg/mL solution (Molecular Probes, Eugene, OR; cat. no. H-3570); store at 4°C. Working solution: dilute in CMF to 1 µg/mL (1/10,000). The working solution can be stored at 4°C and reused several times.
9. Gelvatol aqueous mounting medium for glass or plastic slides: heat 70 mL CMF to 50°C. Add 15 g Celvol (Celanese Ltd., Dallas, TX; cat. no. Celvol 205) while stirring. Maintain temperature (do not boil) until granules swell, then continue stirring at room temperature until mixture clears. Add 30 mL anhydrous glycerol and stir to mix. To preserve fluorescence, add 1 g *N*-propyl gallate (Sigma, cat. no. P-3130) and stir to dissolve. Aliquot into 10-mL syringes and store at 4°C. Warm to room temperature before use to prevent bubbles from forming on the slide as it dries (*see* **Note 7**).
10. 200-Microliter cover wells (Midwest Scientific, St. Louis, MO; cat. no. PC200): These gasketed plastic cover slips can be rinsed thoroughly with distilled water and reused many times. The 200-µL size covers an area about 22 × 40 mm on the slide.
11. DAB staining solution: to make 30 mL, which is enough for a copelin jar that can hold up to 10 slides, dissolve two DAB tablets (Sigma, cat. no. D-5905) in 30 mL CMF. Just before use, add 30 µL 30% H_2O_2 (Sigma, cat. no. H-1009) (*see* **Note 8**).
12. Permount mounting medium (Fisher Scientific, cat. no. SP15-100) (*see* **Note 9**).

2.4. In Vitro Differentiation: Blood and Vasculature

Prepare EB medium as in **Subheading 2.1.**, **item 4**.

2.5. Characterization of ES-Derived Blood Cells and Vasculature

Additional materials beyond those needed for characterization of ES-derived neural cells:

1. Benzidine staining solution: use a needle and 20-mL syringe to add 50 mL 0.5 M glacial acetic acid to the Isopac bottle of benzidine (Sigma, cat. no. B-1883). To make the staining solution (just before use): withdraw stock solution from Isopac bottle with syringe and needle; add 20 µL/mL 30% H_2O_2 and dilute this 1:4 with PBS (*see* **Note 10**).
2. Bovine serum albumin (Sigma, cat. no. A-7906).
3. Zinc fixative: to make 1 L, add 0.5 g calcium acetate, 5 g zinc acetate, and 5 g zinc chloride to 1 L 0.1 M Tris-HCl at pH 7.4 (make from 12.1 g Tris-base, 900 mL distilled water, and 81.5 mL HCl). Do not readjust pH. This can be stored at room temperature.

3. Methods (see Note 11)
3.1. Maintenance of Undifferentiated ES Cells

Maintain ES cells on mitomycin-treated STO feeder layers.

3.1.1. Gelatin Coating for Tissue Culture Dishes

1. Pipet gelatin solution to cover dish bottom.
2. Incubate for 10 min (up to 0.5 h).
3. Remove gelatin and pipet in cell suspension.

3.1.2. Preparation of STO Feeder Layers

STO cells are maintained in the medium described in **Subheading 2.1., item 1**, in 10-cm tissue culture dishes. They should be seeded at a density that requires passing every 3 or 4 d.

1. To treat with mitomycin C, remove medium from confluent dishes and replace with 5 mL fresh medium containing 50 µL mitomycin C solution (10 µg/mL final concentration) (*see* **Subheading 2.1., item 5**).
2. Return to 37°C incubator for 2–4 h.
3. Remove mitomycin C solution, rinse dishes once with PBS, and trypsinize them as usual.
4. Replate into gelatin-coated 6-cm dishes at half the previous density (for example, one 10-cm dish would seed four 6-cm dishes). Feeder layers are only good for 1 wk or so, but the cells can be frozen and stored for several weeks at −70°C (*see* **Subheading 3.1.3.**).

3.1.3. Freezing STO Feeder Cells

1. After treatment with mitomycin C and trypsinization, resuspend the cell pellet in freezing medium (**Subheading 2.1., item 10**).
2. Combine the cells from two 10-cm dishes into each 1-mL vial and freeze at −70°C. Cells remain viable for several weeks at −70°C, possibly longer in liquid nitrogen.
3. To use, thaw one vial into eight gelatin-coated 6-cm tissue culture dishes.

3.1.4. Maintenance of ES Cells

Grow stem cells on STO feeder layers in 6-cm tissue culture dishes. They should be seeded at a density that requires passing every 3–4 d (*see* **Note 12**).

3.1.5. Formation of EBs

1. Trypsinize confluent dish of ES cells as usual (*see* **Note 12**).
2. Resuspend cells from 6-cm dish in 10 mL EB medium. This cell suspension will contain the STO feeder cells as well as the ES cells. It is better to eliminate the feeder cells at this point (*see* **Note 13**).
3. To eliminate feeder cells, return the cell suspension to a tissue culture dish (without gelatin) and incubate at 37°C for 6–7 min.
4. At the end of this time, observe in a microscope to check that a good number of the STO cells have adhered to the dish (they will do this much more quickly than the ES cells).
5. When they have, pipet the floating ES cells into a 10-cm Petri dish. Feed EBs as needed (*see* **Note 14**) by pipeting the medium and cells into a 15-mL conical tube, settling the EBs to the bottom, aspirating the supernatant, and replacing it with fresh medium.

3.1.6. Freezing and Thawing STO and ES Cells

1. To freeze STO or ES cells, trypsinize as usual and resuspend the cell pellet in 0.5 mL complete medium for each dish you are freezing.
2. Add equal volume of 2X freezing medium and store at −70°C. For long-term storage, cells should be moved in a few days to liquid nitrogen. It does not matter that you will be freezing a feeder layer along with the ES cells because most of these will not survive the subsequent thaw. Thaw frozen ES cells onto a 6-cm treated STO feeder layer; thaw STO cells onto a plain 10-cm tissue culture dish.

3.2. In Vitro Differentiation Into Neural Cells

3.2.1. Laminin Coating of Tissue Culture Dishes or Chamber Slides

1. Pipet just enough 10-µg/mL laminin solution to cover the bottom of the dish or well and return to a 37°C incubator overnight.
2. To use, aspirate off the laminin solution and immediately add cell suspension.

3.2.2. RA-Based Neural Differentiation Protocol (27)

1. Grow EBs for 4 d in EB medium, then for 4 d more with the addition of $10^{-7} M$ RA.
2. Change the medium as needed, generally every 1 or 2 d.
3. On the ninth day, plate 30–40 EBs into each well of a four-well chamber slide that has been coated with 10 µg/mL laminin and continue to culture them in EB medium (without RA) for 5–10 d. In this time-period, wild-type cultures, untreated with Hh agonist or antagonist, should show morphological evidence of neuronal and glial differentiation (*see* **Fig. 4B**). Mutant or treated cells can be compared with these controls.

3.3. Characterization of ES Cell-Derived Neural Cells

3.3.1. Histochemical Staining for β-Gal

Using the heterozygous *Ptc1/LacZ* cell line, in which LacZ is driven by the *Ptc1* promoter, the cells responding to an Hh signal can be identified.

1. Rinse the cells in the chamber slide, or the EBs in suspension, in PBS twice.
2. Fix them in 3.7% formaldehyde for 15 min at room temperature.
3. Rinse with CMF twice and incubate with X-gal solution at 30–37°C for several hours to overnight (*see* **Note 15**).

Table 1
Antibodies for Characterization of ES-Derived Neural Cells

Antibody/source	Cells marked	Dilution	Secondary
Nestin (DSHB, Iowa City, IA; cat. no. Rat-401) (see **Note 21**)	Neural progenitors (see **Note 21**)	1/4 for supernatant 1/1000 for ascites	Antimouse
β-III-tubulin (Covance, Berkeley, CA; cat. no. MMS-435P)	Young neurons	1/1000	Antimouse
Microtubule-associated protein 2 (MAP2; Sigma, cat. no. M-4403)	Mature neurons	1/400	Antimouse
Phospho-H3 histone (Upstate, Lake Placid, NY; cat. no. 06-570)	Dividing cells	1/400	Antirabbit
Stage-specific embryonic antigen 1 (SSEA-1; DSHB, cat. no. MC-480)	Embryonic stem cells	1/50	Antimouse immunoglobulin M
Troma 1 (DSHB, cat. no. Troma 1)	Extraembryonic endoderm	1/10	Antirat
RC2 (gift of J. Crandall)	Radial glia	Undiluted	Antimouse immunoglobulin M
Hu (Molecular Probes, cat. no. A-21271)	Neurons (nuclear staining)	1/200	Antimouse

4. When staining is completed, postfix in 3.7% formaldehyde for 30 min at room temperature and rinse in CMF.
5. Dehydrate EBs to methanol, embed, and section, followed by immunocytochemistry (see **Note 16**).
6. Samples in chamber slides may be mounted in an aqueous mounting medium such as Gelvatol (see **Note 5**) or stored at 4°C in CMF for up to a few days prior to antibody staining.

3.3.2. Immunocytochemical Staining for Neural Markers

Immunocytochemical staining for neural markers may be done on recently fixed material in chamber slides or on paraffin sections of EBs.

1. For chamber slides, rinse wells twice with PBS, then fix 12 min in 3.7% formaldehyde.
2. Rinse again twice with CMF. At this point, the material can be stored at 4°C for a few days. This procedure can be done on material that has been previously stained for β-gal expression (see **Subheading 3.3.1.**). To use stored or X-gal-incubated chamber slides, start at **step 3**.
3. Permeabilize with 0.5% Triton X-100 in CMF for 10 min.
4. Rinse again in CMF and block for 1–2 h at room temperature in CMF containing 3% powdered milk.
5. Apply primary antibody diluted in blocker (see **Table 1**) and incubate overnight at 4°C. Four-well chamber slides need about 400 µL antibody solution to cover the bottom of each well.
6. The next morning, wash wells three times for 20 min each with CMF, then incubate in secondary antibody diluted in blocker (see **Table 1**) for 90 min at room temperature. If using fluorescent secondary antibody, protect samples from light.

7. After antibody incubation, rinse twice with CMF for 5 min at room temperature.
8. If using Hoechst to visualize the nuclei (recommended with fluorescent secondaries), incubate 10 min with Hoechst staining solution, then rinse again twice with CMF and cover slip in Gelvatol (*see* **Notes 5** and **17**).

3.3.3. Immunohistochemical Staining on Paraffin Sections of EBs

1. Dewax slides in xylenes three times for 10 min each.
2. Soak off the xylenes in 100% ethanol for 5 min.
3. If using a horseradish peroxidase (HRP)-conjugated secondary antibody (*see* **Note 16**), quench the endogenous peroxidase by soaking for 10 min in 1% H_2O_2 in methanol. Then, rehydrate by soaking for 10 min in distilled water, then 10 min in CMF.
4. If using a fluorescent secondary antibody (*see* **Note 16**), skip **step 3** and rehydrate through an ethanol series to distilled water and then CMF.
5. After the CMF incubation, block approx 1 h in CMF containing 0.1% Triton X-100 and 3% powdered milk.
6. Do the primary antibody incubation under cover wells. Each gasketed cover slip requires 200 µL antibody solution diluted in CMF containing 3% powdered milk (*see* **Table 1** for antibody dilutions). Leave primary antibody overnight at 4°C.
7. In the morning, carefully remove the cover wells, return the slides to a square staining jar, and rinse them three times for 5 min in CMF (the cover wells can be reused many times after rinsing in distilled water and air drying).
8. Incubate slides, again using cover wells, for 1–2 h in secondary antibody.
9. After secondary incubation, rinse again three times for 5 min each in CMF.
10. If using Hoechst to visualize the nuclei (strongly recommended if using a fluorescent secondary antibody), do it now (*see* **Subheading 3.3.2.**, **item 8**). If using fluorescent secondary antibody, cover slip in Gelvatol (*see* **Note 17**).
11. If using HRP-conjugated secondary antibody, incubate rinsed slides in DAB staining solution for a few seconds up to 10 min. It is recommended to monitor staining under a microscope. When staining is done, stop DAB reaction by immersing slides in CMF. Slides can be immediately cover slipped in Gelvatol or dehydrated through an ethanol series into xylenes and cover slipped in Permount (*see* **Note 17**).

3.4. In Vitro Differentiation Into Hematopoietic and Vascular Cells

Grow EBs in EB medium (*see* **Subheading 3.1.5.**). After 8–12 d, a proportion of the EBs will expand into a yolk saclike structure containing blood islands and vasculature (*see* **Note 18** and **Fig. 4A,A'**). EBs from mutant cell lines or those treated with agonists or antagonists of the Hh pathway can be compared with wild-type, untreated controls.

3.5. Characterization of ES Cell-Derived Hematopoietic Cells

3.5.1. Benzidine Staining for Hemoglobin

1. Rinse an aliquot of EBs in PBS.
2. Aspirate off the PBS and replace with 1 mL benzidine staining solution. Hemoglobin-containing blood cells will stain blue-black within a few minutes, sometimes in a few seconds. The samples can be transferred to a 35-mm Petri dish so that staining can be monitored in a dissecting microscope.
3. When staining is complete, return samples to a conical tube and rinse with PBT (*see* **Note 19**).
4. Fix with 3.7% formaldehyde for 15 min at room temperature (*see* **Note 20**).

Table 2
Antibodies for Characterization of ES-Derived Hematopoietic and Vascular Cells

Antibody/source	Cells marked	Dilution	Secondary
PECAM1 (Pharmingen, Los Angeles, CA; cat. no. Mec13.3)	Vascular endothelial cells	1/500	Antirat
Flk-1 (Pharmingen, cat. no. Ly-73)	Vascular endothelial cells	1/50	Antirat
CD34 (Pharmingen, cat. no. RAM 34)	Vascular endothelial cells	1/50	Antirat
αSMA (Sigma, cat. no. 1A4)	Vascular smooth muscle cells	1/500	Antimouse

5. Rinse twice in CMF. At this point, the EBs can be dehydrated into methanol, embedded in paraffin, and sectioned for immunostaining, or further staining can be done on them in whole mount (see **Subheading 3.5.3.**).

3.5.2. Histochemical Staining for β-gal on Ptc+/−LacZ EBs

The Ptc1+/−LacZ cell line allows the identification of cells responding to an Hh signal because these cells will express β-gal. Staining can be done on fresh EBs or on EBs previously stained for hemoglobin, as described in **Subheading 3.3.1**. X-gal-exposed EBs can then be subjected to immunohistochemical staining, either in whole mount or after paraffin embedding and sectioning, to determine the identity of the cells responding to an Hh signal.

3.5.3. Whole-Mount Immunohistochemical Staining for Blood and Vasculogenesis Markers

Whole-mount staining allows visualization of vascular tubelike structures in the EBs.

1. Fix EBs for 30 min in 3.7% formaldehyde (or use EBs previously stained for β-gal and refixed as described in **Subheading 3.3.1.**). Do not dehydrate.
2. Block all day (6–8 h) at 4°C in PBTr plus 3% powdered milk.
3. Incubate overnight on a rocker at 4°C in primary antibody diluted in blocking solution (for antibodies and dilutions, see **Table 2**).
4. Wash five times for 1 h in blocking solution.
5. Incubate overnight at 4°C with secondary antibody diluted in blocker.
6. Perform five 1-h washes in blocking solution, followed by a 20-min incubation in PBTr plus 0.2% bovine serum albumin.
7. Develop color in DAB staining solution plus 0.5% nickel chloride (optional; the nickel chloride imparts a blue-black color instead of a golden-brown color). Monitor staining, which should take several minutes, under a dissecting microscope.
8. When staining is completed, stop reaction by immersing slides in CMF. Photographs can be taken at this point.
9. Dehydrate to methanol, embed in paraffin, and section.

3.5.4. Immunohistochemical Staining on Paraffin Sections

This allows staining with different antibodies to be done on adjacent sections of the same material, facilitating comparisons. However, the EBs must be fixed in a zinc fixative (see **Subheading 2.5., item 3**) if antibody against PECAM is to be used. Fixing in formaldehyde will result in a very weak signal from the PECAM antibody. To fix EBs in zinc fixative:

1. Rinse them in CMF.
2. Incubate in the zinc fixative 8–12 h at room temperature.
3. In the morning, rinse three times in CMF and then dehydrate to 100% methanol. EBs can be stored for several weeks in methanol.
4. Embed in paraffin and section. Antibody staining is done the same as for neural markers (*see* **Subheading 3.3.3.**). For antibody dilutions, *see* **Table 2**.

4. Notes

1. All water used for tissue culture is glass distilled. Glassware is rinsed several times immediately after use with distilled water (no soap or detergent) and autoclaved.
2. Batches of fetal calf serum must be tested to find the best for maintaining the stem cells in an undifferentiated and proliferating state. Atlanta Biologicals has provided the best serum for us in recent years. We test batches for use both in maintaining stem cells and in differentiating EBs. Sometimes, a different batch is needed for each of these applications. At present, we are using Knockout Serum Replacement (Gibco, cat. no. 10828-028) for ES cell growth, but it is not suitable for EBs and does not support the proliferation of STO fibroblasts (although the feeder layer cells will survive in it). Note that serum batches pretested as "ES cell competent" may be very unsatisfactory for supporting the differentiation of EBs.
3. Mitomycin C is very toxic; wear gloves and mix it in the original vial.
4. It is advisable to have at least two lines of heterozygous and homozygous mutant cells because ES cell lines can vary considerably in their growth and differentiation characteristics. For this reason, any peculiarity in a mutant cell line must be confirmed in another mutant cell line before attributing it to the mutation.
5. Permanox is not identical to tissue culture plastic, and cells do not adhere to it as well. However, for further analysis using immunocytochemistry, Permanox does not autofluoresce, and chamber slides provide a smaller volume than a tissue culture dish and better optics than a multiwell plate. *Note that aqueous mounting media must be used.* Organic mounting media such as Permount will soften and warp the plastic slide.
6. Both Triton X-100 and Tween-20 are viscous and difficult to pipet accurately. Make 10% stock solutions in water. To prevent growth of microorganisms, filter-sterilize. Store at room temperature.
7. Gelvatol will harden on the slides after a few days, so they can be stored vertically, but slides should be refrigerated for long storage.
8. DAB tablets generally do not dissolve completely, but this is okay. DAB (3,3′-diaminobenzidine) is hazardous. Use gloves when handling it; buy the tablets instead of the powder, which would need to be measured, and dispose of the solution as hazardous waste (benzidine).
9. Permount should be applied in a fume hood, and slides mounted with Permount should be kept in the fume hood until they have dried; then they can be stored at room temperature.
10. The benzidine stock solution should be wrapped in foil to protect from light, and it is stable for months at room temperature. Benzidine is hazardous, so both the stock solution and the staining solution should be disposed of as hazardous waste, and gloves should be worn when handling them.

11. The cultures described here require a humidified incubator at 37°C and 5% CO_2. Prewarm the medium to 37°C before working with the cells and do not leave cells out of the incubator for any longer than necessary.
12. It is important to trypsinize and triturate cells thoroughly, so achieve a single-cell suspension when passing. It may be necessary to trypsinize for as long as 15 min. Large clumps of stem cells will tend to differentiate into extraembryonic endoderm over time. Use of old feeder layers or those that are too sparse or too dense or failing to pass cells when confluent will also lead to untimely differentiation of stem cells.
13. STO feeder cells that contaminate the culture of EBs can adhere to the Petri dish and reconstitute the feeder layer if they are numerous. This will in turn enable the ES cells to adhere to the dish and delay or prevent their differentiation into EBs. In extreme cases, they can be incorporated into the EBs as they form.
14. There will always be a few STO cells at the beginning of EB culture despite your best efforts (*see* **Note 5**). They will stick to the Petri dish, sometimes also enabling the ES cells to adhere. If ES cells have adhered to the dish, blow them off gently with a serological pipet and transfer them to another dish, leaving the STO cells behind. As the EBs grow, they may become too numerous for one dish. Either discard some or split them into additional dishes.
15. Staining may develop in only 1 h, or it may need to incubate overnight. It will happen faster at the higher temperature, but it may be more convenient to incubate at the lower temperature overnight. If staining becomes too intense, background may develop, and it may be difficult to discern the signal from subsequent immunostaining (generally golden/bronze) against the deep aqua of the X-gal reaction.
16. Fluorescent antibody staining cannot be done on paraffin sections of material that have been previously stained using X-gal. The background will be unacceptably high, completely obscuring the signal. A nonfluorescent secondary such as HRP must be used.
17. Hoechst-stained material must be mounted in aqueous mounting medium. Permount will cause the Hoechst stain to bleed. Samples that are mounted in Permount may also require counterstaining with eosin to render the unstained parts visible. Mounting in Gelvatol will preserve the optical contrast better (at the expense of definition).
18. Some ES cell lines do this better (more rapidly and in greater numbers) than others. Some lines will not differentiate in this manner. We have found R1 ES cells and those derived from R1s generally good for this purpose. It may take as much as 15 d for a good number of EBs to expand. Blood islands may or may not be visible morphologically before staining.
19. It is important to use PBT here and not CMF because the treated EBs tend to float in CMF, which can lead to loss of sample.
20. Do not panic when the staining seems to disappear during fixation. The color turns from blue-black to golden brown in formaldehyde, but it is not gone. The golden brown staining will persist through subsequent manipulations, such as immunohistochemistry or *in situ* hybridization.
21. DSHB, Iowa City, IA, sells nestin antibody as a supernatant and as ascites. The supernatant is harvested from hybridoma cells grown with rabbit serum; therefore, the antibody cannot be used for double labeling with rabbit polyclonals. The ascites is suitable for double labeling, but dilutions must be established for each batch. Note that nestin antibody labels muscle and endoderm precursor cells as well as neuronal precursors, so care must be used in interpreting the results.

Acknowledgment

We would like to thank Joe Carpentino and Hunter Cai for careful reading of the manuscript.

References

1. Hubner, K., Fuhrmann, G., Christenson, L. K., et al. (2003) Derivation of oocytes from mouse embryonic stem cells. *Science* **300**, 1251–1256.
2. Geijsen, N., Horoschak, M., Kim, K., Gribnau, J., Eggan, K., and Daley, G. Q. (2004) Derivation of embryonic germ cells and male gametes from embryonic stem cells. *Nature* **427**, 148–154.
3. Evans, M. J. and Kaufman, M. H. (1981) Establishment in culture of pluripotential cells from mouse embryos. *Nature* **292**, 154–155.
4. Martin, G. R. (1981) Isolation of a pluripotent cell line from early mouse embryos cultured in medium conditioned by teratocarcinoma stem cells. *Proc. Natl. Acad. Sci. USA* **78**, 7634–7638.
5. Ingham, P. W. and McMahon, A. P. (2001) Hedgehog signaling in animal development: paradigms and principles. *Genes Dev.* **15**, 3059–3087.
6. McMahon, A. P., Ingham, P. W., and Tabin, C. J. (2003) Developmental roles and clinical significance of hedgehog signaling. *Curr. Topics Dev. Biol.* **53**, 1–114.
7. Ruiz i Altaba, A. (1999) Gli proteins and Hedgehog signaling: development and cancer. *Trends Genet.* **15**, 418–425.
8. Lum, L., Zhang, C., Oh, S., et al. (2003) Hedgehog signal transduction via Smoothened association with a cytoplasmic complex scaffolded by the atypical kinesin, Costal-2. *Mol. Cell* **12**, 1261–1274.
9. Price, M. and Kalderon, D. (2002) Proteolysis of the Hedgehog signaling effector cubitus interruptus requires phosphorylation by glycogen synthase kinase 3 and casein kinase 1. *Cell* **108**, 823–835.
10. Briscoe, J. and Ericson, J. (1999) The specification of neuronal identity by graded Sonic Hedgehog signaling. *Semin. Cell Dev. Biol.* **10**, 353–362.
11. Pearse, R. V., 2nd, and Tabin, C. J. (1998) The molecular ZPA. *J. Exp. Zool.* **282**, 677–690.
12. Machold, R., Hayashi, S., Rutlin, M., et al. (2003) Sonic hedgehog is required for progenitor cell maintenance in telencephalic stem cell niches. *Neuron* **40**, 189–190.
13. Bale, A. E. (2002) Hedgehog signaling and human disease. *Annu. Rev. Genomics Hum. Genet.* **3**, 47–65.
14. Bale, A. E. and Yu, K. P. (2001) The hedgehog pathway and basal cell carcinomas. *Hum. Mol. Genet.* **10**, 757–762.
15. Goodrich, L. V. and Scott, M. P. (1998) Hedgehog and patched in neural development and disease. *Neuron* **21**, 1243–1257.
16. Matise, M. P. and Joyner, A. L. (1999) Gli genes in development and cancer. *Oncogene* **18**, 7852–7859.
17. Rathjen, J., Haines, B. P., Hudson, K. M., Nesci, A., Dunn, S., and Rathjen, P. D. (2002) Directed differentiation of pluripotent cells to neural lineages: homogeneous formation and differentiation of a neurectoderm population. *Development* **129**, 2649–2661.
18. Maye, P., Becker, S., Sieman, H., et al. (2004) Hedgehog signaling is required for the differentiation of ES cells into neurectoderm. *Dev. Biol.* **265**, 276–290.
19. Byrd, N., Maye, P., Becker, S., et al. (2002) Hedgehog signaling is essential for yolk sac vasculogenesis. *Development* **129**, 361–372.
20. Maye, P., Becker, S., Kasameyer, E., Byrd, N., and Grabel, L. (2000) Indian hedgehog signaling in extraembryonic ectoderm differentiation in ES embryoid bodies. *Mech. Dev.* **94**, 117–132.
21. Goodrich, L. V., Milenkovic, L., Higgins, K. M., and Scott, M. P. (1997) Altered neural cell fates and medulloblastoma in mouse *Patched* mutants. *Science* **277**, 1109–1113.

22. Chen, Y., Gallaher, N., Goodman, R., and Smolik, S. (1998) Protein kinase A directly regulates the activity and proteolysis of cubitus interruptus. *Proc. Natl. Acad. Sci. USA* **95,** 2349–2354.
23. Chen, J. K., Taipale, J., Cooper, M. K., and Beachy, P. A. (2002) Inhibition of Hedgehog signaling by direct binding of cyclopamine to smoothened. *Genes Dev.* **16,** 2743–2748.
24. Chen, J. K., Taipale, J., Young, K. E., Maiti, T., and Beachy, P. A. (2002) Small molecule modulation of Smoothened activity. *Proc. Natl. Acad. Sci. USA* **99,** 14,071–14,076.
25. Frank-Kamenetsky, M., Zhang, X. M., Bottega, S., et al. (2002) Small-molecule modulators of Hedgehog signaling: identification and characterization of Smoothened agonists and antagonists. *J. Biol.* **1,** 10.
26. Wijgerde, H., McMahon, J., Rule, M., and McMahon, A. P. (2002) A direct requirement for Hedgehog signaling for normal specification of all ventral progenitor domains in the presumptive mammalian spinal cord. *Genes Dev.* **16,** 2849–2864.
27. Bain, G., Kitchens, D., Yao, M., Huettner, J., and Gottleib, D. (1995) Embryonic stem cells express neuronal properties in vitro. *Dev. Biol.* **168,** 342–357.

13

Transfection and Promoter Analysis in Embryonic Stem Cells

Sangmi Chung and Kwang-Soo Kim

Summary

Embryonic stem (ES) cells are useful for the molecular analysis of developmental pathways when combined with efficient genetic manipulation and in vitro differentiation procedures. To facilitate the genetic modification of ES cells, we describe how to analyze the transcriptional activities of different promoter systems. The DNA segments from a number of promoters are subcloned into the multiple cloning sites of the green fluorescent protein (GFP) reporter plasmid. The resulting recombinant plasmids are used to transfect ES cells by lipofection. Using luciferase-normalized GFP assays, we observed differing promoter activities in transient transfections of ES cells. The judicious selection of the optimum promoter system for specific genetic modification of ES cells will ensure the successful generation of transgenic ES cell lines.

Key Words: Embryonic stem cells; GFP assay; luciferase assay; promoter strength; transfection; transgene expression.

1. Introduction

Embryonic stem (ES) cells are pluripotent cells derived from the inner cell mass of preimplantation mouse embryos *(1)* and represent embryonic precursor cells that can give rise to any cell type in the embryo *(2,3)*. This developmental potential of ES cells has provided a powerful system to generate experimental animals with specific genetic alterations via homologous recombination *(4,5)*. Moreover, ES cells can also be differentiated into many different cell lineages in vitro *(6,7)*, making them a useful tool to analyze critical steps of early and late events of cell development that would otherwise be difficult to study in experimental animal systems. Because ES cells are easily accessible for genetic modification, they can be used to test certain transgene expression during cellular differentiation *(8,9)*. Another great potential of ES cells is that they may provide an unlimited cell source for transplantation therapies of various brain disorders, such as Parkinson's disease *(10–13)*. For the last application, genetic modification of ES cells may help drive cell type-specific differentiation appropriate for therapeutic purposes *(9)*.

From: *Methods in Molecular Biology, vol. 329: Embryonic Stem Cell Protocols: Second Edition: Volume 1*
Edited by: K. Turksen © Humana Press Inc., Totowa, NJ

For genetic modification of cells by exogenous transgene expression, different viral and cellular promoters have been used (14–16), and one of the most popular choices has been the cytomegalovirus promoter because of its strong activity in most cell lines (17,18). However, promoters show different strength depending on the cellular context in which they operate. Therefore, it is necessary to determine the optimal promoter system for genetic manipulation of ES cells. Thus, here we describe how to transfect ES cells by lipofection and analyze the activity of different promoter systems in an ES cell context using the green fluorescent protein (GFP) reporter and the luciferase assays (see Note 1). The choice of an optimum promoter system in ES cells will ensure successful generation of transgenic ES cell lines.

2. Materials
2.1. Cell Culture
1. Mg^{2+}- and Ca^{2+}-free D-phosphate-buffered saline (PBS; 500-mL bottle; Invitrogen, Carlsbad, CA; cat. no. 14190-144).
2. Dulbecco's modified Eagle's medium (DMEM; 500-mL bottle; Invitrogen, cat. no. 10569-010).
3. Donor herd, heat-inactivated, cell culture-tested horse serum (500-mL bottle; Sigma, St. Louis, MO; cat. no. H1138).
4. Leukemia inhibitory factor (ESGRO; 10^7 U in 1 mL; Chemicon International Inc., Temecula, CA, cat. no. ESG1107).
5. 5-μM β-mercaptoethanol (50-mL bottle; Invitrogen, cat. no. 21985-023).
6. 200 mM L-glutamine (20-mL bottle; Invitrogen, cat. no. 25030-149).
7. 10 mM minimum essential medium nonessential amino acids (50-mL bottle; Invitrogen, cat. no. 11140-050).
8. 1 M HEPES (50-mL bottle; Invitrogen, cat. no. 105630-080).
9. 100X streptomycin-penicillin (50-mL bottle; Invitrogen, cat. no. 15140-122).
10. 10X trypsin-ethylenediaminetetraacetic acid (EDTA; 50-mL bottle; Invitrogen, cat. no. 15400-054).
11. Gelatin (Sigma, cat. no. G1393). For 500-mL 0.1% solution, dissolve 0.5 g gelatin in 500 mL Mg^{2+}- and Ca^{2+}-free PBS in a water bath at 50–65°C for 15–30 min. Filter the solution, preferably while warm, through a 0.2-μm filter unit. Store at 4°C.
12. Falcon polystyrene 100-mm tissue culture dishes (Fisher Scientific, Pittsburgh, PA; cat. no. 353003).
13. Falcon polystyrene 60-mm tissue culture dishes (Fisher Scientific, cat. no. 353002).
14. Falcon polystyrene 35-mm tissue culture dishes (Fisher Scientific, cat. no. 353001).
15. 15-mL polypropylene conical tubes (Fisher Scientific, cat. no. 352097).
16. 0.2-μm filter units (Fisher Scientific, cat. no. 09-740-25A).
17. Dimethyl sulfoxide (DMSO; Sigma, cat. no. D-8779).

2.1.1. Media
1. Media for maintaining ES cells: ES cells are maintained in growth medium consisting of DMEM supplemented with 2 mM glutamine, 0.001% β-mercaptoethanol, 1X nonessential amino acids, 10% donor horse serum, and human recombinant leukemia inhibitory factor (ESGRO; 2000 U/mL). To prepare 500-mL media, combine 429 mL DMEM, 5 mL L-glutamine, 5 mL minimum essential medium nonessential amino acid (NEAA), 5 mL HEPES, 1 mL β-mercaptoethanol, 5 mL penicillin-streptomycin, 50 mL horse serum, and 50 μL ESGRO and then filter through a 0.2-μm filter unit. Store the media at 4°C.

ES Cell Transfection and Promoter Analysis

2. Media for ES cell transfection: for transfection of ES cells, media are prepared in the same way as in **Subheading 2.1.1.**, **item 1**, without the addition of penicillin-streptomycin.
3. Freezing medium: 90% horse serum and 10% DMSO.

2.1.2. Required Equipment for Cell Culture

1. 37°C water bath to warm the media before use.
2. 65°C water bath to heat inactivate serum.
3. Hematocytometer.
4. Laminar flow hood.
5. Humidified incubator with 5% CO_2.
6. Tabletop centrifuge.

2.2. Construction of Reporter DNA

1. Restriction enzymes.
2. pEGFP-1 vector DNA (BD Biosciences, Clontech, Palo Alto, CA; cat. no. 632319).
3. Agarose (Fisher Scientific, cat. no. BP1356-500).
4. TAE buffer (Sigma, cat. no. T6025).
5. Gel extraction kit (Qiagen, Valencia, CA; cat. no. 28704).
6. T4 DNA ligase (New England Biolabs, Beverly, MA; cat. no. M0202S).
7. Competent cells (Stratagene, La Jolla, CA; cat. no. 200230).
8. Qiagen miniprep kit (Qiagen, cat. no. 27106).
9. LB media (Fisher Scientific, cat. no. BP1426-2).
10. Ampicillin (Amresco, Solon, OH; cat. no. 0339-25G).
11. Qiagen midiprep kit (Qiagen, cat. no. 10043).

2.3. Transfection of ES Cells

1. Lipofectamine reagent (1 mL; Invitrogen, cat. no. 18324-012) (*see* **Note 2**).
2. Plus reagent (Invitrogen, cat. no. 11514-015).
3. pGL3 internal control vector (Promega, Madison, WI; cat. no. E1741).
4. Reporter DNA (1 µg/µL each).

2.4. Fluorescence Assay

1. Lysis buffer (Promega, cat. no. E2661).
2. FluoroCount™ Microplate Fluorometer (Packard Instrument Company, now part of PerkinElmer Life and Analytical Sciences Inc., Boston, MA; model BF10000).
3. Corning Costar 96-well plate (Fisher Scientific, cat. no. 9017).

2.5. Luciferase Assay

1. Luciferase assay kit (Promega, cat. no. E1500).
2. Corning Costar 96-well plate (Fisher Scientific, cat. no. 3693).

3. Methods

3.1. ES Cell Culture

ES cells are cultured in a high-glucose medium conditioned with ESGRO (leukemia inhibitory factor) to prevent cell differentiation. To provide a substrate for cell attachment, the cells are cultured on plates coated with 0.1% gelatin. Recommended splitting frequency is every 2–3 d, and the ratio is 1:8 from an 80–90% confluent plate. When cells are passaged, separation of differentiated and undifferentiated cells is done by attaching differentiated cells to uncoated tissue culture dishes for 2 h prior to seeding cells in gelatin-coated dishes. Cells are incubated at all times at 37°C, 5% CO_2, and 100% humidity.

3.1.1. Thawing ES Cells

1. Take out one vial from liquid N_2 and place it in a 37°C water bath for about 2 min (or until contents are completely thawed but not longer than that; *see* **Note 3**).
2. Transfer cells to a 15-mL Falcon tube and add approx 5 mL ES medium (use this to wash out the vial as well).
3. Spin for approx 3 min at 800g.
4. Remove supernatant and resuspend cells by triturating in 2 mL ES medium.
5. Plate in a gelatin-coated (*see* **Subheading 3.1.3.**) tissue culture dish (six-well or 6-cm dish) and incubate.

3.1.2. Freezing ES Cells

1. Wash cells with 1X PBS and leave a little PBS in the plate.
2. Harvest the cells using a cell scraper.
3. Transfer the cells to a 15-mL Falcon tube and spin for approx 3 min at 800g.
4. Discard supernatant and resuspend cells in cold freezing medium (2 mL for 10-cm dish, 6–7 mL for 15-cm dish).
5. Transfer 1 mL aliquots into cryovials.
6. Place at −80°C overnight in a closed styroform box and move to liquid N_2 the next day.

3.1.3. Gelatin Coating of Tissue Culture Plates

1. Add enough gelatin to cover surface (approx 2 mL/15-cm dish, approx 0.5–1 mL/10-cm dish; *see* **Note 4**).
2. Leave at room temperature for 30 min.
3. Remove the gelatin solution and store the plates in their original bag at room temperature, preferably standing (*see* **Note 5**).

3.1.4. Passaging ES Cells

It is recommended to passage cells every 2–3 d because overgrowth of cells seems to induce a higher rate of spontaneous differentiation. To remove differentiated cells, we have established a purification method by which differentiated cells are removed by attachment to uncoated tissue culture dishes prior to seeding cells in gelatin-coated dishes. The purification step is included in the following protocol:

1. Remove medium and wash with 1X PBS (Mg^{2+} and Ca^{2+} free).
2. Add 1X trypsin-EDTA (10X trypsin-EDTA diluted in PBS) and incubate at 37°C for 5 min.
3. Add ES medium to inactivate trypsin.
4. Transfer cells to a 15-mL tube and spin for approx 3 min at 800g.
5. Remove supernatant, resuspend cells in approx 2 mL ES medium, and triturate at least 10–20 times.
6. Seed cells in an uncoated tissue culture dish (the same size as the original dish) and leave in the incubator for 2 h (differentiated cells will attach during this time; ES cells will stay in suspension).
7. Transfer medium containing cells to gelatin-coated (*see* **Subheading 3.3.**) tissue culture dishes. Triturate to make sure that cells are dissociated (*see* **Note 6**). The ATCC recommended splitting ratio is 1:4 to 1:10.

ES Cell Transfection and Promoter Analysis

3.2. Construction of Reporter DNA

1. Cut 1 µg pEGFP-1 reporter plasmid using an appropriate restriction enzyme to insert the promoter of choice in front of EGFP's open reading frame (*see* **Note 7**).
2. Cut out the promoter of choice using an appropriate restriction enzyme or amplify the promoter by polymerase chain reaction (PCR) using primers linked with appropriate enzyme sites that can be cut and ligated to the pEGFP-1 reporter plasmid.
3. Ligate pEGFP-1 reporter plasmid and the promoter using T4 DNA ligase.
4. Transform competent *Escherichia coli* with the 1 µL ligation mixture.
5. Plate the transformed *E. coli* on an LB plate containing ampicillin to facilitate the selection of transformed *E. coli*.
6. Pick single colonies and grow in 1 mL LB media with ampicillin.
7. Prepare plasmid DNA from grown colonies using the Qiagen miniprep kit following the manufacturer's instructions.
8. Cut DNA using an appropriate restriction enzyme and run the cut DNAs on a 1% agarose gel to confirm the insertion of promoter DNA into the GFP reporter plasmid (*see* **Note 8**).
9. Once the correct clone is identified, grow 50 mL of transformed *E. coli* (LB media with ampicillin) overnight at 37°C and prepare plasmid DNA using the Qiagen midiprep kit according to the manufacturer's instructions.

3.3. Transfection of ES Cells

Each EGFP-expressing reporter DNA along with the pGL3 internal control vector, containing the Simian virus 40 enhancer/promoter and the luciferase gene, were transfected into the different cell lines using the lipofectamine-Plus technique. Because ES cell lines transfected with the pGL3 vector yielded reasonable levels of luciferase expression, it was chosen as an internal control of transfection efficiencies.

1. At 1 d before transfection, ES cells are plated on gelatin-coated 35-cm tissue culture dishes at a density of 3×10^5 per well using ES media and are incubated at 37°C in a CO_2 incubator overnight.
2. The next day, when the cells are about 60% confluent, dilute 1 µg of reporter DNA and 1 µg pGL3 internal control vector together in 100 µL DMEM. Then, add 20 µL Plus reagent to the diluted DNA, mix, and incubate at room temperature for 15 min.
3. Dilute 10 µL Lipofectamine reagent into 100 µL DMEM.
4. Combine diluted DNA and diluted Lipofectamine reagent and incubate for 15 min at room temperature.
5. Replace ES cell medium with 1 mL ES cell transfection medium without penicillin-streptomycin.
6. Add DNA-Plus-Lipofectamine complex into the ES cells with fresh medium and incubate for 3 h at 37°C in a CO_2 incubator.
7. Add another 1 mL of ES cell transfection media into the ES cells.
8. The next day, replace media with ES cell media and further incubate the cells until they are ready for assaying (*see* **Subheadings 3.4.** and **3.5.**).

3.4. Fluorescence Assay

1. Harvest cells 36 h after transfection.
2. Bring 1X lysis buffer to room temperature (lysis buffer is also compatible with luciferase assay for normalization).
3. Rinse the cells in the plate with PBS and add 200 µL 1X lysis buffer per 35-mm plate.

4. Freeze the plate in a deep freezer (−80°C) for 5 min and thaw in a 37°C incubator for complete lysis.
5. Scrape the cells and transfer to a microcentrifuge tube.
6. Vortex the tube and centrifuge at 12,000g for 1 min.
7. The supernatant is titrated with different aliquots (e.g., 10, 20, and 50 µL) placed into a 96-well plate and evaluated for GFP fluorescence using a FluoroCount Fluorometer (see **Note 9**).

3.5. Luciferase Assay

To correct for differences in transfection efficiencies and to normalize the GFP fluorescence intensity, measure luciferase activity from the same lysate using the luciferase assay kit from Promega according to the manufacturer's instructions.

1. Add 20 µL cell lysate from **Subheading 3.4.**, **step 6**, into each well of a 96-well plate.
2. Program the luminometer for a 2-s delay and a 10-s read time (see **Note 10**).
3. Normalize fluorescence value using luciferase activity.

4. Notes

1. As a reporter, GFP has an advantage in that it can be visualized in living cells without any special procedure. However, for more sensitive quantification, one can use luciferase instead of GFP as a reporter system. In this case, a double luciferase assay with *Renilla* luciferase and firefly luciferase can be used so that the promoter activity and internal control can be measured in the same well.
2. We compared a number of commercially available lipofection reagents, and so far Lipofectamine Plus is the best for ES cells. Other groups have also successfully used electroporation for transfecting ES cells.
3. Cells are frozen in 90% horse serum, 10% DMSO, to prevent crystal formation, which would damage the cells. However, DMSO is toxic to the cells, and it is therefore important to work fast when you thaw the cells.
4. The volume is not important but covering the entire surface is.
5. This is done to avoid gelatin smearing the lid and the outside of the plate.
6. When passaging ES cells, it is easier to triturate and dissociate the cells if the amount of medium is minimal. ES cells tend to make aggregates, and it is important that they are carefully dissociated when passaged. This seems to reduce spontaneous differentiation.
7. Consult the first chapter of *Molecular Cloning*, a laboratory manual by Sambrook and Russel, which is published by Cold Spring Harbor Laboratory Press, for a more detailed protocol for DNA subcloning.
8. PCR can be used to screen the subclones. In that case, boil grown *E. coli* clones in 100 µL water and use 10 µL for the PCR reaction using one primer from the plasmid and another from the promoter.
9. It is important to titrate the amount of lysate to be assayed to make sure that it is in the linear range of detection because the fluorometer has a narrow range of detection (less than 10^3-fold).
10. The injector adds 100 µL of luciferase assay reagent per well, and the well is read after the time delay.

References

1. Evans, M. J. and Kaufman, M. H. (1981) Establishment in culture of pluripotential cells from mouse embryos. *Nature* **292,** 154–156.
2. Nagy, A., Gocza, E., Diaz, E. M., et al. (1990) Embryonic stem cells alone are able to support fetal development in the mouse. *Development* **110,** 815–821.

3. Nagy, A., Rossant, J., Nagy, R., Abramow-Newerly, W., and Roder, J. C. (1993) Derivation of completely cell culture-derived mice from early-passage embryonic stem cells. *Proc. Natl. Acad. Sci. USA* **90,** 8424–8428.
4. Thomas, K. R. and Capecchi, M. R. (1987) Site-directed mutagenesis by gene targeting in mouse embryo-derived stem cells. *Cell* **51,** 503–512.
5. Hooper, M., Hardy, K., Handyside, A., Hunter, S., and Monk, M. (1987) HPRT-deficient (Lesch-Nyhan) mouse embryos derived from germline colonization by cultured cells. *Nature* **326,** 292–295.
6. Smith, A. G. (1991) Culture and differentiation of embryonic stem cells. *J. Tissue Cult. Methods* **13,** 89–94.
7. Desbaillets, I., Ziegler, U., Groscurth, P., and Gassmann, M. (2000) Embryoid bodies: an in vitro model of mouse embryogenesis. *Exp. Physiol.* **85,** 645–651.
8. Wutz, A. and Jaenisch, R. (2000) A shift from reversible to irreversible X inactivation is triggered during ES cell differentiation. *Mol. Cell* **5,** 695–705.
9. Dinsmore, J., Ratliff, J., Deacon, T., et al. (1996) Embryonic stem cells differentiated in vitro as a novel source of cells for transplantation. *Cell Transplant.* **5,** 131–143.
10. Isacson, O., Costantini, L., Schumacher, J. M., Cicchetti, F., Chung, S., and Kim, K. (2001) Cell implantation therapies for Parkinson's disease using neural stem, transgenic or xenogenic donor cells. *Parkinsonism Rel. Disord.* **7,** 205–212.
11. Deacon, T., Dinsmore, J., Costantini, L. C., Ratliff, J., and Isacson, O. (1998) Blastula-derived stem cells can differentiate into dopaminergic and serotonergic neurons after transplantation. *Exp. Neurol.* **149,** 28–41.
12. Brustle, O., Jones, K. N., Learish, R. D., et al. (1999) Embryonic stem cell-derived glial precursors: a source of myelinating transplants. *Science* **285,** 754–756.
13. Lumelsky, N., Blondel, O., Laeng, P., Velasco, I., Ravin, R., and McKay, R. (2001) Differentiation of embryonic stem cells to insulin-secreting structures similar to pancreatic islets. *Science* **292,** 1389–1394.
14. Najjar, S. and Lewis, R. (1999) Persistent expression of foreign genes in cultured hepatocytes: expression vectors. *Gene* **230,** 41–45.
15. Ramezani, A., Hawley, T., and Hawley, R. (2000) Lentiviral vectors for enhanced gene expression in human hematopoietic cells. *Mol. Ther.* **2,** 458–469.
16. Fan, X., Brun, A., and Karlsson, S. (2000) Adenoviral vector design for high-level transgene expression in primitive human hematopoietic progenitors. *Gene Ther.* **7,** 2132–2138.
17. Keating, A., Horsfall, W., Hawley, R. G., and Toneguzzo, F. (1990) Effect of different promoters on expression of genes introduced into hematopoietic and marrow stromal cells by electroporation. *Exp. Hematol.* **18,** 99–102.
18. Muller, S., Sullivan, P. D., Clegg, D. O., and Feinstein, S. C. (1990) Efficient transfection and expression of heterologous genes in PC12 cells. *DNA Cell Biol.* **9,** 221–229.
19. Okabe, S., Forsberg-Nilsson, K., Spiro, A. C., Segal, M., and McKay, R. D. (1996) Development of neuronal precursor cells and functional postmitotic neurons from embryonic stem cells in vitro. *Mech. Dev.* **59,** 89–102.
20. Johe, K. K., Hazel, T. G., Muller, T., Dugich-Djordjevic, M. M., and McKay, R. D. (1996) Single factors direct the differentiation of stem cells from the fetal and adult central nervous system. Functions of basic fibroblast growth factor and neurotrophins in the differentiation of hippocampal neurons. *Genes Dev.* **10,** 3129–3140.

14

SAGE Analysis to Identify Embryonic Stem Cell-Predominant Transcripts

Kenneth R. Boheler and Kirill V. Tarasov

Summary

The Human Genome Consortium has successfully sequenced the entire human genome (http://www.genome.gov/11006945), but an unfinished goal remains the identification of specific genes responsible for unique cellular processes. With respect to embryonic stem (ES) cells, this includes the identification of factors that govern self-renewal and pluripotentiality. One technique that facilitates this last goal is serial analysis of gene expression (SAGE), a functional genomics technique that identifies and quantifies mRNA transcripts. This technique relies on the preparation and sequencing of complementary DNA concatemers to rapidly generate a comprehensive profile of gene expression within a cell, and unlike microarrays, it does not require prior knowledge of the genes to be assayed. Because SAGE is a sequence-based technique, it can be used to search for ES-restricted genes (i.e., markers) by sequence comparisons among stem cells, differentiated cells, and tissues. These markers can then be genetically manipulated to understand the molecular basis for stem cell biology to help define how transcriptional mechanisms distinguish ES cells from other, less-pluripotent cell types. SAGE is, thus, a powerful technique that permits a comprehensive analysis of mRNA abundance that can define, at a molecular level, fundamental characteristics of ES cells. In this chapter, we illustrate the basic principles of SAGE, describe a complete protocol for the generation of SAGE libraries, and show how this technique can be employed to analyze embryonic stem cells.

Key Words: ES cells; gene expression profiling; mRNA; SAGE; transcriptome.

1. Introduction

Embryonic stem (ES) cell lines are functionally characterized by pluripotentiality and self-renewal capacity, and these properties, more than any others, distinguish stem cells from more differentiated somatic cell types. The molecular factors that are responsible for these functional properties have only partially been elucidated, and the identification of signals that regulate stem cell differentiation to specific cell types remains fundamental to the understanding of ES cell biology (i.e., embryonic stemness). One approach to identifying these regulatory factors is to generate a gene expression profile of ES cells through functional genomic assays such as cDNA and oligonucleotide

From: *Methods in Molecular Biology, vol. 329: Embryonic Stem Cell Protocols: Second Edition: Volume 1*
Edited by: K. Turksen © Humana Press Inc., Totowa, NJ

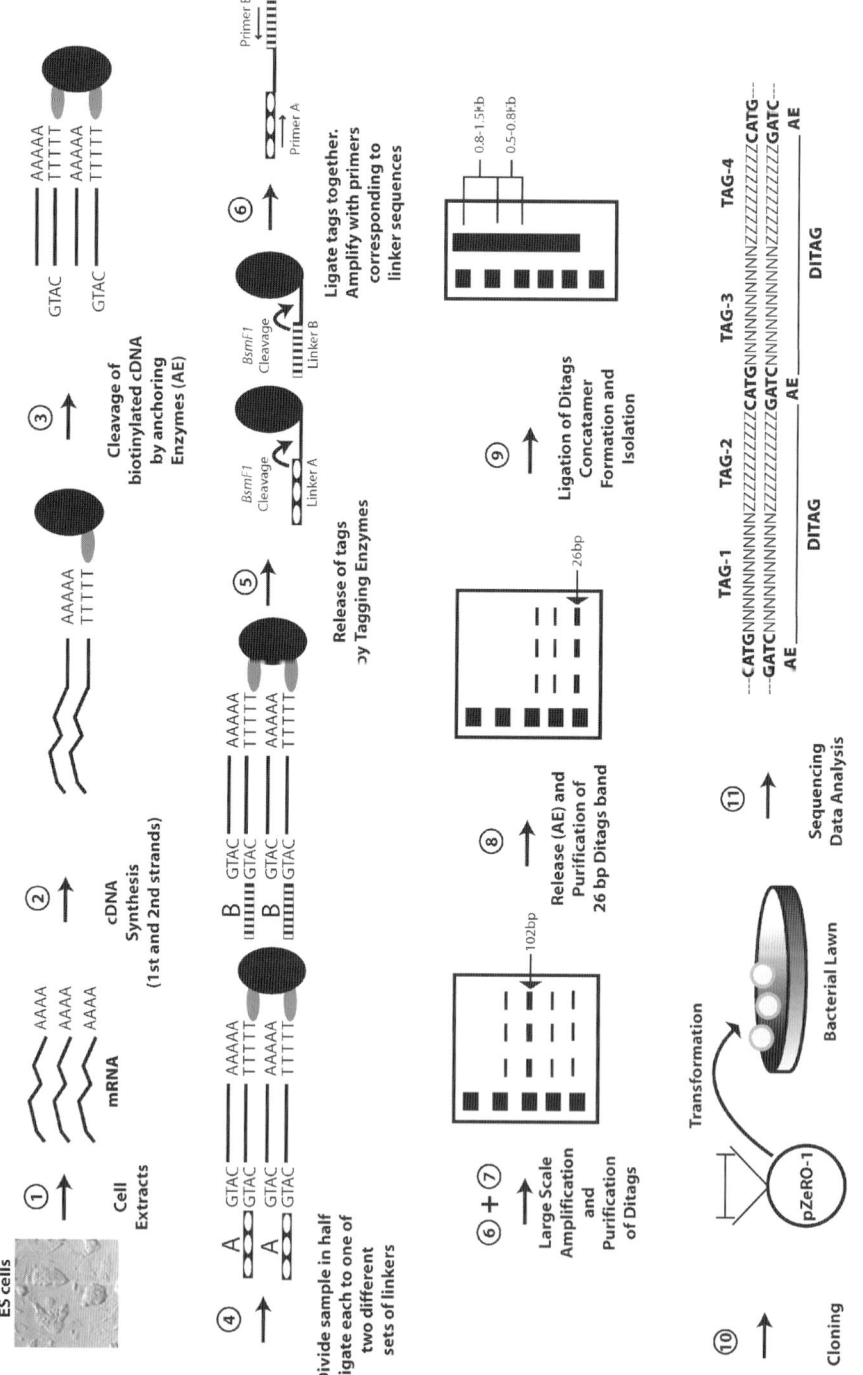

Fig. 1.

arrays (microarrays) or serial analysis of gene expression (SAGE). Both techniques are useful in the identification of quantitative differences among cells and tissues, but the "absolute" quantitative nature and uniformity of SAGE greatly facilitates comparative and comprehensive analyses among laboratories, particularly when large data sets are available in the public domain (http://www.ncbi.nlm.nih.gov/sage/).

SAGE permits the simultaneous evaluation of thousands of transcripts in a single sequencing-based assay, and it is specifically designed to identify both known and unknown (e.g., novel) transcripts. This technique relies on two major principles. First, short DNA sequences are sufficient to identify individual gene products (transcripts or mRNAs) provided that they are derived from a unique region within a transcript; second, concatenation (linking together) of short DNA sequences or tags increases the efficiency of identifying expressed transcripts in a sequencing reaction.

The tag (transcript) profile generated by SAGE relies primarily on 14–21 base nucleotide sequences (i.e., SAGE tags) for gene identification *(1,2)*. The technique generates large numbers of short tags originating from the last (most 3′) unique location of an enzyme recognition site in a single transcript. When the tags are sequenced, the technique can theoretically identify up to 4^{10} (1,048,576) unique transcripts, a number much larger than the estimated 33,000–35,000 genes present in the human genome *(3)*. The newer LongSAGE method (17 base tags) can distinguish 4^{17} different transcripts, a number sufficient to be unique even within the whole genome *(2)*. The potential of SAGE to identify all transcripts present within a cellular transcriptome is limited by the fidelity and cost of sequencing *(1)*; consequently, this technique is perhaps best described as a quantitative sampling technique of a cellular transcriptome that is highly informative for comparative studies.

A general scheme for the generation of SAGE libraries is given in **Fig. 1**. The method involves several major steps:

Fig. 1. Principal steps of serial analysis of gene expression (SAGE). This protocol relies on short DNA sequences (10–14 bp) that are sufficient to identify individual gene products and concatenation (linking together) of these short sequences (i.e., DNA tags) to increase the efficiency of identifying expressed mRNAs in a sequence-based assay. To generate the sequences, purified mRNA from pluripotent embryonic cells passaged in the absence of feeder layers is used to generate double-stranded cDNA. Using streptavidin-coated magnetic beads, the double-stranded cDNA is purified and digested with an anchoring enzyme (AE) like NlaIII, which recognizes specific recognition sites located in the double-stranded DNA sequence (e.g., CATG for NlaIII). The DNA fragment located closest to the biotinylated primer is purified by binding to magnetic beads, and this fraction is divided in half and ligated to two different linker/primer sets to ensure an accurate quantitative representation of the original transcripts in the sample. SAGE tags are generated by digestion of the cDNA molecules with a type IIS restriction enzyme or tagging enzyme like BsmFI, which cleaves DNA sequences several basepairs away from the recognition site. The SAGE tags are joined to form ditags and are amplified by polymerase chain reaction with a set of primers that recognize Linkers A and B. The ditags are separated from the linkers and ligated together to form concatemers of purified ditags. These are then subcloned into a plasmid vector, amplified, and sequenced. The individual tags can then be extracted by identifying the CATG AE sequences. Each individual tag sequence is then run against GenBank databases to identify the corresponding gene product. Once a SAGE library is completed, each individual tag sequence can be run against GenBank databases to identify the corresponding gene product.

1. Isolation and preparation of total RNA and mRNA from pure populations of cells.
2. Synthesis of cDNA.
3. Cleavage of biotinylated cDNA with an anchoring enzyme (AE) and isolation of these fragments by binding to magnetic beads.
4. Linker ligation to the bound cDNA.
5. Release of cDNA tags using a tagging (type IIs) restriction enzyme, followed by blunt ending.
6. Ligation of individual tags to form ditags, followed by a polymerase chain reaction (PCR) amplification.
7. Isolation and ligation of ditags to form concatemers.
8. Cloning of concatemers into plasmids.
9. Transformation of bacteria, plasmid amplification, and isolation.
10. Sequencing of inserts.

Once completed, the sequence files are analyzed to identify the recognition site of the AE. Ditags are electronically extracted to identify matching gene transcripts and to compare among other SAGE libraries. Importantly, SAGE libraries can be compared both within a laboratory and with other publicly available catalogs (http://www.ncbi.nlm.nih.gov/SAGE). In fact, the comparative power of SAGE increases as a function of the number of publicly available libraries, which can be continuously used to confirm or refute the authenticity of putative transcripts with differential gene expression either within a tissue or among tissues, including the identification of predominant or novel ES cell-restricted transcripts.

2. Materials (see Note 1)

2.1. Tissue Culture

2.1.1. Cells and Cell Lines (see *Note 2*)

1. Mouse ES cell lines R1 *(4)* (http://www.ncbi.nlm.nih.gov/sage/, GSM580).
2. Mouse ES cell lines D3 (http://www.ncbi.nlm.nih.gov/sage/, GSM11349).
3. Mouse ES cell line ESF 116 (http://www.ncbi.nlm.nih.gov/sage/, GSM3829).
4. Human ES cell line HES3 *(5)*.
5. Human ES cell line HES4 *(5)*.
6. Buffalo rat liver cells (ATCC, Manassas, VA; cat. no. CRL 1422) *(6)*.
7. Feeder cells *(6,7)*.

2.1.2. Media, Reagents, and Stock Solutions

1. Phosphate-buffered saline (PBS), pH 7.2, without Ca^{2+} and Mg^{2+} (Invitrogen, Carlsbad, CA; cat. no. 10010-023).
2. Dulbecco's modified Eagle's medium (DMEM; Invitrogen, cat. no. 11995-040): contains 4500 mg/L glucose, L-glutamine, and 110 mg/L sodium pyruvate (*see* **Note 3**).
3. Fetal bovine serum (FBS; ES cell qualified): mycoplasma, virus, endotoxin tested (selected batches; *see* **Note 4**).
4. L-Glutamine (200 m*M*; Invitrogen, cat. no. 25030-081).
5. 100X penicillin (5000 U/mL)/streptomycin (5000 µg/mL) solution (pen/strep; Invitrogen, cat. no. 15070-063).
6. 100X minimum essential medium 10 m*M* nonessential amino acid (NEAA) solution (Invitrogen, cat. no. 11140-050).
7. 10 ng/mL leukemia inhibitory factor (LIF; Sigma, St. Louis, MO; cat. no. L5158).
8. 1 mg/mL transferrin (Invitrogen, cat. no. 13008-016).

9. Trypsin solution: 0.2% trypsin (Invitrogen, cat. no. 27250-042) 1:250 in PBS filter-sterilized through a 0.22-μm filter.
10. Trypsin-ethylenediaminetetraacetic acid (EDTA): 0.25% trypsin, 1 mM EDTA-4Na (Invitrogen, cat. no. 25200-056).
11. Gelatin solution: 1% gelatin (Sigma, cat. no. G1890) in double-distilled water (ddH_2O), autoclaved and diluted 1:10 with PBS. Coat tissue culture dishes with 0.1% gelatin solution for 1–24 h at 4°C before use.
12. Mitomycin C (MMC) solution: dissolve 2 mg MMC (Sigma, cat. no. M0503) in 10 mL PBS; filter-sterilize. From this stock solution, dilute 300 μL into 6 mL PBS (0.01 mg/mL final concentration). MMC stock solution should be freshly prepared at weekly intervals and stored at 4°C (*see* **Note 5**).
13. β-Mercaptoethanol (β-ME; Sigma, cat. no. M7522): prepare a stock solution from 7 μL β-ME into 10 mL PBS (10 mM stock concentration). Make fresh at weekly intervals and store at 4°C.
14. Cultivation medium I: DMEM (4.5 g/L glucose) supplemented with 15% FBS for feeder layer cells, 50 U penicillin, and 50 μg/mL streptomycin. Mix together 5 mL pen/strep solution and 75 mL FBS and bring it to a final volume of 500 mL with DMEM. Filter-sterilize through a 0.22-μm filter.
15. Cultivation medium II: DMEM (4.5 g/L glucose) supplemented with 15% FBS, 2 mM L-glutamine, 10 μM β-ME, 0.1 mM NEAA, 50 U/mL penicillin, 50 μg/mL streptomycin. For 500 mL, mix together 5 mL pen/strep solution, 5 mL β-ME stock, 5 mL NEAA, 5 mL L-glutamine, and 75 mL FBS and bring it to a final volume of 500 mL with DMEM. Filter-sterilize through a 0.22-μm filter.
16. Buffalo rat liver (BRL) culture media: DMEM supplemented with 10% FBS, 50 U penicillin, and 50 μg/mL streptomycin. For 500 mL, mix together 5 mL pen/strep solution and 50 mL FBS and bring it to a final volume of 500 mL with DMEM. Filter-sterilize through a 0.22-μm filter.
17. BRL-conditioned media (BRL-CM; DMEM supplemented with BRL-conditioned stock (*see* **Subheading 3.1.1.** for details) containing additives and 10 ng/mL LIF). For 100 mL, mix 60 mL BRL conditioned stock with 9 mL FBS, 1 mL transferrin, 1 mL L-glutamine, 1 mL β-ME, 1 mL NEAA, 1 mL pen/strep, and 100 μL LIF and bring to a final volume of 100 mL with DMEM. Filter-sterilize through a 0.22-μm filter.

2.1.3. Tissue Culture Requirements

1. Tissue culture plates (35, 60, and 100 mm; Nunc, Rochester, NY, or BD Falcon, Franklin Lakes, NJ).
2. 6- and 24-well (Falcon) and 96-well microwell plates (Nunc or Costar, Cambridge, MA).
3. Pipets for tissue culture (2, 5, 10, and 25 mL).
4. Pasteur pipets (2 mL).
5. For feeder layer culture: sterile dissecting instruments, screen or sieve (approx 0.5- to 1-mm diameter), Ehrlenmeyer flasks with stir bars, and centrifuge tubes (*see* **Note 6**).
6. Tissue culture incubator at 37°C and with a 5% CO_2 atmosphere.
7. Laminar flow biosafety cabinet.

2.2. Serial Analysis of Gene Expression

2.2.1. Kits

1. FastTrack 2.0 kit for preparation of mRNA (Invitrogen, cat. no. 1593-02).
2. cDNA Synthesis kit (Invitrogen, cat. no. 18267-013).
3. Deoxynucleotide triphosphate (dNTP) kit (25 mM dNTPs; Amersham, Piscataway, NJ; cat. no. 13-5423-01).

2.2.2. Enzymes (see **Note 7**)

1. T4 DNA ligase (1 U/µL; Invitrogen, cat. no. 15224-017).
2. T4 DNA ligase (5 U/µL; Invitrogen, cat. no. 15224-041) (*see* **Note 8**).
3. T4 polynucleotide kinase (10 U/µL; New England Biolabs [NEB], Beverly, MA; cat. no. M0201S).
4. DNA polymerase I, Klenow fragment (1 U/µL; Pharmacia, Uppsala, Sweden; cat. no. 27-0929-01).
5. Taq DNA polymerase; AmpliTaq Gold polymerase (Applied Biosystems, Foster City, CA; cat. no. 4311814).
6. NlaIII (10 U/µL; NEB, cat. no. R0125) (*see* **Note 9**).
7. BsmFI (2 U/µL; NEB, cat. no. R0572S) (*see* **Note 9**).
8. SphI (5 U/µL; NEB, cat. no. R0182S).

2.2.3. Reagents (see **Note 1**)

1. Absolute ethanol (EtOH; Sigma, cat. no. E7148) and 75% EtOH, prepared in diethylpyrocarbonate (DEPC)-H_2O.
2. Adenosine triphosphate (Amersham, cat. no. 27-1006.01).
3. Agar (Invitrogen, cat. no. 30391-023).
4. Agarose (Invitrogen, cat. no. 15510-027).
5. Ammonium acetate (10 M NH_4OAc; Sigma, cat. no. A1452): dissolve 385.4 g NH_4OAc in DEPC-H_2O and bring to a final volume of 500 mL.
6. Ammonium persulfate (Bio-Rad, Hercules, CA; cat. no. 161-0700).
7. Ammonium sulfate $(NH_4)_2SO_4$ (Sigma, cat. no. A5132).
8. β-ME stock (Sigma, cat. no. M7522).
9. Water treated with 0.1% DEPC (Biosource, Camarillo, CA; cat. no. 395-000).
10. Dimethyl sulfoxide (DMSO; Sigma, cat. no. D2650).
11. 0.5 M disodium ethylenediaminetetraacetic acid (Na_2EDTA), pH 8.0 (Quality Biological, Gaithersburg, MD; cat. no. 351-027-061).
12. Na_2EDTA (Sigma, cat. No. E5134).
13. dNTPs (25 mM; Amersham, cat. no. US77119).
14. Glycogen (Roche Diagnostics, Indianapolis, IN; cat. no. 10901393001).
15. Dynabeads M-280 (Dynal, Brown Deer, WI; cat. no. 112.05).
16. ElectroMAX DH10B cells (Invitrogen, cat. no. 18290-015).
17. Glycerol (Invitrogen, cat. no. 15514-011).
18. Guanidine thiocyanate (Sigma, cat. no. 50990).
19. Isopropanol (J. T. Baker Inc., Phillipsburg, NJ; cat. no. 9084-01).
20. Magnesium chloride ($MgCl_2$; Sigma, cat. no. M8266).
21. Molecular weight markers: 25-bp DNA Ladder (Invitrogen, cat. no. 10597-011); 100-bp DNA Ladder (Invitrogen, cat. no. 15628-050); 1-kb DNA Ladder (Invitrogen, cat. no. 15615-016).
22. *N*-Lauroyl-sarcosine (Sarcosyl; Sigma, cat. no. L-9150).
23. Phenol-chloroform-isoamyl alcohol mix (25:24:1, pH 8.0, Amresco Inc., Solon, OH; cat. no. K169) (*see* **Note 10**).
24. Polyacrylamide solutions (acrylamide:*bis*) 19:1 (Bio-Rad, cat. no. 161-0144).
25. Polyacrylamide solutions (acrylamide:*bis*) 37.5:1 (Bio-Rad, cat. no. 161-0148).
26. Polyacrylamide criterion precast gels (15%, 20% TBE) (Bio-Rad, cat. no. 345-0056).
27. pZeRO-1 (Invitrogen, cat. no. K2500-01).

28. OC/S.O.C media (Biosource, cat. no. 396-110).
29. Sodium chloride (NaCl; Sigma, cat. no. 71383).
30. NaCl (5 M; Quality Biological, cat. no. 351-036-100).
31. Sodium citrate dihydrate (Na-citrate) (Mallinckrodt, Phillipsburg, NJ; cat. no. 0754).
32. Sodium perchlorate ($NaClO_4$) (Sigma, cat. no. S1513).
33. SYBR green I (Molecular Probes, Eugene, OR; cat. no. S-7567).
34. N,N,N',N'-Tetramethylenediamine (TEMED) (Bio-Rad, cat. no. 161-0800).
35. Tris/acetate/EDTA (TAE) (50X; Quality Biological, cat. no. 330-008-161).
36. Tris/borate/EDTA buffer (TBE) (10X, pH 8.3; Biosource, cat. no. 336-000).
37. Tris-HCl (1 M, pH 7.5; Quality Biological, cat. no. 351-006-100).
38. Tris-HCl (1 M, pH 8.0; Quality Biological, cat. no. 351-007-100).
39. Trizma base (Sigma, cat. no. 93362).
40. Tryptone (Sigma, cat. no. 27293).
41. Yeast extract (Sigma, cat. no. Y4000).
42. Zeocin (Invitrogen, cat. no. R250-01).

2.2.4. Solutions

1. RNA lysis buffer: add 23.6 g guanidine thiocyanate to 5 mL 250 mM Na-citrate, pH 7.0, and 2.5 mL 10% Sarcosyl, add DEPC-H_2O to a total volume of 49.5 mL, and mix carefully. Make fresh at monthly intervals. Add 1% β-ME before use.
2. EDTA (0.25 M, pH 7.5): dissolve 93.05 g Na_2EDTA in approx 400 mL ddH_2O, adjust pH to 7.5 with 10 N NaOH, and adjust to 1 L final volume with distilled water (dH_2O). Autoclave to sterilize.
3. Tris-HCl (1 M, pH 8.8): to approx 800 mL ddH_2O, add 121.1 g Trizma base and dissolve. Adjust pH with concentrated HCl to pH 8.8 and then add double-distilled water to 1 L.
4. LoTE: 3 mM Tris-HCl, pH 7.5, 0.2 mM EDTA, pH 7.5, in distilled water. For 100 mL, mix 0.3 mL 1 M Tris-HCl, pH 7.5, with 0.08 mL 0.25 M EDTA. Store at 4°C.
5. 10X PCR buffer: 166 mM $(NH_4)_2SO_4$, 670 mM Tris-HCl, pH 8.8, 67 mM $MgCl_2$, 100 mM β-ME. For 100 mL, mix 2.192 g $(NH_4)_2SO_4$, 67 mL 1 M Tris-HCl, pH 8.8, 0.638 g $MgCl_2$, and 0.78 mL β-ME. Bring to a final volume of 100 mL with sterile water. Prepare stock, distribute into 0.5-mL aliquots, and store at −20°C.
6. 2X binding and washing buffer (B+W; 10 mM Tris-HCl at pH 7.5, 1 mM EDTA, 2.0 M NaCl) (see **Note 11**): mix 1 mL 1 M Tris-HCl, pH 7.5, 0.4 mL 0.25 M EDTA, and 40 mL 5 M NaCl and bring to 100 mL with sterile water. Store at room temperature.
7. 2 M $NaClO_4$: dissolve 12.25 g $NaClO_4$ in 50 mL ddH_2O.
8. 12% polyacrylamide solution (for isolating PCR products and ditags) (44 mL final volume): for a 20 × 16 × 0.1-cm gel, mix 29.4 mL dH_2O, 13.2 mL 40% PAAG mix (19:1 acrylamide:*bis*), 875 µL 50X TAE, 437.5 µL 10% ammonium persulfate, and 37.5 µL TEMED. Stir well.
9. 8% polyacrylamide solution (for separating concatemers) (44 mL final volume): for a 20 × 16 × 0.1-cm gel, mix 33.8 mL dH_2O, 8.8 mL 40% PAAG mix (37.5:1 acrylamide:*bis*), 875 µL 50X TAE, 438 µL 10% ammonium persulfate, and 37.5 µL TEMED. Stir well.
10. LB with 10% glycerol: for 1 L, add 10 g tryptone, 5 g yeast extract, 10 g NaCl, and 100 mL anhydrous glycerol to a 1-L bottle. Fill to 1 L with distilled water and autoclave; cool to 55°C; add Zeocin to final concentration of 50 µg/mL. Store in the dark at 4°C for up to 1 mo.
11. Low-salt LB agar plates or lawns with Zeocin (see **Note 12**): 1% tryptone, 0.5% yeast extract, 0.5% NaCl, 1.5% Bacto-Agar, and 50 µg/mL Zeocin. For preparation, add 5 g

tryptone, 2.5 g yeast extract, and 2.5 g NaCl. Stir, adjust pH of the solution to 7.5 with NaOH, add 7.5 g Bacto-Agar, and bring to a final volume of 500 mL with water. Autoclave, cool to 55°C, add 250 µL Zeocin 100-µg/µL stock, pour plates, and seal with Parafilm. Store in dark at 4°C for up to 1 mo.

2.2.5. Primers for SAGE Library Construction and Sequencing (see **Note 13**)

1. Linker A1 (gel purified): 5′ TTT GGA TTT GCT GGT GCA GTA CAA CTA GGC TTA ATA GGG ACA TG 3′.
2. Linker A2 (gel purified): 5′ TCC CTA TTA AGC CTA GTT GTA CTG CAC CAG CAA ATC C (amino-modified C7) 3′.
3. Linker B1 (gel purified): 5′ TTT CTG CTC GAA TTC AAG CTT CTA ACG ATG TAC GGG GAC ATG 3′.
4. Linker B2 (gel purified): 5′ TCC CCG TAC ATC GTT AGA AGC TTG AAT TCG AGC AG [amino-modified C7] 3′.
5. SAGE primer 1: 5′ GGA TTT GCT GGT GCA GTA CA 3′.
6. SAGE primer 2: 5′ CTG CTC GAA TTC AAG CTT CT 3′.
7. Biotinylated oligo dT (gel purified): 5′ [biotin]T_{18}.
8. M13 forward: 5′ GTA AAA CGA CGG CCA GT 3′.
9. M13 reverse: 5′ GGA AAC AGC TAT GAC CAT G 3′.

2.2.6. SAGE Specialty Requirements

1. SpinX microcentrifuge tubes (Costar, cat. no. 8160).
2. 96-well Costar flat-bottom plates (VWR, West Chester, PA; cat. no. 29442-070).
3. Thermo-Fast® 96-well detection plate and adhesive sealing sheets (Abgene Inc., Rochester, NY; cat. no. AB-1100 and AB-0558).
4. MicroAmp reaction tubes (Applied Biosystems, cat. no. 801-0580).
5. MicroAmp caps (Applied Biosystems, cat. no. 801-0535).
6. 25 × 25-cm bioassay trays (Abgene, Inc, cat. no. QH2216).

2.2.7. Equipment for SAGE

1. Thermocycler (*see* **Note 14**): mastercycler gradient (Brinkmann/Eppendorf, Westbury, NY) or Gene Amp PCR System 9700 (PerkinElmer, Wellesley, MA).
2. Electrophoresis units for polyacrylamide (i.e., Protein II electrophoresis unit [Bio-Rad] or similar) and agarose electrophoresis (i.e., Hoefer HE33 [Pharmacia] or similar); power supply (Power PAC 3000 [Bio-Rad] or similar).
3. Electroporation system Gene-Pulser II/Pulse Controller II (Bio-Rad) or equivalent.
4. 0.2-cm electroporation cuvets (Bio-Rad, cat. no. 165-2086).
5. Magnetic particle concentrator (Dynal, model MPC-E type).
6. GeneQuant RNA/DNA calculator (Pharmacia); Agencourt 2100 Bioanalyzer System (RNA 6000 Nano Assay kit) or equivalent.
7. 0.5- and 1.5-mL microtubes and 20-, 100-, and 1000-µL filtertips.
8. RT6000 centrifuge and H1000B rotor (Kendro Laboratory Products/Sorvall, Asheville, NC) or similar.

2.2.8. Software

1. *SAGE2000*, version 4.12, or better is recommended (available at http://www.sagenet.org).
2. *WinZip* (available at www.winzip.com) for Windows.
3. *MS Access* (Microsoft, Redman, WA).
4. *MS Excel* (Microsoft).

3. Methods

3.1. Tissue Culture

Pluripotent ES cells should be cultivated on monolayers of mouse embryonic fibroblasts (MEFs) that have been mitotically inactivated (i.e., feeder cells) *(8,9)*, followed by passaging on gelatinized tissue culture dishes in the presence of BRL-CM *(10)* or in the presence of LIF *(11,12)*. The generation of MEFs and co-cultivation of ES cells have been previously described *(6,7)*, and the reader is encouraged to review these publications for experimental details. All tissue culture manipulations should be performed in a laminar flow cabinet, and ES cells should be incubated in a humidified incubator (37°C, 5% CO_2). Cultivation of ES cells can be adversely affected by a number of potential factors, including mycoplasm, yeast, or bacterial contamination; changes in CO_2 levels and cellular pH; or inadequate feeding or passaging of ES cells. Ideally, ES cell lines are cultivated without antibiotics, but the addition of penicillin and streptomycin to the culture generally proves useful.

A good transcriptomic analysis ideally should assay every transcript in a homogeneous cell population; therefore, preparation of RNA from a mixed culture system is not ideal for SAGE. For the purposes of SAGE, it is therefore necessary to eliminate contaminating feeder cells. To remove these cells, two techniques have been employed: (1) trypsinization of the mouse ES cell/feeder layer mixture and preplating on tissue culture dishes (fibroblasts generally attach more quickly than ES cells, so that the latter can be isolated as a relatively "pure" population; however, we generally find a relatively high degree of feeder cell contamination with this technique); and (2) trypsinization of the mouse ES cell/feeder layer mixture and plating onto gelatinized plates in the presence of conditioned medium derived from BRL cells supplemented with LIF *(4,6)*. Under these last conditions, mouse ES cell lines grow as colonies, express stem cell markers (e.g., stage-specific embryonic antigen 1, alkaline phosphatase), maintain a high degree of pluripotentiality, and rapidly proliferate, consistent with the growth characteristics of ES cells grown on feeder layers (**Fig. 2**).

By passaging ES cells in the absence of feeder layers for four to five passages, the number of contaminating feeder cells is reduced to less than 0.1% of the total cell number because MEFs, which are highly proliferative at early passages, senesce, go into crisis (i.e., no longer proliferate) *(13)*, and are lost after six to seven passages in culture. We find that these conditions are effective at yielding homogeneous populations of ES cells, particularly if the ES cells used in the experiment were plated on inactivated MEFs that had already been passaged four or five times prior to cultivation in the absence of MEFs. Using this approach, we routinely obtain a relatively pure (>99%) population of ES cells for transcriptome analyses.

3.1.1. Preparation of BRL-CM

In the absence of feeder layers, ES cells can be maintained in an undifferentiated state by cultivation in the presence of BRL-CM *(10)* or in the presence of LIF; however, long-term maintenance of ES cells should involve co-culture with MEFs. The active factor secreted by BRL cells is identical to LIF, but we find that differentiation to some cell lineages is enhanced when ES cell cultures are grown in the presence of both

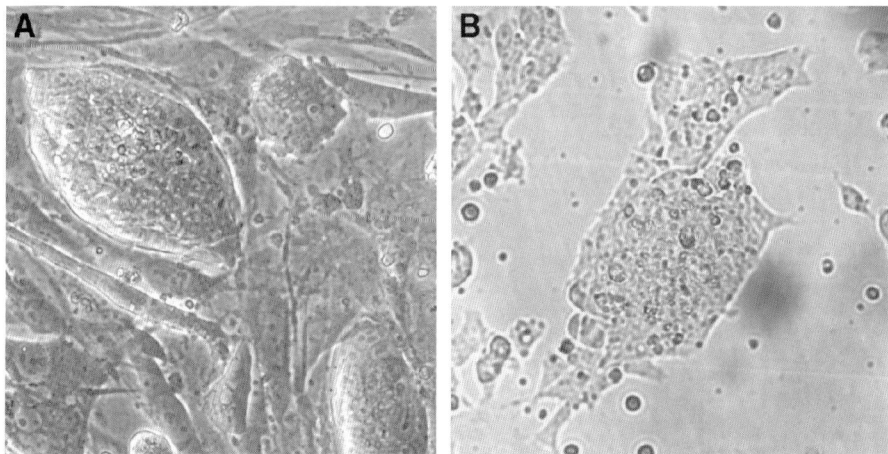

Fig. 2. Embryonic stem (ES) cell line R1 grown under in the presence of mitotically inactivated mouse feeder cell and buffalo rat liver-conditioned media (BRL-CM) growth conditions. (**A**) ES cell colonies grown on feeder cells. (**B**) ES cell colonies grown with BRL-CM plus leukemia inhibitory factor (LIF) in the absence of feeder cells. ES cell colonies grown in the presence of feeder layers or in conditioned media plus LIF test positive for alkaline phosphatase activity and Oct3/4 and Rex-1 (Zfp42) expression. ES cell colonies grown in BRL-CM plus LIF for five passages and reintroduced onto feeder cells containing growth media reacquire a similar morphology of ES cell lines grown on feeder layers; with subsequent passaging, the morphology becomes largely indistinguishable from cells cultivated only on feeder layers *(4)*.

BRL-CM and LIF. When cultivated on gelatinized plates in the absence of feeder cells, ES cells will have a somewhat flattened morphology, but after one to two passages back on feeder layers, the ES colony morphology returns to its original form *(4)*.

1. Cultivate BRL cells in 75-cm^2 tissue culture flasks with 20 mL BRL culture media. Grow until confluent with media changes as needed.
2. Once the BRL cells are confluent, change the media and incubate each flask with 20 mL BRL culture media for 3 d.
3. Collect the media of several plates every 3 d and filter-sterilize. If necessary, centrifuge conditioned medium at 300 g for 5 min before filtration to remove floating cells and debris.
4. Store the filtered medium at 4°C for 1 wk or at −80°C until needed. The BRL monolayer cultures can be used to condition medium for up to 2 wk, and we refer to this medium as BRL stock.
5. The BRL stock is subsequently used to make BRL-CM (*see* **Subheading 2.1.2., item 17**).

3.1.2. Cultivation of Undifferentiated ES Cells

The culture conditions described in this section are routinely employed for cultivation of mouse R1 and D3 ES cells; all these procedures have been previously described *(6,7)*. These protocols are designed to maintain ES cells in an undifferentiated, pluripotent state suitable for passage off of MEFs and subsequent SAGE analyses. MEFs are cultivated in cultivation medium I until the addition of ES cells. It is important to passage

ES cells cultivated on feeder layers every 24–48 h; in general, we do not recommend cultivating these cells longer than 48 h without passaging because the cells may begin to differentiate and be unsuitable for SAGE analyses of undifferentiated cells.

1. Thaw 1 mL (1×10^6 cells) of frozen ES cells (in cultivation medium II containing 8% DMSO) rapidly at 37°C and transfer to 10 mL cultivation medium II.
2. Centrifuge the cells at 300g for 3–5 min to remove supernatant and resuspend the cell pellet in 1 mL cultivation medium II.
3. Transfer the cell suspension to 6-cm tissue culture plates previously coated with 0.1% gelatin and containing either:
   a. A monolayer of MMC-treated feeder cells containing 3–4 mL cultivation medium II (see **Note 15**).
   b. 4 mL cultivation medium II plus 10 ng/mL LIF, BRL-CM, or BRL-CM supplemented with 10 ng/mL LIF (see **Note 15**).
4. Change the medium after the first overnight incubation and passage as needed every 24–48 h.
5. Prior to passaging, change the medium of ES cell cultures (cultivated in either the presence or the absence of feeder cells) 1–2 h before trypsinization.
6. Remove the medium and wash 1X with PBS. Add 1 mL prewarmed (37°C) trypsin:EDTA and incubate at room temperature for 30–60 s. Add 0.5 mL cultivation medium II.
7. Disperse the cell suspension by vigorous trituration through the narrow-bore pipet of a 2-mL Pasteur pipet to obtain a good cell suspension. Pellet the cells by centrifugation at 300g for 5 min and either passage on feeder layers or proceed to **step 3b**.
8. When passaging the cells, resuspend the pellet in cultivation medium II and passage 1:5 to 1:10 onto MMC-treated (6-cm) feeder cells (as in **step 3a**) or passage 1:3 to 1:6 onto gelatin-coated plates (6 cm) containing either BRL-CM plus 10 ng/mL LIF or cultivation medium II plus 10 ng/mL LIF. If removing feeder layers, then passage four to six times until the number of contaminating feeder cells is less than 0.1% (see **Note 15**).

Although embryo-derived stem cell lines cultivated in this manner represent an excellent source of cells for SAGE analysis, cell cultivation techniques are not consistent between laboratories (differences in culturing conditions or reagents, FBS, isolation techniques and lysis, etc.), and some variation is expected. Because serum is such an important component for maintaining ES cell properties (i.e., stemness), different batches of FBS would also be expected to affect the outcome. Spontaneous differentiation of embryo-derived stem cells in vitro can occur even when cultivated with BRL-CM plus LIF, making interpretation of transcriptome analyses challenging in the absence of complimentary data. Finally, the assumption in performing functional genomic analyses of ES cells during differentiation is that self-renewal and pluripotentiality involve a degree of regulation (e.g., transcriptional, posttranscriptional) that alters mRNA abundance. With these limitations in mind, SAGE can be employed to study mouse ES cell lines.

3.2. Serial Analysis of Gene Expression
*3.2.1. Routine Procedures for SAGE (see **Note 16**)*
3.2.1.1. Phenol-Chloroform Extraction

1. Bring sample volume to 200 µL with LoTE and add an equal volume (200 µL) phenol-chloroform-isoamyl alcohol (PC8; 25:24:1, pH 8.0).

2. Gently vortex and separate the aqueous phase, which contains the DNA, from the organic phase by centrifugation at maximum speed for 2 min.
3. Carefully transfer the aqueous phase into a clean Eppendorf tube and ensure that no organic solution is included. If portions of the organic phase are inadvertently transferred, then reextract the solution.

3.2.1.2. EtOH Precipitation

Precipitate DNA in aqueous solution by addition of EtOH in the presence of sodium or ammonium acetate followed by cooling. Because of the small amount of DNA and small fragment sizes, glycogen is added as a carrier.

1. For a 200-µL volume of sample/LoTE, add 3 µL glycogen (20 mg/mL), one-half volume 10 M ammonium acetate (or 1/10 vol 3 M sodium acetate) and 2.5 vol ice-cold ($-20°C$) 95% EtOH and vortex.
2. Pellet the precipitated DNA by centrifugation at maximum speed for 20 min.
3. Remove the EtOH with care and wash with 75% EtOH to remove salt (*see* **Note 17**).
4. Centrifuge at 10,000g for 10 min, dry the pellet, and resuspend in LoTE (*see* **Note 18**).

3.2.1.3. Isopropanol Precipitation

1. Bring sample volume to 450 µL with LoTE, add 3 µL glycogen, 150 µL 2 M NaClO$_4$, and 330 µL isopropanol.
2. Vortex and spin at maximum speed for 10 min at room temperature.
3. Aspirate supernatant, wash twice with 75% EtOH, dry the pellet, and resuspend in specified volume of LoTE (usually 20 µL).

3.2.2. Preliminary Procedures for SAGE (see **Note 19**)

3.2.2.1. Kinasing (Phosphorylation) Reaction for Linkers

1. In two Eppendorf tubes labeled A and B, add 6 µL LoTE, 2 µL 10X kinase buffer, 2 µL 10 mM adenosine triphosphate, and 1 µL T4 polynucleotide kinase (10 U/µL).
2. Add 9 µL Linker A2 (350 ng/µL) to a tube marked Linker A and 9 µL of Linker B2 (350 ng/µL) to tube marked Linker B.
3. Incubate both tubes at 37°C for 30 min, then at 65°C for 10 min.
4. Add 9 µL Linker A1 to Linker A tube and 9 µL Linker B1 to Linker B tube.
5. Incubate both tubes at 95°C for 2 min, then at 65°C for 10 min, at 37°C for 10 min, and at room temperature for 20 min. Kinased linkers are stable at $-20°C$ for up to 1 yr.
6. Prior to use, dilute linkers to 20 ng/µL.

3.2.2.2. Control Reaction for Linkers

Linkers can be tested by self-ligation and then running on a 20% precast acrylamide gel.

1. Prepare four Eppendorf tubes marked 1L, 1N, 2L, and 2N.
2. Add 21 µL LoTE, 6 µL 5X ligation buffer, and 1 µL selected linker (Linker A to the tubes marked 1L and 1N and Linker B to the tubes marked 2L and 2N) in each tube.
3. Incubate all four tubes at 50°C for 2 min, then at room temperature for 15 min.
4. Add 2 µL T4 ligase (1 U/µL) to the tubes marked 1L and 2L and 2 µL LoTE to the tubes marked 1N and 2N.
5. Incubate both tubes at 16°C for 2 h.

SAGE Identification of ES Cell-Predominant Transcripts 207

6. Run half of each sample on 20% precast TBE polyacrylamide gels at 130 V for 3 h.
7. Stain with SYBR Green I (20 µL/200 mL 1X TBE buffer) for 15 min on a shaker. Kinased linkers should form linker-linker dimers (80–100 bp) after ligation; unkinased linkers will not self-ligate. Only linkers dimerized at more than 70% efficiency are acceptable.

3.2.2.3. LINEARIZED pZeRO-1 FOR CLONING

1. Add 14 µL dH$_2$O, 2 µL Buffer 2, 2 µL pZeRO-1 (1 µg/µL), and 2 µL SphI (5 U/µL).
2. Incubate at 37°C for 25 min.
3. Add 60 µL TE buffer (25 ng/µL final concentration) and heat inactivate at 68°C for 15 min.
4. Store on ice until used, ideally the same day. Linearized pZeRO-1 plasmid can be stored at −20°C for a few weeks, but its cloning efficiency will decrease.

3.2.3. Standard SAGE Protocol

The standard SAGE protocol is based on the original protocol described by Velculescu et al. *(1)*, but it has been modified to improve the overall yield and reduce the time necessary to make a SAGE library *(4,14,15)*. The whole procedure (except sequencing and data analysis) takes about 7–9 d; however, this time will vary depending on the experience of the investigator and whether the preliminary procedures for SAGE (**Subheading 3.2.2.**) have been completed.

3.2.3.1. PREPARATION OF CELL EXTRACTS FOR PREPARATION OF RNA (*SEE* **NOTE 20**)

The initial step in creating a SAGE library is the isolation of total RNA from ES cell lysates. Numerous protocols have been successfully employed to isolate total RNA and mRNA for the construction of SAGE libraries; for this reason, a detailed protocol is not given in this chapter.

1. Wash ES cells cultivated in the absence of feeder cells for four to six passages with PBS.
2. Remove PBS and discard.
3. Lyse ES cells by the addition of RNA lysis solution (enough to cover the plate). Either scrape the cells or triturate with a Pasteur pipet until all the cells are lysed.
4. Use 1 mL RNA lysis buffer to wash/recover the maximum amount of RNA/DNA from each of the plates.
5. Transfer the lysis buffer with lysed cells to a polypropylene tube and process the lysates using any standard RNA protocol *(16)*.
6. Dissolve isolated RNA in DEPC-treated water (at neutral pH) and determine its concentration either spectrophotometrically or with the Agencourt 2100 Bioanalyzer.
7. Adjust the samples to give an RNA concentration of approx 0.5 µg/µL, aliquot, and store either at −80°C as an aqueous solution or at −20°C as an EtOH precipitate.
8. For the preparation of Poly(A)$^+$ RNA (i.e., mRNA), use the FastTrack 2.0 kit following the manufacturer's instructions. About 0.5–1 mg total RNA is usually required for mRNA preparation (*see* **Note 21**). Ideally, you should have 5 µg Poly(A)$^+$ RNA for the preparation of a SAGE library using the conditions that we describe here.

3.2.3.2. cDNA SYNTHESIS

At the cDNA synthesis step, we recommend using α^{32}P labeling for test purposes (i.e., to ensure appropriate first- and second-strand synthesis). This is particularly important for researchers who have limited experience in the construction of cDNA libraries. This can be achieved using a small portion of the original mix (**step 2**) for cDNA synthesis, but for the most part, we skip this protocol. If used, nucleotide incorporation into the first strand can be employed to calculate cDNA yield, and isotope labeling allows cDNA tracing up to the tagging enzyme (TE) digestion step.

1. Starting with 5 μg purified mRNA, use the cDNA synthesis kit according to the manufacturer's protocol. However, substitute the 5′-biotinylated oligo dT_{18} primer (500 ng/μL) for the oligo dT primer (supplied with this kit) in the first-strand synthesis reaction.
2. After the second-strand reaction is terminated with 25 μL 0.25 M Na_2EDTA (pH 7.5), extract the cDNA sample with PC8, EtOH precipitate, and resuspend the pellet in 20 μL LoTE (*see* **Subheading 3.2.1.1.**).

3.2.3.3. CLEAVAGE OF BIOTINYLATED cDNA WITH AE AND BINDING TO MAGNETIC BEADS

Once double-stranded cDNA has been successfully synthesized from RNA, the AE (e.g., NlaIII) is used to cleave the cDNA close to the 3′ end. In theory, any restriction enzyme that recognizes and cuts within a four-base recognition sequence may be used for this purpose because, on average, these enzymes will cleave the cDNA once every 256 bp (i.e., 4^4). However, we recommend the use of NlaIII because of the possibility of performing comparisons with other publicly available SAGE catalogs. It should be noted, however, that NlaIII recognition sequences are not located in all transcripts, and these mRNAs will not be detected in the final library.

1. Take 10 μL cDNA sample; add 163 μL LoTE, 2 μL of bovine serum albumin (BSA, 100X), 20 μL 10X restriction enzyme buffer 4, and 5 μL NlaIII (AE, 10 U/μL).
2. Incubate at 37°C for 1 h.
3. Extract the samples with PC8 (*see* **Subheading 3.2.1.1.**), EtOH precipitate, and resuspend in 20 μL LoTE.
4. To two new Eppendorf tubes marked 1 and 2, add 100 μL resuspended M-280 streptavidin magnetic beads.
5. Use the magnetic particle concentrator to bind the beads and then remove the storage buffer. Wash beads once with 200 μL 1X B+W buffer, rebind the beads, and then discard the buffer.
6. Add 100 μL 2X B+W buffer, resuspend, and add 90 μL dH_2O and 10 μL cDNA digestion products to each Eppendorf tube.
7. Incubate at room temperature for 15 min, carefully resuspending the beads intermittently.
8. Wash beads three times with 200 μL 1X B+W, then once with 200 μL LoTE, removing the wash each time. Proceed immediately to the next step.

3.2.3.4. LIGATING LINKERS TO BOUND cDNA

After cleavage by the AE, divide the cDNA mix equally into two tubes into which are added different kinased linkers that contain sequences complementary to the AE cohesive overhangs; these are ligated together within one tube. Linkers are about 41 bp long and contain DNA that is complementary to primers for subsequent amplification steps and that are complementary to type IIS enzyme (e.g., BsmFI) recognition sequences.

1. Resuspend the magnetic beads in 25 µL LoTE and add 8 µL 5X ligase buffer.
2. Add 5 µL annealed Linker A (A1, A2) to the tube marked 1 and 5 µL of annealed Linker B (B1, B2) to the tube marked 2.
3. Incubate both tubes at 50°C for 2 min and then at room temperature for 15 min.
4. Add 2 µL T4 ligase (high concentration, 5 U/µL) to each tube and incubate at 16°C for 2 h.
5. Wash the beads four times with 200 µL 1X B+W, then two times with 1X buffer for BsmFI. Proceed immediately to the next step.

3.2.3.5. RELEASE OF CDNA TAGS BY TE AND BLUNT ENDING

The enzyme BsmFI cleaves 14 bp downstream of its recognition sequence [GGGAC(N_{10})] and generates the 10-bp tag downstream of the NlaIII recognition sequence. This 10-bp sequence is commonly referred to as the SAGE tag.

1. Resuspend the magnetic beads from each tube in 86 µL LoTE.
2. Add 10 µL 10X NEB buffer 4, 2 µL 100X BSA, and 2 µL BsmFI (2 U/µL, TE).
3. Incubate at 65°C for 1 h, mixing intermittently.
4. Collect reaction mixture (separate out beads by taking supernatant).
5. Extract the supernatant with PC8 (*see* **Subheading 3.2.1.1.**), EtOH precipitate, and resuspend the pellets in 10 µL LoTE.
6. To each of two tubes (1 and 2) with 10 µL of TE digestion products, add 32.4 µL dH_2O, 5 µL 10X second-strand buffer (from cDNA synthesis kit), 1 µL 100X BSA, 1 µL dNTPs (25 m*M*), and 0.6 µL Klenow (5 U/µL).
7. Incubate at 37°C for 30 min and add 150 µL LoTE.
8. Extract samples with PC8, EtOH precipitate with 1 mL EtOH and ammonium acetate (*see* **Note 18**), and resuspend pellets in 6 µL LoTE.

3.2.3.6. LIGATION OF DITAGS AND PCR AMPLIFICATION

Using primers complementary to the linker, the ditags are amplified by PCR to facilitate their isolation and identification. The amplified fragments run as a 102-bp band (20-bp ditags + 82-bp linkers) on a gel. These can be readily identified after staining, excised from the gel, and purified.

1. For ditag ligation reactions, take 2 µL blunt-end samples 1 and 2 from the previous step and add 1.2 µL 5X ligase buffer and 0.8 µL high-concentration T4 ligase (5 U/µL). For negative control reactions, do not add enzyme.
2. Incubate at 16°C overnight and then add 14 µL LoTE to the samples and either proceed directly to the next step or store the samples for up to 8 h at +4°C (or frozen at −20°C indefinitely).
3. Use 1-µL aliquots of the samples to make serial LoTE dilutions (*see* **Note 22**) 1/10, 1/25, 1/50, and 1/100 for ditag ligation reaction; 1/10 and 1/25 for negative control reactions; and the pure LoTE sample as an additional negative control.
4. Prepare a PCR mix: 30.5 µL sterile water, 1 µL selected ligation reaction or control dilution, 5 µL 10X PCR Gold buffer, 4 µL $MgCl_2$ (25 m*M*), 3 µL DMSO, 4 µL dNTPs (25 m*M*), 1 µL SAGE primers 1 and 2 (350 ng/µL each), and 0.5 µL AmpliTaq Gold polymerase (5 U/µL). Use one strip (eight tubes) of MicroAmp reaction tubes for PCR.
5. Run cycling conditions as follows: hold initially for 10 min at 95°C; perform 28 cycles as follows: 30 s at 95°C, 1 min at 55°C, and 1 min at 72°C and a final extension at 72°C for 5 min. Hold at 4°C.

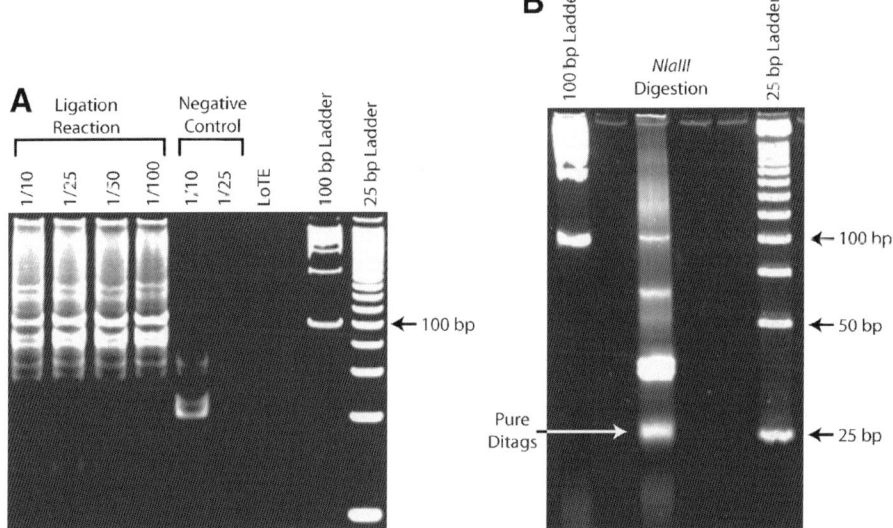

Fig. 3. Ditags produced by serial analysis of gene expression. (**A**) Electrophoresis of polymerase chain reaction-amplified ditags in a 20% Tris/borate/EDTA buffer polyacrylamide gel. The major band of 102 basepairs (20-bp ditags + 82-bp linkers) represents amplification products from ligated and amplified DNA tags. Other bands represent nonspecific amplification products or linker-linker ligation products (80 bp). (**B**) Following purification, the linkers are digested away from the ditags by the AE. The image shows a 12% Tris/acetate/EDTA (TAE) polyacrylamide gel containing NlaIII digestion products. Pure ditags can be seen as 24- to 26-bp products (arrow). Other bands represent products of incomplete AE digestion (linker-ditag structures) or undigested linker-ditag-linkers.

6. Load 20 µL PCR products mixed with 4 µL any 6X loading buffer (containing bromophenol blue) into the wells of a 10-well 20% preready Novex TBE polyacrylamide gel. Load 100-bp DNA Ladder to the outside wells of the gel. Use 1X TBE as a running buffer and run gels initially at 30 V until all the samples have fully entered the gel. Reset the voltage to 120 V and run until the bromophenol blue dye reaches the bottom of the gel (approx 5 h).
7. Stain the gel with SYBR Green I (20 µL/200 mL 1X TBE buffer) for 15 min while gently shaking.

Ditag ligation reaction-PCR-amplification products should be seen as a major 102-bp band with another bright band at approx 80 bp (linker-linker ligation). Any other bands should be considerably less bright. Negative control reactions should lack the 102-bp band, and LoTE (negative control) should not produce any bands (**Fig. 3A**). The optimal dilution of the ditag ligation reaction products for large-scale PCR is determined based on the brightness of the 102-bp band relative to background. The highest dilution that gives the greatest brightness-to-background ratio is considered optimal. Typically, 1/50 to 1/100 dilutions produce good results.

3.2.3.7. Large-Scale Amplification

The large-scale amplification process is necessary to generate enough ditags for concatemerization and subsequent cloning. Specifically, the test amplification is scaled up in size to generate relatively large amounts of ditags.

1. Perform large-scale PCR amplifications similarly to the test-PCR conditions but with the selected LoTE dilutions as a template (e.g., 1/100). Use a Thermo-Fast 96-well detection plate and a cover plate with adhesive sealing sheets and spin (see **Note 23**) to eliminate bubbles in the wells. Prepare 120 reactions in 50 µL in 96-well plates.
2. Collect PCR products in 12 Eppendorf tubes (500 µL in each), extract samples with PC8, EtOH precipitate (using 1 mL absolute EtOH), and resuspend pellets in 18 µL LoTE (216 µL total).
3. Add 54 µL 6X loading buffer (270 µL total) and load PCR products onto eight wells of 12% polyacrylamide gel (19:1). Load 100-bp DNA Ladder to the outside wells of the gel. Use 1X TAE as running buffer. Set voltage to 30 V until the entire sample has entered the gel, then reset the voltage to 130 V.
4. At the end of the electrophoresis, stain the gel with SYBR Green I (20 µL/200 mL 1X TAE buffer) for 15 min while shaking.
5. Pierce the bottoms of eight 0.5-mL Eppendorf tubes with a needle to form a hole with a diameter of about 0.8–1.0 mm in each and place them in 1.5-mL Eppendorf tubes.
6. Using a surgical blade, cut the 102-bp bands out of the gel and put them into the individual 0.5-mL tubes.
7. Spin tubes at maximum speed for 2 min at room temperature and then discard 0.5-mL tubes.
8. Add 300 µL LoTE to each Eppendorf tube, vortex, and incubate at 65°C for 15 min, mixing intermittently.
9. Transfer the contents of each Eppendorf tube to individual SpinX tubes; spin at maximum speed for 2 min.
10. Discard SpinX cartridges, EtOH precipitate the samples (liquid phase) using 940 µL absolute EtOH, and resuspend pellets in 14.8 µL LoTE (118.5 µL total).

3.2.3.8. Purification of Ditags

Following purification, the linkers are digested from the ditags by the AE. This releases a 26-bp band that consists of two tags or a single ditag. We use modified conditions, relative to the original protocol, to reduce potential problems associated with incomplete digestion during the step of ditag isolation.

1. Combine all samples into one Eppendorf tube and add 15 µL 10X restriction buffer 4, 1.5 µL 100X BSA, and 15 µL NlaIII (10 U/µL).
2. Incubate digestion reactions at 37°C for 1 h and 15 min, extract sample with PC8, then add 50 µL LoTE to bring sample volume to 200 µL. During this time, prepare a 12% polyacrylamide gel (19:1).
3. EtOH precipitate on dry ice/EtOH bath, spin at full speed at 4°C for 20 min in an Eppendorf centrifuge, and resuspend the pellet in 30 µL LoTE.
4. Add 6 µL 6X loading buffer and load samples onto two wells of a 12% polyacrylamide gel (19:1). Two lanes are reserved for 100-bp and 25-bp DNA Ladders. Use 1X TAE as the running buffer. Set voltage to 30 V until the entire sample has entered the gel, then reset voltage to 130 V. At the end of the electrophoresis, stain the gel with SYBR Green I (20 µL/200 mL 1X TAE buffer) for 15 min while shaking.
5. Using a surgical blade, cut 24- to 26-bp bands off the gel (**Fig. 3B**) and place them into individual bottom-pierced 0.5-mL tubes positioned in 1.5-mL Eppendorf collection tubes.
6. Spin tubes at 12,000g for 2 min to shred the gel slice and discard the 0.5-mL tubes.
7. Add 300 µL LoTE to each Eppendorf tube and vortex.
8. Incubate at 37°C for 15 min, mixing intermittently.
9. Transfer the contents of each Eppendorf tube to individual SpinX tubes; spin at 12,000g for 2 min.

10. Discard SpinX cartridges, distribute liquid phases from two SpinX tubes into three Eppendorf tubes; EtOH precipitate on dry ice/EtOH bath. Resuspend and combine the pellets in 7 µL LoTE.

3.2.3.9. Ligation of Ditags

Using T4 DNA ligase, ditags are linked to form concatemers, which can be subcloned into a plasmid vector system. The length of time necessary to ligate ditags into catemers of suitable average length (approx 800 bp) depends on the quality of the purified ditags. Although concatenation times can vary from 45 min to overnight, depending on the volume of ditags purified, we find that a 3-h concatenation generally produces good results.

1. To the purified and pooled ditags (7 µL total), add 2 µL 5X ligation buffer and 1 µL high-concentration T4 ligase (5 U/µL).
2. Ligate ditags to form concatemers by incubation at 16°C for 3 h. During this time, prepare an 8% polyacrylamide gel.
3. At the end of the 3-h incubation, add 90 µL LoTE and reincubate samples at 60°C for 5 min.
4. Extract sample with PC8, EtOH precipitate, and resuspend pellet in 10 µL LoTE.
5. Incubate ditag concatemer sample (*see* **Note 24**) at 60°C for 5 min.
6. Add 190 µL LoTE, extract sample with PC8, EtOH precipitate, and resuspend pellet in 10 µL dH$_2$O.
7. Incubate at 60°C for 5 min and immediately load sample onto one well of an 8% polyacrylamide gel (37.5:1). Load 100-bp DNA ladder and 1-kb DNA ladder to distal wells of the gel. Run and stain the gel as previously described.
8. Using a surgical blade, cut concatemers with a length of 600–2500 bp out of the gel and divide into two parts (**Fig. 4**).
9. Place each portion into two labeled, bottom-pierced 0.5-mL tubes as described previously.
10. Spin tubes at maximum speed in microfuge for 2 min and then discard 0.5-mL tubes.
11. Add 300 µL LoTE to each Eppendorf tube; vortex.
12. Incubate at 65°C for 15 min, mixing intermittently.
13. Transfer the contents of each Eppendorf tube to two SpinX tubes; spin at maximum speed for 2 min.
14. Discard SpinX cartridges; transfer liquid phases from two SpinX tubes to one Eppendorf tube.
15. EtOH precipitate using 1 mL absolute EtOH and resuspend pellets in 6 µL total volume of LoTE.

3.2.3.10. Cloning Concatemers and Sequencing

1. To 6 µL purified concatemers, add 2 µL 5X ligase buffer; 1 µL pZeRO-1 previously linearized (*see* **Subheading 3.2.2.3.**) with SphI (25 ng/µL); and 1 µL T4 ligase (1 U/µL) (*see* **Subheading 3.2.2.3.**).
2. Incubate at 16°C overnight.
3. Add 190 µL LoTE to the samples, extract with PC8, EtOH precipitate, wash pellets three times with 75% EtOH, and resuspend pellets in 6 µL LoTE. Use self-ligated pZeRO-1 vector as a negative control.
4. Transform competent bacteria by the addition of 2 µL of each DNA sample to 50 µL freshly thawed ElectroMAX DH10B cells on ice. Mix gently with the pipet tip in a test tube rather than by pipetting.

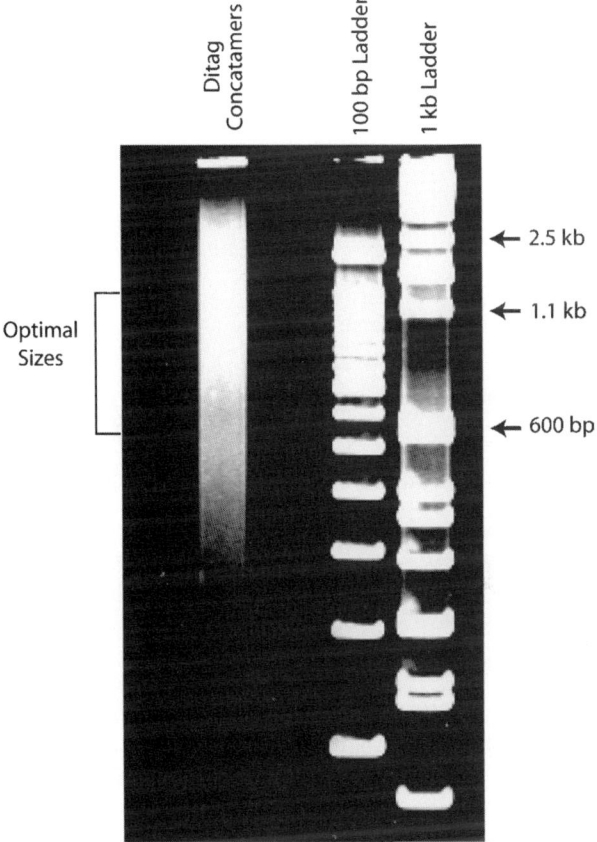

Fig. 4. Concatemerized ditag products observed in an 8% TAE polyacrylamide gel. Using T4 DNA ligase, ditags are ligated together to form concatemers that can be subcloned into a vector. Depending on the length of ligation time and the initial starting products, ditags concatemerize with varying degrees of efficiency. Ideally, cloning of the concatemers should be performed with DNA strands longer than 500 bp. The ideal cloning products (600–1200 bp) are shown in the figure.

5. Transfer the cold mixture to the bottom of 0.2-cm electroporation cuvets, avoiding bubbles, and pulse-electroporate using the following constants: 2.5 kV, 25-μF capacitor, and 200 Σ resistance.
6. After the pulse, immediately transfer bacterial cells to a 15-mL bacterial tube with 1 mL prewarmed SOC media (no Zeocin), place in a shaker incubator, and let the bacteria recover from the electroporation at 37°C for 40 min while rotating at 200g.
7. Plate 100 mL transformed bacteria on two or three 10-cm^2 Zeocin-containing plates and save the remaining contents at 4°C overnight. If the transformation and colony PCR are successful, then plate the remaining contents onto 25-cm^2 plates (*see* **step 13**).
8. Use about 15 individual colonies from the Petri dishes for each sample to check the cloning efficiency and ditag concatemer structure in the clones by PCR. For this, run 25-μL PCR reactions with 18 μL dH$_2$O, 2.5 μL 10X PCR buffer, 1.5 μL DMSO, 1.5 μL dNTPs

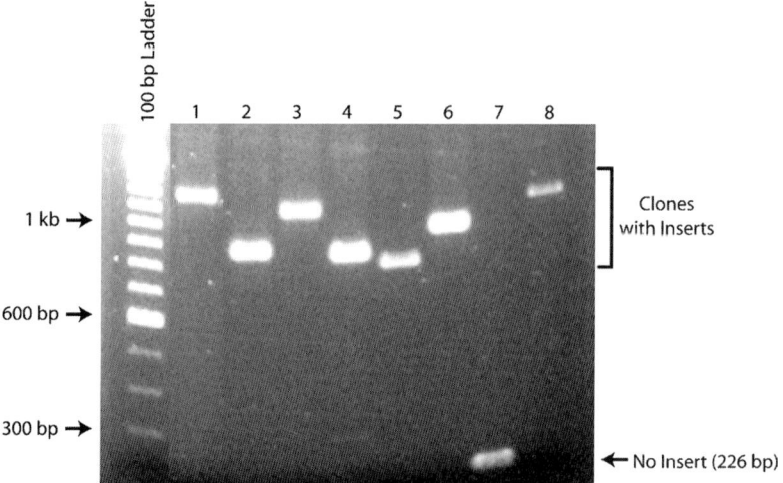

Fig. 5. Polymerase chain reaction (PCR) products of DNA inserts from test clones. To determine the quality of a serial analysis of gene expression library, inserts are amplified by PCR and assessed. Clones with inserts demonstrate a variety of insert sizes that depend on the concatemer length (e.g., a 15-bp ditag product runs at >600 bp). Bands of 226 bp represent amplification products of colonies that do not contain inserts.

(25 mM), 0.5 µL of each M13 forward and reverse primers (350 ng/µL each), and 0.1 µL of thermostable Taq DNA polymerase (0.5 U) inoculated with a single bacterial colony picked by a sterile toothpick. Cycling conditions are as follows: the initial hold is 2 min at 95°C, followed by 25 cycles of 30 s at 95°C, 1 min 30 s at 56°C, 1 min 30 s at 70°C, and a final extension at 70°C for 5 min. Hold at 4°C.
9. To the 5-µL aliquots of control PCR reactions, add 1 µL 6X loading buffer and load onto the wells of 1.2% agarose gels with 1 µg/mL ethidium bromide. Load 100-bp DNA ladder to the outside wells of the gels and use 0.5X TBE as a running buffer. Run electrophoresis reactions at a constant voltage setting of 120 V.
10. Estimate the cloning efficiency by the ratio of clones with large inserts to the total number of clones amplified with PCR. Background clones with no inserts will appear on the gel as 226-bp bands (**Fig. 5**).
11. To the remaining 20 µL PCR products for 6–10 clones with inserts, add 180 µL LoTE, extract sample with PC8, add 250 run microliters of LoTE (to the total volume of 450 µL for each sample), precipitate samples with isopropanol, and resuspend pellet in 10 µL LoTE.
12. For in-house sequencing, use forward or reverse M13 primers (*see* **Note 25**). If the inserts of the tested clones demonstrate correct concatemer structure (**Fig. 6**) and the cloning efficiency is satisfactory, then the SAGE library should be sequenced.
13. From **step 7**, plate 350-µL aliquots of the cell/SOC solution onto the 25-cm^2 bioassay trays with low-salt LB agar plates supplemented with Zeocin (50 ng/µL) and 80 µL cell/SOC solution onto the 10-cm Petri dishes (*see* **Note 26**).
14. Incubate plates for about 20 h at 37°C or send out for sequencing (*see* **Note 27**).

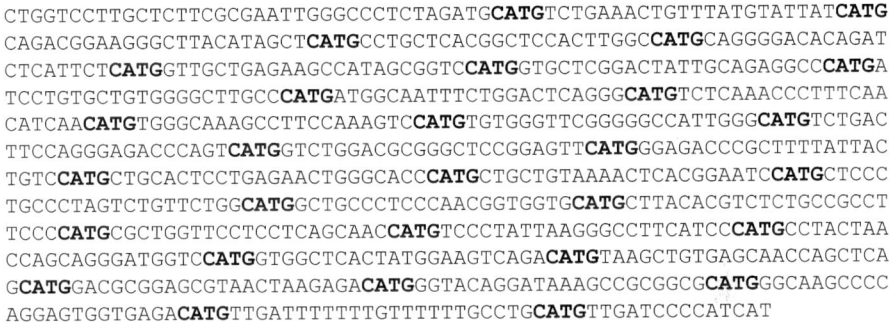

Fig. 6. Correct concatemer structure of sequenced plasmid DNA. Fragments of 24–26 bp, which are flanked by NlaIII sites (in bold), represent authentic ditags. These ditags can be extracted from the raw data files and used to identify transcripts, both quantitatively and qualitatively. Comparisons among serial analysis of gene expression libraries (http://www.ncbi.nlm.nih.gov/sage/) facilitate the identification of factors present or differentially expressed in the cell/tissue of interest.

15. Alternatively and for submission of libraries in glycerol stock format, inoculate wells of 96-well plates filled with 250 µL LB with 10% glycerol/Zeocin media by colonies from Petri dishes using sterile toothpicks. Incubate plates for about 12 h at 37°C in the darkness. Seal with adhesive SealPlate film, freeze at –80°C, and ship on dry ice for sequencing (*see* **Note 27**).

3.3. SAGE Data Collection and Analysis (see Note 28)

A good SAGE library can yield up to more than 60 tags per sequencing reaction in a majority of clones. The number and identity of tags in each reaction can be analyzed using software designed by the Kinzler/Vogelstein group (available at www.sagenet.org). Data collection should be performed as described in the guide "SAGE Analysis Software," version E, available at http://invitrogen.com/Content/sfs/ manuals/ SAGE software_man.pdf (*see* **Note 28**), but to minimize sequencing errors, we recommend the exclusion of any tag that is not observed to be fivefold greater than the expected occurrence of chance sequencing errors *(17)*. Once collected, the entire data set can be archived electronically, and we encourage investigators to deposit their SAGE data with the public gene expression repository known as SAGEMap. The major goals of this repository are to allow for easy handling, analysis, and exchange of expression data in the public forum. At this Web site, one can also find SAGEGenie, which is a set of tools designed to process SAGE data and compare data between various tissues/catalogs (i.e., SAGE Anatomic Viewer). In summary, SAGE analysis software extracts tag sequences from raw sequence data, quantifies tag frequency, identifies the gene transcripts encoded by the tag sequences (using a reference sequence database), and can be useful to compare tag abundances between SAGE libraries. For comparisons among SAGE libraries, we use *MS Access*, and a number of useful SAGE-related Web links are as follows:

1. SAGE home page: http://www.sagenet.org/
2. SAGEmap home page: http://www.ncbi.nlm.nih.gov/SAGE/
3. EntrezGene: http://www.ncbi.nlm.nih.gov/gquery/gquery.tcgi.
4. SAGE Genie: http://cgap.nci.nih.gov/SAGE

Finally, and as noted in **Subheading 2.1.1.**, three mouse ES SAGE catalogs have been deposited to SAGEMap. These libraries serve as valuable reference sets that can be used to identify or confirm transcript abundances in ES cells relative to other cell types.

3.4. SAGE Reference Libraries for Mouse ES Cells

SAGE library comparisons are very useful in the elucidation of gene transcripts that are restricted to or differ among ES cell lines. These differences may be critical to our understanding of potential epigenetic changes that occur with time of cultivation and, more important, in defining those transcripts that may underlie ES cell pluripotentiality and self-renewal properties. In our initial attempts to analyze the transcriptomes of ES cells, we prepared SAGE libraries from R1 ES cell and secondarily P19 embryonic carcinoma cell lines *(4,18)*. At that time, only two other mouse SAGE libraries were available for comparative purposes, precluding a clear analysis of the molecular basis for the ES cell phenotype.

Since our initial SAGE analysis of mouse R1 ES cells, over 40 mouse SAGE libraries from various tissues and cell lines, including two additional ES cell lines (D3 and ESF 116) and one from an EG cell line (EG-1), have been deposited to the public domain. This has allowed us to identify numerous transcripts with expression patterns similar to that of Oct3/4 (Tarasov et al., unpublished data). We have also been able to exploit the comparative power of SAGE (http://www.ncbi.nlm.nih.gov/SAGE), which increases as a function of the number of publicly available libraries, to confirm or refute the authenticity of other stemness-associated transcripts. As an example, we have taken a subset of known and putative stemness factors identified from recently published microarray analyses *(19,20)* and compared the abundance (tags per million) of each transcript among 40 SAGE libraries.

Based on these analyses, we would conclude that Mdr1 and the LIF receptor are not stemness-restricted factors, but factors like UTF-1, Dppa-5, Sox2, and Tdgf (in addition to Oct3/4 and Nanog) are authentic embryonic stemness-related transcripts. Recent comparisons between the mouse R1 ES cell SAGE library with the recently published SAGE analyses of human ES cells, however, have pointed out a number of important differences *(5)* that may account for the role of LIF in mouse ES cells but is not required for cultivation of human ES cells. Specifically, members of the LIF signaling pathway (STAT3, LIFR, and gp130) are highly expressed in mouse ES cells, but are either absent or poorly expressed in human ES cells. In addition, both Oct3/4 and Sox2 were more highly abundant in human than in mouse ES cells.

Because SAGE data are quantitative in nature, we were also able to use the R1 mouse SAGE data set to estimate the total number of transcripts present in ES cells. For statistical reasons, it proved difficult to estimate accurately the total number of unique transcripts, but a simple correction indicated that greater than 54,000 unique transcripts must be present, and model simulations indicated that 130,000 unique transcripts were compatible with the R1 ES cell sampling profile *(21)*. Because approx 10% of the tags in the R1 SAGE library did not map with any previously described EST data set, we estimated that the number of unique transcripts (splice variants or novel gene transcripts) that have not yet been identified in ES cells remain quite high (approx 6000–13,000), underscoring a potential limitation in our ability to define the molecular basis of ES cell identity.

4. Notes

1. All materials and reagents for tissue culture should be kept as sterile as possible, by filtration or autoclaving if appropriate. Similarly, for SAGE, all materials and reagents should be ribonuclease- and deoxyribonuclease-free until the cloning step, either through the purchase of molecular biology-grade reagents or sterilization when required.
2. Several ES cell lines have been successfully employed for SAGE, including two (D3 and R1) that were analyzed using the conditions described in this chapter. SAGE catalogs (i.e., the list of all identified SAGE tags) prepared from these cell lines are indicated by GSM numbers.
3. Either DMEM or Iscove's modified Dulbecco's medium (IMDM) can be employed for cultivation of undifferentiated ES cells. Differentiation with IMDM may require fewer starting cells than differentiation with DMEM, but this tendency is sometimes subtle. In general, and when starting with the same number of starting cells, IMDM promotes cardiomyocyte formation and differentiation more rapidly than DMEM.
4. FBS testing: the most convenient tests for sera include: (1) comparative plating efficiencies at 10, 15, and 30% serum concentrations; (2) alkaline phosphatase activity in undifferentiated ES cells; and (3) in vitro differentiation capacity of ES cells to cardiomyocytes after five passages in selected batches of serum. The last is essential for studies of cardiomyocyte differentiation.
5. Use caution because MMC is carcinogenic.
6. We use CD-1 mice, approx 16 dpc, for the preparation of primary feeder layers or, alternatively, frozen stocks of feeder cells (primary frozen MEF aliquots or STO cells).
7. For restriction enzyme digestions and ligase reactions, we employ buffers provided by the respective manufacturer.
8. It is easy to be confused by the T4 ligase (5 and 1 U/µL) concentrations for a particular step. Please pay attention to the concentrations of each reagent.
9. SAGE uses two major enzymes: BsmFI and NlaIII. The most commonly used AE (NlaIII) has a half-life of only about 3 mo at $-20°C$. Be sure to use fresh batches of enzyme for each SAGE project. The TE BsmFI is a type IIs restriction enzyme that cuts approx 10–14 basepairs downstream of the $GGGAC(N_{10})$ recognition sequence, whereas the AE NlaIII is a type II restriction enzyme that yields a 3′ overhang.
10. Optionally, add 8-hydroxyquinoline to a final concentration of 0.1%.
11. It is easy to become confused by B+W solutions (2X or 1X). Please note if the text refers to 1X or 2X.
12. Zeocin is light sensitive and comparatively unstable. Zeocin-containing media should be kept in the dark for a period of no longer than 2 wk. Low-salt LB agar/Zeocin plates should be protected from light immediately after pouring and until bacterial colonies have grown to an appropriate size.
13. Oligonucleotides (Linkers A1, B1, A2, B2, and biotinylated oligo dT) should be gel purified.

14. We have successfully employed several thermocyclers (GeneAmp PCR Systems 9600, 9700 from Applied Biosystems and Master Gradient from Eppendorf) for the construction of SAGE libraries. If you use other thermocyclers, then we recommend that you test the system before large-scale amplification.
15. Ideally, ES cells are passaged at least one time on feeder layers after thawing. Once ES cells have been passaged off feeder cells, these cells should be cultivated in BRL-CM containing LIF. This protocol seems to be the best for maintaining pluripotentiality. Finally, contaminating feeder cells can be distinguished by their relatively large size when compared with ES cells.
16. Ribonuclease-/deoxyribonuclease-free conditions are important for the successful construction of SAGE libraries. Ensure that all pipets are accurate and solutions are prepared properly and use DEPC-treated glassware and plasticware (during the early steps of library construction) and aerosol-barrier pipet tips throughout the procedure.
17. EtOH precipitation in the presence of ammonium acetate or sodium acetate works equally well; however, precipitation in the presence of sodium acetate is preferable prior to ligation steps. Ammonium ions inhibit the ligase activity; therefore, sodium acetate should be used for all steps prior to ligation reactions. Regardless, remove salt from precipitates by washing in 75% EtOH to ensure appropriate reaction conditions in subsequent steps.
18. In certain instances, an increased volume of EtOH or a 15-min dry ice/EtOH incubation may be required at the precipitation step as indicated in the text.
19. All incubations for SAGE steps are performed in a water bath, which is preferable to a dry block. All centrifugation steps, unless otherwise noted, are performed in an Eppendorf microcentrifuge at room temperature.
20. We usually add 2 mL RNA lysis buffer per 60-mm plate and transfer this solution to lyse cells from up to five plates. We routinely purify RNA by ultracentrifugation through a $CsCl_2$ gradient and use guanidine isothiocyanate, a chaotropic agent that inactivates ribonucleases; however, RNeasy Mini Kits (from Qiagen) or other suitable kits also yield good-quality RNA. For SAGE libraries, we have usually found the methods of Chirgwin to be superior to that of kits. Short RNA sequences that frequently are isolated by other common RNA techniques tend to adversely affect the final SAGE libraries. If analyzed spectrophotometrically, then the absorbance ratio of 260–280 is expected to be about 1.8–2.0. Total RNA must be of good quality before proceeding with the isolation of mRNA.
21. If you have small volumes of starting material, then you can use the microSAGE *(22,23)* and miniSAGE *(24)* modifications to the SAGE protocol.
22. We recommend always preparing fresh dilutions of the ligation reactions for the PCR amplification step; do not use dilutions that have been frozen.
23. To spin 96-well plates, we use a RT6000 centrifuge with an H1000B rotor (Sorvall) at 600*g* for 3 min.
24. A modification of the original SAGE protocol allows better resolution of concatemers on polyacrylamide gel, which improves the overall yield of ditags in a SAGE library: use 65°C for 15 min, chill on ice for 10 min, then load onto the gel. We recommend using the "hot" variation of the same modification: use 60°C for 5 min, PC8 extract, EtOH precipitate, keep at 60°C for 5 min, and load immediately.
25. For large-scale sequencing, we suggest the use of M13 forward primers. For clones with long inserts, an additional M13 reverse-primed sequencing could be beneficial, yielding

more tags, but it also leads to rapid accumulation of duplicate dimers because of the partial overlapping of the sequences.

26. Selection based on X-Gal/IPTG blue/white coloring of colonies with or without inserts has been suggested to increase the cloning efficiency. Although this modification can enhance the overall yield of clones with inserts, the use of pZeRO vectors obviates this requirement.

27. Currently, some commercial sequencing companies provide automated picking of colonies from agar lawns on 25-cm^2 bioassay trays (e.g., Agencourt Biosciences Corporation, http://www.agencourt.com/services/sage/). We currently send all our libraries to be sequenced commercially directly from colonies on bacterial plates or after picking and growing each colony. For the latter, we amplify the library in glycerol stock format, through the use of LB media with 10% glycerol, according to the recommendations of Agencourt. Other companies may require purification of PCR products. To do this, isolate the bands, remove the DNA from the gel in 200 µL LoTE, extract with PC8, isopropanol precipitate (add 3 µL glycogen, 150 µL 2 M NaClO$_4$, and 330 µL 2-propanol). Vortex gently and centrifuge at maximum speed for 10 min. Wash twice with 75% EtOH and resuspend the pellets in 20 mL LoTE. Sequence according to standard protocols. It is recommended that one well in each plate be filled with sterile LoTE to serve as a negative control in sequencing reactions. We recommend that you duplicate each plate and store these duplicates at −80°C until the library has been successfully sequenced, after which time these plates can be discarded.

28. When you perform your data analyses, keep in mind three major limitations to SAGE: (1) the occurrence of multiple tag hits (it is possible that one tag can be associated with several genes or several tags with one gene); (2) there are sequences that lack NlaIII digestion sites *(25)*; and (3) some tags are not informative, such as when the CATG site is located immediately upstream of the poly(A$^+$) tail. We have previously detailed procedures for downloading GenBank databases from the National Center of Biotechnology Information Web site and step-by-step instructions for tag extraction from GenBank sequences for identification of "tag-to-gene matches." But because the National Center of Biotechnology Information Web site has been significantly improved, it is now preferable to use the SAGEmap_tag_ug-full or SAGEmap_tag_ug-rel files as described in the chapter "Downloading a UniGene Reference database" in the *SAGE Analysis Software*, version E, release date July 14, 2003.

References

1. Velculescu, V. E., Zhang, L., Vogelstein, B., and Kinzler, K. W. (1995) Serial analysis of gene expression. *Science* **270**, 484–487.
2. Saha, S., Sparks, A. B., Rago, C., et al. (2002) Using the transcriptome to annotate the genome. *Nat. Biotechnol.* **20**, 508–512.
3. Lander, E. S., Linton, L. M., Birren, B., et al. (2001) Initial sequencing and analysis of the human genome. *Nature* **409**, 860–921.
4. Anisimov, S. V., Tarasov, K. V., Tweedie, D., et al. (2002) SAGE identification of gene transcripts with profiles unique to pluripotent mouse R1 embryonic stem cells. *Genomics* **79**, 169–176.
5. Richards, M., Tan, S. P., Tan, J. H., Chan, W. K., and Bongso, A. (2004) The transcriptome profile of human embryonic stem cells as defined by SAGE. *Stem Cells* **22**, 51–64.

6. Boheler, K. R. (2003) ES cell differentiation to the cardiac lineage. *Methods Enzymol.* **365**, 228–241.
7. Wobus, A. M., Guan, K., Yang, H.-T., and Boheler, K. R. (2002) Embryonic stem cells as a model to study cardiac, skeletal muscle and vascular smooth muscle cell differentiation, in *Methods in Molecular Biology* (Turksen, K., ed.), Humana Press, Totowa, NJ, pp. 127–156.
8. Martin, G. R. (1981) Isolation of a pluripotent cell line from early mouse embryos cultured in medium conditioned by teratocarcinoma stem cells. *Proc. Natl. Acad. Sci. USA* **78**, 7634–7638.
9. Doetschman, T. C., Eistetter, H., Katz, M., Schmidt, W., and Kemler, R. (1985) The in vitro development of blastocyst-derived embryonic stem cell lines: formation of visceral yolk sac, blood islands and myocardium. *J. Embryol. Exp. Morphol.* **87**, 27–45.
10. Smith, A. G. and Hooper, M. L. (1987) Buffalo rat liver cells produce a diffusible activity which inhibits the differentiation of murine embryonal carcinoma and embryonic stem cells. *Dev. Biol.* **121**, 1–9.
11. Smith, A. G., Heath, J. K., Donaldson, D. D., et al. (1988) Inhibition of pluripotential embryonic stem cell differentiation by purified polypeptides. *Nature* **336**, 688–690.
12. Williams, R. L., Hilton, D. J., Pease, S., et al. (1988) Myeloid leukaemia inhibitory factor maintains the developmental potential of embryonic stem cells. *Nature* **336**, 684–687.
13. Parrinello, S., Samper, E., Krtolica, A., Goldstein, J., Melov, S., and Campisi, J. (2003) Oxygen sensitivity severely limits the replicative lifespan of murine fibroblasts. *Nat. Cell Biol.* **5**, 741–747.
14. Kenzelmann, M. and Muhlemann, K. (1999) Substantially enhanced cloning efficiency of SAGE (serial analysis of gene expression) by adding a heating step to the original protocol. *Nucleic Acids Res.* **27**, 917–918.
15. Anisimov, S. V., Tarasov, K. V., Stern, M. D., Lakatta, E. G., and Boheler, K. R. (2002) A quantitative and validated SAGE transcriptome reference for adult mouse heart. *Genomics* **80**, 213–222.
16. Chirgwin, J. M., Przybyla, A. E., MacDonald, R. J., and Rutter, W. J. (1979) Isolation of biologically active ribonucleic acid from sources enriched in ribonuclease. *Biochemistry* **18**, 5294–5299.
17. Lash, A. E., Tolstoshev, C. M., Wagner, L., et al. (2000) SAGEmap: a public gene expression resource. *Genome Res.* **10**, 1051–1060.
18. Anisimov, S. V., Tarasov, K. V., Riordon, D., Wobus, A. M., and Boheler, K. R. (2002) SAGE identification of differentiation responsive genes in P19 embryonic cells induced to form cardiomyocytes in vitro. *Mech. Dev.* **202**, 25–74.
19. Tanaka, T. S., Kunath, T., Kimber, W. L., et al. (2002) Gene expression profiling of embryo-derived stem cells reveals candidate genes associated with pluripotency and lineage specificity. *Genome Res.* **12**, 1921–1928.
20. Sharov, A. A., Piao, Y., Matoba, R., et al. (2003) Transcriptome analysis of mouse stem cells and early embryos. *PLoS Biol.* **1**, 410–419.
21. Stern, M. D., Anisimov, S. V., and Boheler, K. R. (2003) Can transcriptome size be estimated from SAGE catalogs? *Bioinformatics* **19**, 443–448.
22. Datson, N. A., van der Perk-de Jong, J., van den Berg, M. P., de Kloet, E. R., and Vreugdenhil, E. (1999) MicroSAGE: a modified procedure for serial analysis of gene expression in limited amounts of tissue. *Nucleic Acids Res.* **27**, 1300–1307.

23. Angelastro, J. M., Klimaschewski, L. P., and Vitolo, O. V. (2000) Improved NlaIII digestion of PAGE-purified 102 bp ditags by addition of a single purification step in both the SAGE and microSAGE protocols. *Nucleic Acids Res.* **28,** E62.
24. Ye, S. Q., Zhang, L. Q., Zheng, F., Virgil, D., and Kwiterovich, P. O. (2000) miniSAGE: gene expression profiling using serial analysis of gene expression from 1 microgram total RNA. *Anal. Biochem.* **287,** 144–152.
25. Pleasance, E. D., Marra, M. A., and Jones, S. J. (2003) Assessment of SAGE in transcript identification. *Genome Res.* **13,** 1203–1215.

15

Utilization of Digital Differential Display to Identify Novel Targets of Oct3/4

Yoshimi Tokuzawa, Masayoshi Maruyama, and Shinya Yamanaka

Summary

Embryonic stem (ES) cells proliferate infinitely while maintaining pluripotency. The POU family transcription factor Oct3/4 is specifically expressed in ES cells and early embryos and plays a critical role in self-renewal of ES cells. However, only a few examples of Oct3/4 target genes have been identified. In this chapter, we describe our strategy to isolate novel Oct3/4 target genes. We first identify genes that are specifically expressed in ES cells by means of digital differential display of expressed sequence tag databases. Reporter gene and gel mobility shift assays are used to confirm the role of Oct3/4. Identification of novel Oct3/4 targets will facilitate our understanding of pluripotency.

Key Words: ES cell; homeobox; homeoprotein; pluripotency.

1. Introduction

Embryonic stem (ES) cells are derived from the inner cell mass of mammalian blastocysts. ES cells proliferate infinitely while maintaining pluripotency, an ability to differentiate into all types of somatic and germ cells *(1)*. ES cells were developed from mouse blastocysts in 1981 *(2,3)* and resulted in the development of the knockout mouse technology. Human ES cells were established in 1998 and are considered promising sources for cell transplantation therapy *(4)*.

The POU transcription factor Oct3/4 is expressed specifically in pluripotent cells, including ES cells, early embryos, and germ cells *(5,6)*. Targeted disruption of the mouse Oct3/4 gene results in early embryonic lethality *(7)*. Inner cellular mass of these embryos differentiates exclusively into trophoblasts. Furthermore, conditional deletion of Oct3/4 in ES cells leads to spontaneous differentiation into trophectoderm, whereas its overexpression results in differentiation into primitive endoderm and mesoderm *(8)*, demonstrating its essential role for self-renewal of pluripotent cells.

Only a few Oct3/4 target genes have been identified. These include FGF-4 *(9)*, UTF-1 *(10)*, and Rex1 *(11)*, in which Oct3/4 binds to an octamer motif, ATT (T/A) GCAT, located in their regulatory elements. In FGF-4 and UTF-1, the sex-determining region Y-related transcription factor Sox2 binds to a motif adjacent to the octamer sequence and synergistically activates transcription *(10)*. The Oct3/4 binding site in UTF-1 is one nucleotide different from the canonical octamer sequence *(10)*. In Rex-1, the hypothetical factor ROX1 functions in a similar manner *(12)*. In addition to these genes, we identified Fbx15 as an Oct3/4 and Sox2 target *(13)*.

In this chapter, we summarize our method of identifying novel target genes of Oct3/4. This method can be utilized to identify target genes of other transcription factors and genes specifically expressed in tissues or cells of interest.

2. Materials
2.1. Cell Culture
Please refer to Chapter 30.

2.2. Luciferase Reporter Assay
1. Lipofectamin 2000 (1.5 mL; Invitrogen Corp., Carlsbad, CA; cat. no. 11668-019).
2. Dual Luciferase Reporter Assay System (Promega, Madison, WI; cat. no. E1960).

2.3. Gel Mobility Shift Assay
1. Fugene6 transfection reagent (Roche, Tokyo, Japan; cat. no. 1 814 443).
2. F9 nuclear extract (Active Motif, Carlsbad, CA; cat. no. 36007).
3. [α-^{32}P] adenosine triphosphate (ATP; Amersham, Tokyo, Japan; cat. no. AA 3368).
4. Whole-extract buffer: for 50 mL, mix 2 mL 0.5 M HEPES pH 7.9, 40 µL 0.5 M ethylenediaminetetraacetic acid, 250 µL 0.1 M dithiothreitol (DTT), 4.5 mL 5 M NaCl, 12.5 mL glycerol, 125 µL 0.2 M PMSF, and 30.6 mL H$_2$O.
5. Protease inhibitor cocktail (Nakalai Tesque, Kyoto, Japan; cat. no. 25955-11).
6. Microdialyzer (Nippon Genetics, Tokyo, Japan; cat. no. TOR-50).
7. Dialysis buffer: for 500 mL, mix 20 mL 0.5 M HEPES pH 7.9, 400 µL 0.5 M ethylenediaminetetraacetic acid, 2.5 mL 0.1 M DTT, 125 mL glycerol, 1.25 mL 0.2 M PMSF, and 350.9 mL H$_2$O.
8. 100 pmol/µL sense and antisense oligonucleotides.
9. 10X annealing buffer: for 1 mL, mix 200 µL 1 M Tris-HCl pH 7.5, 100 µL 1 M MgCl$_2$, 200 µL 5 M NaCl, and 500 µL H$_2$O.
10. Megalabel DNA 5′-End labeling kit (Takara, Kyoto, Japan; cat. no. 6070).
11. Micro BioSpin column (Bio-Rad, Hercules, CA; cat. no. 732-6250).
12. 10 U poly(dG-dC) (Amersham; cat. no. 27-7910-01).
13. 2X binding buffer: for 1 mL, mix 80 µL 0.5 M HEPES pH 7.9, 4 µL 0.5 M MgCl$_2$, 400 µL 20% Ficoll, 10 µL 0.1 M DTT, and 506 µL H$_2$O.
14. 30% acrylamide: in 200 mL H$_2$O, dissolve 58 g acrylamide monomer and 2 g N,N'-methyrenbisacrylamide.
15. 0.5X TBE buffer: for 500 mL, add 50 mL 10X TBE to 450 mL H$_2$O.
16. 4% acrylamide gel: for 30 mL, mix 4 mL 30% acrylamide, 1.5 mL 10X TBE, 30 µL TEMED, and 100 µL 40% APS.

DDD Identification of Oct3/4 Targets 225

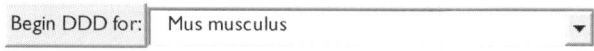

Fig. 1. The initial table of digital differential display for selection of an organism.

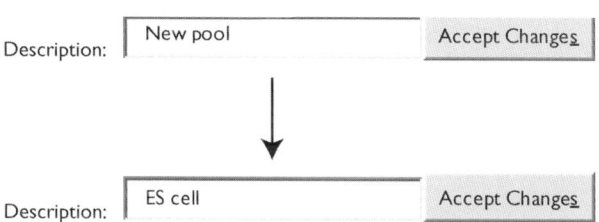

Fig. 2. The text box to name each pool of libraries.

3. Methods

3.1. Identification of Candidates of Oct3/4 Target Genes With Digital Differential Display

Because Oct3/4 is specifically expressed in pluripotent cells and germ cells, it is likely that its target genes are expressed in a similar manner. Digital differential display (DDD) analysis of UniGene can be utilized to identify such genes. UniGene assigns all the expressed sequence tag (EST) sequences that meet minimal standards of quality to distinct "clusters," each representing a unique expressed gene. DDD takes advantage of UniGene by comparing the number of times sequences from different libraries were assigned to a particular UniGene cluster. This has the advantage that DDD will only report on sequences that are likely to represent *bona fide* expressed genes (*see* **Note 1**). The program is freely available through the UniGene Web site (http://www.ncbi.nlm.nih.gov/UniGene/info_ddd.html; *see* **Note 2**).

1. Select organism and start DDD in the table at the top (**Fig. 1**). Another table appears at the top summarizing your choices for comparison (**Fig. 2**). The unit of DDD is the *pool*, which is a collection of libraries (minimum = 1) that you want to group together on the basis of your own criteria for comparison with other pools you will define. Enter the name of the pool in the text box.
2. To select the library or libraries to include in the pool, scroll down and check the box to the left of each library you want to include (**Fig. 3**). Libraries that have been "normalized" are indicated by **N**, whereas those that have been subtracted are indicated by **S** (*see* **Note 3**).
3. Click the button Accept Changes (*see* **Note 4**) The column labeled ESTs indicates the number of sequences from the pool that have been assigned to a UniGene cluster.
4. Repeat the process because you must have a minimum of two pools to compare. Click on New to define the new pool. Repeat the previous steps. On clicking Accept Changes this second time, DDD compares the two pools. For each sequence in a pool that has been mapped to a UniGene cluster, DDD will compare the number of sequences from that cluster that appears in every other pool. DDD will report when there is a significant difference

Fig. 3. The table to define libraries for each pool.

between the number of sequences that belong to a cluster in one pool vs another pool (*see* **Note 5**).
  a. To add another pool, click on NEW and repeat the previously mentioned process.
  b. To change one of your previous choices, click on EDIT.
  c. The description of the pools will remain at the top of the DDD page, whereas the results will be found below in a table (**Fig. 4**).
5. Save the data as text. Open and analyze the data with spreadsheet programs (e.g., Excel™).
6. Confirm specific expression of candidate genes identified by DDD by Northern blot or reverse transcriptase polymerase chain reaction.

3.2. Identification of Putative Oct3/4 Binding Sites in Candidate Genes
  1. Use Blat search (http://genome.ucsc.edu/cgi-bin/hgBlat) to obtain chromosome localization, gene structure, and the nucleotide sequences of the flanking regions of candidate genes.
  2. Use TESS (http://www.cbil.upenn.edu/tess/) to identify putative binding sites of Oct3/4 and other transcription factors in candidate genes.

DDD Identification of Oct3/4 Targets

A ES cell	B Somati..	Gene index	Gene description
200 0.00542 ● A>B	2555 0.00220 B<A	Mm.335315	Eukaryotic translation elongation factor 1 alpha 1 (Eef1a1)
139 0.00377 ● A>B	1056 0.00091 B<A	Mm.336743	Heat shock protein 8 (Hspa8)
89 0.00241 ● A>B	 0.00000 B<A	Mm.10205	Undifferentiated embryonic cell transcription factor 1 (Utf1)
106 0.00287 ● A>B	618 0.00053 B<A	Mm.290774	MRNA fragment for heat shock

Fig. 4. The table showing the digital differential display (DDD) result. Columns labeled with letters (A, B, ...) represent their respective pool and will contain at least part of the name. Each row represents a UniGene cluster: the column Gene Description contains the name of the cluster, whereas Gene Index refers to the numerical ID of the cluster. Clicking on this ID will link you to a summary report for the UniGene cluster. The DDD results table provides several pieces of information for each gene. The first numerical value is the total number of expressed sequence tag clones within the pool that mapped to the cluster shown. The second numerical value is the fraction of sequences within the pool that mapped to the cluster shown. The dot is merely a visual aid that reflects the numerical values. If any pool participates in a statistically significant pairwise comparison with another pool, the relationship is indicated at the bottom.

In the case of Fbx15, we identified a putative binding site of the Sox family of transcription factors between nucleotide positions -534 and -528 from the transcription initiation site and an octamer-like sequence, TTTATCAT, immediately upstream of the Sox-binding site.

3.3. Reporter Gene Analysis to Characterize the Regulatory Elements in the Candidate Genes

3.3.1. Construction of Reporter Plasmids

Two types of reporter plasmids are used, pGV-B and pGB-P. pGV-B contains the firefly luciferase (Fluc) gene but lacks promoter sequences. A DNA fragment of various lengths from the 5′-flanking region of the genes of interest is subcloned upstream of the Fluc gene. This method can be used when the putative *cis*-element is located in the 5′-flanking region of the candidate gene. pGV-P contains the Fluc gene driven by a minimum thymidine kinase promoter. DNA fragments derived from any positions of candidate genes can be tested for enhancer activity with this type of reporter plasmid. Reporter plasmids can be constructed by standard molecular biology techniques or by Gateway cloning technology (Invitrogen).

3.3.2. Transfection of Reporter Plasmids Into ES Cells and NIH3T3 Cells

1. Seed undifferentiated RF8 cells (2.5×10^5 cells/well), differentiated RF8 ES cells (0.5×10^5 cells/well), or NIH3T3 cells (0.5×10^5 cells/well) to gelatin-coated 24-well plates.
2. Culture overnight at 37°C with 5% CO_2.
3. The next day, dilute plasmid DNA (1 μg reporter plasmid and 0.025 μg pRL-TK; *see* **Note 6**) and 2 μL Lipofectamine 2000 each in 150 μL DMEM.
4. Combine the diluted DNA/Lipofectamine 2000, mix by vortex briefly, and incubate for 20 min at room temperature.
5. During incubation, remove the medium from cells and wash once with DMEM.
6. After incubation, apply the DNA/Lipofectamine 2000 mixture to ES cells and incubate for 4 h at 37°C with 5% CO_2. pGV-B should be used as a negative control. pGV-C, in which the Fluc gene is driven by the Simian virus 40 promoter/enhancer, should be used as a positive control.
7. After 4 h, add 500 μL ES medium and incubate at 37°C for 17–20 h.

3.3.3. Measure of Luciferase Activity

1. Remove medium and wash cells with phosphate-buffered saline (PBS).
2. Add 100 μL 1X lysis buffer and shake at room temperature for 20 min.
3. Add 25 μL Fluc substrates to the 5-μL aliquot and measure the Fluc activity with a luminometer. Add 25 μL of Rluc substrates to the mixture and Rluc activity.
4. Measure Fluc and Rluc of cells that were mock transfected as background values.
5. Calculate relative luciferase activity of each sample (*see* **Note 6**).

3.4. Gel Mobility Shift Assay to Study the Binding of Oct3/4 to the Enhancer Elements of Candidate Genes

3.4.1. Preparation of Expression Plasmids

pCAGIP-HA-Oct3/4 and pCAGIP-HA-Sox2 were constructed by a recombination reaction between the pDONR-Sox2 or pDONR-Oct3/4 (entry vectors) and pCAGIP-HA-gw (destination vectors) containing Gateway *ref*A cassette (M. Maruyama et al., unpublished data).

3.4.2. Preparation of Cell Lysates

To express Oct3/4 and Sox2 in Cos7, pCAGIP-HA-Oct3/4 and pCAGIP-HA-Sox2 were introduced into Cos7 cells with Fugene6.

1. Seed Cos7 cells to 100-mm dishes (1×10^6 cells per dish) and culture overnight.
2. Mix 500 μL DMEM and 15 μL FuGene 6 and incubate at room temperature for 5 min.
3. Add 5 μg of vectors to the mixture and incubate at room temperature for 15 min.
4. Add the mixture to Cos7 cells drop by drop and incubate at 37°C for 2 d.
5. After 2 d, aspirate medium and wash cells with PBS.

6. Add 1 mL PBS and remove the cells with a cell scraper. Transfer the cells to a new tube.
7. Spin down at 14,000g for 1 min at 4°C.
8. Remove the supernatant and wash the pellet with PBS.
9. Resuspend the pellet in 150 μL whole-extract buffer containing 3 μL protease inhibitor cocktail.
10. Shake gently.

3.4.3. Dialysis

To desalt the cell lysate, we used a microdialyzer at 4°C.

1. To swell the dialysis membrane, soak it in water for 5 min.
2. Remove water from the dialysis chamber and transfer the cell lysate into the chamber. Set collection tubes and floats.
3. Put on the dialysis buffer and stir for 2 h. Spin down at 14,000g for 30 s and collect the dialyzed cell lysates.
4. To check the expression, carry out Western blot analysis.

3.4.4. Gel Mobility Shift Assay

3.4.4.1. Probe Preparation

1. Prepare the following reaction mixture: 10 μL 100 pmol/μL sense-oligo, 10 μL 100 pmol/μL antisense-oligo, 3 μL 10X annealing buffer, and 7 μL H_2O.
2. Incubate at 85°C for 5 min.
3. Switch off the heat block and cool the sample gradually to room temperature.
4. Add 75 μL 100% ethanol to the sample and centrifuge at 14,000g for 30 min at 4°C.
5. Wash the pellet with 70% ethanol.
6. Resuspend the sample in TE buffer.

3.4.4.2. Labeling Reaction

1. Prepare the following reaction mixture: 1 μL 10 pmol/μL double-stranded oligo, 1 μL 10X phosphorylation buffer, 1 μL T4 polynucleotide kinase, 5 μL [α-^{32}P] ATP, and 2 μL H_2O.
2. Incubate at 37°C for 30 min with Megalabel DNA 5′-End labeling kit.
3. After incubation, add 40 μL H_2O and purify the sample with Micro BioSpin columns.
4. Dilute 1 μL purified probes in 19 μL H_2O.

3.4.4.3. Binding Reaction

1. Prepare the following reaction mixture: 12–15 μL cell lysate (dialyzed), 10 μL 2X binding buffer, 1 μL polydG-dC (2 μg/μL), and 1 μL labeled oligo (10 fmol/μL); adjust volume to 20 μL with water.
2. Incubate on ice for 40 min.

3.4.4.4. Electrophoresis

1. Prerun the gel at 150 V for 5–10 min.
2. Apply the samples on 4% polyacrylamide gel (0.5X TBE) and run for 2 h at 150 V and 4°C.
3. Dry the gel on filter paper at 80°C for 40 min with a gel dryer.
4. Analyze the gel by autoradiography.

A

	Pool	Lib ID(s)	Clustered ESTs
Edit...	A. ES cell	15703, 10023, 14556, 2512, 1882	36879
New...			

B

	Pool	Lib ID(s)	Clustered ESTs
Edit...	A. ES cell	15703, 10023, 14556, 2512, 1882	36879
Edit...	B. New pool	no libraries selected	0
New...			

Fig. 5. The table showing the name and libraries selected for the first pool. To define the second pool, click on New in table **A**. Another table (**B**) will appear.

4. Notes

1. Among many differences in the number of sequences that are assigned to a particular UniGene cluster contained between libraries, only some reflect biological reality. In DDD, significant differences are identified by the Fisher exact test. One important factor in determining statistical relevance is the absolute number of sequences in each library that have been successfully assigned to a UniGene cluster. In many cases, there are not enough sequences in dbEST libraries to meet the threshold of significance employed in the Fisher exact test. Because DDD will only yield a report if there are differences that exceed this threshold, it is expected that many comparisons will yield nothing.
2. The Web address of DDD may and did change but can be easily identified from the Web site of Unigene.
3. Each library has a numerical ID; clicking on the ID will link you to a summary report of that library.
4. The table at the top of the page indicates the numerical library ID of those you chose to include in pool A as well as the name of the pool (**Fig. 5**).
5. DDD only reports on a cluster if at least one pairwise comparison results in a statistically significant difference. However, when such a difference is found, the DDD report includes information on the number of times that UniGene cluster was represented in all pools, although the report distinguishes those differences that are statistically significant. You may find it useful to include a control pool that contains a large number of sequences because the library source is unlikely to share sequences with your libraries of interest. Because there will be many statistically significant differences between the control and other pools, DDD will report on many more genes, and although many of the differences between your libraries of interest will not be statistically significant, you may find them biologically interesting nevertheless.
6. Relative luciferase activity = (Fluc − background)/(Rluc − background).

Acknowledgments

We thank Eiko Kaiho, Tomoko Ichisaka, and other members in Yamanaka laboratory for valuable discussion; Yukiko Ikeguchi, Junko Iida, and Masako Shirasaka for technical and administrative support; and Dr. Hitoshi Niwa for MG1.19 ES cells.

References

1. Smith, A. G. (2001) Embryo-derived stem cells: of mice and men. *Annu. Rev. Cell Dev. Biol.* **17,** 435–462.
2. Evans, M. J. and Kaufman, M. H. (1981) Establishment in culture of pluripotential cells from mouse embryos. *Nature* **292,** 154–156.
3. Martin, G. R. (1981) Isolation of a pluripotent cell line from early mouse embryos cultured in medium conditioned by teratocarcinoma stem cells. *Proc. Natl. Acad. Sci. USA* **78,** 7634–7638.
4. Thomson, J. A., Itskovitz-Eldor, J., Shapiro, S. S., et al. (1998) Embryonic stem cell lines derived from human blastocysts. *Science* **282,** 1145–1147.
5. Okamoto, K., Okazawa, H., Okuda, A., Sakai, M., Muramatsu, M., and Hamada, H. (1990) A novel octamer binding transcription factor is differentially expressed in mouse embryonic cells. *Cell* **60,** 461–472.
6. Scholer, H. R., Dressler, G. R., Balling, R., Rohdewohld, H., and Gruss, P. (1990) Oct-4: a germline-specific transcription factor mapping to the mouse t-complex. *EMBO J.* **9,** 2185–2195.
7. Nichols, J., Zevnik, B., Anastassiadis, K., et al. (1998) Formation of pluripotent stem cells in the mammalian embryo depends on the POU transcription factor Oct4. *Cell* **95,** 379–391.
8. Niwa, H., Miyazaki, J., and Smith, A. G. (2000) Quantitative expression of Oct-3/4 defines differentiation, dedifferentiation or self-renewal of ES cells. *Nat. Genet.* **24,** 372–376.
9. Dailey, L., Yuan, H., and Basilico, C. (1994) Interaction between a novel F9-specific factor and octamer-binding proteins is required for cell-type-restricted activity of the fibroblast growth factor 4 enhancer. *Mol. Cell Biol.* **14,** 7758–7769.
10. Ben-Shushan, E., Thompson, J. R., Gudas, L. J., and Bergman, Y. (1998) Rex-1, a gene encoding a transcription factor expressed in the early embryo, is regulated via Oct-3/4 and Oct-6 binding to an octamer site and a novel protein, Rox-1, binding to an adjacent site. *Mol. Cell Biol.* **18,** 1866–1878.
11. Damjanov, I. (1978) Teratoma and teratocarcinoma in experimental animals. *Natl. Cancer Inst. Monogr.* **49,** 305–306.
12. Nishimoto, M., Fukushima, A., Okuda, A., and Muramatsu, M. (1999) The gene for the embryonic stem cell coactivator UTF1 carries a regulatory element which selectively interacts with a complex composed of Oct-3/4 and Sox-2. *Mol. Cell Biol.* **19,** 5453–5465.
13. Tokuzawa, Y., Kaiho, E., Maruyama, M., et al. (2003) Fbx15 is a novel target of Oct3/4 but is dispensable for embryonic stem cell self-renewal and mouse development. *Mol. Cell Biol.* **23,** 2699–2708.
14. Winston, J. T., Koepp, D. M., Zhu, C., Elledge, S. J., and Harper, J. W. (1999) A family of mammalian F-box proteins. *Curr. Biol.* **9,** 1180–1182.

16

Gene Silencing Using RNA Interference in Embryonic Stem Cells

J. Matthew Velkey, Nicole A. Slawny, Theresa E. Gratsch, and K. Sue O'Shea

Summary

Pluripotent embryonic stem (ES) cells are an important model system to examine gene expression and lineage segregation during differentiation. One powerful approach to target and inhibit gene expression, RNAi, has been applied to ES cells with the goal of teasing out the cascades of gene expression/repression that shape the early embryo. In this chapter, we describe the current understanding of the mechanisms of gene silencing by small hairpin RNAs, as well as controls and caveats to using this approach in ES cells. A consideration of synthetic vs plasmid-based RNAi vectors, design of targeting constructs, transfection of ES cells, and flow sorting of targeted cells is followed by methods for the analysis of phenotype and behavior of targeted cell populations using immunohistochemistry, reverse transcriptase polymerase chain reaction, Western blotting, and scanning electron microscopy.

Key Words: Differentiation; flow cytometry; immunohistochemistry; RISC; RT-PCR; scanning electron microscopy; siRNA; small hairpin RNA (shRNA); transfection; Western blotting.

1. Introduction

Embryonic stem (ES) cells are a powerful model system to tease out the successive waves of gene expression and repression that produce specific cell and tissue types in the early embryo because their gene expression profile is similar to the inner cell mass and epiblast of the preimplantation embryo. In addition, because they retain the capacity for multilineage differentiation, if that differentiation could be controlled, then they would also form an unsurpassed source of cells for tissue and cell replacement therapies. ES cells are particularly appropriate for studies of lineage choice in early development not only because they are pluripotent but also because they can be easily transfected to express genes of interest, and loss of function experiments can be carried out using gene targeting and homologous recombination or more recently using RNAi.

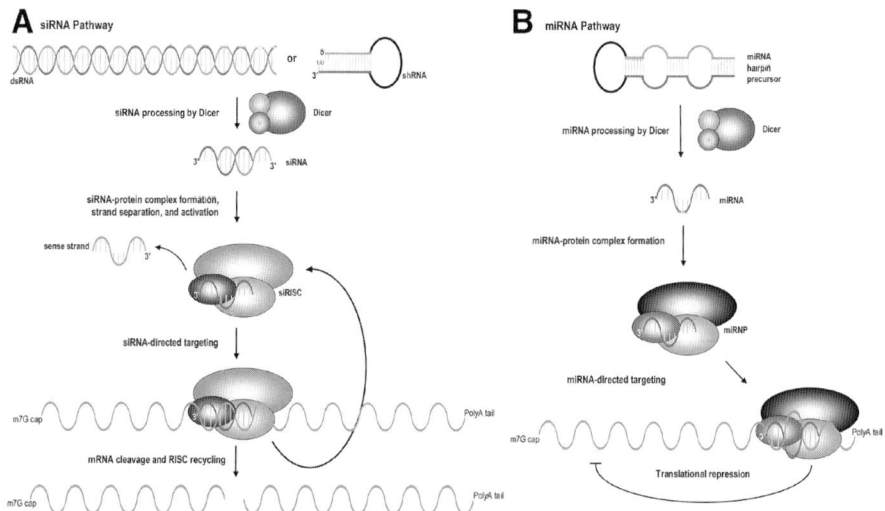

Fig. 1. siRNA and miRNA pathways. (**A**) The siRNA pathway. Cytosolic double-stranded RNA (dsRNA) or short-hairpin RNA (shRNA) is processed into short-interfering RNA (siRNA) molecules by an RNase III enzyme-containing complex (Dicer). The resulting siRNAs consisting of a 19-nucleotide (nt) duplexed region with 2-nt unpaired 3′ overhangs are recognized and incorporated into an siRNA-inducing silencing complex (siRISC) in which the sense and antisense strands are unwound. The antisense strand is retained in the complex and guides the targeting of mRNA containing complementary sequence for endonucleolytic cleavage and degradation. The siRISC remains activated and can mediate several cycles of mRNA targeting before the antisense strand dissociates and is replaced by another siRNA. (**B**) The miRNA pathway. MicroRNAs (miRNAs) are derived from approx 70-nt miRNA hairpin precursors that are processed by Dicer into approx 22-nt miRNAs. These miRNAs are single stranded and are incorporated into an miRNA-protein complex (miRNP) in which they guide the targeting of mRNA containing partial sequence homology—typically at the 3′ untranslated region—for translational repression.

Gene silencing by double-stranded, small-interfering RNA (siRNA) molecules that bind and cleave sequence-specific cognate messenger RNA (mRNA) transcripts is also a powerful method for analyzing gene function in development. RNAi has been widely applied, initially to invertebrates and plants, more recently to mammalian cells, and specifically to ES cells *(1–3)*. Targeted ES cells can then be employed to create chimeric embryos *(4,5)*, wholly ES-derived embryos via tetraploid complementation *(2)*, or cells can be differentiated in vitro or as teratomas. Interestingly, evidence suggests that many noncoding RNAs contain antisense RNAs, and it has been estimated that at least 20% (and likely more) of all human genes have antisense sequences associated with them *(6)*. Naturally occurring microRNAs (miRNA) are short (21–22 nucleotide [nt]), single-stranded RNA molecules that have widespread roles in human disease, in meiosis and mitosis, and in normal development *(7)*.

The mechanism of action of both miRNAs and siRNAs is not entirely understood, although there are striking similarities in each pathway (**Fig. 1**). Both siRNAs and

RNAi Gene Silencing in ES Cells

miRNAs are small, 19-nt RNA molecules produced from longer double-stranded (ds) or hairpin RNA precursors by ribonuclease (RNase) III activity of the Dicer enzyme *(8)*, and both mediate silencing by forming an activated RNA-induced silencing complex (RISC). In the siRNA pathway, dsRNA (either in linear or hairpin form) is cleaved by Dicer into 19-nt duplexes that are then incorporated into a RISC. The siRNA duplex is then unwound, and the antisense strand is retained to guide the activated RISC to the target mRNA for silencing. In the miRNA pathway, the miRNAs produced by Dicer are single-stranded molecules that are immediately incorporated into a similar RISC. Once activated, RISCs mediate gene silencing in the siRNA pathway via the cleavage and degradation of target mRNA, whereas activated RISCs in the miRNA pathway typically silence genes by binding to target mRNA and inhibiting translation *(9)*.

Interestingly, some miRNAs can direct mRNA cleavage, much like siRNAs *(10)*, and conversely, siRNAs can be engineered to function like miRNAs in repressing translation *(11)*. A general hypothesis that emerges from these findings is that RISCs containing small RNAs with perfect homology will induce mRNA degradation, and those with imperfect matches lead to translational repression *(12)*. However, it is unclear whether sequence homology alone is the key determinant in the regulation of mRNA degradation vs translational repression because siRNAs containing 1-nt or 2-nt mismatches to the target mRNA have been used as negative controls in many RNAi experiments (e.g., **ref. 13**) to test for sequence-specific gene silencing. These constructs were found to have no effect on gene expression at either the mRNA or protein level.

Much more remains to be learned about the regulation of the siRNA and miRNA pathways, and the "rules" that determine what effect a particular small RNA will have when introduced into cells are still a work in progress. For now, investigators interested in silencing gene expression via RNAi must be aware of the possibility that a particular targeting construct may have nonspecific, variable, or possibly even no effects on gene expression, and this should be carefully controlled and assessed in any experiment. Some of the strategies and caveats for the application of this technology to study gene function in ES cells are described here.

1.1. Developing Effective Targeting Constructs

The first choice to be made in the design of an RNAi is whether to use a synthetic siRNA or plasmid-based short-hairpin RNA (shRNA). Several companies will design and synthesize siRNA to target specific genes, among them Dharmacon, Oligoengine, Pharmacia, and Invitrogen. The synthetic siRNAs are limited in the duration of their effect as they are depleted with each cell division, which may be an advantage in some experiments. We use a plasmid-based system (*see* description in **Subheading 1.2.**) for the delivery of shRNA that allows for the sustained expression in the transfected ES cells.

The sequence specificity of the RNAi designed is critical because it must recognize a single target. To select unique sequences, we start with the Gene Fisher primer program *(14)* (*see* **Note 1**). The general parameters for shRNA design are 19–25mer (we have made functional shRNAs as large as 35-mer); a melting temperature (t_m) of 42–55°C, a G:C content of 40–55%, and an A nucleotide at the 3′ end. Each of the primers is then analyzed *in* *silico* with the Basic Local Alignment Search Tool (BLAST) *(15)* to minimize potential off-target silencing effects. Only oligomers that exclusively identify the gene of interest are considered. Also, if the selected gene has been knocked out, then you may want to target your RNAi to the same region.

Reynolds et al. *(16)* provide some additional considerations for RNAi design that address G:C content (36–52%), internal instability of the 3′ end of the sense strand (increase A/T at last five positions), and several other nucleotide-specific positional effects. We have designed a variety of shRNAs that target either the 5′ or 3′ regions on the mRNA and have not found that a particular region is more or less susceptible to downregulation. Once the nucleotide sequences for the shRNA templates are selected, they are synthesized by Invitrogen as two long complementary oligomers comprised of the gene-specific sense and antisense cDNA (19–35 nt), RNA polymerase type III (pol III) start and stop signals, and additional sequences to form the hairpin structure (**Fig. 2**). It is a good idea to test several different shRNAs to compare the level of gene-specific knockdown; we usually test at least three, individually and in combination, to determine the most efficient target sequences. Combinations of shRNAs can be employed to target several regions of the same gene or multiple genes.

In all RNAi experiments, whether using synthetic or plasmid-based methods, it is essential to design a negative control shRNA that contains one or more nucleotide substitutions from the wild type. These shRNAs should not disrupt the expression of the target mRNA. These sequences must also be analyzed *in silico* to be certain they will not target other mRNAs. There is a report that missense mutations are most effective (i.e., that do not knockdown) if targeted to the 3′ end *(17)*. We have designed shRNA with three nucleotide changes; generally, the three-nucleotide substitutions are targeted to the 5′ end, middle (i.e., at the cleavage site), and 3′ end. We have used the plasmid-based method to successfully knockdown expression of several genes, including Bmp4, Oct4 *(3)*, wnt8a, and wnt8b in ES cells.

1.2. Expression Plasmids

Constitutive expression of shRNAs can be achieved by using vectors containing promoters for RNA pol III, which is preferred over pol II because of the precision of its initiation and termination signals. Attempts to generate shRNAs using pol II promoter-based expression vectors have not been particularly successful because of imprecise initiation or excessive read-through after the termination signal, resulting in hairpins with long, single-stranded 5′ or 3′ tails that render them ineffective as siRNAs. In contrast, pol III initiates transcription at a defined distance from sequence elements in the promoter region and precisely terminates transcription at sequences containing five successive thymidines, so that the transcript is cleaved after the second uridine, producing a hairpin with a 2-nt 3′ overhang ideal for activity as an siRNA (**Fig. 2**).

The most widely used pol III promoters thus far have been the U6 small nuclear RNA promoter (U6) *(18,19)* and the H1 promoter for the RNA component of the RNase P ribozyme *(13)*. Both U6 and H1 are compact, constitutively active promoters that are amenable to modification by cloning based on polymerase chain reaction (PCR), so the general strategy for the construction of shRNA expression vectors has been to amplify these promoters from genomic DNA using primers that insert unique restriction enzyme

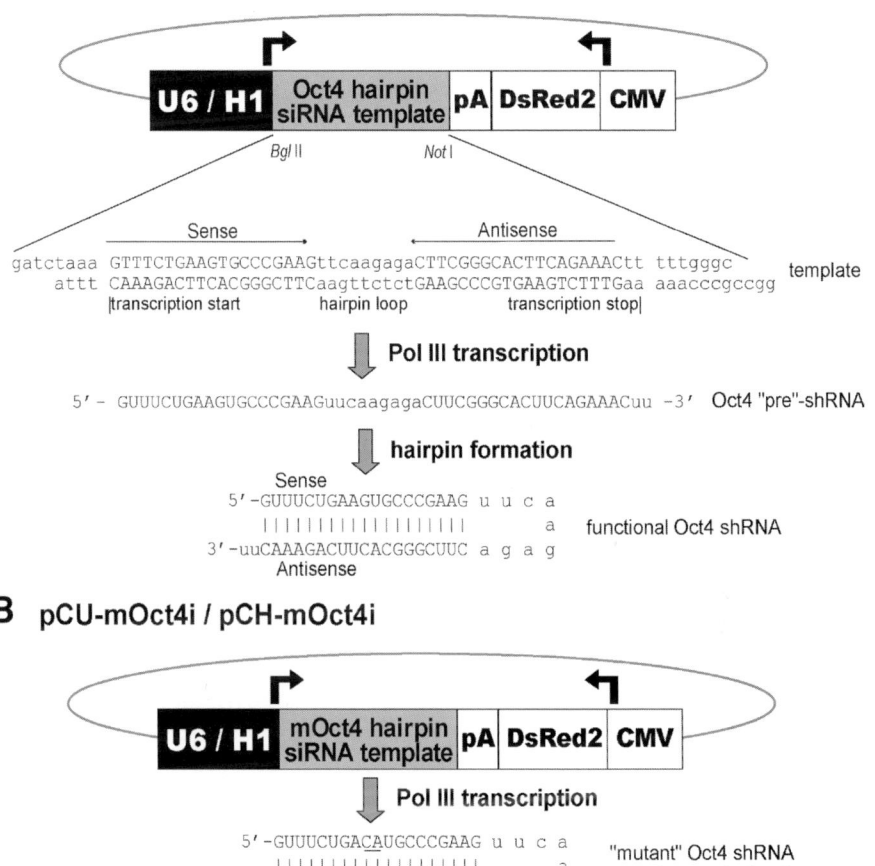

Fig. 2. shRNA expression constructs. (**A**) RNA pol III promoter-based shRNA expression vectors used to target Oct4 in ES cells. The U6 or H1 promoter with an engineered *Bgl*II site at position -9 to -4 has been cloned into a DsRed reporter plasmid. Templates encoding hairpin siRNAs with 19 nucleotides of homology to Oct4 are synthesized as 64-mer DNA oligonucleotides, annealed in vitro, and inserted as shown. Pol II transcription from the cytomegalovirus (CMV) promoter (left arrow) produces DsRed, whereas pol III transcription from U6 or H1 (right arrow) produces nascent shRNAs targeting Oct4. Basepairing between sense and antisense strands of the transcript is predicted to produce a functional shRNA with a 9-nt loop. (**B**) "Mutant" Oct4 RNAi (mOcti) constructs used as negative controls are identical to the functional Octi constructs except for a 2-nt substitution (underlined) in the middle of the duplexed stem of the shRNA.

sites at the 5' and 3' ends of the promoters (e.g., *EcoRI* and *BamHI*, respectively) to allow cloning into a vector polylinker site. In addition, another unique restriction site (e.g., *BglII*) is engineered just upstream of the initiation site to allow insertion of the shRNA oligonucleotide template (*see* **Fig. 2**).

The H1 promoter is slightly more flexible for expression of shRNAs because any nucleotide can be at the transcription initiation site (+1), whereas the U6 promoter shows highest activity with transcripts containing a guanosine at +1 *(20)*. We have found both promoters to be effective at silencing gene expression in mouse ES cells, although generally we and others have had better results using H1-based plasmids *(3,21)*. Rigorous quantitative analysis of the relative activities of these promoters in vector-based RNAi applications has yet to be done, so it is unknown whether the differences observed reflect inherent differences in the actual promoters or merely a reflection of variations in transfection or integration efficiency of the more compact H1 promoter, as has been proposed *(21)*.

Nonetheless, we recommend the use of H1-based vectors for RNAi because of their flexibility in terms of the shRNA template sequence and their slightly higher overall efficacy. It should also be noted that although many plasmid-based RNAi systems exist, numerous viral vectors have been developed (e.g., **ref. 22**), particularly lentiviral-based systems, which have the advantage of efficient integration even in noncycling cells and resistance to transgene silencing.

1.2.1. Inducible RNAi

Although the use of constitutively active pol III promoters in shRNA expression vectors has made it possible to achieve stable gene knockdown in cells, there are instances when it is more desirable to suppress genes in a regulated fashion. This is particularly important when using ES cells in which knocking down a gene of interest may result in differentiation, thereby making the establishment of stable cell lines difficult. Many inducible expression systems (Tet-on, Cre-LoxP, etc.) have been developed; however, these have typically utilized pol II promoters. To generate conditional RNAi expression vectors, much interest has been directed at developing inducible pol III promoters. As illustrated in **Fig. 3**, one of the strategies utilizes Cre-LoxP technology in which a 5-T strong termination signal or a stuffer fragment (usually containing an antibiotic selection cassette) flanked by LoxP sites is placed upstream of the shRNA template such that transcription is prematurely terminated *(23,24)* or not initiated because of disruption of spacing between the distal and proximal sequence elements of the promoter region *(25)*. Conditional expression of Cre recombinase from a drug-inducible or tissue-specific promoter results in the excision of the disrupting fragment, thereby permanently restoring expression of a functional shRNA. To date, numerous such Cre expression constructs and transgenic mouse lines have been developed to which this form of regulatable RNAi can be applied; however, it should be noted that, once the RNAi is induced in this manner, there is no way to turn it off.

To achieve inducible and reversible RNAi, other investigators have modified pol III promoters such that they could be induced by tetracycline *(26–29)* or ecdysone analogs *(30)*. The ecdysone-inducible system that has been described uses a U6 promoter modified so that it is no longer constitutively active but can be induced by treatment with muristerone A, an analog of ecdysone *(30)*. In the tet-inducible systems that have been developed, the pol III promoters remain constitutively active but are inhibited by transcriptional repressors.

RNAi Gene Silencing in ES Cells

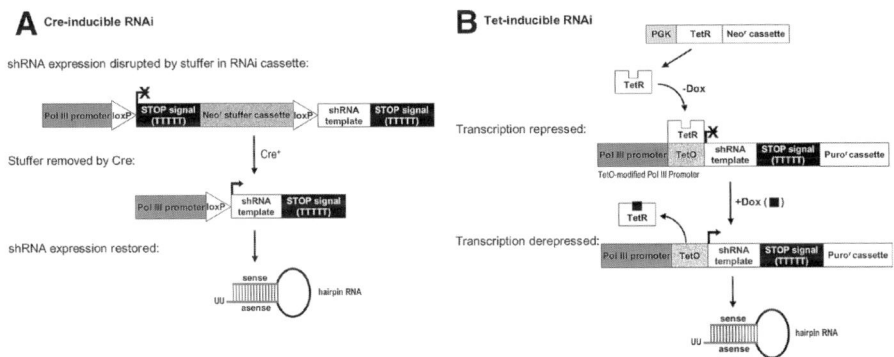

Fig. 3. Inducible RNAi systems. (**A**) Cre-inducible RNAi. Cells are transfected with a construct containing an RNAi expression cassette that has been disrupted by the insertion of a "stuffer" fragment flanked by LoxP sites. The stuffer fragment inactivates the RNAi cassette by the presence of a 5-thymidine strong stop signal or by disrupting spacing between key sequence elements in the pol III promoter. An antibiotic (e.g., neomycin) resistance cassette contained within the stuffer allows for selection of cell lines. Expression of Cre results in the excision of the stuffer and restoration of shRNA expression. After excision, a single LoxP site persists in the cassette, so care must be taken to ensure that it is located within a region that will not interfere with basal activity of the RNAi cassette. (**B**) Tet-inducible RNAi. Cells are transfected with a construct driving expression of a Tet repressor protein (TetR) as well as an antibiotic (e.g., neomycin) resistance cassette to allow selection of stable cell lines. A second construct is then introduced into the cells that contain an RNAi cassette with a TetO recognition sequence engineered into the promoter region and a different antibiotic resistance cassette (e.g., puromycin) to allow selection of double-stable cell lines. In these cells, TetR binds to the TetO recognition sequence in the RNAi cassette and represses transcription. Treatment with doxycyline (Dox) inhibits the binding of TetR to its recognition sequence, thus restoring (or "derepressing") constitutive activity of the RNAi cassette.

To achieve this, transgenic cell lines are generated that express (either constitutively or via a tissue-specific promoter) the classic tet repressor (TetR) protein or a hybrid tet suppressor (tTS) protein in which TetR is fused to a strong silencer such as the KRAB transcriptional repression module found in many zinc finger proteins *(31)*. Into this background, one introduces an shRNA expression vector that has been modified such that a tetO recognition sequence is inserted in frame between the TATA box and transcription initiation site. In this context, the TetR or tTS protein binds to the tetO sequence and represses transcription (**Fig. 3**). The presence of tetracycline or its analog doxycycline results in "derepression," the TetR or tTS proteins are released from the tetO sequence, and expression of the shRNA is thus permitted.

1.3. Markers: Fluorophores, Selectables

One difficulty in using RNAi in ES cells is the need to obtain relatively pure populations of cells expressing the targeting construct. Analysis of mixed cultures can

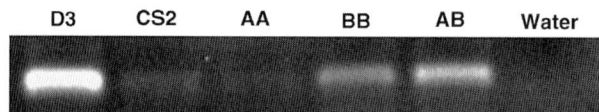

Fig. 4. RT-PCR analysis of interferon response. D3 ES cells were transfected with empty vector (CS2) or with Wnt8 RNAi constructs in combination: two targeting wnt8a (AA), two targeting wnt8b (BB), or one targeting wnt8a and one targeting wnt8b (AB). After a 48-h recovery in complete medium without LIF, differentiation medium containing puromycin was added to the cells. RNA was harvested 7 d posttransfection and analyzed using RT-PCR for OAS1g expression. All transfected cells expressed less OAS1g than the parent D3 ES cells, indicating that a nonspecific interferon response was not stimulated by these constructs.

be misleading because of target gene expression in untransfected cells (*see* **Figs. 4** and **5**). We have found that the best method to obtain relatively pure populations of transfected cells is to combine flow cytometry (to enrich for targeted cells) with antibiotic selection (to prevent overgrowth of untargeted cells that inevitably remain after flow sorting). There are several choices of both fluorochromes and antibiotics. Both EGFP and DsRed 2.1 (Clontech) work well as fluorescent markers. Although DsRed 2.1 can be easier to visualize by fluorescence microscopy, it needs approx 48 h to mature (because of tetramer formation) and tends to be packaged and degraded by ES cells, leading to signal loss after 7–14 d. There is a relatively new monomeric red fluorescent protein (mRFP1) that matures in less than 24 h *(32)*; however, in our experience this protein still appears packaged after a few days in ES cells.

We use two different antibiotic resistance cassettes, neomycin and puromycin. For ES cells, we use 350 µg/mL (active) G418 or 1 µg/mL puromycin to prevent overgrowth of untransfected cells. Not all ES cell lines have the same tolerance for antibiotics, even when they express a gene conferring resistance. The best way to determine the right dose of antibiotic for an ES cell line is to expose transfected cells to a range of concentrations and choose the one that kills untransfected cells fairly rapidly but does not kill a large number of transfected cells.

1.4. Delivery to ES Cells

Initially, siRNAs were added to the medium of cells in culture like antisense oligonucleotides *(33)*, infused into tissues or delivered to adult animals via tail vein *(34)*, where the efficiency of gene knockdown can be variable and may depend on the specific construct employed. Without transfection, delivery to tissues has often been ineffective *(35)*, and ultimately the effectiveness of an RNAi may be determined by the route of delivery (for a review, *see* **ref. 36**). Liposome delivery has also been successfully employed *(33)*.

We have employed both electroporation and Lipofectamine to deliver the shRNA constructs. We prefer Lipofectamine because of the high transfection efficiency, the lack of cell death that can result from the electroporation itself, and the fact that there is no need to control the copy number of plasmid DNA in transfected cells.

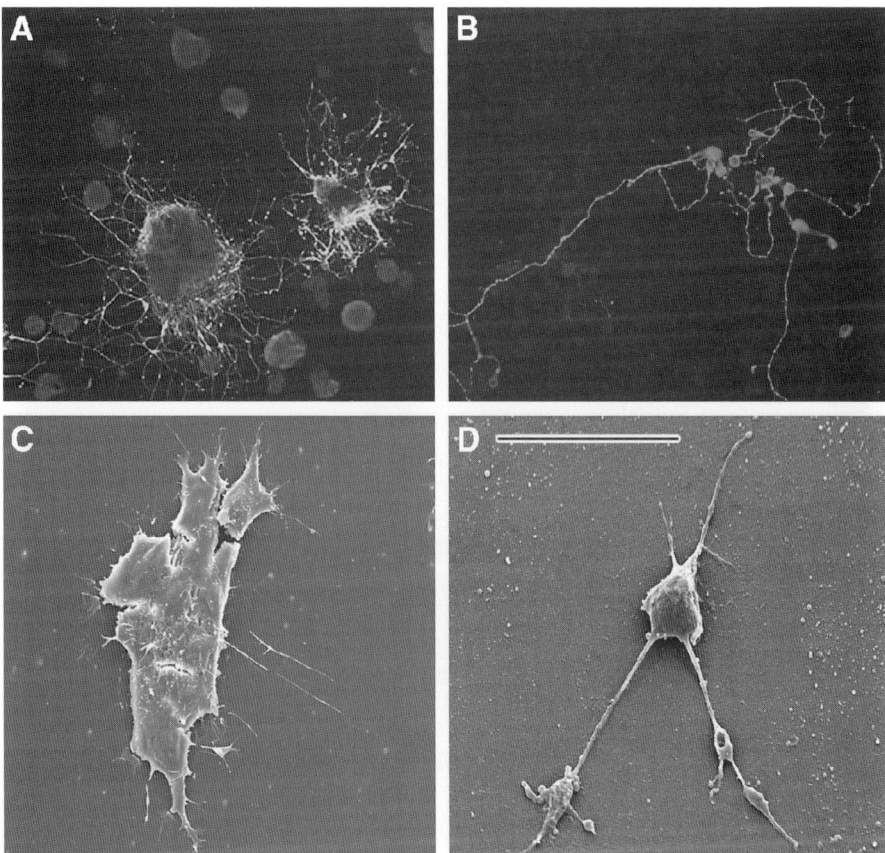

Fig. 5. Immunohistochemistry and SEM evaluations. Cells were transfected with Wnt8 RNAi constructs: (**A**) wnt8a and wnt8b; (**B**) two targeting wnt8b. After a 48-h recovery in complete medium without LIF, differentiation medium with puromycin was added to the cells. At 10 d posttransfection, cells were fixed with 2% paraformaldehyde, washed with PBS, and stored at 4°C prior to immunohistochemical localization of neuronal tubulin (anti-TUJ1 primary antibody, BABCO 1:100), secondary antibody conjugated to fluorescein isothiocyanate. Hoechst 23187 was added to the PBS wash to visualize nuclei (blue). (**C**) and (**D**) SEM view of control D3: (**C**) ES cells illustrating a compact cluster of 10–12 cells. Note their stellate appearance, presence of multiple filopodia, and pronounced cell–cell adhesion. (**D**) SEM of single neuron differentiated from ES cells. Numerous cell processes arise from the large cell body. Scale bar = 100 µm. (Please *see* the companion CD for the color version of this figure.)

2. Materials
2.1. Tissue Culture
1. ES cells: D3, B5, A9 (*37–39*).
2. Complete ES culture medium: Dulbecco's modified Eagle's medium (DMEM; Invitrogen, Carlsbad, CA; cat. no. 11965-092) supplemented with 10% ES tested fetal bovine serum

(FBS; Atlanta Biologicals, Atlanta, GA; cat. no. S 1150), 10^{-4} M β-mercaptoethanol (Sigma, St. Louis, MO; cat. no. M6250), 0.224 µg/mL L-glutamine (Invitrogen, cat. no. 35050-061), 1.33 µg/mL HEPES (Invitrogen, cat. no. H4034), with and without 500 U/mL human recombinant leukemia inhibitory factor (LIF; Oncogene Research Products, San Diego, CA; cat. no. PF044). Add 50 µL LIF to 100 mL medium.
3. Six-well plates, 35 mm/well (Fisher, Pittsburgh, PA; cat. no. 07-200-80), 10-cm dishes (Fisher, cat. no. 08-772E).
4. 0.1% gelatin (Sigma, cat. no. G1890), autoclave solution.
5. Tissue culture-grade water (Sigma, cat. no. W3500).
6. Trypsin/EDTA (ethylenediaminetetraacetic acid): 2.5% (10X) trypsin (Invitrogen, cat. no. 15090-046) diluted to 1X with Hank's balanced salt solution (HBSS) (Invitrogen, cat. no. 14185-052) and 1 mM EDTA (Invitrogen, cat. no. 15575-038).
7. Genticin, G418 (Invitrogen, cat. no. 11811-023) or puromycin (Sigma, cat. no. P8833).

2.2. Transfection Reagents
1. Purified plasmid DNA (Plasmid prep kit; Qiagen, Valencia, CA; cat. no. 27106).
2. Lipofectamine (Invitrogen, cat. no. 18324-012).
3. Plus reagent (Invitrogen, cat. no. 11514-015).
4. 5-mL polystyrene tubes (Fisher, cat. no. 14-959-1A).
5. DMEM (*see* **Subheading 2.1., item 2**).

2.3. Flow Cytometry
1. 10-cm dish of RNAi-transfected ES cells and 6-cm dish of untransfected ES cells.
2. Ca^{2+}-/Mg^{2+}-free HBSS and trypsin (*see* **Subheading 2.1., item 6**).
3. DMEM without phenol red, with HEPES, with high glucose, high L-glutamine (Invitrogen, cat. no. 21063-029).
4. Complete ES cell media (containing serum and LIF) (*see* **Subheading 2.1., item 2**).
5. Hemacytometer or Coulter counter.
6. Penicillin/streptomycin solution (Invitrogen, cat. no. 15140-122).
7. Flourescence activated cell sorter.
8. Gelatin-coated six-well plates or 3.5-cm dishes (*see* **Subheading 2.1., item 3**, and **Subheading 3.1.1., step 1**).
9. G418 (ILT, cat. no. 11811-027) or puromycin (Sigma, cat. no. P8833).
10. ES cell differentiation or complete medium (*see* **Subheading 2.1., item 2**).

2.4. RNA Analysis
2.4.1. RNA Extraction and Purification
1. Trizol reagent (Invitrogen, cat. no. 15596-018).
2. Chloroform (Sigma, cat. no. C2432).
3. Isopropanol.
4. Diethylpyrocarbonate (DEPC)-H_2O/RNase-free water (DEPC, Sigma, cat. no. D5758; water, Sigma, cat. no. W4502).
5. Amplification-grade deoxyribonuclease (DNase) I (10 U/µL) and 10X buffer (Invitrogen, cat. no. 18068-015).
6. 70 and 75% ethanol (EtOH) in DEPC-H_2O.
7. 0.6- or 1.5-µL microtubes.
8. Agarose (Invitrogen, cat. no. 15510-019).
9. 10X Tris/borate/EDTA buffer (Sigma, cat. no. T4415).
10. Ethidium bromide (EtBr) (10 mg/mL; Invitrogen, cat. no. 15585-011).

RNAi Gene Silencing in ES Cells 243

11. TE buffer (10 mM Tris-HCl, 1 mM EDTA): to make 100 mL TE, combine 1 mL 1 M Tris-Cl, 200 μL 0.5 M EDTA, and 99 mL RNase-/DNase-free water.
12. Gel loading buffer: 30% glycerol, 10 mM EDTA, 0.25% bromophenol blue (Sigma, cat. no. 11-439-1), 0.25% xylene cyanol (Sigma, cat. no. X4126) in RNase-/DNase-free water.

2.4.2. Reverse Transcription

1. Total RNA, 1–5 μg per reaction.
2. DEPC-H$_2$O.
3. Oligo-dT$_{12-18}$ primer (0.5 μg/μL; Invitrogen, cat. no. 18418-012).
4. 10 mM dNTP mix: 10 mM each of dATP, dCTP, dGTP, and dTTP (Invitrogen, cat. no. 10297-018).
5. Powerscript (200 U/μL), 5X first-strand buffer, and 0.1 M dithiothreitol (Clontech, Palo Alto, CA; cat. no. 639501).

2.4.3. Polymerase Chain Reaction

1. Reverse transcriptase (RT) template, 1/20 total volume of the RT reaction.
2. 0.2- to 0.6-μL thin-wall PCR tubes.
3. Taq DNA polymerase (5 U/μL), 10X Taq DNA polymerase buffer, 50 mM MgCl$_2$ (Invitrogen, cat. no. 10342-046), and 5 M Betaine (Sigma, cat. no. B2629) or dimethyl sulfoxide (DMSO; Sigma, cat. no. D2650).
4. 10 mM dNTP mix (*see* **Subheading 2.4.2.**, **item 4**).
5. Gene-specific primers, 10–20 pM/reaction.
6. Thermocycler (M. J. Research, Waltham, MA; model no. MJR PTC100).
7. 100-bp or 1-kb plus DNA ladders, (Invitrogen, cat. no. 15628-019 and 15615-016).
8. Gel loading buffer (*see* **Subheading 2.4.1.**, **item 12**).

2.5. Western Blotting

2.5.1. Sample Collection and Protein Isolation

1. Phosphate-buffered saline (PBS) (Invitrogen, cat. no. 14190-044).
2. RIPA buffer: 50 mM Tris-HCl, pH 7.4 (Fisher, cat. no. BP153-500), 1% IGEPAL (Sigma, cat. no. I-3021), 0.25% Na-deoxycholate (Sigma, cat. no. D-6750), 150 mM NaCl (Fisher, cat. no. S271-3), and complete protease inhibitor tablet with EDTA (Roche, Nutley, NJ; cat. no. 1 836 153) (*see* **Note 2**). To prepare 100 mL RIPA, add 790 mg Tris-HCl to 75 mL distilled water. Add 900 mg NaCl and stir the solution until all solids are dissolved. Using HCl, adjust to pH 7.4. Add 10 mL 10% IGEPAL to the solution. Add 1 mL 100 mM EDTA to the solution. Adjust the volume of the solution to 100 mL using a graduated cylinder. Store RIPA at 2–8°C until ready to use.
3. Cell scrapers (Fisher, cat. no. 08-773-1).
4. Microcentrifuge tubes (Eppendorf-Brinkman, Westbury, NY; cat. no. 22 36 320-4).
5. Sonicator with a microtip (Sonic Dismembrator; Fisher, model no. 500).
6. Bradford protein assay kit (Bio-Rad, Hercules, CA; cat. no. 500-0006).
7. Spectrophotometer (Bio-Rad, model no. SmartSpec 3000).

2.5.2. Electrophoresis

1. Tris-HCl precast gels (Bio-Rad or Invitrogen) or acrylamide, ammonium persulfate (ILT, cat. no. 15523-012), $N,N,N´,N´$-tetramethylenediamine (Bio-Rad, cat. no. 161-0800), Tris-HCl (ILT, cat. no. 15804-020), and sodium dodecylsulfate (SDS).
2. Laemmli sample buffer (Bio-Rad, cat. no. 161-0737).

3. β-Mercaptoethanol (Sigma, cat. no. M7522).
4. Miniprotean electrophoresis cell (Bio-Rad, cat. no. 165-3301).
5. Tris-glycine electrophoresis buffer: 25 mM Tris (ILT, cat. no. 15804-020), 250 mM glycine (pH 8.3) (Bio-Rad, cat. no. 161-0717), 0.1% SDS (ILT, cat. no. 15525-017). For a 5X stock solution, dissolve 15.1 g Tris-base and 94 g glycine in 900 mL of double-distilled water (ddH_2O) (do not adjust the pH). Add 50 mL of a 10% (w/v) stock solution of SDS and adjust the volume to 1 L with water.
6. Power supply for electrophoresis.

2.5.3. Immunoblotting

1. Minitrans blot (Bio-Rad, model no. 170-3935).
2. Transfer buffer: 39 mM glycine (Bio-Rad, cat. no. 161-0717), 48 mM Tris-base (Roche, cat. no. 604 205), and 0.037% SDS. To prepare 1 L transfer buffer (pH 8.3), add 2.9 g glycine, 5.8 g Tris-base, and 0.37 g SDS to 800 mL ddH_2O.
3. Methanol (MeOH).
4. Whatman filter paper.
5. Nitrocellulose (*see* **Note 3**).
6. Biosafe Coomassie blue stain (Bio-Rad, cat. no. 161-0786).
7. Ponceau red stain (Gelman Sciences, Ann Arbor, MI; cat. no. 51284).
8. TBST (Tris-buffered saline plus Tween-20): 50 mM Tris (ILT, cat. no. 15504 020) at pH 7.5 (should be adjusted before adding detergent), 150 mM NaCl, and 0.2% Tween-20 (Sigma, cat. no. F1379). To prepare 1 L TBST, add 6.1 g Tris and 8.8 g NaCl to 900 mL distilled water. Stir the solution until all solids are dissolved. Adjust to pH 7.5. Add 2 mL Tween-20 and stir until the solution is clear. Adjust the volume of the solution to 1 L using a graduated cylinder. Store TBST at room temperature until ready to use. To make blocking buffer simply add 5% w/v powdered milk to TBST.
9. Primary antibody to bind to the target protein.
10. Horseradish peroxidase-conjugated secondary antibody to bind to primary antibody.
11. SuperSignal West Pico chemiluminescent substrate (Pierce, Rockford, IL; cat. no. 34077).
12. Lightproof film holder.
13. Bio-Max film (Kodak, Rochester, NY; cat. no. 868 9358).

2.6. Immunohistochemistry

1. Fixative (phosphate-buffered 2% paraformaldehyde): combine 10 mL 16% formaldehyde (Electron Microscopy Sciences, Hatfield, PA; cat. no. 15700), 8 mL 10X Dulbecco's phosphate-buffered saline (D-PBS) with Ca^{2+} and Mg^{2+} (Invitrogen, cat. no. 14080-055) and 62 mL distilled water (dH_2O). Adjust to pH 7.4; store at 4°C up to 6 mo.
2. Permeabilization solution: combine 50 mL dH_2O, 50 μL Triton X-100 (Sigma, cat. no. T-9284), 50 mg sodium citrate ($Na_6H_5Na_3O_7$-2H_2O; Sigma, cat. no. C-0909). Store at 4°C up to 3 mo.
3. Blocking solution: combine 1 mL total normal serum (*see* **Note 4**), 8.85 mL 1X D-PBS (ILT, cat. no. 14040-133), 50 μL Triton X-100, and 100 μL 10% sodium azide (Sigma, cat. no. S-8032; dissolve 1 g in 10 mL dH_2O) (*see* **Note 5**). Store at 4°C up to 1 mo.
4. Antibody dilution buffer: combine 100 μL total normal serum (*see* **Note 4**), 9.75 mL D-PBS, 50 μL Triton X-100, and 100 μL 10% sodium azide. Store at 4°C up to 1 mo.
5. Serum wash: combine 100 μL total normal serum, 9.8 mL D-PBS, and 100 μL 10% sodium azide. Store at 4°C up to 1 mo.
6. 1000X Hoechst 33258 stock solution: mix 25 mg Hoechst 33258 (*bis*-benzamide; Sigma, cat. no. B1155) and 50 mL dH_2O. The product will arrive as a powder in a sealed amber

bottle. Add 10 mL dH$_2$O directly to the bottle and dissolve by gentle agitation. Transfer solution to a clean amber bottle and add 40 mL dH$_2$O. Store at 4°C in the dark up to 1 yr. To make working solution, dilute 100 µL in 100 mL dH$_2$O and store at 4°C (in dark).
7. 4′,6-Diamidino-2-phenylindole (DAPI; Sigma, cat. no. D-8417): working solution 100 µg/mL in water.
8. ProLong Gold without DAPI (Molecular Probes, Eugene, OR; cat. no. 36934).

2.7. Phase Contrast Microscopy and Scanning Electron Microscopy

1. 13-mm Thermanox plastic cover slips (Nalge Nuc International, Rochester, NY; cat. no. 174950) coated with 0.1% gelatin (*see* **Subheading 2.1.**, **item 4**, and **Subheading 3.1.1.**, **step 1**).
2. 0.2 *M* pH 7.4 2X buffer solution (either Sorensen's phosphate or sodium cacodylate) (*see* **Note 6**). For Sorensen's phosphate buffer, prepare 0.2 *M* stock solutions of dibasic sodium phosphate (Na$_2$HPO$_4$·2H$_2$O, 35.61 g/L) and monobasic sodium phosphate (NaH$_2$PO$_4$·2H$_2$O, 31.21 g/L). Mix mono- and dibasic solutions to obtain desired pH (for pH 7.2, 72 mL dibasic sodium phosphate plus 28 mL monobasic sodium phosphate; for pH 7.4, 81 mL dibasic sodium phosphate plus 19 mL monobasic sodium phosphate). Buffer solution will be 2X; mix 1:1 with 2X fixative solution to obtain 1X working solution of buffered fixative. For sodium cacodylate buffer, prepare 0.2 *M* solution of sodium cacodylate (C$_2$H$_6$AsNaO$_2$·3H$_2$O, 42.80 g/L). Adjust pH by adding 0.2 *M* HCl (*see* **Note 6**) (for pH 7.2, add 8.4 mL 0.2 *M* HCl per 100 mL cacodylate solution; for pH 7.4, add 5.6 mL 0.2 *M* HCl per 100 mL cacodylate solution). Buffer solution will be 2X; mix 1:1 with 2X fixative solution to obtain 1X working solution of buffered fixative.
3. Primary fixative-buffered 2.5% glutaraldehyde: mix 1 mL 25% glutaraldehyde (Electron Microscopy Sciences, cat. no. 16200), 5 mL 0.2 *M* pH 7.4 2X buffer solution, and 4 mL dH$_2$O.
4. Primary fixative (alternative) of buffered modified Karnovsky's fixative (2% formaldehyde/ 2.5% glutaraldehyde): combine 1 mL 16% formaldehyde or paraformaldehyde (Electron Microscopy Sciences, cat. no. 15700), 1 mL 25% glutaraldehyde, 5 mL 0.2 *M* pH 7.4 2X buffer solution, and 3 mL dH$_2$O (*see* **Note 7**).
5. Secondary fixative of 1% osmium tetroxide: mix 1 mL 2% OsO$_4$ (Electron Microscopy Sciences, cat. no. 19172) and 1 mL 0.2 *M* pH 7.4 2X buffer solution (*see* **Note 8**).
6. Graded EtOH series (50, 70, and 95%, 2X 100%).
7. Hexamethyldisilizane (HMDS; Electron Microscopy Sciences, cat. no. 16700).
8. Double-sided tape or conductive adhesive tabs (Electron Microscopy Sciences, cat. no. 76762-01).
9. SEM specimen stubs (we prefer 1/2-in. diameter aluminum with slotted heads; Electron Microscopy Sciences, cat. no. 75220).

3. Methods
3.1. Transfection
3.1.1. Day 1

1. At 24 h prior to transfection, plate cells in six-well plates, or in 10-cm dishes or cover slips previously coated with gelatin for at least 20 min at 4°C. Prior to use, gelatin is removed and medium added.
2. Remove the medium from a confluent T-75-cm flask and rinse the cell monolayer twice with 1X HBSS.

3. Add 3 mL 0.25% trypsin/EDTA to the flask and distribute evenly across the cells, incubate for 2 min at 37°C, 5% CO_2; cells should lift easily from the flask.
4. Inactivate the trypsin by adding 6 mL ES medium (–) LIF, triturate gently, place cell suspension into a 15-mL conical tube, and centrifuge at approx 1500g for 5 min.
5. Remove the supernatant and resuspend cells in a total volume of 1 mL; triturate gently to resuspend the pellet.
6. Determine the cell number per milliliter using a hemacytometer or Coulter counter.
7. Plate 4×10^5 cells/well for a six-well plate 2 mL complete ES medium; for a 10-cm dish, plate 3×10^6 cells in 10 mL complete ES medium.

3.1.2. Day 2

The cells should be 60–70% confluent for optimal transfection.

1. Prepare DNA/Lipofectamine complexes: for mixture A, for a 35-mm dish, combine 2 µg plasmid DNA (10 cm = 4 µg) and 100 µL DMEM (10 cm = 0.75 mL) in a 5-mL polystyrene tube, then add 12 µL Plus reagent (10 cm = 20 µL); incubate 15 min at room temperature. For mixture B, in a separate tube combine 8 µL Lipofectamine (10 cm = 30 µL) with 100 µL DMEM (10 cm = 0.75 mL); incubate 15 min at room temperature (*see* **Note 9**).
2. Add mixture B to mixture A; incubate another 15 min at room temperature. Remove the ES medium from the cells and replace it with 0.8 mL (35-mm dish) or 5 mL (10-cm dish) DMEM. Transfection volume for 35-mm dish is approx 1 mL; it is approx 6.5 mL for a 10-cm dish.
3. Use a pipet to apply the DNA/Lipofectamine complexes to the cells in a dropwise manner. Incubate at 37°C with 5% CO_2 for 4 h.
4. Remove the transfection medium and replace it with ES medium (–) LIF (for differentiation); for a 35-mm dish, the volume is 2 mL, and it is 10 mL for a 10-cm dish. Continue the incubation at 37°C, 5% CO_2.
5. If using an antibiotic for selection, then start treatment at 24 h posttransfection.
6. Sort the cells by flow cytometry at 24–48 h posttransfection when using a fluorophore.

3.2. Nonspecific/Interferon Response

It was initially believed that large (>30-nt) shRNAs would elicit a nonspecific, interferonlike response as an innate cellular antiviral reaction. Although this appears to be essentially correct, nonspecific responses to targeting constructs have been widely reported *(40–43)* and appear to be plasmid dependent. ES, embryonal carcinoma, and mouse embryos prior to d 7 appear to lack the ability to mount an interferon response to large dsRNAs *(44–47)*. However, once ES cells begin to differentiate, this response can occur nonspecifically, halting transcription and translation. Plasmid-based RNAi, which is constitutively expressed, could cause an interferonlike response in differentiating cell cultures that could initiate cell death or otherwise affect the experimental results. Therefore, it is crucial to monitor interferon-responsive target gene expression to ensure specificity. Other important controls include monitoring the expression of one or more genes outside the targeted pathway and use of a missense RNAi construct to ensure target specificity of the RNAi.

We use RT-PCR to monitor the expression of 2´5´-oligoadenylate synthetase1 (OAS1g), a classic interferon response gene. The primer sequences, F-GGA GAC CCA GGA AGC TCC AGA and R-TTG AGG AAG GCT TGT GAT TG *(48)* are used in

RNAi Gene Silencing in ES Cells

PCR. The PCR conditions used for these primers are: 95°C/3 min, 95°C/30 s, 50°C/30 s, 72°C/1 min for 35 cycles; 72°C/7 min; and 4°C hold (*see* **Note 10**). An example of the results obtained using this primer pair is shown in **Fig. 4**.

3.3. Evaluation

Gene expression must be carefully monitored to determine the efficiency of targeting. This should be done at multiple levels (RNA and protein) as well as by observing the morphology of the ES cells. Because nontargeted cells present in ES cultures will rapidly overgrow the cultures and render the evaluation of RNA and protein meaningless, ES cell cultures should be either selected in antibiotic (include a neo^r in the targeting plasmid) or enriched by flow cytometry (include a fluorochrome in the plasmid), or both, so that resulting cultures are composed largely of the transfected cell population.

3.3.1. Flow Cytometry

This protocol describes how to enrich for ES cells transfected with an RNAi construct containing both a fluorochrome and an antibiotic resistance cassette. One important control for flow sorting equipment setup is untransfected ES cells. These cells allow the flow technicians to set the gates of the machine properly as well as helping to determine which population of cells expresses the fluorescent protein. Not all ES cell lines seem to tolerate the stress of sorting equally. Therefore, it is important to carry out at least one small-scale test sort with each of your cell lines to determine which are the most tolerant of the sorting. One critical parameter to improve survival of ES cells is to keep the excitation laser power to less than 100 mW. Because no flow cytometry procedure can eliminate 100% of the untransfected cells, we recommend including an antibiotic resistance cassette in your RNAi plasmid and postsort growth/differentiation of your cells in medium with antibiotic. The Purdue University Cytometry Laboratory has an excellent Web site (http:www.cyto.purdue.edu) to help with flow cytometry questions and troubleshooting.

3.3.1.1. Preparing Cells for Flow Cytometery

1. Remove the medium from the cells and discard.
2. Wash the cells once with Ca^{2+}-/Mg^{2+}-free HBSS to remove serum proteins, then add 2 mL 2.5% trypsin and incubate for 5 min at 37°C.
3. To terminate trypsin activity, add 2 mL complete (with serum, can be LIF–) medium, then triturate thoroughly to remove cells and break up clumps.
4. Transfer the cell suspension to a 15-mL conical tube and centrifuge for 2 min at room temperature, approx 1500g.
5. Resuspend the cells in 1 mL DMEM without phenol red.
6. Count the cells with a hemacytometer or a Coulter counter and adjust the volume until there are 5×10^6 cell/mL. Always include 100 U/mL penicillin/streptomycin in all media used before and directly after cell sorting to reduce the potential for contamination (*see* **Note 11**). Place the cells on ice for delivery to the flow facility.
7. Trypsinize and resuspend the untransfected ES cells (which will serve to set the gates of the flow machine) exactly as for the transfected cells.

8. Prepare 1- or 2.5-mL polystyrene snap-cap collection tubes with 0.5 mL collection media (the media you plan to use for differentiation) and 100 U/mL penicillin/streptomycin and place these tubes on ice.
9. Deliver the cells and collection tubes to your flow facility.

3.3.1.2. Maintaining Cell Purity Postsorting

1. Plate the sorted cells with no fewer than 5×10^5 cells in a 10-cm dish. Add 100 U/mL penicillin/streptomycin in the media to prevent bacterial growth.
2. Allow the cells to recover for 24 h.
3. Add 350 µg/mL G418 or 1 µg/mL puromycin to the cultures to prevent overgrowth of untransfected cells.

3.3.2. RNA Analysis: RT-PCR

The most critical variable in this protocol is having a pure population of undifferentiated and differentiated ES cells as your starting material (see **Subheadings 3.1.** and **3.3.**). Primers are then designed to identify downregulation of the shRNA targeted gene, genes directly downstream, as well as control genes such as β-actin or glyceraldehyde-3-phosphate dehydrogenase. It is also important to identify transcripts restricted to pluripotent stem cells and other tissue- or cell-type-specific markers.

3.3.2.1. RNA Extraction and Purification

This procedure involves the extraction of total RNA from wild-type shRNA and mutant shRNA-transfected mouse ES cells, followed by DNase I treatment to remove the chromosomal DNA. The presence of chromosomal DNA in the RNA preparation will interfere with the PCR and can yield products independent of mRNA expression. One way around this is to design primers that span intron/exon borders so that they distinguish a genomic vs an mRNA product. Many of the genes we study have only a single exon, so it is critical to remove all genomic DNA from the RNA sample (see **Note 12**). There are a variety of reagents and kits that are available for the isolation of total RNA. We prefer the Trizol reagent and follow the manufacturer's protocol.

To harvest RNA from adherent ES cells, remove the culture medium and discard. Add 500 µL Trizol to the cells in a 35-mm dish (5×10^5 cells).

1. Gently rock the plate so the Trizol covers the cell monolayer; with a pipet triturate the cell lysate and transfer it to 1.5-mL tubes and place on ice immediately (see **Note 13**).
2. Incubate tubes at room temperature for 5 min.
3. Add 100 µL chloroform to each tube (see **Note 14**) and shake tubes by hand for 15 s.
4. Incubate samples at room temperature for 3 min.
5. Centrifuge at 12,000g for 10 min at 4°C.
6. Carefully pipet off the upper, aqueous phase (300 µL) and put in a fresh 1.5-mL tube.
7. Add 250 µL isopropanol to each sample; mix by inversion (see **Note 15**) and incubate at room temperature for 10 min.
8. Centrifuge at 4°C for 10 min at 12,000g.
9. Pipet or pour off the isopropanol from the side opposite the pellet (see **Note 16**).

RNAi Gene Silencing in ES Cells

10. To wash the pellet, add 500 µL cold 75% EtOH (in DEPC-H$_2$O) and centrifuge at 4°C for 10 min at 9500g.
11. Pipet or pour off the EtOH, keeping the tube inverted; place on a clean Kimwipe and let the remainder of the EtOH drain for about 10–15 min. Do not let the pellet dry completely because it will be difficult to resuspend. Alternatively, after the EtOH is poured off, recentrifuge the samples for 1 min. Carefully pipet off the residual EtOH (~100 µL) but do not disturb the pellet.
12. Resuspend RNA pellets in 20–100 µL DEPC-H$_2$O and keep on ice (*see* **Note 17**).
13. Determine the concentration and integrity of the RNA by ultraviolet (UV) spectrophotometry and gel electrophoresis (*see* **Note 18**).
14. Electrophorese 2 µL of each RNA in a 1% agarose gel with 0.3 µg EtBr/mL and 1X Tris/borate/EDTA buffer (standard agarose gel). To each 2 µL RNA sample, add 3 µL loading buffer and heat denature at 65°C for 10 min. Put tubes on ice and load directly. Run gel for 2 h at 100 V. Examine the gel on a UV transilluminator and photograph/record.
15. Treat the remainder of the RNA with DNase I. To 1 µg RNA, add the following: 1 µL 10X DNase reaction buffer, 1 µL DNase I (1 U/µL), and DEPC-H$_2$O to a total volume of 10 µL. Incubate at room temperature for 15 min (*see* **Note 19**).
16. Stop the reaction with 1 µL 25 m*M* EDTA and heat at 65°C for 10 min. The RNA is ready for the RT reaction; no phenol:chloroform extraction is necessary.

3.3.2.2. REVERSE TRANSCRIPTION

RT uses the purified RNA as a template to synthesize a first-strand cDNA in the presence of an oligo d-T primer. The amount of total RNA used in the RT reactions can range from 1 to 5 µg (an RT reaction with 1 µg RNA will yield enough template to test 18 primer pairs in PCR, but the same amount of RNA must be used in each RT). To verify that the RNA is free of chromosomal DNA, it is important always to include a RT (–) control. This reaction will include the RNA and all the components except the RT. We often do not do this because we carry out 25 cycles of PCR with β-actin primers and the RNA following DNase I treatment; this avoids having to do RT for each RNA sample to be analyzed, which is especially useful if you have small amounts of RNA from several samples (i.e., a time-course study).

1. Denature the RNA template as follows: combine 11 µL (1 µg) DNased RNA and 0.5 µL oligo d-T (0.5 µg/µL) in a 0.6-µL tube; total volume is 11.5 µL.
2. Heat at 70°C for 10 min in a thermal cycler, then place immediately on ice.
3. To the denatured RNAs, add 1.5 µL 0.1 *M* dithiothreitol, 4 µL 5X first-strand buffer, 2 µL 10 m*M* dNTP mix, and 1 µL Powerscript RT (*see* **Notes 20** and **21**).
4. Incubate at 48°C for 1 h in a thermal cycler (*see* **Note 22**).
5. Follow by incubation at 70°C for 15 min to inactivate the RT. The templates are ready to use in PCR.

3.3.2.3. POLYMERASE CHAIN REACTION

The DNA/RNA hybrids serve as templates with gene-specific primers to amplify selected targets in PCR. Many parameters in PCR must be determined for each gene-specific primer pair and template; they include primer and MgCl$_2$ concentrations, primer annealing temperatures, and cycle number. Another concern is PCR contamination. A reagent (buffer, water, dNTPs, etc.) can become contaminated with a DNA template. This is caused by non-sterile techniques such as not changing pipet tips between tubes. Therefore, it is essential that a reagent only (without template) control be included with each PCR.

To analyze the functionality of each shRNA, select primers that amplify the target gene (should be downregulated); also include primers for genes downstream/directly induced by the gene of interest and genes that should not be affected by downregulation of the target gene. Because our interest is in neural differentiation, we have primers designed to identify pluripotent stem cells (Oct4), pan-neural (nestin), early neurons (musashi1), ectoderm (cytokeratin, BMP-2, and BMP-4), mesoderm (brachyury), endoderm (Gata4, Foxb3), and a positive control (β-actin). The positive control primers represent a gene that is expressed at the same level by all cells. Genes that show changes in expression patterns by semiquantitative RT-PCR can be further analyzed by quantitative/real-time PCR.

1. For each 50-μL reaction, combine 1 μL of the total RT reaction, 5 μL 10X Taq DNA polymerase buffer, 1.5–2.5 μL (1.5–2.5 mM) 50 mM $MgCl_2$, 0.5 μL 10 mM dNTP mix, 1–2 μL (10–20 pM) of each primer (i.e., β-actin forward and reverse), and 0.25 μL Taq DNA polymerase (5 U/μL); bring total volume to 50 μL with RNase-/DNase-free water (*see* **Note 23**).
2. Test the RT templates first with the positive control primer pair (i.e., β-actin primers) before using the other gene-specific primers. Analyze the β-actin products in a 1.5% agarose gel in the presence of EtBr; the intensity of these products for all templates should be uniform. Standard PCR conditions are 94°C/3 min (initial denaturation); 94°C/30 s to 1 min (denaturation/cycle); 50–60°C/30 s to 1 min (annealing, primer dependent); 72°C/1–2 min (extension, depends on the size of the PCR product; generally, synthesis occurs at a rate of 500 bp/30 s) for 20–35 cycles; 72°C/10 min (final extension); and 4°C hold (*see* **Note 24**).
3. Typically, 20% of the PCR products are electrophoresed in 1.2–2% agarose gels in the presence of EtBr (0.3 μg/mL) and photographed/recorded using a transilluminator.

3.4. Protein Analysis: Western Blotting and Immunohistochemistry

When targeting a gene by RNAi, it is important to monitor levels of target protein as well as mRNA. Immunohistochemistry allows for the detection/localization of protein within the cellular environment. Western blotting can be used for a quantitative comparison to determine the level of knockdown by RNAi targeting at the protein level. These methods can also be used to monitor RNAi-targeted cells for off-target effects or the interferon response (*see* above).

3.4.1. Western Blotting

3.4.1.1. SAMPLE COLLECTION

1. Remove and discard media, then wash the ES cells twice with ice-cold PBS (*see* **Note 25**).
2. Add 100–500 μL collection buffer (*see* **Notes 26** and **27**) and scrape the cells from the substrate with a cell lifter.
3. Transfer the buffer/cells to a microfuge tube and store them on ice (*see* **Notes 28** and **29**).

3.4.1.2. PROTEIN ISOLATION

There are three alternative methods to lyse cells for protein extracts: (1) samples can be sonicated; (2) samples can be drawn through a 21-gage needle two or three times (this step shears DNA and breaks up cell membranes); or (3) samples can be frozen and

Table 1
Effective Range of Separation of SDS-Polyacrylamide Gels (49)

Acrylamide concentration (%)	Linear range of separation (kDa)
15	12–43
10	16–68
7.5	39–94
5	57–212

thawed several times using a dry ice/EtOH bath and 37°C water bath (*see* **Note 30**).

1. Leave the lysed cells on ice for 10 min, vortexing frequently.
2. Centrifuge for 15 min at 4°C, 14,000g.
3. Transfer the supernatant to a new microfuge tube and determine the protein concentration using a protein assay.
4. Store samples at –80°C.

3.4.1.3. Electrophoresis

Precast Tris-HCl polyacrylamide gels can be purchased from Bio-Rad or Invitrogen. Alternatively, gels can be cast by hand using the protocols and recipies found in *Molecular Cloning (49)*. Use **Table 1** to determine the acrylamide percentage of the gel based on the size of the protein you are detecting.

1. Calculate the volume of each sample needed to load 15–50 µg of protein per well (*see* **Note 31**) and disperse that volume into a new microfuge tube.
2. Calculate the amount of Laemmli sample buffer needed to dilute the samples 1:1 with buffer. Make up fresh Laemmli sample buffer by adding 5% β-mercaptoethanol to the buffer sold by Bio-Rad (*see* **Note 32**).
3. Boil the samples for 5 min (*see* **Note 33**).
4. Quick cool samples on ice and centrifuge briefly; samples are ready to load.
5. Prepare the gel and tank for loading samples (*see* **Note 34**). Fill the interior chamber of the electrode assembly with 1X Tris-glycine running buffer and check for leaks. Then, remove the comb from the stacking gel.
6. Using a syringe, wash each well with 1X Tris-glycine buffer to remove unpolymerized acrylamide.
7. Fill the outer chamber of the gel tank with 1X Tris-glycine buffer (*see* **Note 35**).
8. Load the protein samples and protein ladder in separate wells. Run the gel at a maximum of 15–25 mA for one gel and 30–50 mA for two gels.
9. While the gel is running, cut a piece of nitrocellulose the size of the running gel and soak it in double-distilled water (*see* **Note 36**).
10. Stop the gel when the dye front approaches or exits the bottom of the plates (*see* **Note 37**). Remove the electrode assembly from the gel tank.

3.4.1.4. Immunoblotting

1. Cut six pieces of Whatman filter paper exactly the size of the nitrocellulose.
2. Add 100 mL MeOH to 400 mL of transfer buffer.

3. Fill the transfer tank about half full with transfer buffer/MeOH and place the transfer cassette holder in the transfer tank.
4. Place the nitrocellulose in transfer buffer/MeOH for at least 10 min before assembly of the transfer stack.
5. Put two sheets of Whatman paper into transfer buffer/MeOH, then remove a transfer cassette from the transfer tank (*see* **Note 38**).
6. Open the transfer cassette on the benchtop (*see* **Note 39**). Place two pieces of transfer buffer-soaked Whatman paper on the positive side of the cassette.
7. Place the nitrocellulose paper on top of the Whatman paper. Add a thin layer of transfer buffer/MeOH to the top of the paper (*see* **Note 40**).
8. Gently pry the glass plates of the gel sandwich apart. Cut off and discard the stacking gel. Place the glass plate with the gel into the transfer buffer/MeOH, then wait 1–2 min.
9. Carefully lift the gel off the glass plate and lay it flat and bubble free on the nitrocellulose. Add a thin layer of transfer buffer/MeOH to the top of the gel and gently smooth any bubbles with a gloved finger.
10. Put the two remaining sheets of Whatman paper into transfer buffer/MeOH and then place them on top of the gel. Add a small volume of transfer buffer/MeOH to the top of the transfer stack and smooth the stack by rolling a pipet across it in two directions.
11. Close the transfer cassette, place it into the tank (be sure the nitrocellulose is toward the positive pole), and then attach the tank lid.
12. Transfer the protein at 200 mA for 45–75 min depending on the size of your protein.
13. Take the transfer cassette out of the tank, open the cassette, and gently remove the top two sheets of Whatman paper from the stack. Carefully lift the gel off the nitrocellulose.

3.4.1.5. MONITOR TRANSFER EFFICIENCY

There are now two optional steps used to monitor transfer efficiency prior to probing the blot with antibodies: biosafe Coomassie stain of the gel posttransfer and Ponceau red stain of nitrocellulose membrane.

3.4.1.5.1. Biosafe Coomassie Stain of the Gel Posttransfer

1. Place the gel in biosafe Coomassie and stain overnight at room temperature with rocking.
2. Remove the stain (can be reused multiple times) and wash the gel with double-distilled water until blue bands of untransferred protein become clearly visible.

3.4.1.5.2. Ponceau Red Stain of Nitrocellulose Membrane

1. Pour Ponceau red stain over the membrane and rock by hand to cover the blot completely.
2. Remove excess stain (can be reused multiple times) and wash the blot with double-distilled water until red bands of transferred protein become clearly visible (*see* **Note 41**).
3. Enclose the blot in plastic wrap and photocopy to capture an image of the stained membrane (*see* **Note 42**).
4. After recording the image, wash the blot with double-distilled water until the majority of the red stain has been removed from the membrane.

3.4.1.6. WESTERN BLOTTING

1. Block the membrane in 5% milk powder/TBST (blocking buffer) for a minimum of 1 h at room temperature or overnight at 4°C.
2. Dilute the primary antibody to the appropriate concentration in 5% milk powder/TBST (*see* **Note 43**).

3. Place the blot into a container as close to the size of the membrane as possible, add a minimum of 5 mL diluted primary antibody, and incubate from 1 h at room temperature to overnight at 4°C depending on the antibody and protein sample (*see* **Note 44**).
4. Wash the blot in 5% milk powder/TBST for 5–10 min with at least four changes of 5% milk powder/TBST.
5. Dilute the appropriate horseradish peroxidase-conjugated secondary antibody to the appropriate concentration in 5% milk powder/TBST (*see* **Note 45**).
6. Add diluted secondary antibody and incubate for 1 h at room temperature (*see* **Note 46**).
7. Wash the blot in TBST for 10 min with at least four changes of TBST.
8. Prepare the Pierce SuperSignal West Pico chemiluminescent substrate by mixing 500 µL Luminol/enhancer solution with 500 µL stable peroxide solution in a clean and dry blot container (*see* **Note 47**).
9. Place the blot protein side up into the freshly mixed substrate, shake the container to evenly spread the substrate across the blot, and incubate for 5 min at room temperature. Remove the blot and place it protein side up under a clear sheet of plastic in a film holder.
10. Proceed to a darkroom and turn on the safe light before opening the film box. Place a piece of Kodak Bio-Max film on top of the blot and expose for 30 s, then develop the film. Exposure time can be adjusted to enhance the signal; however, if there is significant background after 30 s, blocking and wash steps may need to be adjusted.

3.4.2. Immunohistochemistry

Although Western blotting allows quantitative analysis of protein levels, it is important to note that the analysis is performed on populations of cells that, because of inefficient selection, may contain untransfected cells, the presence of which may produce confounding results. Therefore, it can be helpful to include immunohistochemical assays that, although more qualitative rather than quantitative, are still advantageous because the detection and localization of proteins can be performed at the cellular level. Moreover, the presence of untransfected cells can actually be advantageous as they will serve as an internal control for immunostaining. In addition to assessing the gene product that is knocked down, it can also be informative to look for the expression of downstream products as well as markers of ES cell pluripotency (e.g., Oct4) or lineage-specific differentiation (e.g., Tuj1 for neurons) to help determine the effects of gene knockdown.

1. Remove medium from cells and rinse with 1X D-PBS.
2. Fix cells in 2 mL buffered 2% formaldehyde for 15 min at room temperature (*see* **Note 48**).
3. Rinse cells twice with 2 mL 1X D-PBS. The cells can be stored at this point at 4°C for later processing or proceed immediately to **step 4**.
4. Remove D-PBS and treat cells with 2 mL of permeabilization solution for 10 min at room temperature.
5. Remove permeabilization solution and rinse cells with 1 mL serum wash.
6. Remove serum wash and apply 500 µL blocking solution. Incubate for 30 min at room temperature in a humidity chamber made by lining the bottom of an airtight chamber with wet paper towels.
7. Remove blocking solution and incubate with 350 µL primary antibody in dilution buffer (*see* **Note 49**).
8. Recover primary antibody and store at 4°C for up to 1 mo. Rinse cells twice with 1 mL serum wash (*see* **Note 50**).

9. Remove serum wash and apply 350 μL fluorophore-labeled secondary antibody diluted 1:100 in dilution buffer. Incubate for 5 min at room temperature in a humidity chamber (*see* **Note 51**).
10. Recover secondary antibody and store at 4°C for up to 1 mo. Rinse cells with 1 mL serum wash (*see* **Note 50**).
11. Nuclear labeling (optional): rinse cells with distilled water and apply one of the following nuclear dyes and incubate for 30 min at room temperature: 5 μg/mL Hoechst 33258 diluted in distilled water or 100 μg/mL DAPI diluted in distilled water.
12. Rinse cells with distilled water.
13. Replace distilled water with D-PBS and assess fluorescence using an inverted fluorescence microscope. If signal is low, then **steps 5–10** can be repeated (there is usually no need to repeat nuclear labeling).
14. If signal is good, then samples can be stored at 4°C (wrapped in foil to protect from light) and viewed at any time up to 6 mo.
15. Alternatively, samples can be mounted in antifade medium, cover slipped with a 25-mm no. 1 circle, and stored at –20°C up to 6 mo (*see* **Note 52**).

3.4.3. Morphological Analysis: Phase Contrast Microscopy and SEM

In addition to changes in protein and mRNA expression profiles, gene silencing by RNAi can induce overt morphological changes in ES cells. Undifferentiated ES cells grow as compact clusters, and they are typically small cells with a high nucleus:cytoplasm ratio and multiple nucleoli. Undifferentiated cells are attached via numerous filopodia and sometimes a few lamellipodia, giving them a stellate appearance overall. On differentiation, ES cells adopt a variety of morphologies (e.g., epithelioid, mesenchymal, neuronal). Such changes can be routinely assessed using a phase contrast microscope; however, the morphological features of RNAi-treated cells can be appreciated in much finer detail by SEM.

1. Purify RNAi-treated cells by antibiotic selection or fluorescent-activated cell sorting and seed them onto 13-mm Thermanox plastic cover slips coated with 0.1% gelatin (*see* **Note 53**). The cell density, time in culture, and culture media depend on the experiment, keeping in mind that the cells require at least 12 h to attach, and that each cover slip will accommodate about 3×10^5 cells at confluence.
2. At the desired time to fix the cells, remove growth media and gently rinse cells with 1X buffer (1 part 2X buffer to 1 part distilled water), then gently apply the primary fixative and fix for 15 min at room temperature in a fume hood.
3. Remove primary fixative and rinse cells with 1X buffer.
4. Apply osmium tetroxide secondary fixative and fix for 30 min at room temperature in a fume hood. Remove osmium fixative and rinse samples twice with 1X buffer (*see* **Note 54**).
5. Rinse samples twice with distilled water (to remove buffer salts that may precipitate during dehydration) and dehydrate by running through a graded EtOH series at room temperature (5 min each in 30, 50, 70, and 95% EtOH, then twice in 100% EtOH).
6. Rinse samples twice (5 min each) with 100% HMDS.
7. Remove cover slips from HMDS and allow to dry by placing on filter paper (sample side up) in a fume hood for at least 30 min (*see* **Note 55**).
8. To mount, take a cleaned stub, place a conductive double-sided adhesive tab on the surface, and then place the cover slip gently on the tab (*see* **Note 56**).

9. Sputter coat the cover slips with 20 nm Au-Pd (*see* **Note 57**).
10. View samples in an SEM (*see* **Note 58**) set at a relatively low accelerating voltage (5–10 keV) to prevent specimen damage and charging.

3.5. Discussion

RNAi as a method to silence gene expression is particularly powerful in creating graded knockdown of gene expression and has the additional advantages that multiple genes can be targeted, labeled targeting vectors have been developed to identify successfully transfected cells, and inducible systems have been developed. Use of RNAi in "indicator" ES cell lines that express a lineage-restricted gene driving the expression of a fluorochrome such as Oct4-EGFP, t-EGFP, Sox1-EGFP, or Tau-EGFP allows rapid assessment of the role of the targeted gene in lineage differentiation.

A number of caveats with this approach are unique to ES cells. First is the fact that targeted cells must somehow be enriched in the culture, which contains both transfected and untransfected cells; otherwise, group measures such as protein or RNA analysis will be uninformative. When the targeting plasmid contains a fluorochrome marker, it is also possible to visually identify and monitor differentiation of successfully transfected cells. This is particularly useful when targeting a transcription factor rather than a secreted growth factor or signaling molecule that may be provided by the untransfected, normal ES cells in the culture, which also produce LIF *(50)*. It is also informative to grow targeted ES cells in complete medium. In these conditions, we have shown that RNAi targeted to Oct4 produces trophectoderm differentiation even in the presence of inhibitors of differentiation (LIF, serum).

ES cells have a number of advantages as a starting material for RNAi, including creation of embryos from transfected cells and knockdown of inhibitors of differentiation at selected stages of differentiation. Fertilized eggs have also been transfected with siRNAs *(51)* as an alternative to gene targeting via homologous recombination, producing stable, heritable gene knockdown. This approach has tremendous untapped potential to examine the role of maternal RNAs in early development, recently using human ES cell lines.

4. Notes

1. There are several different programs that can be used to design your own siRNA or shRNA. The following are just a few of the companies that offer free design programs: BD Clontech, Invitrogen, and OligEngine. Regardless, be sure to carry out BLAST analyses of the selected sequences to be sure you will only target the gene of interest.
2. Prepare collection buffer by addition of complete protease inhibitor (follow package insert) just prior to use; once inhibitor is added, buffer must be used within 24 h.
3. PVDF membrane can also be used for Western blotting. However, this membrane must be handled differently from nitrocellulose; consult the package insert.
4. Species choice is based on secondary antibodies; for instance, use goat serum if secondary antibody is goat derived. If two secondaries from two different species will be used, then add 0.5 mL serum from each species.
5. Sodium azide is poisonous and should be handled with care.
6. In contrast to cacodylate buffers, phosphate buffers are more prone to precipitate and must be kept sterile; however, cacodylate buffers contain arsenic and should be handled and disposed of properly. *Addition of acid will produce arsenic gas, so perform this step in a fume hood.*

7. For electron microscopy, modified Karnovksy's is the fixative of choice for most animal tissues because it combines the rapid penetration of formaldehyde with the superior protein crosslinking of glutaraldehyde. However, for cells grown as a monolayer, penetration is not an issue, so glutaraldehyde will suffice.
8. All work with OsO_4 *must* be done in a fume hood and with *extreme care*. All contaminated waste should be disposed of properly.
9. Master mixes should be set up by increasing the amount of each reagent proportionally for multiple wells/dishes.
10. The OAS1g primer pair functions best when 1 M Betaine is included in the PCR master mix.
11. The cells must be taken to our flow cytometry core in 17 × 100-mm snap-cap polystyrene tubes; you should check with your facility to determine what you will need.
12. The precautions for working with RNA are listed in many references (e.g., **ref. *49***). Be sure always to wear gloves and use only sterilized tips and glassware at minimum. It is also a good idea to have reagents that are only used with RNA. In all RNA protocols, DEPC- or RNase-free water is always used in reagents and reactions and for resuspension of RNAs.
13. If there are multiple plates to harvest, then just be sure that all cell lysates are kept on ice until all plates have been harvested. At this point, the samples can be stored at –80°C for at least 1 mo or proceed to **step 2**. To finish the extractions at a later time, thaw tubes on ice and then proceed with the protocol.
14. Chloroform:isoamyl alcohol can also be used.
15. Usually, you see a small white pellet.
16. The pellet will be clear and gel-like but usually is visible.
17. For 5×10^5 cells (one 35-mm dish), the pellet is resuspended in 20 µL and yields 1–3 µg RNA/µL.
18. UV spectrophotometry is done to determine the concentration of RNA or DNA. Make a 1:100 dilution of each sample in TE and obtain A_{260} and $A_{260/280}$ ratio. The A_{260} is used to determine concentration for RNA (A_{260} × 40 µg/mL × dilution factor) and DNA (A_{260} × 50 µg/mL × dilution factor). The $A_{260/280}$ ratio indicates the purity of the sample and should be in the range of 1.7–2.0. A ratio less than 1.7 indicates that the prep contains contaminants that could interfere with the RT-PCR.
19. Do not exceed 15 min.
20. Master mixes should always be made to decrease pipetting errors.
21. Master mixes contain all the common reagents (without templates) for a particular reaction. It is a good rule to make at least enough of the master mix for one additional reaction (i.e., if you have 9 reactions, then the master mix is made for 10 reactions). Be sure to mix well (vortex) prior to use.
22. Powerscript RT has activity at temperatures from 37 to 55°C (higher temperatures increase specificity and keep the RNA denatured during the reaction).
23. All PCR reactions are set up in 600 µL, RNase-/DNase-free, thin-wall tubes. Make sure tops are closed tightly and place tubes in a thermocycler.
24. Troubleshooting PCR: if more than one PCR product is present, then the reaction conditions need to be optimized for the primer pair (i.e., alter annealing temperature, primer concentration, $MgCl_2$, etc.). If none of the templates yield PCR products with the control primer pair (i.e., β-actin), then it is possible one of the reagents could have been left out of either the RT reaction or the PCR or the RNA may be degraded. Repeat the PCR with only one or two of the templates. If a product is obtained, then some component of the PCR was not present in the initial PCR. If a product of the same size is amplified with all of the templates including

the PCR reagent control, then some component in the reaction is contaminated with an extraneous DNA template. The best way to resolve the contamination problem is to discard the reagents that were used and start with fresh PCR reagents. Electrophorese 2 μL of the RNAs in a 1% agarose/EtBr gel to check for RNA degradation (degraded RNA will appear as a smear when viewed with UV light). Be sure to select good positive control templates for each primer pair. Betaine or DMSO can be added to the PCR (20% and 5–10%, respectively); this helps disrupt the secondary structure of the template, and it is useful with G:C-rich templates. We test each new primer pair with and without Betaine or DMSO with a positive control template.

25. Be sure to remove all of the PBS washes before adding collection buffer.
26. Protein collection buffers vary widely based on subcellular location and composition of the target protein.
27. The amount of buffer added depends on the size of the tissue culture dish and the number of cells. RIPA buffer is a good general-purpose collection buffer for proteins that reside in the cytosol of cells.
28. Always thaw and store protein samples on ice to reduce degradation.
29. The cells can be frozen at –80°C and stored at this point, or you can continue with the protocol.
30. If there are any proteases present in your sample, then the freeze/thaw lysis method can result in protein degradation.
31. The amount of total protein necessary to detect immunoblotted target protein will vary based on the quality of the primary antibody used for detection.
32. For best results, do not store sample buffer with β-mercaptoethanol.
33. The tubes must have locking caps or cap holders, or the tubes will pop open during boiling and fill with water, causing sample loss.
34. If you are running only one gel, then you must place a buffer dam or glass plate on the other side of the electrode assembly.
35. Check the buffer level inside the electrode assembly. It will be very difficult to add/remove buffer after the samples have been loaded without losing some protein sample.
36. *Do not touch the nitrocellulose with your bare hands.*
37. This depends on the size of your protein and the gel percentage.
38. Note the orientation of the transfer cassette; be sure that the gel is placed toward the negative pole and the nitrocellulose toward the positive pole.
39. Make sure that the sponges are completely soaked with transfer buffer/MeOH.
40. Make sure to identify the side of the nitrocellulose that will have protein by cutting off one upper corner and then noting which side is trimmed. Alternatively, you can make pencil marks on the nitrocellulose to indicate the gel lanes either before or after protein transfer.
41. Excess washing can result in a loss of staining; if this happens, then just repeat staining.
42. Do not allow the blot to dry out as this can adversely affect protein detection with some antibodies.
43. The titer of the primary antibody should be determined experimentally. In general, a 1:1000 dilution is a good starting concentration for primary antibody.
44. Be sure the blot is protein side up when placed on the rocker.
45. In general, a 1:5000 dilution is a good starting concentration for secondary antibody.
46. The incubation time can be lengthened to increase detection, but this will often increase background as well.
47. Be sure to change tips when going from the substrate bottle to the peroxide bottle to prevent cross-contamination and loss of substrate strength.

48. Volumes given are per well in a six-well plate.
49. Dilution values are antibody dependent and generally range from 1:50 to 1:1000. Incubation times also vary but try 2 h at room temperature (in a humidity chamber). If signal is low, then switch to an overnight incubation at 4°C (also in a humidity chamber) with rocking.
50. Generally, the prepared antibodies (both primary and secondary) can be reused up to three times without loss of activity.
51. Fluorophores are light sensitive, so samples should be protected from light in all subsequent steps.
52. Noncover slipped samples can be viewed in D-PBS. However, cover slipped samples mounted in antifade media typically yield better signal and resolution.
53. These cover slips are preferred because they will fit four to a well in a six-well plate or singly in a 24-well plate; they are treated on one side to promote cell adhesion, and they are resistant to many of the solvents used to prepare samples for electron microscopy.
54. Dispose of contaminated waste appropriately.
55. From this point, samples should be stored in a desiccator at all times.
56. The goal is to establish good electrical contact between the sample and the stub. Any portions of the cover slip that hang over the stub can be "grounded" by painting a thin strip of conductive paint along the underside from the stub to the edge of the cover slip.
57. We coat our samples for 80 s in a Polaron E5100 sputter coater with a discharge voltage of 1.5 kV and sputter current of 20 mA.
58. We use an AMRAY 1000b.

Acknowledgments

This research was supported by National Institutes of Health grants NS-039438, NS-048187, and GM-069985. J. M. V. was supported by DE-07057.

References

1. Yang, S., Tutton, S., Pierce, E., and Yoon, K. (2001) Specific double-stranded RNA interference in undifferentiated mouse embryonic stem cells. *Mol. Cell Biol.* **21,** 7807–7816.
2. Kunath, T., Gish, G., Lickert, H., Jones, N., Pawson, T., and Rossant, J. (2003) Transgenic RNA interference in ES cell-derived embryos recapitulates a genetic null phenotype. *Nat. Biotechnol.* **21,** 559–561.
3. Velkey, J. M. and O'Shea, K. S. (2003) Oct4 RNA interference induces trophectoderm differentiation in mouse embryonic stem cells. *Genesis* **37,** 18–24.
4. Carmell, M. A., Zhang, L., Conklin, D. S., Hannon, G. J., and Rosenquist, T. A. (2003) Germline transmission of RNAi in mice. *Nat. Struct. Biol.* **10,** 91–92.
5. Rubinson, D. A., Dillon, C. P., Kwiakowski, A. V., et al. (2003) A lentivirus-based system to functionally silence genes in primary mammalian cells, stem cells and transgenic mice by RNA interference. *Nat. Genet.* **33,** 401–406.
6. Kiyosawa, H., Yamanaka, I., Osato, N., Kondo, S., and Hayashizaki, Y. (2003) Antisense transcripts with FANTOM2 clone set and their implications for gene regulation. *Genome Res.* **13,** 1324–1334.
7. Mattick, J. S. (2004) RNA regulation: a new genetics? *Nat. Rev. Genet.* **5,** 316–323.
8. Grishok, A., Pasquinelli, A. E., Conte, D., et al. (2001) Genes and mechanisms related to RNA interference regulate expression of the small temporal RNAs that control *C. elegans* developmental timing. *Cell* **106,** 23–34.

9. Dykxhoorn, D. M., Novina, C. D., and Sharp, P. A. (2003) Killing the messenger: short RNAs that silence gene expression. *Nat. Rev. Mol. Cell Biol.* **4,** 457–467.
10. Llave, C., Xie, Z., Kasschau, K. D., and Carrington, J. C. (2002) Cleavage of Scarecrow-like mRNA targets directed by a class of *Arabidopsis* miRNA. *Science.* **297,** 2053–2056.
11. Doench, J. G., Petersen, C. P., and Sharp, P. A. (2003) siRNAs can function as miRNAs. *Genes Dev.* **17,** 438–442.
12. Denli, A. M. and Hannon, G. J. (2003) RNAi: an ever-growing puzzle. *Trends Biochem. Sci.* **28,** 196–201.
13. Brummelkamp, T. R., Bernards, R., and Agami, R. (2002) A system for stable expression of short interfering RNAs in mammalian cells. *Science* **296,** 550–553.
14. Giegerich, R., Meyer, F., and Schleiermacher, C. (1996) GeneFisher-software support for the detection of postulated genes, in *Proceedings of the Fourth International Conference on Intelligent Systems for Molecular Biology,* AAAI Press.
15. Altschul, S. F., Madden, T. L., Schaffer, A. A., et al. (1997) Gapped BLAST and PSI-BLAST: a new generation of protein database search programs. *Nucleic Acids Res.* **25,** 3389–3402.
16. Reynolds, A., Leake, D., Boese, Q., Scaringe, S., Marshall, W. S., and Khvorova, A. (2004) Rational siRNA design for RNA interference. *Nat. Biotechnol.* **22,** 326–330.
17. Amarzguioui, M., Holen, T., Babaie, E., and Prydz, H. (2003) Tolerance for mutations and chemical modifications in a siRNA. *Nucleic Acids Res.* **31,** 589–595.
18. Yu, J.-Y., DeRuiter, S. L., and Turner, D. L. (2002) RNA interference by expression of short-interfering RNAs and hairpin RNAs in mammalian cells. *Proc. Natl. Acad. Sci. USA* **99,** 6047–6052.
19. Paul, C. P., Good, P. D., Winer, I., and Engelke, D. R. (2002) Effective expression of small interfering RNA in human cells. *Nat. Biotechnol.* **20,** 505–508.
20. Tuschl, T. (2002) Expanding small RNA interference. *Nat. Biotechnol.* **20,** 446–448.
21. Zhang, L., Yang, N., Mohamed-Hadley, A., Rubin, S. C., and Coukos, G. (2003) Vector-based RNAi, a novel tool for isoform-specific knock-down of VEGF and anti-angiogenesis gene therapy of cancer. *Biochem. Biophys. Res. Commun.* **303,** 1169–1178.
22. Mittal, V. (2004) Improving the efficiency of RNA interference in mammals. *Nat. Rev. Genet.* **5,** 355–365.
23. Kasim, V., Miyagishi, M., and Taira, K. (2004) Control of siRNA expression using the Cre-loxP recombination system. *Nucleic Acids Res.* **32,** e66.
24. Tiscornia, G., Tergaonkar, V., Galimi, F., and Verma, I. M. (2004) CRE recombinase-inducible RNA interference mediated by lentiviral vectors. *Proc. Natl. Acad. Sci. USA* **101,** 7347–7351.
25. Coumoul, X., Li, W., Wang, R. H., and Deng, C. (2004) Inducible suppression of Fgfr2 and Survivin in ES cells using a combination of the RNA interference (RNAi) and the Cre-LoxP system. *Nucleic Acids Res.* **32,** e85.
26. Wiznerowicz, M. and Trono, D. (2003) Conditional suppression of cellular genes: lentivirus vector-mediated drug-inducible RNA interference. *J. Virol.* **77,** 8957–8961.
27. Czauderna, F., Santel, A., Hinz, M., et al. (2003) Inducible shRNA expression for application in a prostate cancer mouse model. *Nucleic Acids Res.* **31,** e127.
28. van de Wetering, M., Oving, I., Muncan, V., et al. (2003) Specific inhibition of gene expression using a stably integrated, inducible small-interfering-RNA vector. *EMBO Rep.* **4,** 609–615.
29. Ohkawa, J. and Taira, K. (2000) Control of the functional activity of an antisense RNA by a tetracycline-responsive derivative of the human U6 snRNA promoter. *Hum. Gene Ther.* **11,** 577–585.

30. Gupta, S., Schoer, R. A., Egan, J. E., Hannon, G. J., and Mittal, V. (2004) Inducible, reversible, and stable RNA interference in mammalian cells. *Proc. Natl. Acad. Sci. USA* **101,** 1927–1932.
31. Margolin, J. F., Friedman, J. R., Meyer, W. K., Vissing, H., Thiesen, H. J., and Rauscher, F. J., 3rd. (1994) Kruppel-associated boxes are potent transcriptional repression domains. *Proc. Natl. Acad. Sci. USA* **91,** 4509–4513.
32. Campbell, R. E., Tour, O., Palmer, A. E., et al. (2002) A monomeric red fluorescent protein. *Proc. Natl. Acad. Sci. USA* **99,** 7877–7882.
33. Makimura, H., Mizuno, T. M., Tastaitis, J. W., Agami, R., and Mobbs, C. V. (2002) Reducing hypothalamic AGRP by RNA interference increases metabolic rate and decreases body weight without influencing food intake. *BMC Neurosci.* **3,** 18.
34. Lewis, D. L., Hagstrom, J. E., Loomis, A. G., Wolff, J. A., and Herweijer, H. (2002) Efficient delivery of siRNA for inhibition of gene expression in postnatal mice. *Nat. Genet.* **32,** 107–108.
35. Isacson, R., Kull, B., Salmi, P., and Wahlestedt, C. (2003) Lack of efficacy of "naked" small interfering RNA applied directly to rat brain. *Acta Physiol. Scand.* **179,** 173–177.
36. Holen, T. and Mobbs, C. V. (2004) Lobotomy of genes: use of RNA interference in neuroscience. *Neuroscience* **126,** 1–7.
37. Doetschman, T. C., Eistetter, H., Katz, M., Schmidt, W., and Kemler, R. (1985) The in vitro development of blastocyst-derived embryonic stem cell lines: formation of visceral yolk sac, blood islands and myocardium. *J. Embryol. Exp. Morphol.* **87,** 27–45.
38. Hadjantonakis, A. K., Gertsenstein, M., Ikawa, M., Okabe, M., and Nagy, A. (1998) Generating green fluorescent mice by germline transmission of green fluorescent ES cells. *Mech. Dev.* **76,** 79–90.
39. Wernig, M., Tucker, K. L., Gornik, V., et al. (2002) Tau EGFP embryonic stem cells: an efficient tool for neuronal lineage selection and transplantation. *J. Neurosci. Res.* **69,** 918–924.
40. Bridge, A. J., Pebernard, S., Ducraux, A., Nicoulaz, A.-L., and Iggo, R. (2003) Induction of an interferon response by RNAi vectors in mammalian cells. *Nat. Genet.* **34,** 263–264.
41. Jackson, A. L., Bartz, S. R., Schelter, J., et al. (2003) Expression profiling reveals off-target gene regulation by RNAi. *Nat. Biotechnol.* **21,** 635–637.
42. Scacheri, P. C., Rozenblatt-Rosen, O., Caplen, N. J., et al. (2004) Short interfering RNAs can induce unexpected and divergent changes in the levels of untargeted proteins in mammalian cells. *Proc. Natl. Acad. Sci. USA* **101,** 1892–1897.
43. Sledz, C. A., Holko, M., de Veer, M. J., Silverman, R. H., and Williams, B. R. G. (2003) Activation of the interferon system by short-interfering RNAs. *Nat. Cell Biol.* **5,** 834–839.
44. Barlow, D. P., Randle, B. J., and Burke, D. C. (1984) Interferon synthesis in the early postimplantation mouse embryo. *Differentiation* **27,** 229–235.
45. Burke, D. C., Graham, C. F., and Lehman, J. M. (1978) Appearance of interferon inducibility and sensitivity during differentiation of murine teratocarcinoma cells in vitro. *Cell* **13,** 243–248.
46. Francis, M. K. and Lehman, J. M. (1989) Control of β-interferon expression in murine embryonal carcinoma F9 cells. *Mol. Cell Biol.* **9,** 3553–3556.
47. Harada, H., Willison, K., Sakakibara, J., Miyamoto, M., Fujita, T., and Taniguchi, T. (1990) Absence of the type I iFN system in EC cells: transcriptional activator (IRF-1) and repressor (IRF-2) genes are developmentally regulated. *Cell* **63,** 303–312.
48. Mashimo, T., Glaser, P., Lucas, M., et al. (2003) Structural and functional genomics and evolutionary relationships in the cluster of genes encoding murine 2′,5′-oligoadenylate synthetases. *Genomics* **82,** 537–552.

49. Sambrook, J., and Russell, D. S. (2001) *Molecular Cloning: A Laboratory Manual*, Cold Spring Harbor Laboratory Press, Cold Spring Harbor, NY.
50. Rathjen, P. D., Toth, S., Willis, A., Heath, J. K., and Smith, A. G. (1990) Differentiation inhibiting activity is produced in matrix associated and diffusible forms that are generated by alternate promoter usage. *Cell* **62,** 1105–1114.
51. Hasuwa, H., Kaseda, K., Einarsdottir, T., and Okabe, M. (2002) Small interfering RNA and gene silencing in transgenic mice and rats. *FEBS Lett.* **532,** 227–230.

III

GENETIC MANIPULATION OF EMBRYONIC STEM CELLS

17

Efficient Transfer of HSV-1 Amplicon Vectors Into Embryonic Stem Cells and Their Derivatives

Dieter Riethmacher, Filip Lim, and Thomas Schimmang

Summary

The derivation of specialized differentiated cells from embryonic stem (ES) cells is now a major focus for future therapies involving cell and organ replacement in humans. To obtain populations of differentiated cells for transplantation into the human body, highly optimized protocols are needed that allow direction of the development of ES cells and their derivatives. These protocols mostly include the use of a combination of growth factors that control the differentiation of ES cells into highly specialized cells found in adult organs. The introduction of different growth factors into ES cells and their derivatives via viral gene transfer may greatly facilitate the optimization of these protocols. In this chapter, we describe a method based on herpes simplex virus type 1, which allows efficient gene transfer during several steps of a protocol, directed to obtain neural progenitors from ES cells. This protocol therefore may allow the study of potential factors influencing the cell fate or differentiation of ES cells and their derivatives.

Key Words: Amplicon; embryonic stem cell; gene transfer; Herpes; nestin; neural progenitor.

1. Introduction

Embryonic stem (ES) cells have been shown to differentiate in vitro into a large variety of cell types (for review, *see* **ref. *1***). The elaboration of defined protocols permitting the production of highly specialized cells may provide donor cells, which could be used for transplantation therapies. For example, protocols have been described that increase the amount of neurons derived from ES cells using specific growth factors *(2–4)*. In this context, the efficient introduction of growth factors via viral vectors may greatly improve the efficiency of protocols and allow screening for the influence of novel factors during ES cell differentiation. Several viral vectors have already been introduced into ES cells, including retrovirus, adenovirus, adeno-associated virus, and herpes saimiri *(5–9)*.

From: *Methods in Molecular Biology, vol. 329: Embryonic Stem Cell Protocols: Second Edition: Volume 1*
Edited by: K. Turksen © Humana Press Inc., Totowa, NJ

The use of herpes simplex virus type 1 (HSV-1), the viral vector applied in the present study, has several advantages, including a large capacity for transgenes. The insert capacity of the vector may be used to integrate large promoter sequences, which contain essential promoter elements required during ES cell differentiation. Although HSV-1 amplicons infect mitotic and postmitotic cells, they do not integrate into the host genome and thus do not bear the risk of affecting functions of the host cell via insertional mutagenesis *(10)*. This point may become essential once ES cell-based therapies enter the clinic. On the other hand, the nonintegrating nature of amplicon vectors will preclude the maintenance of its genetic information in cells that are derived from the cell in which the gene transfer has occurred *(10)*. The transient presence and expression of a gene transfer vector may yet provide an ideal signal for cell specification of an ES cell *(11)*.

Amplicons are plasmid-based vectors derived from HSV-1 *(12,13)*. They are replication defective and contain both an *Escherichia coli* and HSV origin of replication and HSV packaging sequences. Amplicon vectors are amplified in bacteria and subsequently packaged into HSV-1 particles using a complementing cell line. We have used HSV-1 mutant helper viruses harboring deletions of the immediate-early (IE)2 gene, together with an appropriate cell line; the protocols described here refer to the use of the IE2 deletion mutant 5dl1.2 grown on the complementing cell line 2-2. A detailed protocol of how to prepare HSV amplicon vectors can be found in **ref. *13***.

In this chapter, we describe a protocol that allows the introduction of amplicon vectors into ES cells and their derivatives. In comparison to the use of recombinant growth factors, the introduction via amplicons has been shown to increase the amount of differentiated cells obtained *(11)*. The application of vectors expressing growth factors instead of the direct use of recombinant factors is essential if the latter cannot be obtained. In this context, amplicon vectors offer a flexible vehicle to transfer growth factors into ES cells.

2. Materials

2.1. Tissue Culture

1. 1X phosphate-buffered saline (PBS): for 1 L, dissolve 0.185 g NaH_2PO_4, 0.385 g Na_2HPO_4, and 9 g NaCl and bring to pH 7.4 with concentrated KOH.
2. 1X Dulbecco's modified Eagle's medium (DMEM) with Glutamax-I (500 mL; Gibco BRL/Invitrogen Corporation, Karlsruhe, Germany; cat. no. 31966-021).
3. 100X 10 m*M* minimum essential medium (MEM) nonessential amino acids (NEAA) solution (100 mL; Gibco BRL, cat. no. 11140-050).
4. 100X penicillin-streptomycin (100 mL; Gibco BRL, cat. no. 15140-122).
5. Fetal bovine serum (FBS) characterized and screened for ES cell growth (100 mL; Gibco BRL, cat. no. 16141-061).
6. 100X insulin-transferrin-selenium (ITS) supplement (10 mL; Gibco BRL, cat. no. 41400-045).
7. DMEM/F12 (1:1) (500 mL; Gibco BRL, cat. no. 31331-028).
8. N2 represents a chemically defined system that allows maintenance of neuronal progenitor cells (100X) (5 mL; Gibco BRL, cat. no. 17502-048).
9. Laminin (1 mg; Gibco BRL, cat no. 23017-015). Stock solution is 1 mg/mL.

10. Polyornithine (0.01%) (50 mL; Sigma, Seelze, Germany; cat. no. P4957).
11. Trypsin-ethylenediaminetetraacetic acid (EDTA; 500 mL; Gibco-BRL, cat. no. 25300-062).
12. 1 M HEPES (100 mL; Gibco BRL, cat. no. 15630-056).
13. 50 mM, 1000X 2-mercaptoethanol (20 mL; Gibco BRL, cat. no. 31350-010).
14. Leukemia inhibitory factor (LIF; 10^7 U/mL; Chemicon, Hofheim, Germany; cat. no. ESG1107) (*see* **Note 1**).
15. Fibronectin (5 mg; Gibco BRL, cat. no. 33016-015). Stock solution is 5 mg/mL.
16. Mitomycin (2 mg; Sigma, cat. no. M-0503). Stock solution is 1 mg/mL.
17. Dimethyl sulfoxide (DMSO) (100 mL; Sigma, cat. no. C-8919).
18. Gelatin (100 g; Sigma, cat. no. G9391).

2.1.1. Media

1. Medium for embryonic fibroblast cells: murine embryonic fibroblast cells that serve as a feeder for ES cells are grown in DMEM with Glutamax-I supplemented with 10% FBS, 1% penicillin-streptomycin, 1% NEAA, and 100 µM 2-mercaptoethanol. For 1 L, combine 878 mL DMEM with Glutamax-I and 100 mL FBS, 10 mL penicillin-streptomycin, 10 mL NEAA, and 2 mL 2-mercaptoethanol. Store at 4°C for up to 4 mo.
2. Media for ES cells: ES cells are maintained in supplemented DMEM with Glutamax-I. DMEM with Glutamax-I is supplemented with 15% ES-tested FBS, 1% NEAA, 1% penicillin-streptomycin, 10 mM HEPES, 0.1 mM 2-mercaptoethanol, and 1000 U/mL LIF. For 1 L, combine 818 mL DMEM with Glutamax-I and 150 mL FBS, 10 mL penicillin-streptomycin, 10 mL NEAA, 10 mL HEPES, 2 mL 2-mercaptoethanol, and 0.1 mL LIF. Store at 4°C for up to 2 mo.
3. For differentiation of ES cells and enrichment of neural precursors, ES cells were grown in ITSFn medium. ITSFn medium is DMEM/F12 supplemented with 1% penicillin-streptomycin, 5 µg/mL insulin, 50 µg/mL transferrin, 30 nM selenium chloride, and 5 µg/mL fibronectin. For 100 mL, combine 98 mL DMEM/F12, 1 mL penicillin-streptomycin, 1 mL ITS supplement, and 0.1 mL fibronectin. Store at 4°C for up to 1 mo.
4. After selection of neuronal progenitors, cells are grown in DMEM/F12 supplemented with N2 and containing 2 mM HEPES and 1 µg/mL laminin. For 100 mL, combine 90 mL DMEM/F12, 10 mL N2, 200 µL HEPES, and 100 µL laminin. Filter-sterilize and store at 4°C for up to 1 mo.

2.2. β-*Galactosidase Staining*

1. 4% paraformaldehyde, pH 7.0: to prepare 4% paraformaldehyde, add 20 g paraformaldehyde (1 kg; Fluka, Seelze, Germany; cat. no. 76240) to 300 mL H_2O and heat to 55–60°C. Slowly add 1 M NaOH dropwise over about 10 min until the solution becomes clear. Cool the solution to room temperature. Use pH paper to check that the pH is 7.0–7.5 (add more NaOH if necessary). Add 100 mL 0.5 M sodium phosphate buffer at pH 7.0 and then water to a final volume of 500 mL. The final pH should be 7.0–7.5. Store at 4°C.
2. Tris-buffered saline (TBS): dissolve 8 g NaCl, 0.2 g KCl, and 3 g Tris-base in 800 mL distilled water. Adjust to pH 7.4 with HCl and add water to 1 L.
3. 5-Bromo-4-chloro-3-indolyl-β-D-galactoside (X-gal; 1 g; Roche, Mannheim, Germany; cat. no. 745740): a 50-mg/mL stock in DMSO can be stored in aliquots at −20°C.
4. *N*-*N*-Dimethylformamide (250 mL; Sigma, cat. no. 4551).
5. Staining solution (kept at 4°C in the dark): to prepare a 200-mL solution of 10 mM PBS (pH 7.4) containing 0.332 g potassium ferrocyanide (5 mM; Sigma, cat. no. P-8131), add

0.424 g potassium ferrocyanide (5 mM; Sigma, cat. no. P9387) and add 0.2 mL 1 M MgCl$_2$ (1 mM). Store at 4°C in a darkened container because this reagent is light sensitive. To the staining solution, add at a concentration of 1 mg/mL X-gal from a 50-mg/mL stock in N-N-dimethylformamide.

2.3. Immunofluorescence

1. 4% paraformaldehyde (see **Subheading 2.2.**, **item 1**).
2. TBS (pH 7.5) (see **Subheading 2.2.**, **item 2**).
3. Ethanol.
4. Acetic acid.
5. Bovine serum albumin (BSA) (100 mL; Gibco BRL, cat. no. 10106-151).
6. Phalloidin-fluorescein isothiocyanate (0.1 mg; Sigma, cat. no. P5282).
7. Mowiol (500 g; Polysciences, Eppelheim, Germany; cat. no. 17951).
8. Nestin antibodies: rat-401 mouse monoclonal immunoglobulin (Ig) G$_1$ obtained from the Developmental Studies Hybridoma Bank (Iowa City, IA).
9. Secondary antibodies: alexa fluor 546 goat antimouse IgG$_1$ (250 µL; Molecular Probes, Karlsruhe, Germany; cat. no. A-21123).
10. Tween-20 (100 mL; Sigma, cat. no. P1379).

3. Methods

3.1. Maintenance of Undifferentiated ES Cells and HSV-1-Mediated Gene Transfer

3.1.1. Maintenance of ES Cells

The particular characteristics of murine ES cells include the possibility to expand them almost indefinitely without apparent morphological changes (see **Fig. 1**, **step 1**). However, it is necessary to exchange the culture medium daily, and the ES cell culture has to be split regularly; otherwise, ES cell differentiation will be initiated. Two additional factors are important to prevent ES cell differentiation: addition of LIF to the culture medium and growth of ES cells on a feeder layer of mouse embryonic fibroblasts (MEFs). To prevent growth of the latter cells, fibroblasts are treated with mitomycin C, which stops their division. ES cells are maintained on 10-mm tissue culture dishes containing a mitomycin C-treated MEF layer. To split ES cell cultures, cells are trypsinized.

1. Wash cells three times with PBS.
2. Add 5 mL 0.25% trypsin-EDTA.
3. Incubate cells for 5 min at 37°C.
4. Gently pipet cells up and down to obtain a single-cell suspension.
5. Centrifuge the suspension at 700g for 5 min.
6. Resuspend the pellet in ES cell medium (see **Subheading 2.1.1.**, **item 2**) and distribute onto fresh culture dishes with feeder layers (see **Note 2**).

3.1.2. Infection of ES Cells and Analysis of HSV-1-Mediated Gene Transfer

This part describes transfer of an amplicon vector expressing β-galactosidase (pHSV*lacZ*) into undifferentiated ES cells. The β-galactosidase gene may be used as a reporter from bicistronic amplicon vectors and thus facilitates the monitoring of gene transfer into ES cells (see **Note 3**).

HSV-1 Amplicon Vector Transfer Into ES Cells

Step 1: Routine culture of ES cells in ES cell medium+LIF

Step 2: Generate EB: ES medium without LIF, non-adherent culture dishes
(4 days)

Plate whole EB on adhesive tissue culture surface
(1 day)

Step 3: Selection of neuronal progenitors by serum-free culture in ITSFn medium
(8 days)

Trypsinize and seed on polylysin+laminin treated coverslips

Step 4: Expansion of neuronal progenitors and culture in serum-free N2 medium+laminin

Step 5: Differentiation

Fig. 1. Embryonic stem cell differentiation.

1. ES cells grown in 24-well plates are infected with the amplicon vector expressing β-galactosidase by adding different amounts of viral particles directly into the ES cell culture medium. For instance, from a stock solution of 2×10^7 infectious vector units (ivu)/mL different amounts corresponding to 0.2 to 2×10^6 ivu/mL are added to ES cell cultures (*see* **Note 4**).
2. After 24 h, cells are fixed in 4% PFA for 3–5 min.
3. Cells are washed twice with TBS.
4. Fresh staining solution (*see* **Subheading 2.2.**, **item 3**) is added and left for 30 min at 37°C in the dark.
5. Reaction is stopped by washing twice with PBS.

3.2. Differentiation of ES Cells and Gene Transfer Using Amplicon Vectors

3.2.1. Embryoid Body Formation

The development of embryoid bodies from stem cells is the first step of all protocols aiming to differentiate ES cells in vitro (*see* **Fig. 1, step 2**).

1. Plate ES cells at a high density (>10^6 cells per 100-mm plate) on tissue culture dishes without feeder cells.
2. Expand cells in ES cell culture medium containing LIF (*see* **Subheading 2.1.1.**, **item 2**) for 2–3 d.
3. Dissociate these almost-confluent cultures by the addition of trypsin (0.05%) and EDTA (0.04%) in PBS.
4. Carefully transfer cell aggregates to nonadherent culture dishes, in which they are grown in suspension in ES cell medium without LIF. Embryoid bodies are formed after 4 d.

3.2.2. Selection of Neuronal Progenitor Cells

For further differentiation, a selection for neuronal progenitors is initiated (*see* **Fig. 1**, **step 3**). The 4-d-old embryoid bodies (from **Subheading 3.2.1.**) are transferred to culture dishes with an adhesive surface (treated with 0.5% gelatin in PBS for 1 h at room temperature) and incubated for 24 h in ES cell medium (*see* **Subheading 2.1.1.**, **item 2**) without LIF. This allows the embryoid bodies to settle at the bottom of the culture dish. Selection of neuronal progenitors is then initiated by replacing the ES cell medium with ITSFn medium (*see* **Subheading 2.1.1.**, **item 3**) During the following days, a massive amount of cell death is observed. Therefore, during 8 d the medium is exchanged daily. After 8 d, cells are treated according to the following protocol:

1. Wash cells three times in PBS.
2. Dissociate cells with 0.05% trypsin/0.02% EDTA for 2–3 min.
3. Centrifuge at 700*g* for 5 min.
4. Resuspend cells in DMEM/F12 medium supplemented with N2 medium (*see* **Subheading 2.1.1.**, **item 4**) containing 1 µg/mL laminin and replate them at a cell density of 7×10^5 cells/cm^2 on culture dishes (*see* **Note 5**) covered with polyornithine (50 µg/mL diluted in water, incubated overnight at room temperature) and laminin (10 µg/mL diluted in PBS, incubated at 4°C for 1–2 h).

3.2.3. Infection of Neural Progenitors With Amplicon Vectors

After the selection of neural progenitors, amplicon vectors may be used to transfer genes into these cells with the objective to amplify (*see* **Fig. 1**, **step 4**) or direct them to further differentiate (*see* **Fig. 1**, **step 5**). In the present protocol, we used amplicon vectors expressing FGF2 under the control of the IE 415 promoter, termed pHSV*fgf2*, to amplify the number of neuronal progenitors. Infection with pHSV*fgf2* leads to a transient expression of FGF-2 for 6 d, leading to the selective proliferation of neuronal progenitors.

1. Plate cultures at a cell density of 3×10^5 cells/cm^2 in 24-well dishes (*see* **Note 6**). Cells are grown in DMEM/F12 supplemented with N2 and containing 1 µg/mL laminin.
2. Cultures are infected 48 h after plating with different concentrations of pHSV*fgf2*. A corresponding volume of virus solution is added to the medium of the ES cell cultures (*see* **Note 7**).
3. Cells are grown for 6 d, changing the culture medium daily; after this time, neuronal progenitors have been amplified (*see* **Note 8**).

3.2.4. Detection of Neuronal Progenitors by Staining With Nestin Antibodies

Neuronal progenitors may be identified by antibodies against nestin *(2)*. The determination of the amount of neuronal progenitors after selection and amplification (*see* **Fig. 1**, **step 4**) allows the researcher to estimate the percentage of neuronal progenitors obtained before continuing with further differentiation steps to obtain neurons (*see* **Fig. 1**, **step 5**; **refs.** *2–4*).

1. Fix cultures of neuronal progenitors derived from ES cells (*see* **Fig. 1**, **step 4**) grown on glass cover slips (22 × 22 mm) with 4% PFA in PBS for 3–5 min.
2. Rinse cultures once in TBS.

3. Permeabilize cells in 95% ethanol and 5% acetic acid at $-20°C$ for 20 min.
4. Wash cells three times for 5 min in TBS.
5. Block cells in 1% BSA in TBS for 15 min.
6. Incubate cells with Nestin antibodies (Rat-401) at a 1:1 dilution in PBS containing 0.1% BSA at 4°C overnight.
7. Rinse cells once and wash four times in PBS.
8. Incubate cells with secondary antibody Alexa fluor 546 goat antimouse IgG_1 diluted 1:400 in PBS containing 0.1% BSA for 1 h at room temperature (see **Note 9**).
9. Rinse cells once and wash four times in PBS.
10. For counterstaining of the cytoplasm, add phalloidin-fluorescein isothiocyanate at 12 µg/mL in PBS containing 0.05% Tween-20 and 1% BSA for 20 min.
11. Wash cells five times with PBS containing 0.05% Tween-20.
12. Rinse once in water.
13. Mount glass cover slips in Mowiol.

3.2.5. Differentiation Into Neurons

After the amplification of neuronal progenitors (see **Subheading 3.2.3.**), neuronal progenitors are further differentiated into neurons. This is usually achieved by the removal of FGF-2 from the culture medium. In the present protocol, the transient expression of pHSV*fgf2* (see **Subheading 3.2.3.**) allows amplification of neuronal progenitors. Differentiation of neuronal progenitors is therefore achieved by continuing the culture of neuronal progenitors in DMEM/F12 supplemented with N2. Cells are cultured for 2–3 wk, changing the medium daily or every second day (see **Note 10**).

4. Notes

1. LIF may also be supplied by the LIF-producing cell line M1, which secretes LIF into the supernatant. The supernatant is used at a 1:2500 dilution.
2. Alternatively, cells may be frozen in culture medium containing 10% DMSO.
3. Alternatively, amplicon vectors that express green fluorescent protein may be used; this allows direct visualization of the reporter in the fluorescent microscope. In this case, ES cells are plated on glass cover slips (22 × 22 mm).
4. High amounts of virus may cause cytotoxic effects. Therefore, it is essential to test different concentrations of virus to infect stem cells.
5. In case an immunofluorescence study is planned, plate cells on glass cover slips (22 × 22 mm) treated with polyornithine (50 µg/mL) and laminin (10 µg/mL).
6. Also, different cell densities may be plated. This will need an adjustment of the amount of virus used for infection (see **Subheading 3.2.3.**, **step 2**). In case an immunofluorescence study is planned, plate cells on glass cover slips (22 × 22 mm).
7. We have obtained efficient infection at 1.2×10^6 ivu/mL pHSV*fgf2* using a stock solution that contained a concentration of 2×10^6 ivu/mL.
8. This can be monitored and quantified by immunofluorescence using nestin antibodies (see **Subheading 3.2.4.**).
9. A nuclear counterstain may be performed using Hoechst 33258 (Sigma, cat. no. H6024) during the incubation of the secondary antibody at 5 µg/mL, added together with the secondary antibody.
10. Neuronal differentiation may be analyzed by immunofluorescence with neuron-specific

antibodies, for example, β-tubulin type III (mouse monoclonal IgG$_{2a}$; Babco represented by Hiss Diagnostics, Freiburg, Germany; cat. no. MMS-435P). The same protocol as specified under **Subheading 3.2.4.** may be used, applying an antibody dilution of 1:1000. As a secondary antibody, Alexa fluor 488 goat antimouse IgG$_{2a}$ (Molecular Probes, cat. no. A-21131) is used.

References

1. Blau, H. M., Brazelton, T. R., and Weimann, J. M. (2001) The evolving concept of a stem cell: entity or function? *Cell* **105,** 829–841.
2. Okabe, S., Forsberg-Nilsson, K., Cyril Spiro, A., Segal, M., and McKay, R. D. G. (1996) Development of neuronal precursor cells and functional postmitotic neurons from embryonic stem cells in vitro. *Mech. Dev.* **59,** 89–102.
3. Lee, S. -H., Lumelsky, N., Studer, L., Auerbach, J. M., and McKay, R. D. G. (2000) Efficient generation of midbrain and hindbrain neurons from mouse embryonic stem cells. *Nat. Biotechnol.* **18,** 675–679.
4. Rolletschek, A., Chang, H., Guan, K., Czyz, J., Meyer, M., and Wobus, A. M. (2001) Differentiation of embryonic stem cell-derived dopaminergic neurons is enhanced by survival-promoting factors. *Mech. Dev.* **105,** 93–104.
5. Stevenson, A. J., Clarke, D., Meredith, D. M., Kinsey, S. E., Whitehouse, A., and Bonifer, C. (2000) Herpes virus saimiri-based gene delivery vectors maintain heterologous expression throughout mouse embryonic stem cell differentiation in vitro. *Gene Ther.* **7,** 464–471.
6. Ketteler, R., Glaser, S., Sandra, O., Martens, U. M., and Klingmuller, U. (2002) Enhanced transgene expression in primitive hematopoietic progenitor cells and embryonic stem cells efficiently transduced by optimized retroviral hybrid vectors. *Gene Ther.* **9,** 477–487.
7. Gropp, M., Itsykon, P., Singer, O., et al. (2003) Stable genetic modification of human embryonic stem cells by lentiviral vectors. *Mol. Ther.* **7,** 281–287.
8. Ma, Y., Ramezani, A., Lewis, R., Hawley, G., and Thomson, J. A. (2003) High-level sustained transgene expression in human embryonic stem cells using lentiviral vectors. *Stem Cells* **21,** 111–117.
9. Smith-Arica, J. R., Thomson, A. J., Ansell, R., Chiorini, J., Davidson, B., and McWhir, J. (2003) Infection efficiency of human and mouse embryonic stem cells using adenoviral and adeno-associated viral vectors. *Cloning Stem Cells* **5,** 51–62.
10. Burton, E. A., Wechuk, J. B., Wendell, S. K., Goins, W. F., Fink, D. J., and Glorioiso, J. C. (2001) Multiple applications for replication-defective herpes simplex virus vectors. *Stem Cells* **19,** 358–377.
11. Vicario, I. and Schimmang, T. (2003) Transfer of FGF-2 via HSV-1 based amplicon vectors promotes efficient formation of neurons from embryonic stem cells. *J. Neurosci. Meth.* **123,** 55–60.
12. Lim, F., Hartley, D., Starr, P., et al. (1996) Generation of high-titer-defective HSV-1 vectors using an IE2 deletion mutant and quantitative study of expression in cultured cortical cells. *Biotechniques* **20,** 460–469.
13. Lim, F., Hartley, D., Starr, P., et al. (1997) Use of defective herpes-derived plasmid vectors. *Methods Mol. Biol.* **62,** 223–232.

18

Lentiviral Vector-Mediated Gene Transfer in Embryonic Stem Cells

Masahiro Oka, Lung-Ji Chang, Frank Costantini, and Naohiro Terada

Summary

The major limitations in gene transduction to embryonic stem (ES) cells are (1) low efficiency of gene delivery and (2) suppression of gene expression after integration into the host genome. A human immunodeficiency virus type 1 (HIV-1)-based lentiviral vector has been demonstrated to be an excellent tool for stable and efficient gene expression in ES cells. Here, we introduce a protocol for lentiviral vector-mediated transgene expression in murine ES cells. Using lentiviral vectors expressing LacZ, green fluorescent protein, and Cre recombinase, we demonstrate the efficiency and utility of the vectors in ES cell study.

Key Words: Embryonic stem cell; differentiation; gene expression; gene transfer; lentivirus.

1. Introduction

Embryonic stem (ES) cells have the capacity to differentiate into all three germ layers. Furthermore, the pluripotent ES cells can be maintained undifferentiated and proliferate indefinitely in vitro. These distinct characteristics of ES cells make the cells a valuable source for diverse applications, ranging from basic developmental studies to transplantation therapy. Accordingly, the development of methods for efficient gene manipulation in ES cells is worthwhile.

The major limitations in gene transduction to ES cells are (1) low efficiency of gene delivery and (2) suppression of gene expression after integration into the host genome. Low efficiency of gene delivery into ES cells using a plasmid vector with a cationic lipid transfection reagent or electroporation method requires a selection of transduced cells using drugs for selective markers. Besides the low transduction efficiency of ES cells, gene silencing poses another problem for stable gene expression in ES cells. Exogenous genes integrated into the ES cell genome are frequently suppressed through epigenetic gene modifications (e.g., genomic structure alteration, histone modification, DNA methylation, etc.). Gene silencing occurs more frequently, particularly when ES cells are differentiated, because dramatic epigenetic changes occur during ES cell differentiation. To remedy this, a polyoma-based episomal vector has been successfully applied to achieve stable gene expression devoid of epigenetic effect *(1–3)*. However, one possible limitation for this method is the requirement for ES cells to express polyoma large antigen.

A human immunodeficiency virus type 1 (HIV-1)-based lentiviral vector has been demonstrated to be an excellent tool for stable and efficient gene expression in ES cells *(4,5)*. They can transduce into ES cells with high efficiency, and the integrated genes can escape the silencing of gene expression when cells were both undifferentiated and differentiated *(4,5)*. Furthermore, the lentiviral vector-derived gene expression can be maintained during germline transmission *(5,6)*. Therefore, lentiviral vector-based gene manipulation is a valuable tool for in vitro and in vivo ES cell applications.

Here, we introduce a protocol for lentiviral vector-mediated transgene expression in murine ES cells. The lentiviral vector used here possesses modified U3 and U5 long-terminal repeat (LTR) sequences to eliminate the transcriptional interference between the LTR and the internal promoter, allowing efficient transgene expression *(7)*. The HIV-1 central DNA flap sequence *(8,9)* was also inserted to further improve the transduction efficiency. Using lentiviral vectors expressing LacZ, green fluorescent protein (GFP), and Cre recombinase, we demonstrate the efficiency and utility of the vectors in ES cell study.

2. Materials
2.1. Tissue Culture

1. Embryonic stem-leukemia inhibitory factor (ES-LIF) medium: Knock-out Dulbecco's modified Eagle's medium (Knock-out DMEM; Invitrogen, Carlsbad, CA; cat. no. 10829-018) containing 10% Knockout SR (Invitrogen; cat. no. 10828-028), 1% fetal bovine serum (FBS; Atlanta Biologicals, Norcross, GA; cat. no. S11550), 2 mM L-glutamine, 100 U/mL penicillin, 100 µg/mL streptomycin (Invitrogen, cat. no. 10378-016), 25 mM HEPES (Invitrogen, cat. no. 15630-080), 300 µM monothioglycerol (Sigma, St. Louis, MO; cat. no. M6145), and 1000 U/mL recombinant mouse LIF (Chemicon International, Temecula, CA; cat. no. ESG-1107). For 1 L, to 855 mL DMEM, add 100 mL Knockout SR, 10 mL FBS, 10 mL 100X penicillin/streptomycin/L-glutamine, 25 mL 1 M HEPES, 26 µL 11.5 M monothioglycerol, and 100 µL LIF.
2. ES differentiation medium: Iscove's modified Dulbecco's medium (IMDM; Invitrogen, cat. no. 12440-084) containing 20% FBS, 2 mM L-glutamine, 100 U/mL penicillin, 100 µg/mL streptomycin, 25 mM HEPES, and 300 µM monothioglycerol. For 1 L, to 765 mL IMDM, add 200 mL FBS, 10 mL 100X penicillin/streptomycin/L-glutamine, 25 mL 1 M HEPES, and 26 µL 11.5 M monothioglycerol.
3. 0.1% (w/v) gelatin (Specialty Media, Phillipsburg, NJ; cat. no. ES-006-B).
4. Six-well cell culture dish (Nunc, Rochester, NY; cat. no. 140675).
5. 10-cm cell culture dish (Nunc, cat. no. 172958).
6. Trypsin-ethylenediaminetetraacetic acid (Mediatech, Herndon, VA; cat. no. 25-053-CI).
7. Phosphate-buffered saline (PBS) (Mediatech; cat. no. 21-031-CV).

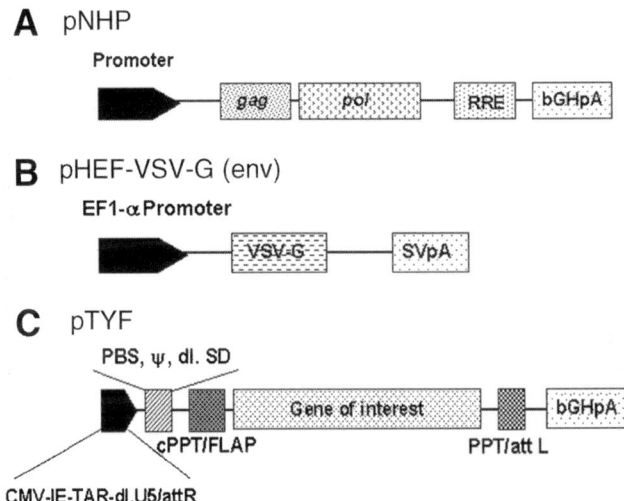

Fig. 1. Maps of the plasmid constructs for HIV-1-based lentiviral system used in this study. The basic vector system consists of three components: (**A**) helper construct pNHP, which encodes HIV-1 *Gag-Pol* and regulatory proteins *Tat* and *Rev*, with all of the accessory genes deleted; (**B**) envelope construct pHEF-VSV-G, which encodes G protein of vesicular stomatitis virus to pseudotype viral particles for expanded tropism; (**C**) transducing vector pTYF, which contains modified 5′ long-terminal repeat (LTR) and deletion of U3 in the 3′ LTR as self-inactivating vector, packaging signal (ψ), minimal HIV-1 *gag* sequence, an internal promoter (P) of non-HIV-1 origin, and the gene of interest (transgene).

2.2. Lentiviral Vector Transduction

2.2.1. Lentiviral Vector Production

1. pHP (packaging plasmid) (*see* **Fig. 1A**).
2. pHEF-VSVG (envelope-coding plasmid) (*see* **Fig. 1B**).
3. pTYF (self-inactivating lentiviral transducing plasmid) (*see* **Fig. 1C**).
4. 293T (transformed human primary embryonal kidney) cell.
5. Biosafety hood (meeting the criteria established by the National Institutes of Health biosafety level 2/3 laboratory safety guidelines).
6. SuperFect Transfection Reagent (Qiagen, Valencia, CA; cat. no. 301305).
7. Six-well cell culture plates.
8. 24-well cell culture plates (Nunc, cat. no. 143982).
9. Dulbecco's modified Eagle's medium (DMEM).
10. FBS.
11. 0.45-μm low-protein binding filter (Millipore, Billerica, MA; cat. no. SLHVR25LS).
12. Centricon (Millipore, cat no. UFC910096).

2.2.2. Lentiviral Vector Infection

1. Polybrene (Sigma, cat no. 107689).
2. Six-well cell culture plates.

3. 24-well cell culture plates.
4. Biosafety hood (meeting the criteria established by the National Institutes of Health biosafety level 2/3 laboratory safety guidelines).
5. *In Situ* β-Galactosidase Staining Kit (Stratagene, La Jolla, CA; cat. no. 200384).
6. Inverted fluorescence microscope (Olympus, Pittsburgh, PA; IX-70).
7. Magnafire image-processing system (Optronics, Goleta, CA).

3. Methods
3.1. Tissue Culture
3.1.1. Maintenance of Undifferentiated ES Cells
3.1.1.1. Gelatin Coating of Tissue Culture Plates

1. Precoat cell culture dishes with 0.1% gelatin (1 mL per well for a six-well cell culture dish; 3 mL for a 10-cm cell culture dish) at least 10 min up to 1 h prior to use.
2. Remove the excess gelatin solution and use immediately.

3.1.1.2. Maintenance of Undifferentiated ES Cells

1. Prewarm culture medium and trypsin-ethylenediaminetetraacetic acid at room temperature or in a 37°C water bath.
2. Remove medium from the ES cell culture dish and rinse once with PBS (2 mL per well for a six-well cell culture dish; 8 mL for a 10-cm cell culture dish).
3. After removing PBS, add trypsin and incubate at 37°C for 5 min or until the cells are detached from the plate by gentle agitation.
4. Gently pipet up and down to dissociate the ES cell colonies further (*see* **Note 1**).
5. Add ES-LIF medium (2 mL per well for a six-well cell culture dish; 10 mL for a 10-cm cell culture dish) to inactivate trypsin and pipet gently.
6. Transfer the cell suspension to a 15-mL conical tube and centrifuge at 350g for 5 min.
7. Remove the supernatant and resuspend the cells in ES-LIF medium (2 mL per well for a six-well cell culture dish; 10 mL per well for a 10-cm cell culture dish).
8. Count cell number and plate the ES cells on gelatin-coated cell culture dish (5×10^4 to 1×10^5 per well for six-well cell culture plate; 2×10^5 to 5×10^5 for 10-cm cell culture dish) that contains ES-LIF medium (2 mL per well for a six-well cell culture dish; 10 mL for a 10-cm cell culture dish) and distribute cells on the plate evenly.
9. Change culture medium every day.

3.1.2. ES Cell Differentiation on Cell Culture Dish

1. Remove medium from maintained undifferentiated ES cell culture and rinse once with PBS (2 mL per well for a six-well cell culture dish; 10 mL for a 10-cm cell culture dish).
2. After removing PBS, add trypsin and incubate at 37°C for 5 min or until the cells are detached from the plate by gentle agitation.
3. Gently pipet up and down to dissociate the ES cell colonies further.
4. Add ES differentiation medium (2 mL per well for a six-well cell culture dish; 10 mL for a 10-cm cell culture dish) and gently resuspend the cell by pipetting.
5. Transfer the cell suspension to a 15-mL conical tube and centrifuge at 350g for 5 min.
6. Remove the supernatant and add ES differentiation medium to resuspend the cells.
7. Harvest the cells by centrifugation at 350g for 5 min.

Lentiviral Vector-Mediated Gene Transfer in ES Cells

8. Remove the supernatant and resuspend the cells (2 mL per well for a six-well cell culture dish; 8 mL per well for a 10-cm cell culture dish) in ES differentiation medium.
9. Count the cell number and plate at a density of 2×10^4 to 1×10^5 cells per well for a six-well cell culture dish that contains 2 mL ES differentiation medium in each well.

3.2. Lentiviral Vector Transduction

3.2.1. Lentiviral Vector Preparation

1. Plate 293T cells on a six-well cell culture dish at a density of 5×10^5 cells per well 18 h prior to DNA transfection.
2. Replace media with fresh DMEM supplemented with 10% FBS (600 µL per well) just prior to transfection.
3. Dilute plasmid DNA (**Fig. 1A,B,C**) with 75 µL (per well) of serum-free DMEM and vortex briefly.
4. Add 7 µL (per well) of SuperFect Transfection Reagent and incubate at room temperature for 7–10 min.
5. DNA-SuperFect mixture is added dropwise to each culture well and incubated for 4–5 h at 37°C.
6. Incubated cells are washed once with PBS and fed with fresh DMEM supplemented with 10% FBS or appropriate media suitable for the later target cells.
7. Harvest virus-containing supernatant at 12-h intervals three times (up to 48 h).
8. Collected supernatant is cleared of cell debris by centrifugation (1000g, 5 min) and filtered using 0.45-µm low-protein binding filter (*see* **Note 2**).
9. Virus is further concentrated by centricon filtration.

3.2.2. Lentiviral Vector Infection for GFP and LacZ Expression

1. ES cells were plated on a gelatin-coated 6- or 24-well cell culture plate (at a density of 5×10^4 cells per well for 6 wells, 1×10^4 cells per well for 24 wells) 24 h prior to infection.
2. Remove medium by aspiration, add various multiplicities of infection (MOI) of lentivirus (MOI 0.1–1000) directly to the cells and incubate for 1 min.
3. Add 100 µL of ES-LIF medium containing 10 µg/mL polybrene and incubate for 1 h at 37°C.
4. Add 2 mL medium per well for a six-well plate and 500 µL medium for a 24-well plate.
5. Virus-containing medium is replaced with fresh ES-LIF medium at 24–48 h after infection.
6. For the titration of LacZ-expressing lentivirus, cells are fixed with 3.7% formaldehyde at 48 h after infection.
7. Cells are washed twice with PBS and incubated with 5-bromo-4-chloro-3-indolyl-β-D-galactoside (X-gal) staining reagent overnight at 37°C.
8. Stained cells are observed using an inverted microscope (*see* **Fig. 2**).
9. For further observation of LacZ or GFP marker gene expression, infected cells are passaged and maintained undifferentiated or differentiated on cell culture plates for indicated times (*see* **Fig. 3** for LacZ and GFP marker gene expression).

3.2.3. Lentiviral Vector Infection for CRE-Mediated Recombination

1. R26R-EYFP (enhanced yellow fluorescent protein) ES cells (*10*) are plated and infected with lentivirus expressing a nuclear-localized form of Cre recombinase (*11*) as in **Subheading 3.2.2.**
2. YFP expression is examined at 24-h intervals under an inverted fluorescence microscope. Fluorescent images are acquired and processed with Magnafire image-processing system (*see* **Fig. 4**).

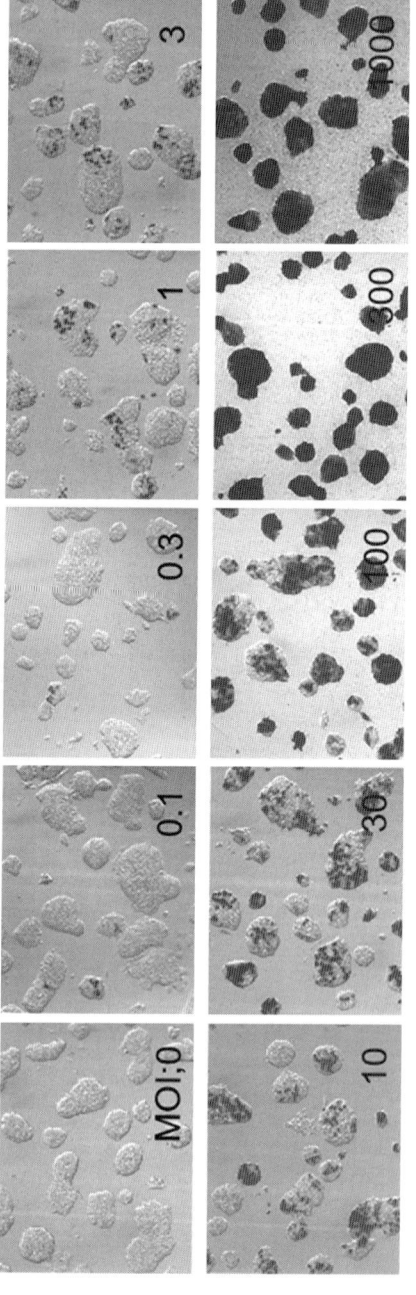

Fig. 2. Undifferentiated J1 embryonic stem (ES) cells were infected with different multiplicities of infection (MOI) of lentivirus-expressing *LacZ* gene under the control of EF1-α promoter. After 2 d, ES cells were fixed and stained with X-gal to examine the expression of β-galactosidase. Note that virus infection was observed at an MOI of 0.1. Almost 100% infection was achieved with an MOI of 300. (Please *see* the companion CD for the color version of this figure.)

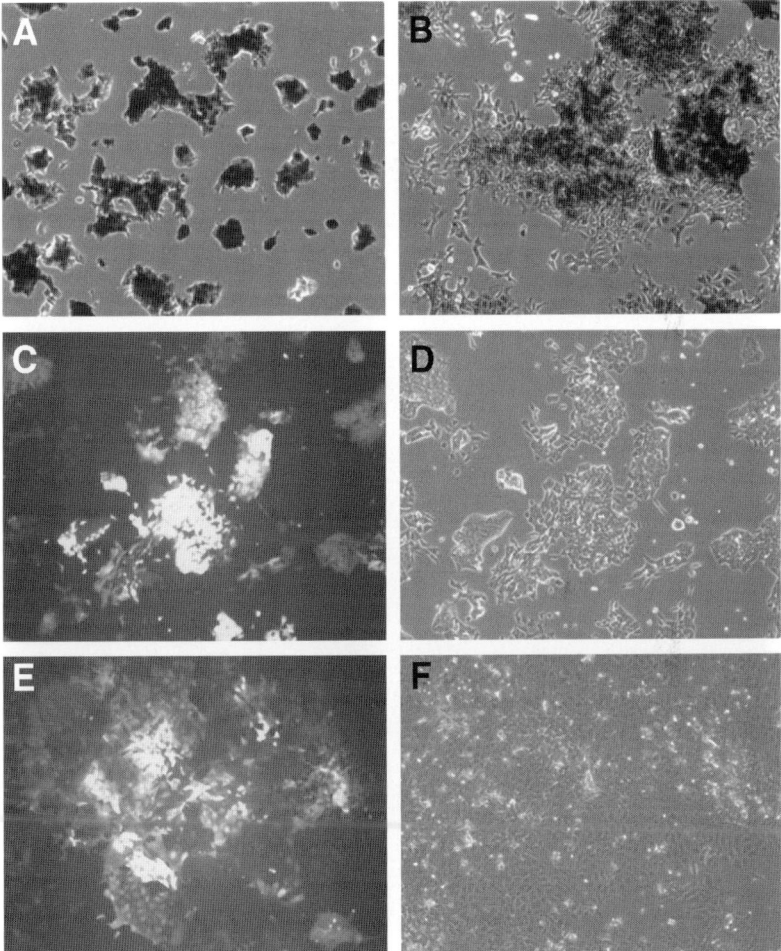

Fig. 3. Undifferentiated R1 embryonic stem (ES) cells maintained in ES-leukemia inhibitory factor (LIF) medium were infected with lentivirus expressing *LacZ* gene (**A,B**) or enhanced green fluorescence protein (EGFP) (**C–F**) under the control of EF1-α promoter (multiplicities of infection of 100). After 2 d, infected ES cells were passaged into a new dish and further maintained in ES-LIF medium for another 7 d (**A**) or 20 d (**C,D**). Infected cells were also induced to differentiate in the absence of LIF for 5 d (**B**) or 7 d (**E,F**). Expression of reporter gene was detected using X-gal staining (**A,B**) or observed under a fluorescence microscope (**C,E**). Phase contrast images are shown for EGFP-expressing cells (**D,F**). Expression of β-galactosidase or EGFP was sustained regardless of the differentiation status of ES cells. (Please *see* the companion CD for the color version of this figure.)

4. Notes

1. For most of the cells, 5 min is sufficient.
2. The transfected cells normally produce lentiviral vectors with titers ranging from 10^7 to 10^8 transducing units per milliliter of culture medium.

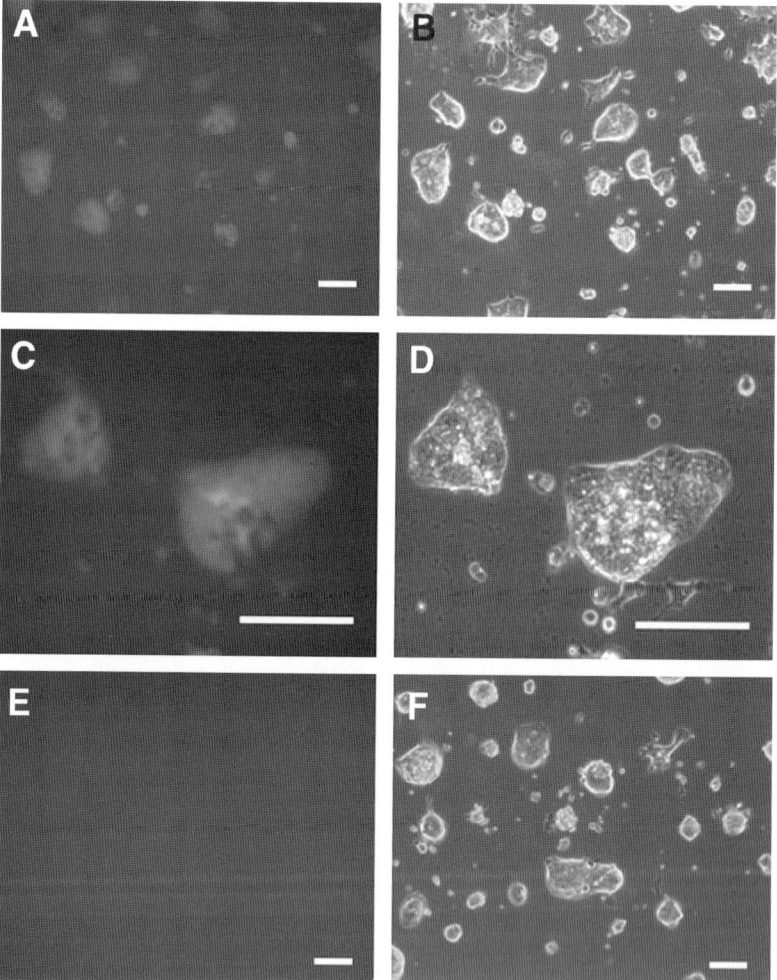

Fig. 4. R26R-EYFP embryonic stem cell, which possesses a *YFP* gene preceded by a loxP-flanked transcriptional termination sequence *(10)*, were infected with lentivirus expressing a nuclear-localized form of Cre recombinase (**A–D**) or control lentivirus (**E,F**) (multiplicities of infection of 100). After 2 d of infection, YFP expression was observed under a fluorescence microscope (**A,C,E**). Phase contrast images of the same field are shown (**B,D,F**). Cre-mediated DNA recombination was monitored by YFP expression under a fluorescence microscope. Highly efficient DNA recombination was observed when Cre recombinases were introduced by lentiviral vector. Bar = 100 μm. (Please *see* the companion CD for the color version of this figure.)

References

1. Gassmann, M., Donoho, G., and Berg, B. (1995) Maintenance of an extrachromosomal plasmid vector in mouse embryonic stem cells. *Proc. Natl. Acad. Sci. USA* **92,** 1292–1296.
2. Camenisch, G., Gruber, M., Donoho, G., Van Sloun, P., Wenger, R. H., and Gassmann, M. (1996) A polyoma-based episomal vector efficiently expresses exogenous genes in mouse embryonic stem cells. *Nucleic Acids Res.* **24,** 3707–3713.

3. Niwa, H., Burdon, T., Chambers, I., and Smith, A. (1998) Self-renewal of pluripotent embryonic stem cells is mediated via activation of STAT3. *Genes Dev.* **12,** 2048–2060.
4. Hamaguchi, I., Woods, N. B., Panagopoulos, I., et al. (2000) Lentivirus vector gene expression during ES cell-derived hematopoietic development in vitro. *J. Virol.* **74,** 10,778–10,784.
5. Pfeifer, A., Ikawa, M., Dayn, Y., and Verma, I. M. (2002) Transgenesis by lentiviral vectors: lack of gene silencing in mammalian embryonic stem cells and preimplantation embryos. *Proc. Natl. Acad. Sci. USA* **99,** 2140–2145.
6. Lois, C., Hong, E. J., Pease, S., Brown, E. J., and Baltimore, D. (2002) Germline transmission and tissue-specific expression of transgenes delivered by lentiviral vectors. *Science* **295,** 868–872.
7. Iwakuma, T., Cui, Y., and Chang, L. J. (1999) Self-inactivating lentiviral vectors with U3 and U5 modifications. *Virology* **261,** 120–132.
8. Zennou, V., Petit, C., Guetard, D., Nerhbass, U., Montagnier, L., and Charneau, P. (2000) HIV-1 genome nuclear import is mediated by a central DNA flap. *Cell* **101,** 173–185.
9. Sirven, A., Pflumio, F., Zennou, V., et al. (2000) The human immunodeficiency virus type-1 central DNA flap is a crucial determinant for lentiviral vector nuclear import and gene transduction of human hematopoietic stem cells. *Blood* **96,** 4103–4110.
10. Srinivas, S., Watanabe, T., Lin, C. S., et al. (2001) Cre reporter strains produced by targeted insertion of EYFP and ECFP into the ROSA26 locus. *BMC Dev. Biol.* **1,** 4.
11. Zaiss, A. K., Son, S., and Chang, L. J. (2002) RNA 3′ readthrough of oncoretrovirus and lentivirus: implications for vector safety and efficacy. *J. Virol.* **76,** 7209–7219.

19

Use of the Cytomegalovirus Promoter for Transient and Stable Transgene Expression in Mouse Embryonic Stem Cells

Katie M. Barrow, Flor M. Perez-Campo, and Christopher M. Ward

Summary

Embryonic stem (ES) cells are pluripotent cells derived from the epiblast of preimplantation embryos. These cells are emerging as a key model system for elucidating mechanisms involved in development and disease as well as having a unique potential as a source of unlimited somatic cells for transplantation therapies. ES cells can be easily manipulated at the DNA level, allowing both transient and stable expression of complementary DNA encoding transgenes of interest. The human cytomegalovirus (CMV) immediate-early enhancer and promoter is commonly used for transient expression of transgenes in ES cells. However, its use in the formation of stable cell lines is less common. We demonstrate an electroporator transformation technique that results in up to 90% transfection efficiency of CMV-encoding vectors in ES cells. Furthermore, we describe the design of vectors and cloning techniques that allow stable expression of transgenes under control of the CMV promoter and a fluorescent microscopy method for detecting protein expression in ES cells *in situ*.

Key Words: Bicistronic vectors; CMV promoter; ES cells; pluripotency; stable gene expression; transgene; transient gene expression.

1. Introduction

Mouse embryonic stem (ES) cells are isolated from preimplantation embryos and can be maintained for prolonged periods in culture in an undifferentiated state. The ability of ES cells to differentiate into all three germ layers provides a unique model to study cell lineage differentiation in vitro *(1)*. ES cells are easily manipulated at the DNA level by transient expression of protein-encoding complementary DNA (cDNA) and the human cytomegalovirus (CMV) immediate-early promoter and enhancer is commonly used for such approaches *(2)*. However, expression of cDNA under control of the CMV promoter is often decreased within several population doublings (**Fig. 1**). Therefore, the CMV promoter is generally not used to establish stable ES cell lines. However, correct vector design can lead to the use of CMV to establish stable ES cell lines, and we describe such a method in this chapter.

From: *Methods in Molecular Biology, vol. 329: Embryonic Stem Cell Protocols: Second Edition: Volume 1*
Edited by: K. Turksen © Humana Press Inc., Totowa, NJ

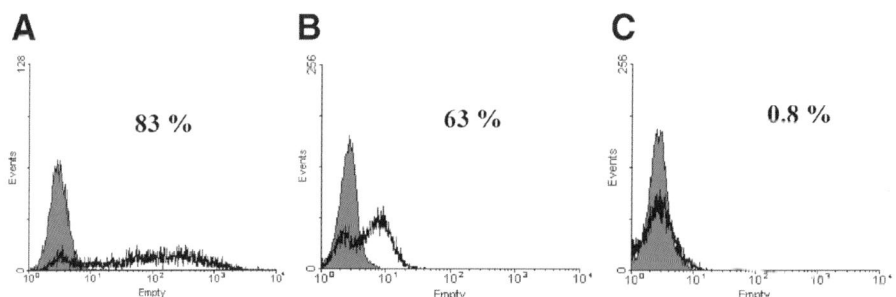

Fig. 1. Transient expression of a cytomegalovirus-enhanced green fluorescence protein (EGFP) vector in E14TG2a embryonic stem (ES) cells. pEGFP-N1 was electroporated into ES cells and assayed for EGFP expression in a Becton Dickenson FACScan after 48 h (**A**). G418-resistant clones (350 µg/mL) were picked after 8-d exposure to the selection agent and cultured as described in **Subheading 3**. After further culture (three passages), the clone exhibited significantly decreased EGFP expression (**B**) that was reduced to background levels after two additional passages (**C**). (Please *see* the companion CD for the color version of this figure.)

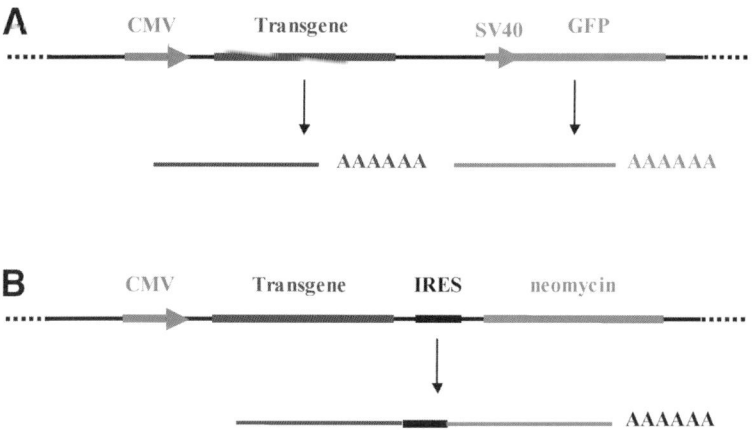

Fig. 2. Example of cytomegalovirus expression vectors for use in (**A**) transient and (**B**) stable transgene expression studies and the formation of subsequent mRNA transcripts. IRES, internal ribosome entry site. (Please *see* the companion CD for the color version of this figure.)

Many CMV-containing vectors are available for transient transfection of ES cells. For example, pCMS-enhanced green fluorescence protein (EGFP) (BD Biosciences, Clontech) allows cloning of cDNA downstream of CMV and isolation of transformants using EGFP expression transcribed via an independent SV40 promoter (**Fig. 2A**). Other vectors are available that allow isolation of transformed cells by antibiotic selection; however, these are not particularly useful for transient expression because the transgene is usually inactive before antibiotic-selected clones can be cultured (**Fig. 1**). Alternatively, transformants can be identified using a dual transfection approach in which the transgene is electroporated with a second vector encoding a fluorescent gene (e.g., pEGFP-N1; BD Biosciences). However, for ease of use and interpretation, it is recommended that single vectors containing a fluorescent marker gene are used for transient expression in ES cells (**Fig. 2**).

Such vectors, however, are not useful for stable expression of transgenes because the construct gives no growth advantage to transformed cells. Indeed, it is likely that the vector represents a significant growth disadvantage, which may explain the rapid decrease in transgene expression observed in ES cell cultures (**Fig. 1**). Thus, to form stable ES cell lines it is necessary to give transformed cells a growth advantage over untransformed cells. This can be achieved using a CMV promoter encoding bicistronic messenger (m)RNA from which both the transgene and an antibiotic resistance gene are transcribed (e.g., pIRESneo, pIREShyg; BD Biosciences) (*see* **Fig. 2** for basic illustration of vectors recommended for transient and stable gene expression). Addition of antibiotic to the medium results in survival of only those cells maintaining expression of the antibiotic-resistance gene; as a result, transgene expression is maintained.

Before embarking on the construction of an engineered ES cell line, it is highly advisable to ensure that the parental line is capable of germline transmission. Many mouse ES cell lines are available, but they vary considerably in their germline transmission abilities *(3)*. Furthermore, ES cell lines exhibit significant variations in transcript expression and lineage differentiation properties when cultured under identical conditions in vitro *(3)*. Thus, if the ES cells are to be utilized in differentiation studies, then it is good practice to assess several cell lines for their efficiency of lineage formation under appropriate conditions.

2. Materials
2.1. Tissue Culture

1. 10X phosphate-buffered saline (PBS) at pH 7.4 (Invitrogen, Paisley, UK; cat. no. 70011-044) diluted to 1X with double-distilled water (ddH$_2$O).
2. Gelatin from porcine skin (type A cell culture tested; Sigma, Dorset, UK; cat. no. G1890). Make 0.1% solution in ddH$_2$O, autoclave to dissolve, and store at 4°C. Use within 1 mo.
3. Dimethyl sulfoxide (500 mL; BDH Laboratory Supplies, Surrey, UK; cat. no. 103234L).
4. Fetal bovine serum (FBS). FBS for use with ES cells must be screened prior to use to ensure low toxicity and ability to maintain the undifferentiated state of the cells *(1)*. Alternatively, serum can be purchased prescreened from serum suppliers. We have used Hyclone ES screened serum (Fisher Scientific, Loughborough, UK; cat. no. SH3007003E) to successfully maintain germline competence in the E14TG2a ES cell line. FBS should be heat inactivated before use at 56°C for 30 min and then aliquoted into sterile containers under aseptic conditions. Store at −20°C for future use.
5. Knockout™ DMEM (500 mL; Invitrogen, cat. no. 10829-018).
6. 100X nonessential amino acids (NEAA) solution (Invitrogen, cat. no. 11140-050).
7. Nucleoside solution: dissolve the following in 100 mL ddH$_2$O at 37°C, filter-sterilize through 0.2-μm filter, and store at 4°C for a maximum of 1 mo: 80 mg adenosine (Sigma, cat. no. A4036), 85 mg guanosine (Sigma, cat. no. G6264), 73 mg cytidine (Sigma, cat. no. C4654), 73 mg uridine (Sigma, cat. no. U3003), and 24 mg thymidine (Sigma, cat. no. T1895). Heat to 37°C prior to each use.

8. L-Glutamine (100 mL; Sigma, cat. no. G7513). This solution should be stored at −20°C in 5-mL aliquots.
9. 1000X 2-mercaptoethanol (Invitrogen, cat. no. 21985-023).
10. 1X trypsin-EDTA (ethylenediaminetetraacetic acid): 0.05% trypsin, 0.02% EDTA·4Na (Sigma, cat. no. T3924).
11. ESGRO (leukemia inhibitory factor [LIF]) (10^7 U; Chemicon International, Temecula, CA; cat. no. ESG1107).
12. Colcemid (Invitrogen, cat. no. 15212-046).
13. Tissue culture dishes (100 mm; Nunc, Hereford, UK; cat. no. 172958).
14. 96-, 24-, and 6-well tissue culture plates (Nunc).
15. 0.2-µm syringe filters.
16. 1.8-mL cryotubes (Scientific Laboratory Supplies, Manchester, UK; cat. no. 368632K).
17. Electroporator (Gene Pulser II; Bio-Rad, Hemel Hempstead, UK).
18. 4-mm electroporation cuvets (Bio-Rad, cat. no. 165-2088).

2.1.1. ES Cell Media and Culture

1. ES cells are maintained in Knockout DMEM™ supplemented with heat-inactivated FBS, NEAA, nucleoside solution, L-glutamine, 2-mercaptoethanol, and LIF (ESGRO). For 550 mL medium, add 50 mL heat-inactivated FBS, 6 mL 100X NEAA (1% v/v), 6 mL nucleoside solution (1% v/v), 6 mL L-glutamine (1% v/v), 0.6 mL 1000X 2-mercaptoethanol, and 0.06 mL ESGRO (10^7 U/mL) to a 500-mL bottle of Knockout DMEM (*see* **Note 1**).
2. Freezing medium: for 10 mL, mix 1 mL dimethyl sulfoxide with 9 mL ES medium.
3. 37°C water bath.
4. Hemocytometer.
5. Plastic disposable 10- and 25-mL pipets.
6. Humidified incubator at 37°C/7.5% CO_2.
7. Inverted phase contrast microscope with ×10 and ×25 objectives.
8. Class II laminar flow cabinet.
9. Liquid nitrogen storage facilities.
10. 10-, 20-, 200-, and 1000-µL pipets and tips designated for tissue culture use only.
11. Refrigerator (4°C) and freezers (−20 and −80°C).
12. Centrifuge suitable for 20-mL universal containers.

2.2. Karyotyping of ES Cells

1. Colcemid (*see* **Subheading 2.1., item 12**).
2. Trypsin-EDTA solution (*see* **Subheading 2.1., item 10**).
3. Potassium chloride (0.56% w/v).
4. ddH_2O.
5. Absolute methanol (Sigma, cat. no. 534013).
6. Glacial acetic acid (Sigma, cat. no. A6283).
7. Fixative: 3:1 absolute methanol/glacial acetic acid.
8. Pasteur pipets.
9. 15-mL centrifuge tubes.
10. Giemsa (500 mL; BDH Laboratory Supplies; cat. no. 350864X).

2.3. Plasmid Propagation and Purification

1. One-Shot Top 10F' chemically competent *Escherichia coli* (Invitrogen, cat. no. C3030-03).
2. Selection antibiotic (e.g., ampicillin, kanamycin).

3. Tryptone.
4. Yeast extract.
5. Sodium chloride.
6. For 500-mL LB broth, add 5 g tryptone (1% w/v), 2.5 g yeast extract (0.5% w/v), and 2.5 g sodium chloride (0.5% w/v) in ddH$_2$O. This solution must be sterilized before use and selection antibiotic added just prior to use (see **Note 2**).
7. Agar.
8. 10-cm^2 plastic bacteriological Petri dishes.
9. LB agar plates: prepare in advance by addition of 2 g agar to 100 mL LB broth. LB agar will be solid at room temperature and must be melted (microwave or water bath) before plates can be poured. Add selection antibiotic when LB agar is at hand temperature and then pour roughly 10 mL into each 10-cm plate using aseptic technique (see **Note 3**). Plates should be left to set at room temperature and thereafter stored inverted at 4°C.
10. Qiagen plasmid maxi kit (Qiagen, Crawley, UK; cat. no. 12163).

2.4. Fluorescent-Activated Cell Sorting

1. 1X PBS (see **Subheading 2.1.**, **item 1**).
2. Bovine serum albumin (BSA) (100 g; Sigma, cat. no. A-2153).
3. Sodium azide (100 g; Sigma, cat. no. S-8032).
4. Fluorescent-activated cell sorting (FACS) buffer: this is used as the antibody diluent and for all washes. For 500 mL FACS buffer, add 1 g BSA (0.2% w/v) and 0.5 g sodium azide (0.1% w/v) to 500 mL PBS. Store at 4°C.
5. Formaldehyde (Sigma).
6. Cell fixative: 1% v/v solution of formaldehyde in PBS.
7. 96-well V-bottom plate (Alpha Laboratories, Eastleigh, UK; cat. no. FX9100S).
8. FACS machine (FACScan; Becton Dickinson, CA).
9. Antibodies (various).
10. 96-well plate centrifuge (New Brunswick Scientific, St. Albans, UK).

2.5. Immunofluorescence of ES Cells

1. 1X PBS (see **Subheading 2.1.**, **item 1**).
2. Paraformaldehyde (500 g; Sigma, cat. no. P-6148).
3. 4% (w/v) paraformaldehyde: dissolve 4 g paraformaldehyde in 100 mL PBS by heating at 60°C with occasional agitation for 3 h or until dissolved. This can be stored at 4°C for 1 mo.
4. ImmEdge pen™ (Vector Laboratories, Peterborough, UK; cat. no. H4000).
5. Antibodies (various).
6. Normal goat serum (10 mL; Dako, Ely, UK; cat. no. X0907).
7. Triton X-100 (Sigma).
8. BSA (100 g; Sigma; see **Subheading 2.4.**, **item 2**).
9. Blocking buffer: blocking buffer is used as the antibody diluent. For 20 mL, add 200 µL normal goat serum (1% v/v) and 20 µL Triton X-100 (0.1% v/v) to 20 mL 0.1% (w/v) BSA (0.5 g BSA in 500 mL PBS; store at 4°C). Filter-sterilize (0.2 µm) and store at 4°C for up to 1 mo.
10. Hot knife (handicraft shop).
11. 15-mm diameter cover slips (Fisher Scientific, cat. no. 15-183-85).
12. Nail polish for sealing.
13. Vectashield mountant containing 10 mL 4',6-diamidino-2-phenylindole (DAPI; Vector Laboratories, cat. no. H-1200).

3. Methods
3.1. ES Cell Culture
3.1.1. Thawing Cells
1. Remove cryovial from liquid nitrogen and leave at room temperature for 1 min (*see* **Note 4**).
2. Thaw vial at 37°C in a water bath.
3. Transfer the contents of the cryotube to 10 mL medium in a 20-mL screw-capped bottle and then pellet the cells by centrifugation.
4. Gently resuspend the cell pellet in the required volume of medium (9 mL for a 10-cm diameter dish) and transfer to a gelatin-treated tissue culture plate (*see* **Subheading 3.1.2.**).

3.1.2. Preparation of Gelatin-Treated Tissue Culture Plates
1. Prepare a 0.1% (w/v) solution of gelatin in 1X ddH_2O and sterilize by autoclaving (*see* **Note 5**).
2. In a laminar flow cabinet, add enough of the 0.1% gelatin solution to cover the bottom of the plate or flask (for a 10-cm^2 dish, this is typically about 6–7 mL).
3. Store dishes containing this solution overnight at 4°C or at 37°C for 1 h.
4. Remove the gelatin solution from the plates and air-dry in a laminar flow cabinet until all solution has evaporated.
5. Store the plates at 4°C for a maximum of 1 mo.

3.1.3. Maintenance of ES Cells
To subculture ES cells:

1. Aspirate the medium.
2. Gently wash the cells twice with 1X PBS.
3. Add 4 mL trypsin-EDTA to a 10-cm^2 dish, remove after 10 s, and place in a tissue culture incubator (37°C and 7.5% CO_2).
4. After 60 s, remove the plate and gently tap to detach the cells.
5. Resuspend cells in 3 mL medium and transfer 0.5 mL of this to a fresh gelatin-treated tissue culture dish (approx 3×10^6 cells) containing 8 mL medium, making a 1:6 dilution of the original culture.
6. Gently move the plate from side to side to ensure an even distribution of the cells and place in a tissue culture incubator at 37°C and 7.5% CO_2.
7. Observe ES cells daily and passage at approx 70% confluency, typically every 2 d (*see* **Note 6**).

3.1.4. Freezing Cells (see **Note 7**)
1. Trypsinize cells from an approx 70% confluent 10-cm dish (*see* **Subheading 3.1.3., steps 1–4**).
2. Pellet cells by centrifugation.
3. Resuspend cell pellet in 1 mL freezing medium and aliquot 250 µL into individual cryotubes.
4. To freeze cells slowly, wrap vials in several layers of paper towels, taking care to keep the tubes upright.
5. The wrapped vials should be stored upright overnight at $-70°C$ and transferred to liquid nitrogen the following day for long-term storage.

3.2. Plasmid Propagation and Purification
We routinely use One Shot Top 10F' chemically competent cells from Invitrogen Corporation for plasmid propagation according to the manufacturer's protocol.

1. Add plasmid DNA directly to a vial of thawed competent cells and mix. Do not pipet up and down.
 2. Incubate cells on ice for 30 min.
 3. Heat shock the cells at 42°C in a water bath for exactly 30 s and immediately return to ice.
 4. Add 250 µL prewarmed SOC medium (supplied with cells) to each vial.
 5. Incubate vials at 37°C for 1 h in a rotary shaker (225 rpm).
 6. After incubation, plate out three different volumes (5, 20, and 40 µL) transformed cells onto pre-prepared LB agar plates with selection antibiotic and spread the cells over the agar surface.
 7. Invert plates and incubate at 37°C overnight.
 8. The following day, select four individual colonies to bulk up. Pick these colonies by stabbing with a yellow pipet tip and place the tip into a 15-mL conical tube containing 5 mL LB broth with appropriate selection antibiotic.
 9. Incubate tubes at 37°C on a rotary shaker at 225 rpm for approx 8 h (*see* **Note 8**.) After incubation, the solution should be cloudy because of bacterial growth (*see* **Note 9**).
 10. Pipet 100 µL of this subculture into 100 mL LB broth with antibiotic in a 250-mL conical flask and incubate overnight at 37°C on a rotary shaker at 225 rpm (*see* **Note 10**).
 11. Follow the manufacturer's protocol for plasmid purification.
 12. Sequence or analyze samples by restriction digest to ensure that the clone is faithful to the original sequence (*see* **Note 11**).

3.3. Transfection of ES Cells (see Note 12)

There are several methods of transfection that may be used; however, we have found electroporation to be the most efficient with our cell lines and growth conditions. We routinely transfect pEGFP-N1 or pdsRed2-N1 into ES cells to assess transfection efficiency. In our experience, it is essential to split the cells the day before electroporation to ensure a rapidly dividing population (*see* **Subheading 3.1.3.**).

3.3.1. Transient Expression of Transgenes Under the Control of the CMV Promoter

 1. Culture cells as described in **Subheading 3.1.** and split 1 d prior to electroporation.
 2. Place required number of cuvets on ice until required.
 3. Trypsinize the cells, collect in media, and pellet by centrifugation.
 4. Wash the pellet twice in PBS and resuspend in PBS to give a final concentration of 1×10^7 cells/mL.
 5. Aliquot 800 µL cell suspension into each 4-mm cuvet.
 6. Add 20 µg DNA to each cuvet and mix by pipetting up and down a few times. If cotransfecting a gene of interest and fluorescent marker protein (FP), then the 20 µg can be in the ratio of 15 µg gene of interest, 5 µg FP. However, this should be optimized for individual cell lines and vectors.
 7. Electroporate each cuvet with one 250-V pulse, 475 µF, and place the cuvet on ice for 15 min.
 8. Transfer the cell suspension into a gelatin-treated 10-cm dish, add 9 mL medium, and place in an incubator at 37°C at 7.5% CO_2.
 9. Assay for transgene expression using FACS analysis (*see* **Subheading 3.4.**). Expression of the transgene should be maintained for at least 3 d (**Fig. 1**).

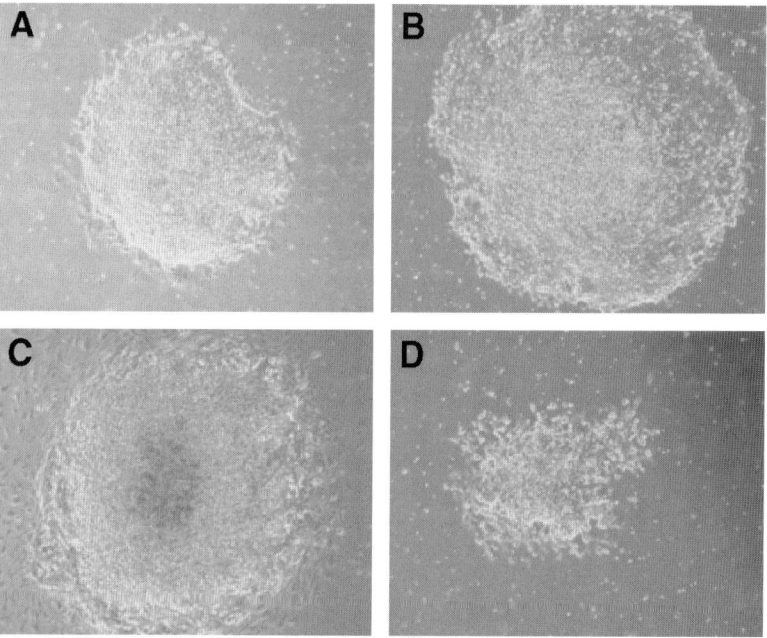

Fig. 3. A range of embryonic stem (ES) cell colonies obtained following selection of E14TG2a ES cells transfected with pEGFP-N1 in 350 µg/mL G418. (**A–C**) Ideal colonies that should be picked. Note that colony C contains differentiated cells at the periphery, but the main colony is comprised of undifferentiated cells. The last will rapidly outgrow the differentiated cells on replating. (**D**) Differentiated or slow-growing colonies such as this should be avoided.

3.3.2. Stable Expression of Transgenes Under the Control of the CMV Promoter

Because CMV-encoded transgenes are rapidly downregulated in ES cells, it is advisable to use bicistronic mRNA vectors encoding both the gene of interest and an antibiotic resistance gene.

1. Carry out the method as described for transient transfection in **Subheading 3.3.1., steps 1–8**.
2. After 48 h, add selection antibiotic: G418 at 350 µg/mL; hygromycin at 150 µg/mL; or puromycin at 1 µg/mL. Significant cell death will occur. After approx 8–10 d, individual colonies will be apparent.
3. When the majority of the colonies reach about 2-mm diameter (**Fig. 3**), pick each colony (using a microscope in a laminar flow hood) and place in 20 µL EDTA/trypsin solution for 1 min.
4. Add 80 µL culture medium, pipet up and down to disaggregate the colony, and transfer to a fresh gelatin-treated 96-well plate.
5. When the cultures are almost confluent, split 1:6 to five fresh gelatin-treated 96-well plates, and FACS analyze (*see* **Subheading 3.4.**) the remaining cells to identify colonies expressing the transgene (**Fig. 1**).

6. When the cultures are nearly confluent, freeze cells from four of the plates to ensure a stock of early passage cells.
7. Some clones will lose expression of the transgene; others will maintain expression. Discard colonies that lose expression.
8. Expand the clones by transfer into 48-, then 24-, and then 6-well plates.
9. Karyotype the cells to ensure the presence of 40 chromosomes (*see* **Subheading 3.6.**).

3.4. Assaying Transgene Expression and Selection of Cell Phenotypes Using FACS

FACS analysis is a quick and reliable method to detect levels of transgene expression within individual cells and to isolate appropriate cellular phenotypes following transgene manipulation. Cells can be labeled using an antibody to the protein of interest encoded by the transgene, which is then detected with a fluorophore-conjugated secondary antibody. Alternatively, transgene-expressing cells can be identified solely by fluorescent marker gene expression (**Fig. 2**).

For detection of transgene expression in ES cells:

1. Trypsinize cells as described in **Subheading 3.1.3.**, collect in media, and pellet by centrifugation.
2. Wash cells twice in PBS and resuspend in required volume of FACS buffer to give a final concentration of 1×10^7 cells/mL.
3. Aliquot 100 µL cells into a well of a 96-well V-bottom plate. An antibody isotype control is essential, so at least two wells will be required per sample to be analyzed.
4. Pellet cells in a benchtop plate centrifuge.
5. Resuspend the pellet in 100 µL antibody diluted to the required concentration in FACS buffer. Incubate plate on ice for 30 min (*see* **Note 13**).
6. Repeat **step 4** and wash the cells three times in 150 µL FACS buffer.
7. Incubate the cells with 100 µL fluorophore-conjugated secondary antibody (diluted in FACS buffer) on ice in the dark for 30 min (*see* **Note 14**).
8. Wash the cells twice with 150 µL FACS buffer.
9. Resuspend the cell pellet in fixative and transfer cells to a 5-mL tube containing 200 µL 1% formaldehyde. Cells can be stored at 4°C in the dark for up to 1 wk prior to analysis, although rapid analysis is recommended.

To detect transgene expression by expression of fluorescent protein rather than by antibody labeling, cells should be trypsinized, washed in PBS, and then fixed and stored in 1% formaldehyde.

3.5. Immunofluorescence (in Tissue Culture Plates)

Mouse ES cells are difficult to grow on glass cover slips because they do not adhere sufficiently. We have developed a simple technique to allow the labeling and visualization of ES cells on plastic gelatin-treated tissue culture dishes (for example, *see* **Fig. 4**).

1. Aspirate medium from cells and wash twice in 1X PBS.
2. Add 5 mL 4% paraformaldehyde to a 10-mm dish (ensuring cells are covered) for 15 min to fix the cells (*see* **Note 15**).
3. Remove paraformaldehyde and wash cells three times in PBS, taking care not to wash cells from the plate. Fixed dishes can be stored in PBS at 4°C for up to 4 wk.

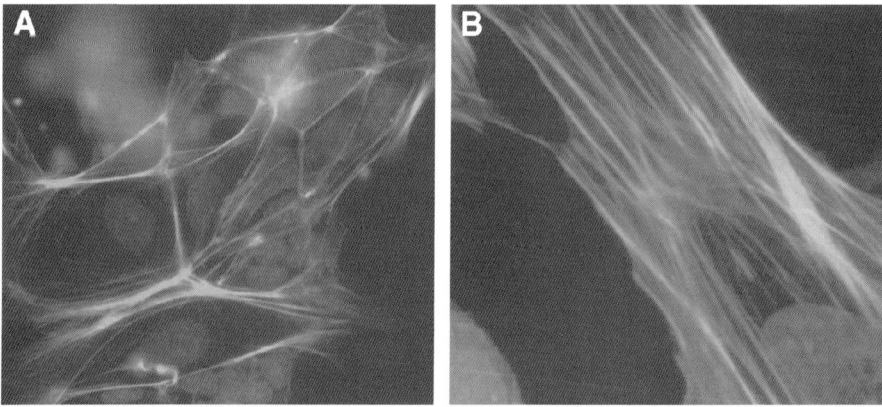

Fig. 4. Example of embryonic stem cell immunofluorescence on plastic tissue culture dishes. Actin filaments of undifferentiated cells were labeled with directly conjugated rhodamine-phalloidin. Cells were then mounted in Vectashield containing DAPI to label nuclei and visualized using an Olympus BX-51 microscope at (**A**) ×400 and (**B**) ×1000 magnification.

4. Remove PBS and score around some cells in the center of the dish with a cotton bud to form an isolated circle of cells approx 2 cm in diameter (*see* **Note 16**).
5. Circumscribe these cell circles with the ImmEdge pen to create a hydrophobic barrier (*see* **Note 17**).
6. Add blocking buffer for 30 min at room temperature (*see* **Note 18**).
7. Add primary antibody diluted in blocking buffer and incubate for 1–2 h at room temperature (*see* **Note 19**).
8. Wash cells three times in blocking buffer for 5 min each.
9. Add secondary antibody containing conjugated fluorophore (e.g., goat antirabbit Alexa Fluor 488 at 1:500) in blocking buffer and incubate at room temperature for 1 h in the dark. For cells expressing green fluorescent protein, 546-nm AlexaFluors can be used; for RFP-expressing cells, use 488-nm AlexaFluors (*see* **Note 20**).
10. Wash cells twice in blocking buffer for 5 min each, then twice for 15 min, followed by twice for 5 min in the dark. Remove all blocking buffer.
11. Add two drops of Vectashield mountant containing DAPI (labels nuclei blue under appropriate excitation wavelength) to each cell circle.
12. Place a cover slip on the cells and apply gentle pressure, taking care not to damage the cells. Remove excess mountant from the edges of the cover slip with a tissue.
13. Seal the cover slip in place with nail polish and leave to set.
14. Excise an area containing the labeled circles of cells in the shape of a standard microscope slide (this can be marked on the base of the dish before **step 4** for ease) using the hot knife.
15. Plastic slides with cover slipped cells can be stored at 4°C in the dark for up to 1 wk (*see* **Note 21**).

3.6. Karyotyping

Transfected cells must be karyotypically assayed to ensure that they have maintained the correct number of chromosomes. Cell numbers required for this method range from 5×10^6 to 10^7 and at the minimum should be 50% confluent, ideally 70% confluent.

1. Add colcemid at 0.02 µg/mL in tissue culture medium to the cells.
2. Incubate for 2 h at 37°C and 7.5% CO_2.
3. Trypsinize the treated cells and collect in 15-mL centrifuge tubes in PBS.
4. Pellet cells by centrifugation.
5. Remove the supernatant and disrupt cell pellet.
6. Using a plastic Pasteur pipet, add 1 mL KCl drop by drop and then an excess up to 6 mL. Invert the tube several times to ensure thorough mixing.
7. Leave for 10 min for cells to swell.
8. Pellet by centrifugation.
9. Remove the supernatant and disrupt the pellet.
10. Add ice-cold fixative a drop at a time and mix thoroughly to prevent cell clumping.
11. Add the fixative to give a final volume of 6 mL and leave for 5 min at room temperature.
12. Pellet the cells by centrifugation.
13. Repeat **steps 9–12** an additional three times, leaving a final volume of 1 mL of cells in fixative.
14. Spread cells onto a slide. Prepare four or five spreads per specimen and stain overnight with Giemsa (*see* **Note 22**).

4. Notes

1. All work should be carried out in a laminar flow class II cabinet using disposable sterile plasticware. All media should be warmed to 37°C prior to use, with the exception of the trypsin/EDTA solution, which should be at room temperature.
2. Heating inactivates the antibiotic, and efficiency is rapidly lost on storage in dilute form at 4°C. For this reason, store 50-mg/mL aliquots of antibiotic at −20°C; defrost and add to LB broth when required.
3. This must be done quickly because LB agar can rapidly resolidify.
4. Cell thawing must be performed quickly to minimize cell death. Care must be taken when removing vials from liquid nitrogen as these can explode when exposed to temperature changes. Always wear appropriate safety wear (including goggles) when handling cryovials.
5. The gelatin will not dissolve until the solution is autoclaved.
6. It is possible to split ES cells grown in medium containing 10% ES cell screened FBS at less than 1:6. However, this is cell line and serum dependent, so care must be taken when determining the optimum dilution for subculture of the user's own cell line. Undifferentiated ES cells form colonies with few spreading cells, although the exact morphology is cell line and serum dependent.
7. To maintain cell integrity, cells should be frozen slowly in medium containing a cryopreservant.
8. Start in the morning so the culture is ready by late afternoon.
9. If the culture is not cloudy, then something has gone wrong. A common mistake is use of the wrong selection agent (i.e., using ampicillin instead of kanamycin).
10. **Notes 9** and **10** are detailed in the protocol provided with the Qiagen maxiprep kit.
11. For detailed methods on DNA cloning, refer to **ref. 4**. *Molecular Cloning: A Laboratory Manual* (Samrook, Fritsch and Maniatis: Cold Spring Harbour Press).
12. DNA prepared with a Qiagen midi/maxiprep kit does not require any further treatment prior to transfection into ES cells.
13. If the protein of interest is not cell surface, then it may be necessary to include a permeabilization step (10 min in 0.1% Triton X-100 diluted in FACS buffer) prior to incubation with primary antibody.

14. Not all fluorophores can be detected by a FACScan.
15. Other fixatives can also be used if the antibody required is not compatible with paraformaldehyde. Acetone degrades plastic and must never be used as a fixative with this technique.
16. Cells within the circles must not dry out. Up to three circles can be made on the same plate.
17. This ensures that solutions added to the cells do not spread over the entire dish.
18. Triton X-100 can destroy some epitopes.
19. Primary antibody can also be left on the cells overnight. Optimal antibody concentration and incubation times should be determined by the end user.
20. AlexaFluors (Molecular Probes Inc.) are preferred to other fluorescein-conjugated secondary antibodies because of their increased stability and resistance to photobleaching.
21. We use DAPI to label cell nuclei for simplicity; however, this can also be achieved using standard phase images, so DAPI is not essential. Care must be taken to ensure that the fluorophores selected as secondary antibodies will emit at a wavelength that can be detected on the microscope to be used.
22. It is best to prepare the spreads the same day. If not, store fixed cells at $-20°C$ and change the fix immediately before preparing spreads.

References

1. Smith, A. G. (2001) Embryo-derived stem cells: of mice and men. *Annu. Rev. Cell Dev. Biol.* **17,** 435–462.
2. Ward, C. M. and Stern, P. L. (2002) The human cytomegalovirus immediate-early promoter is transcriptionally active in undifferentiated mouse embryonic stem cells. *Stem Cells* **20,** 472–475.
3. Ward, C. M., Barrow, K. M., and Stern, P. L. (2004) Significant variations in differentiation properties between independent mouse ES cell lines cultured under defined conditions. *Exp. Cell Res.* **293,** 229–238.
4. Sambrook, J., Fritsch, E. F., and Maniatis, T. (1989) *Molecular Cloning: A Laboratory Manual,* Cold Spring Harbour Press, Cold Spring Harbour, New York.

20

Use of Simian Immunodeficiency Virus Vectors for Simian Embryonic Stem Cells

Takayuki Asano, Hiroaki Shibata, and Yutaka Hanazono

Summary

The ability to stably introduce genetic material into primate embryonic stem (ES) cells could allow broader application. In this chapter, we describe a method of gene transfer into simian (cynomolgus macaque) ES cells using a simian immunodeficiency virus-based lentivirus vector. When cynomolgus ES cells are transduced with a simian immunodeficiency virus vector encoding the green fluorescent protein (GFP) gene, a large fraction of cells (greater than 50%) fluoresce, and high levels of GFP expression persist for months as assessed by flow cytometry and real-time polymerase chain reaction. Thus, the use of GFP as a reporter gene allows direct and simple detection of successfully transduced ES cells and facilitates monitoring of ES cell proliferation and differentiation both in vitro and in vivo. In addition, this highly efficient gene transfer method allows faithful gene delivery to primate ES cells with potential for both research and therapeutic applications.

Key Words: Flow cytometry; gene transfer; green fluorescent protein; lentivirus vector; primate embryonic stem cells; real-time PCR; simian immunodeficiency virus vector.

1. Introduction

Nonhuman primate embryonic stem (ES) cells have remarkable similarities to human ES cells in all aspects, including morphology and surface marker expression. On the other hand, primate (both human and nonhuman) ES cells are quite distinct from mouse ES cells, for instance, in their growth velocity, feeder and leukemia inhibitory factor (LIF) dependency, and their morphology and surface marker expression. Therefore, experimental results using mouse ES cells may not be predictive of those in primates. These discrepancies stimulated us to use nonhuman primate (simian) ES cells as a predictive model to more closely reflect human ES cell characteristics and behavior *(1,2)*.

The lentivirus vector was first established from human immunodeficiency virus (HIV)-1 *(3)*. It can transduce quiescent cells such as neurons and hematopoietic stem cells *(3,4)*. Non-HIV lentivirus vectors have also been established by modifying feline

From: *Methods in Molecular Biology, vol. 329: Embryonic Stem Cell Protocols: Second Edition: Volume 1*
Edited by: K. Turksen © Humana Press Inc., Totowa, NJ

immunodeficiency virus, equine infectious anemia virus, simian immunodeficiency virus (SIV), or bovine immunodeficiency virus *(5–9)*. Among primate lentivirus vectors, the merit of SIV vectors over HIV-1 vectors is safety. The sequence homology between HIV-1 and SIV is considerably low (approx 50%) *(10)*. The generation of replication-competent virus by recombination between SIV vectors and HIV-1 in human subjects is therefore highly unlikely. This provides a great advantage in safety over HIV vectors, especially when target cells are already infected with HIV or permissive to HIV infection.

HIV-1-based lentivirus vectors can efficiently transduce human cells but not those of Old World monkeys *(11)*. A species-specific cytoplasmic component confers the innate postentry restriction to HIV-1 infection in simian cells *(12)*. Unlike HIV-1 vectors, SIV vectors can efficiently transduce simian embryonic and hematopoietic stem cells *(13,14)*. In this chapter, we describe a method to use a SIV-based lentivirus vector for efficient gene transfer into simian (cynomolgus macaque) ES cells.

2. Materials

2.1. Cells

1. Simian (rhesus or cynomolgus) ES cells *(1,2)*.
2. Mouse embryonic fibroblasts (MEFs) from CD-1 (also referred to as ICR [Institute of Cancer Research]) (Charles River, Wilmington, MA) or BALB/c mice (Charles River).
3. 293T human embryonic kidney cell line (ATCC, Manassas, VA; cat. no. 11268).

2.2. Culture Media and Reagents

1. Dulbecco's modified Eagle's medium (DMEM) (Sigma, St. Louis, MO; cat. no. D-6429).
2. DMEM nutrient mixture F-12 1:1 mixture (DMEM/F12) (Invitrogen, Carlsbad, CA; cat. no. 11330-032).
3. ES cell-qualified fetal bovine serum (FBS; Invitrogen, cat. no. 10439-024).
4. 10,000 IU/mL penicillin-10,000 µg/mL streptomycin (100X; Invitrogen, cat. no. 15070-063).
5. 200 mM L-glutamine (100X; Invitrogen, cat. no. 25030-081).
6. 2-Mercaptoethanol (Sigma, cat. no. M3148).
7. FBS (Sigma, cat. no. F-2442).
8. Phosphate-buffered saline (PBS) (Invitrogen, cat. no. 10010-023).
9. Hanks balanced salt solution (HBSS) (Invitrogen, cat. no. 14025-092).
10. 0.25% trypsin-ethylenediaminetetraacetic acid (Invitrogen, cat. no. 25200-056).
11. 2.5% trypsin (Invitrogen, cat. no. 15090-046).
12. Polybrene (Sigma, cat. no. S2667).
13. Culture medium for primate ES cells: DMEM/F12 containing 15% ES cell-qualified FBS, 2 mM L-glutamine, 100 IU/mL penicillin-100 µg/mL streptomycin, and 0.1 mM 2-mercaptoethanol.
14. Culture medium for 293T cells: DMEM containing 10% FBS and 100 IU/mL penicillin-100 µg/mL streptomycin.
15. Post-transfection medium: DMEM containing 20% FBS.

2.3. SIV Vectors

1. pVSV-G (sold as a part of the pantropic retroviral expression system; BD Biosciences Clontech, San Jose, CA; cat. no. 631512 and 631530).

… SIV Vectors for Simian ES Cells

2. SIV packaging plasmid and SIV gene transfer plasmid (for plasmid construction, *see* **ref. 7**).
3. Lipofectamine reagent (Invitrogen, cat. no. 18324-111).
4. Plus reagent (Invitrogen, cat. no. 11514-015).
5. Opti-MEM (Invitrogen, cat. no. 11058-021).
6. Stericup filters (Millipore, Billerica, MA; cat. no. SCHV U01RE).

2.4. Flow Cytometry

1. A flow cytometer equipped with an argon-ion laser (Becton Dickinson FACScan, FACS Caliber, or an equivalent).
2. Cell strainers (BD Falcon, San Jose, CA; cat. no. 352350).
3. Round-bottom test tubes with cell strainer caps (BD Falcon, cat. no. 352235).
4. Fluorescent-activated cell sorting (FACS) medium: 2% FBS and 0.1% NaN_3 (Wako, Osaka, Japan; cat. no. 197-11091) in PBS.
5. Fixing medium: 1% paraformaldehyde (Wako, cat. no. 064-00406) in PBS.
6. Phycoerythrin (PE)-conjugated antimouse-H-$2K^d$ monoclonal antibody (BD PharMingen, San Jose, CA; cat. no. 553566).

2.5. Real-Time Polymerase Chain Reaction

1. A real-time thermal cycler (ABI-PRISM 7000 sequence detection system or an equivalent).
2. A QIAamp DNA minikit (Qiagen, Hilden, Germany; cat. no. 51104).
3. A Quantitect SYBR green polymerase chain reaction (PCR) kit (Qiagen, cat. no. 204143).
4. MicroAmp optical 96-well reaction plates (Applied Biosystems, Foster City, CA; cat. no. N801-0560) and MicroAmp caps (Applied Biosystems, cat. no. N801-0535).
5. A spectrophotometer (Beckman Coulter DU 7500 or an equivalent).

3. Methods
3.1. Construction of SIV Vector

We have used the SIV vector derived from SIV African green monkey (SIVagm) *(7)* to transduce simian ES cells. SIV vectors can transduce simian ES cells more efficiently than adenovirus, adeno-associated virus, or oncoretrovirus vectors *(13)*. In addition, SIV vectors can efficiently transduce nondividing cells, for instance, the ocular tissue and adipocytes *(15,16)*.

Instead of depending on specific SIV entry via CD4 and other co-receptors, the vesicular stomatitis virus (VSV)-G envelope has generally been used to pseudotype SIV vectors. Because the cellular receptors for VSV-G, including phosphatidylserine, phosphatidylinositol, and GM3 ganglioside, appear to be very abundant and ubiquitous membrane components of most mammalian cells, VSV-G-enveloped viruses can infect a wide variety of cells and tissues. In addition to the broader range, VSV-G-pseudo-typed viruses are physically more stable than naturally occurring lentiviruses and can be concentrated by centrifugation (*see* **Subheading 3.1.2.**).

3.1.1. Transfection

1. Dissociate exponentially growing 293T cells with 0.25% trypsin-ethylenediaminetetra-acetic acid solution and plate 5×10^6 293T cells in a 100-mm plate (60–80% confluent) 1 d prior to transfection (*see* **Note 1**).
2. On the day of transfection, mix 4.5 µg of the gene transfer plasmid, 1.3 µg of the packaging plasmid, and 0.5 µg of the envelope plasmid (pVSV-G) in 750 µL of Opti-MEM.

3. Prepare the Plus reagent just prior to use and add 20 µL Plus reagent to the DNA solution (from **step 2**). Vortex gently and incubate the mixture at room temperature for 15 min.
4. Dilute 30 µL of the Lipofectamine reagent into 750 µL of OptiMEM in a separate tube.
5. Mix the DNA/Plus solution (770 µL; from **step 3**) and the Lipofectamine solution (780 µL; from **step 4**) followed by incubation at room temperature for 15 min.
6. During the incubation, replace the medium of 293T cells with 6.5 mL OptiMEM.
7. After the incubation, evenly add the DNA/Plus/Lipofectamine solution (1.55 mL total; from **step 5**) onto 293T cells and incubate the plate at 37°C, 5% CO_2. At 4 h after the transfection, add 8 mL DMEM containing 20% FBS.

3.1.2. Harvest and Concentration of Vector

1. Incubate the plate (from **Subheading 3.1.1.**) overnight and replace medium with 10 mL regular 293T growth medium.
2. At 24 h after media replacement, harvest the supernatant (which contains the vector) and filter it though a 0.45-µm pore membrane. The titer of vector will be 10^5–10^6 transducing units (TU) per milliliter (*see* **Note 2**).
3. Concentrate the vector supernatant at 42,500*g* for 2 h with a high-speed centrifuge.
4. After centrifugation, carefully discard the supernatant and resolve the pellet with PBS containing 5% FBS. The suspension volume should be 1/1000 to 1/100 of the initial volume. The final titer of vector will be 10^8–10^9 TU/mL (*see* **Note 3**).

3.2. Transduction

1. Plate 1.5×10^5 ES cells on an MEF (5×10^5 cells) feeder layer in a 35-mm dish and incubate the dish at 37°C, 5% CO_2, for 12–24 h.
2. Gently wash ES cells with HBSS and add 1 mL (half of the regular volume) of the growth medium.
3. Thaw a viral stock without foaming in a water bath at 37°C and add it to the culture (*see* **Note 4**).
4. After 10 h, aspirate the medium, gently wash ES cells once with HBSS, and replace with 2 mL fresh medium.
5. At 2–3 d after transduction, evaluate the transduction efficiency (*see* **Subheading 3.3.** and **Note 5**).

3.3. Assessment of Transduction Efficiency

After transduction, it is important to assess the transduction efficiency, usually 2–3 d after exposure to the vector. If a marker gene such as green fluorescent protein (GFP) is included in the vector, then you can assess the transduction efficiency by examining the marker gene expression. GFP expression can be easily monitored under a fluorescent microscope or by flow cytometry (*see* **Subheading 3.3.1.**). Another method to assess the transduction efficiency is to examine the SIV-provirus (vector integrated into the host genome) by real-time DNA-PCR (*see* **Subheading 3.3.2.**). It is particularly useful when marker genes are not available or marker gene expression levels are not high enough.

When cynomolgus ES cells are transduced once or twice with an SIV vector encoding the GFP gene, more than 50% of cells fluoresce, and the GFP expression persists for months. In addition, high levels of GFP expression are observed during embryoid body formation *(13)*. On the other hand, transduction of cynomolgus ES cells with an oncoretrovirus vector results in lower gene transfer rates (less than 20%), suggesting that simian lentivirus vectors can transduce simian ES cells more efficiently than oncoretrovirus vectors *(13)*.

3.3.1. Flow Cytometry

1. Aspirate old medium from the culture and rinse cells with HBSS (from **Subheading 3.2., step 5**). Add 2 mL 0.25% trypsin-HBSS to the dish and incubate for 5 min at 37°C. Detach ES cell colonies from the bottom by tapping with your fingers. Add 3 mL ES medium to the dish, disperse the cells into single cells using a 1-mL tip, and transfer the cell suspension to a 15-mL conical tube.
2. Spin cells in a centrifuge at 140g for 4–5 min. Aspirate the medium and resuspend the pellet in FACS medium. Pass the cell suspension through a cell strainer to remove cell clusters (*see* **Note 6**). Count a cell number and adjust it at $1-2 \times 10^6$ cells/mL.
3. Transfer 100 μL cell suspension ($1-2 \times 10^5$ cells) into a 1.5-mL tube. Add 0.1 μg (1 μL) of PE-conjugated antimouse H-2K^d monoclonal antibody solution to the tube and incubate it for 30–60 min on ice.
4. After incubation, add 1 mL FACS medium to the tube and spin cells at 800g for 5 min at 4°C. Aspirate medium and wash the pellet with FACS medium. Spin the cell suspension at 800g for 5 min at 4°C again.
5. Resuspend the pellet with 200–500 μL fixing medium. The cell suspension can be left at 4°C overnight until flow cytometric analysis.
6. Transfer the cell suspension to a round test tube through a strainer cap.
7. Perform flow cytometric analysis using a flow cytometer with excitation at 488 nm. The fluorescence data of GFP and PE can be obtained via FL1 and FL2 parameters, respectively. **Figure 1** shows a typical profile of cynomolgus ES cells transduced with an SIV vector expressing GFP. Cynomolgus ES cells are negative for antimouse H-2K^d, but co-cultured MEF feeder cells (derived from BALB/c mice) are positive for it; thus, you can distinguish both ES and MEF cells.

3.3.2. Real-Time PCR

1. Extract DNA from a culture pellet (containing both ES and MEF cells from **Subheading 3.2., step 5**) using a QIAamp DNA minikit (*see* **Note 7**). Assess the purity of DNA by checking a 260/280-nm absorbance ratio with a spectrophotometer. Preferably, it is higher than 1.75. Adjust the concentration of DNA stocks (dilute with DNase-free water) to 50 μg/mL.
2. Prepare a master mix for real-time PCR as shown in **Table 1** (*see* **Note 8**). Dispense 45 μL into each well of a MicroAmp optical 96-well reaction plate.
3. Add 5 μL (250 ng) template DNA to each well and seal the plate with MicroAmp caps.
4. Place the plate in a real-time thermal cycler and start a PCR program.
5. Analyze data according your software package (*see* **Note 9**).

4. Notes

1. Because 293T cells were established from 293 cells after transfection with the SV40 large T antigen and neomycin resistance genes, it is recommended to treat 293T cells with 800 μg/mL (active) of G418 for 1 wk once a month so the transgenes are not lost. It is, however, important to passage 293T cells several times without G418 before virus production to avoid contamination of G418 in the viral supernatant.

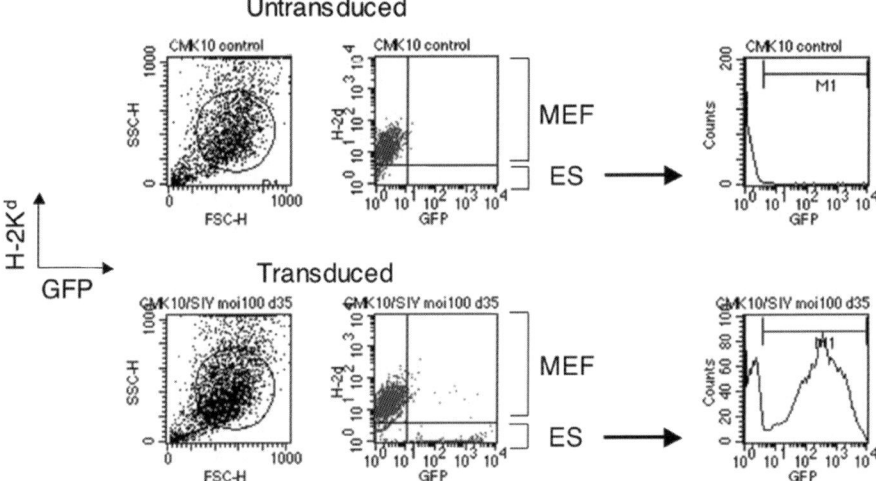

Fig. 1. Assessment of transduction efficiency by flow cytometry. The transgene green fluorescent protein (GFP) expression was analyzed on a FACScan using *CellQuest* software 2–3 d after transduction with a GFP-expressing simian immunodeficiency virus vector. The co-cultured BALB/c-derived feeder cells could be distinguished from the cynomolgus embryonic stem cells using PE-conjugated mouse antimouse H-2K^d monoclonal antibody, which does not react to cynomolgus cells but does react to BALB/c cells.

2. The titer (transducing units, TU) of vector is defined as the ability to transduce target cells. For instance, 10^5 TU/mL indicates that 1 mL vector solution is able to transduce 10^5 cells. We usually use 293T cells as targets to assess the titer. The titer of virus can also be assessed in terms of genomic copies (often designated gc). Genomic copy number of SIV vector can be evaluated by RNA dot-blot or quantitative RNA-PCR.
3. The vector solution can be stored at −80°C at least for several months. The titer will decrease even at −20°C. Frozen stocks should be thawed quickly in a water bath at 37°C just prior to use. Avoid repeated freezing and thawing, or the titer will decrease.
4. The passage of ES cells before and after lentiviral transduction is the same as the routine passage; 30–100 cells per clump is the best. You do not have to disperse clumps for transduction. Vectors are added at 10–50 TU per target cell. We sometimes add polybrene (final concentration 4–8 µg/mL) in the transduction culture and other times do not add it. It does not seem that polybrene improves the transduction efficiency with SIV vectors unlike the case with oncoretrovirus vectors. It is suggested that ES cell exposure to lentivirus solution is no longer than 12 h. Longer exposure may result in a large decrease in ES cell number, presumably because of the toxicity of the pseudotyped envelope VSV-G protein. Serum may greatly hamper lentiviral transduction. If you do not obtain good gene transfer efficiency, then it is suggested to remove the serum from your transduction culture.
5. The transgene expression in ES cells can be enhanced by changing the promoter or adding *cis*-acting elements in the vector. The *cis*-acting sequences include the central polypurine and termination tract (cPPT) to facilitate nuclear import of the viral complex and the woodchuck posttranscriptional regulatory element (WPRE) to increase transgene expression *(17)*. **Figure 2** shows variable GFP expression in cynomolgus ES cells transduced with SIV vectors containing various promoters and cPPT/WPRE sequences.

Table 1
Real-Time PCR Reaction Mixture

Master mix	Volume per reaction	Final concentration
2X QuantiTect SYBR green PCR master mix	25 µL	1X
Forward primer (10 µM)	2.5 µL	0.5 µM
Reverse primer (10 µM)	2.5 µL	0.5 µM
Water	15 µL	NA
Total volume of master mix	45 µL	NA
Template DNA sample (50 µg/mL)	5 µL (250 ng DNA)	5 µg/mL
Total volume of reaction mixture	50 µL	NA

GFP sequence primer set: 5'- CGT CCA GGA GCG CAC CAT CTT C-3' and 5'- GGT CTT TGC TCA GGG CGG ACT-3'. Internal control cynomolgus β-actin sequence primer set: 5'-CAT TGT CAT GGA CTC TGG CGA CGG-3' and 5'-CAT CTC CTG CTC GAA GTC TAG GGC-3'. NA, not applicable.

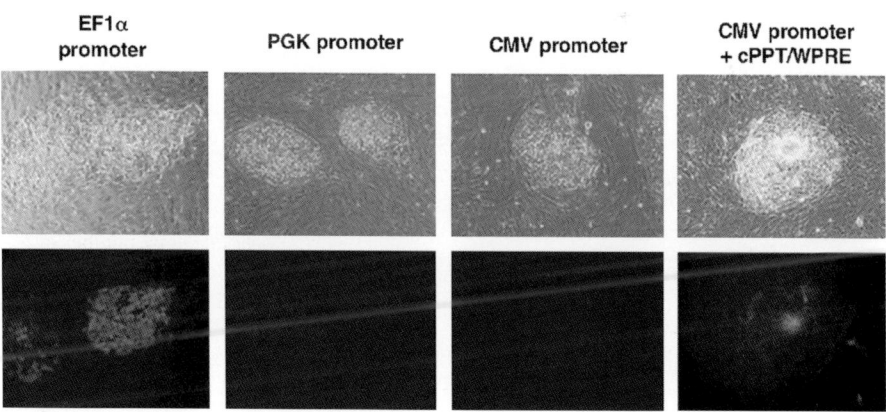

Fig. 2. Promoters and *cis*-acting sequences in simian immunodeficiency virus (SIV) vectors affect transgene expression. Cynomolgus embryonic stem (ES) cells were transduced with green fluorescent protein (GFP)-expressing SIV vectors at 30 TU per target cell. The vectors contain the elongation factor (EF) 1α, phosphoglycerate kinase (PGK), or cytomegalovirus promoter (CMV). The transduced ES cells were observed at d 5 with a fluorescent microscope under a bright field (upper) or dark field (lower). In this cynomolgus ES cell line (CMK6), the usage of the EF1α promoter resulted in the highest GFP expression. In addition, the GFP expression could be enhanced by the inclusion of two *cis*-acting sequences, the central polypurine and termination tract (cPPT) and the woodchuck posttranscriptional regulatory element (WPRE) (rightmost panel).

6. As an ES cell number considerably decreases after passing cells through a strainer before flow cytometry, start experiments with a sufficient number of cells.
7. MEF cells are cotransduced with SIV vector together with ES cells. Therefore, it is suggested to passage transduced ES cells onto untransduced MEF cells several times before DNA

extraction to avoid contamination of transduced MEF cells. In addition, because ES cells are cultured on MEF cells, it is difficult to extract DNA separately from ES or MEF cells. Thus, it is important to know the fraction (percent) of ES cells in total cultured cells (ES plus MEF cells) before DNA extraction in order to calculate the transduction efficiency of ES cells. The fraction (ES vs total cells) can be assessed by flow cytometry (*see* **Subheading 3.3.1.** and **Fig. 1**).

8. We usually use a SYBR green method (Qiagen Quantitect SYBR green PCR kit) rather than a probe method. The former is easier. For the SYBR green method, you do not have to develop specific primers or a probe; rather, regular primer sets are used. It is, however, important to confirm that the PCR does not generate nonspecific bands on an agarose gel because the SYBR green method quantifies all PCR products, including nonspecific ones, if any.

9. The positive control is genomic DNA extracted from cells that contain a known copy number of the target sequence per cell. Dilute the DNA with genomic DNA from naive control monkeys to make a series of diluted positive controls (100, 10, 1, 0.1, 0.01%). The quantitative PCR should be certified each time to yield linear amplifications in the range of the intensity of positive control series (0.01–100%, correlation coefficient >0.98). To certify equal amounts of loaded sample DNA, an internal control sequence (for instance, β-actin) in the same sample should be subjected to real-time PCR. Calculated transduction efficiency (percent) indicates a fraction of cells successfully transduced with SIV vector given that each vector-positive cell contains one copy of the provirus.

References

1. Thomson, J. A., Kalishman, J., Golos, T. G., et al. (1995) Isolation of primate embryonic stem cell line. *Proc. Natl. Acad. Sci. USA* **92**, 7844–7848.
2. Suemori, H., Tada, T., Torii, R., et al. (2001) Establishment of embryonic stem cell lines from cynomolgus monkey blastocysts produced by IVF or ICSI. *Dev. Dyn.* **222**, 273–279.
3. Naldini, L., Blomer, U., Gallay, P., et al. (1996) In vivo gene delivery and stable transduction of nondividing cells by a lentiviral vector. *Science* **272**, 263–267.
4. Miyoshi, H., Smith, K. A., Mosier, D. E., Verma, I. M., and Torbett, B. E. (1999) Transduction of human CD34+ cells that mediate long-term engraftment of NOD/SCID mice by HIV vectors. *Science* **283**, 682–686.
5. Poeschla, E. M., Wong-Staal, F., and Looney, D. J. (1998) Efficient transduction of nondividing human cells by feline immunodeficiency virus lentiviral vectors. *Nat. Med.* **4**, 354–357.
6. Olsen, J. C. (1998) Gene transfer vectors derived from equine infectious anemia virus. *Gene Ther.* **5**, 1481–1487.
7. Nakajima, T., Nakamaru, K., Ido, E., Terao, K., Hayami, M., and Hasegawa, M. (2000) Development of novel simian immunodeficiency virus vectors carrying a dual gene expression system. *Hum. Gene Ther.* **11**, 1863–1874.
8. Schnell, T., Foley, P., Wirth, M., Munch, J., and Uberla, K. (2000) Development of a self-inactivating, minimal lentivirus vector based on simian immunodeficiency virus. *Hum. Gene Ther.* **11**, 439–447.
9. Berkowitz, R., Ilves, H., Lin, W. Y., et al. (2001) Construction and molecular analysis of gene transfer systems derived from bovine immunodeficiency virus. *J. Virol.* **75**, 3371–3382.
10. Hanazono, Y., Asano, T., Ueda, Y., and Ozawa, K. (2003) Genetic manipulation of primate embryonic and hematopoietic stem cells with simian lentivirus vectors. *Trends Cardiovasc. Med.* **13**, 106–110.

11. Owens, C. M., Yang, P. C., Gottlinger, H., and Sodroski, J. (2003) Human and simian immunodeficiency virus capsid proteins are major viral determinants of early, postentry replication blocks in simian cells. *J. Virol.* **77,** 726–731.
12. Stremlau, M., Owens, C. M., Perron, M. J., Kiessling, M., Autissier, P., and Sodroski, J. (2004) The cytoplasmic body component TRIM5alpha restricts HIV-1 infection in Old World monkeys. *Nature* **427,** 848–853.
13. Asano, T., Hanazono, Y., Ueda, Y., et al. (2002) Highly efficient gene transfer into primate embryonic stem cells with a simian lentivirus vector. *Mol. Ther.* **6,** 162–168.
14. Hanawa, H., Hematti, P., Keyvanfar, K., et al. (2004) Efficient gene transfer into rhesus repopulating hematopoietic stem cells using a simian immunodeficiency virus-based lentiviral vector system. *Blood* **103,** 4062–4069.
15. Ikeda, Y., Goto, Y., Yonemitsu, Y., et al. (2003) Simian immunodeficiency virus-based lentivirus vector for retinal gene transfer: a preclinical safety study in adult rats. *Gene Ther.* **10,** 1161–1169.
16. Ogata, K., Mimuro, J., Kikuchi, J., et al. (2004) Expression of human coagulation factor VIII in adipocytes transduced with the simian immunodeficiency virus agmTYO1-based vector for hemophilia A gene therapy. *Gene Ther.* **11,** 253–259.
17. VandenDriessche, T., Thorrez, L., Naldini, L., et al. (2002) Lentiviral vectors containing the human immunodeficiency virus type-1 central polypurine tract can efficiently transduce nondividing hepatocytes and antigen-presenting cells in vivo. *Blood* **100,** 813–822.

21

Generation of Green Fluorescent Protein-Expressing Monkey Embryonic Stem Cells

Tatsuyuki Takada, Yutaka Suzuki, Nae Kadota, Yasushi Kondo, and Ryuzo Torii

Summary

Monkey embryonic stem (ES) cells are a useful tool for studying early human development and evaluating the efficacy of stem cell therapy. Monkey ES cells show closer similarity to human ES cells than their mouse counterparts regarding morphology, cell surface markers, and the maintenance of pluripotency, including the leukemia inhibitory factor requirement. The generation of genetically modified monkey ES cells with a biomarker such as green fluorescent protein, which allows noninvasive monitoring of progeny ES cells, is invaluable for the development of cell transplantation therapy and the study of differentiation mechanisms in primates. Here, we describe the generation of green fluorescent protein-expressing monkey ES cells using a conventional electroporation method.

Key Words: Electroporation; gene transfer; genetic modification; GFP; monkey ES cell.

1. Introduction

Establishment of human embryonic stem (ES) cells (*1,2*) and the development of nuclear transfer are the fundamental technologies for the clinical application of stem cell therapy in regenerative medicine. Generation of autologous ES cells by human therapeutic cloning through somatic cell nuclear transfer has been proposed for cell transplantation therapy (*3*). However, ethical concerns are the major obstacle to using human ES cells, especially to perform human therapeutic cloning.

To circumvent ethical issues and evaluate the efficacy and safety of cell transplantation therapy, an accurate animal model is required. Although mouse ES cells have been widely used for differentiation studies, there are significant differences between mouse and human early development and their ES cells (*4*).

Monkey ES cells, on the other hand, share many characteristics with human ES cells *(4–8)*. For example, monkey ES cells show similar morphology and expression pattern of the cell surface markers, such as stage-specific embryonic antigens (SSEAs) (e.g., SSEA-3, SSEA-4) and tumor rejection antigens (TRAs) (e.g., TRA-1-60, TRA-1-81) to human ES cells. Maintenance of pluripotency and self-renewal of monkey and human ES cells largely depends on the unknown factors provided by mouse embryonic fibroblasts (MEFs), and unlike mouse ES cells, their pluripotency cannot be maintained with leukemia inhibitory factor (LIF) in the absence of MEFs. In addition, early development of nonhuman primate and humans is basically quite similar. Therefore, monkey ES cells are considered the ideal model for studying early human development and the development of stem cell therapy.

Genetic modification of ES cells by introducing disease-related genes and gene targeting is also key for the clinical application of stem cell therapy *(9)*. However, monkey ES cells tend to differentiate when dissociated into single cells, which is required for most of transfection procedures. This results in inefficient transfection and has limited their application in biomedical research *(10)*. We found that electroporation is a reliable method to introduce foreign genes into monkey ES cells without jeopardizing their pluripotency *(11)*, although it is not as efficient as that of mouse ES cells.

In this chapter, we describe protocols to introduce the enhanced green fluorescent protein (EGFP) gene into monkey ES cells. The EGFP-expressing monkey ES cells provide a powerful tool to study pluripotency in ES cells by monitoring their differentiation and tissue regeneration processes. The capability to distinguish ES cells from the host cells should allow the accurate validation of clinical applications of cell transplantation therapy.

2. Materials

2.1. Tissue Culture

1. Phosphate-buffered saline (PBS): dissolve 9.6 g Dulbecco's PBS (−) powder (Nissui, Tokyo, Japan; cat. no. 05913) into 1 L distilled water (dH$_2$O) and autoclave.
2. 0.25% trypsin ethylenediaminetetraacetic acid solution (Nacalai, Kyoto, Japan; cat. no. 35554-64).
3. 0.1% collagenase solution: dissolve 0.1 g collagenase (Wako, Osaka, Japan; cat. no. 034-10533) in 100 mL Dulbecco's modified Eagle's medium (DMEM), filter, aliquot into 10-mL portions, and store at −20°C.
4. 0.1% gelatin: dissolve 0.5 g gelatin (Sigma, St. Louis, MO; cat. no. G1890) in 500 mL dH$_2$O and autoclave. Store at 4°C.
5. Mitomycin C (MMC; Nacalai, cat. no. 23305-94): dissolve 10 mg powder in 10 mL dH$_2$O, filter, and store at −20°C protected from light.
6. DMEM (Nacalai, cat. no. 14246-25).
7. DMEM/F12 Ham (Sigma, cat. no. D6421).
8. 100 mM sodium pyruvate solution (Sigma, cat. no. S8636).
9. Nonessential amino acids (NEAA) solution (Sigma, cat. no. M7145).
10. 200 mM L-glutamine solution (Nacalai, cat. no. 16948-04).
11. 2-Mercaptoethanol (Sigma, cat. no. M7522).
12. Knockout serum replacement (KSR; Invitrogen, Carlsbad, CA; cat. no. 10828-028).

13. Penicillin-streptomycin solution, 5000 U/mL penicillin, 5 mg/mL streptomycin (100X; Nacalai, cat. no. 26252-94).
14. Fetal bovine serum (FBS; JRH Biosciences, Lenexa, KS; cat. no. 12103-500M).
15. Falcon cell strainer (100 µm; BD Biosciences, Bedford, MA; cat. no. 352360).
16. Iwaki polystyrene 100-mm Petri dishes (Asahi Techno Glass, Tokyo, Japan; cat. no. 1020-100X).
17. Iwaki polystyrene 100-mm tissue culture dishes (Asahi Techno Glass, cat. no. 3020-100X).
18. Iwaki polystyrene 60-mm tissue culture dishes (Asahi Techno Glass, cat. no. 3010-060X).
19. Iwaki polystyrene 35-mm tissue culture dishes (Asahi Techno Glass, cat. no. 3000-035X).
20. Nunc four-well multidish (Nalge Nunc, Rochester, NY; cat. no. 176740).
21. MEF medium: MEF cells are maintained in DMEM with 10% heat-inactivated FBS, 2 mM L-glutamate, and 1% penicillin-streptomycin. To prepare a bottle of MEF medium, combine 57 mL FBS, 6 mL L-glutamate, 6 mL penicillin-streptomycin, and 500 mL DMEM. Store at 4°C.
22. ES medium: cynomolgus monkey ES cells (CMS-A2, CMK6) *(7,11)* are maintained in DMEM/F12 Ham containing 20% knockout serum replacement, 1% NEAA, 1 mM sodium pyruvate, 2 mM L-glutamine, 0.1 mM 2-mercaptoethanol, and 1% penicillin-streptomycin. To prepare a bottle of monkey ES medium, add 5 mL NEAA, 5 mL sodium pyruvate, 5 mL L-glutamine, 5 mL penicillin-streptomycin, 4 µL 2-mercaptoethanol, and 100 mL knockout serum replacement into 400 mL DMEM/Ham F12. This medium is stable at least 3 wk when stored at 4°C.

2.2. Electroporation

1. QIAfilter Plasmid Midi kit (Qiagen, Valencia, CA; cat. no. 12243).
2. Alw44 I (ApaL I) (Toyobo, Osaka, Japan; cat. no. 031000).
3. Electroporation cuvet (Bio-Rad, Hercules, CA; cat. no. 165-2088).
4. Neomycin-resistant primary MEF cells (Specialty Media, Phillipsburg, NJ; cat. no. PMEF-N).
5. G-418 50-mg/mL solution (Nacalai, cat. no. 165-13).
6. ES cell characterization kit (Chemicon, Temecula, CA; cat. no. SCR 001).
7. Freezing vessel (Nihon Freezer, Tokyo, Japan; cat. no. BICELL).

3. Methods

3.1. Preparation of MEFs

Monkey ES cells require a mitotically inactivated MEF layer for the maintenance of pluripotency and self-renewal activity. The quality of MEFs critically affects the condition of monkey ES cells (*see* **Note 1**).

1. Sacrifice pregnant mice at 12.5 d postcoitum (dpc).
2. Remove the uterus to a 10-cm sterile Petri dish containing 10 mL PBS.
3. Dissect embryos from the uterus and place them in a new 10-cm sterile Petri dish containing 10 mL PBS.
4. Remove the upper part of the head and internal organs. Wash embryos with PBS in a new 10-cm dish to remove as much blood as possible.
5. Place embryos in a disposable 2.5-mL syringe attached to a 23-gage needle. Mince embryos by inserting plunger into the syringe and passing the embryos through the needle into a new 10-cm tissue culture dish containing 10 mL MEF medium (*see* **Note 2**). Aspirate partially minced embryos into the syringe and expel them again using the

plunger. Repeat this step until embryos are well disrupted into small pieces, then dispersed them in the medium uniformly. Incubate at 37°C, 5% CO_2, in a humidified atmosphere.
6. When cells reach confluence, trypsinize, filter tissue clumps out using a 100-μm cell strainer, and freeze cells at a concentration of 5×10^6 cells/mL.

3.2. MMC Treatment of MEFs

1. Thaw a frozen vial of MEF cells in a 37°C water bath quickly, transfer into a 15-mL tube containing 5 mL MEF medium, and centrifuge (190g for 3 min). Aspirate medium, resuspend the cell pellet with fresh MEF medium, and plate in four new 10-cm dishes.
2. When confluent, adjust the medium to 5 mL and add 50 μL MMC (10 μg/mL final concentration).
3. Return dishes to the incubator for 2–3 h.
4. Wash the dishes three times with 10 mL PBS and trypsinize cells.
5. Either use MMC-treated cells immediately by plating $1–1.5 \times 10^6$ cells/6-cm dish or freeze for future use.

3.3. Culture of Monkey ES Cells

MEF feeder should be prepared 1–2 d before plating ES cells and used within 3–4 d of preparation (*see* **Note 3**).

1. Plate MEFs ($1–1.5 \times 10^6$ cells) in a gelatinized 6-cm dish the day before plating monkey ES cells.
2. Thaw a frozen vial of monkey ES cells (1×10^7) in a 37°C water bath quickly, transfer into a 15-mL tube containing 9 mL monkey ES medium, and centrifuge (190g for 3 min). Aspirate medium, gently resuspend the cell pellet with 5 mL fresh monkey ES medium, and plate into one 6-cm dish.
3. Change medium every day.
4. When the cultures reach confluence or if large colonies become predominant, then passage ES cells.

3.4. Passage of Monkey ES Cells

The timing of passage is important. Pass ES cells at about 50% confluence or before they form multilayer or donutlike (cells located in the center of the colony become flat and thin) colonies.

1. Remove medium and wash with 3 mL PBS.
2. Add 1 mL 0.1% collagenase to cover cell surface, then aspirate and return the dish to the incubator for 3–5 min (*see* **Note 4**).
3. Monitor ES colonies under the microscope. Undifferentiated ES colonies should shrink and separate from MEFs.
4. Add 2 mL ES medium and dissociate ES cells by gentle pipetting (*see* **Note 5**). The MEF layer also detaches from the dish by this procedure. Wash the MEF sheets by gentle pipetting to recover as much as ES colonies.
5. Transfer medium into a 15-mL tube; try to leave MEF clumps in the dish and centrifuge (190g, 3 min).
6. Aspirate medium and resuspend in 4 mL ES medium.

7. Seed the cells onto new feeder plates in the ratio of 1:2 to 1:4. Gently tilt the plates in both the x and y directions to distribute cells evenly. Then, incubate at 37°C. Cells will reach confluence in 3–4 d.

3.5. Vector Preparation

Use the GFP expression vector, which has EGFP gene under the control of the cytomegalovirus immediate-early enhancer and human elongation factor 1α promoter *(11,12)* to express EGFP ubiquitously.

1. Purify supercoiled plasmid DNA with Qiagen column.
2. Linearize plasmid (100 μg) using 100 U Alw44 I (ApaL I) at 37°C overnight.
3. Confirm digestion by agarose gel electrophoresis.
4. Extract DNA with the same volume of phenol/chloroform/isoamylalcohol (25:24:1). After brief centrifugation (9000g, 2 min), transfer the aqueous phase to a new tube and precipitate DNA by adding 1/10 vol of 3 M sodium acetate (pH 5.2) and 2.2 vol of ethanol.
5. Recover DNA by centrifugation (18,000g, 15 min), wash with 70% ethanol, and dissolve in 10 mM Tris-HCl and 1 mM ethylenediaminetetraacetic acid (pH 8.0) at a concentration of 1 mg/mL. This DNA solution can be stored at −20°C.

3.6. Electroporation of Monkey ES Cells

The condition of ES cells affects transfection efficiency, especially survival after electroporation. Exponentially growing ES cells should be used.

1. Plate neomycin-resistant MEFs (1–1.5 × 10^6) in a gelatinized 6-cm dish the day before electroporation of monkey ES cells.
2. Change ES cell media 2 h prior to electroporation.
3. Dissociate ES cells (approx 1 × 10^7) (*see* **Note 6**) with collagenase as described in **Subheading 3.4.**
4. Disperse ES cells into small clumps consisting of approx 5–10 cells by mechanical pipetting using a Pipetman (P-1000) (*see* **Note 7**).
5. Wash cells with ES medium by centrifugation (190g, 3 min). Aspirate supernatant and resuspend with 5 mL ES medium.
6. Transfer cell suspension to a gelatinized plate and incubate at 37°C for 10–20 min (*see* **Note 8**) to eliminate MEF feeders.
7. Collect floating monkey ES cell clumps by centrifugation (190g, 3 min), wash with 10 mL PBS (190g, 3 min), and resuspend in 0.8 mL PBS.
8. Mix the cell suspension with 20–40 μg linearized plasmid DNA and transfer to an electroporation cuvet (4-mm gap).
9. Place the electroporation cuvet on ice for 10 min.
10. Set up the conditions for electroporation (500 μF, 250 V, Gene Pulser II, Bio-Rad).
11. Transfer the cuvet into the cuvet holder of electroporator, then deliver electronic pulse (500 μF, 250 V).
12. Remove the cuvet from the holder and incubate on ice for 10 min.
13. Resuspend cells with 8–12 mL of ES medium and plate into two or three 6-cm dishes with neomycin-resistant MEFs.
14. Change the media the next day.
15. At 48 h after electroporation, switch the medium to ES medium containing G418 (200 μg/mL) for drug selection.

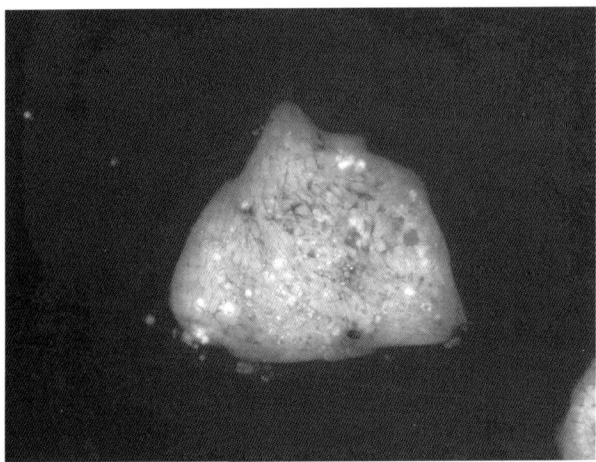

Fig. 1. Green fluorescent protein (GFP)-expressing monkey embryonic stem (ES) cell line. Fluorescent image of an undifferentiated monkey ES cell colony that stably expresses GFP. Strong and uniform GFP expression was observed. Individual cells and their shape were clearly identified by GFP expression.

16. Change the medium every day for 5–8 d. G418-resistant and GFP-expressing colonies will grow to a suitable size for pick up.

3.7. Isolation of GFP-Expressing ES Colonies

1. Plate neomycin-resistant MEFs (1×10^5) in a gelatinized four-well plate the day before picking monkey ES colonies.
2. Monitor GFP-expressing colonies under a fluorescent microscope. Mark the colonies to be picked on the bottom of the dish (see **Note 9**).
3. Place the plate on a dissecting microscope and pick up colonies using a flame-drawn Pasteur pipet.
4. Transfer the colonies into a drop (20–50 µL) of ES medium prepared in a 3.5-cm dish and dissect the colony into several clumps using a 22–26-gage needle or flame-drawn glass capillary (see **Note 10**).
5. Transfer cell clumps into a four-well MEF plate containing 0.5 mL ES medium with G418.
6. When the culture reaches confluence, dissociate ES cells with collagenase and expand culture size to a 3.5-cm dish, then a 6-cm dish (see **Fig. 1**).
7. Confirm the expression of cell surface markers (alkaline phosphatase, SSEA-4, TRA-1-60, TRA-1-81) characteristic of monkey ES cells using an ES cell characterization kit according to the manufacturer's instructions.

3.8. Freezing of Transfected Cells

1. Collect confluent GFP-expressing ES cells from two 6-cm dishes by collagenase treatment followed by centrifugation (190g, 3 min).
2. Resuspend cells in 1 mL freezing medium consisting of 90% KSR and 10% dimethyl sulfoxide.

3. Transfer cell suspension into a cryotube and put it in a freezing vessel precooled at 4°C (*see* **Note 11**).
4. Put the freezing vessel in an −80°C freezer for 24 h.
5. Transfer cryotubes into liquid nitrogen.

4. Notes

1. It is advisable that a newly prepared MEF layer should be tested before use.
2. We usually use one or two embryos per 10-cm dish.
3. We routinely maintain monkey ES cells on MMC-treated MEF layer in 6-cm dishes. Culture and passage of monkey ES cells require some experience.
4. 0.25% trypsin containing 1 mM $CaCl_2$ and 20% KSR also works *(7)*.
5. This step is very important and critical. Do not dissociate ES cell colonies into single cells as they differentiate or die. Check the dissociation under the microscope periodically. We usually dissociate into cell clumps consisting of approx 50 cells.
6. We routinely use three or four confluent 6-cm dishes or one or two 10-cm dishes for each electroporation experiment.
7. Try to minimize pipetting and bubble formation as it may damage ES cells.
8. Check periodically. Do not leave the plate in the incubator for a long time. Monkey ES cell colonies adhere easier to the plate than mouse ES cells do.
9. Try to pick up well-grown colonies with uniform GFP expression. Do not pick up colonies showing weak or mosaic GFP expression.
10. Dissected clumps consist of about 30–50 cells. If a colony is not big enough to dissect, then just transfer the colony into the new four-well plate and culture for 4–7 d more. When it grows to a suitable size, pick it up again and do mechanical dissection as described.
11. A Styrofoam tube stand for 15-mL culture tubes can be used as a freezing box.

References

1. Thomson, J. A., Itskovitz-Eldor, J., Shapiro, S. S., et al. (1998) Embryonic stem cell lines derived from human blastocysts. *Science* **282,** 1145–1147
2. Reubinoff, B. E., Pera, M. F., Fong, C. Y., Trounson, A., and Bongso, A. (2000) Embryonic stem cell lines from human blastocysts: somatic differentiation in vitro. *Nat. Biotechnol.* **18,** 399–404.
3. Hwang, W. S., Ryu, Y. J., Park, J. H., et al. (2004) Evidence of a pluripotent human embryonic stem cell line derived from a cloned blastocyst. *Science* **303,** 1669–1674.
4. Thomson, J. A. and Marshall, V. S. (1998) Primate embryonic stem cells. *Curr. Top. Dev. Biol.* **38,** 133–165.
5. Thomson, J. A., Kalishman, J., Golos, T. G., et al. (1995) Isolation of a primate embryonic stem cell line. *Proc. Natl. Acad. Sci. USA* **92,** 7844–7848.
6. Thomson, J. A., Kalishman, J., Golos, T. G., Durning, M., Harris, C. P., and Hearn, J. P. (1996) Pluripotent cell lines derived from common marmoset (*Callithrix jacchus*) blastocysts. *Biol. Reprod.* **55,** 254–259.
7. Suemori, H., Tada, T., Torii, R., et al. (2001) Establishment of embryonic stem cell lines from cynomolgus monkey blastocysts produced by IVF or ICSI. *Dev. Dyn.* **222,** 273–279.
8. Marshall, V. S., Waknitz, M. A., and Thomson, J. A. (2001) Isolation and maintenance of primate embryonic stem cells. *Methods Mol. Biol.* **158,** 11–18.
9. Zwaka, T. P. and Thomson, J. A. (2003) Homologous recombination in human embryonic stem cells. *Nat. Biotechnol.* **21,** 319–321.

10. Eiges, R., Schuldiner, M., Drukker, M., Yanuka, O., Itskovitz-Eldor, J., and Benvenisty, N. (2001) Establishment of human embryonic stem cell-transfected clones carrying a marker for undifferentiated cells. *Curr. Biol.* **11,** 514–518.
11. Takada, T., Suzuki, Y., Kondo, Y., et al. (2002) Monkey embryonic stem cell lines expressing green fluorescent protein. *Cell Transplant.* **11,** 631–635.
12. Takada, T., Iida, K., Awaji, T., et al. (1997) Selective production of transgenic mice using green fluorescent protein as a marker. *Nat. Biotechnol.* **15,** 458–461.

22

DNA Damage Response and Mutagenesis in Mouse Embryonic Stem Cells

Yiling Hong, Rachel B. Cervantes, and Peter J. Stambrook

Summary

Mutation in embryonic stem (ES) cells can potentially compromise multiple cell lineages and affect the well-being of subsequent generations. Thus, ES cells require sensitive mechanisms to maintain genomic integrity. One mechanism involves suppression of mutation. A complementary mechanism is to regulate the cell cycle checkpoint and facilitate cell death. Here, we describe the detailed protocols we have used to investigate DNA damage response and mutagenesis in mouse ES cells.

Key Words: Apoptosis; cell cycle regulation; DNA damage response; embryonic stem cell; mutagenesis.

1. Introduction

The maintenance of genomic integrity is important to ensure proper cellular function and viability. DNA, however, is under constant threat of incurring damage both from endogenous sources such as metabolism, replication, and recombination, and from exogenous sources such as genotoxic chemicals and ultraviolet or ionizing radiation. Any of these sources can contribute to mutation and chromosomal aberrations *(1)*. To protect genomic integrity, most eukaryotic cells have discrete cell cycle checkpoints and several DNA repair mechanisms *(2–4)*.

Embryonic stem (ES) cells have functions and requirements that are very different from those of somatic cells *(5)*. ES cells retain the potential to give rise to any cell type in the body, including germ cells, whereas somatic cells have restricted patterns of gene expression that are characteristic of their specific differentiated lineage. The consequences of mutation and genomic instability in a somatic cell are limited to the activity of that particular cell lineage and only affect the well-being of the individual. After a latency period, collective mutations in somatic cells may result in somatic disease (e.g., cancer), but the incurred mutations cannot be passed on to the progeny. In contrast, genomic instability in the germline or in ES cells can potentially compromise multiple cell lineages, including germ cells, which may affect not only the majority of cells of the individual, but also the well-being of any progeny that derive from the original mutant cell.

Thus, it is imperative that ES cells have mechanisms that protect the integrity of their genomes. One mechanism involves suppression of mutation via mitotic recombination. The mutant frequency at the endogenous adenine phosphoribosyltransferase (*Aprt*) reporter gene when heterozygous in somatic cells is very high, approaching 10^{-4} in vivo *(6,7)*. In contrast, the mutant frequency in ES cells is suppressed by about two orders of magnitude *(8)*. A complementary mechanism to preserve ES cell genomic integrity would be a facilitated cell death that rids the population of cells with a mutational burden. Consistent with this possibility is that ES cells lack a G1 checkpoint *(9,10)* and are hypersensitive to DNA damage *(11)*.

In this protocol, we detail methods that we have used to investigate cellular response to DNA damage in mouse ES cells, including the measurement of mutation frequency, cell cycle regulation, and apoptosis.

2. Materials

2.1. Measurement of Spontaneous and DNA Damage-Induced Mutation Frequencies in ES Cells

2.1.1. Tissue Culture

1. Dulbecco's modified Eagle's medium (DMEM) (1X; Gibco BRL, Carlsbad, CA; cat. no. 11965-092).
2. Heat-inactivated fetal bovine serum (FBS; Gibco BRL, cat. no.26140-079).
3. ES-qualified FBS (Gibco BRL, cat. no. 16141-079).
4. 10 m*M* minimum essential medium nonessential amino acid (NEAA) solution (100X; Gibco BRL, cat. no. 11140-050).
5. 10 m*M* GlutaMax™-1 solution (100X; Gibco BRL, cat. no. 35050-061).
6. Penicillin-streptomycin (100X; Gibco BRL, cat. no. 15140-122) (optional).
7. 1X phosphate-buffered saline (PBS): to prepare 1X PBS, combine the following: 1.15 g Na_2HPO_4, 0.2 g KH_2PO_4, 8 g NaCl, and 0.2 g KCl in 1 L distilled water.
8. β-Mercaptoethanol (Sigma, St. Louis, MO; cat. no. M-6250).
9. Mitomycin C (Sigma, cat. no. M0503).
10. Trypsin-EDTA (ethylenediaminetetraacetic acid): 025% trypsin and 1X 1 m*M* EDTA, (Gibco BRL, cat. no. 25200-072).
11. Leukemia inhibitory factor (LIF; Chemicon International Inc., Temecula, CA; cat. no. ESG1106).
12. Gelatin (Fisher, Fairlawn, NJ; cat. no. G8).
13. Dimethyl sulfoxide (Sigma, cat. no. D8418).
14. Corning polystyrene 100-mm tissue dishes (Corning Inc., Corning, NY; cat. no. 430167).
15. Corning polystyrene six-well culture cluster (Corning, cat. no. 3516).
16. 5-, 10-, and 25-mL nonpyrogenic serological pipet (Fisher, cat. no. 13678-11, 13678-11E, 13678-11D).

ES Cell DNA Damage Response and Mutagenesis

2.1.1.1. MEDIA

1. Medium for embryonic fibroblast (EF) cells. Supplement DMEM with 10% heat-inactivated FBS and 100 U/mL penicillin-streptomycin. To prepare EF medium, to one 500-mL bottle of DMEM, combine 55 mL heat-inactivated FBS and 5 mL 100X penicillin-streptomycin.
2. Medium for ES cells J11: J11 ES cells are maintained in DMEM supplemented with 15% ES-quality FBS, 1X NEAA, 1X GlutaMax-1, 100 U/mL penicillin-streptomycin, 0.1 μM β-mercaptoethanol, and 50 μM recombinant LIF. To prepare ES medium, to one 500-mL bottle of DMEM add the following: 90 mL ES-qualified FBS, 6 mL NEAA, 6 mL GlutaMax, 6 mL 100 U/mL penicillin-streptomycin, 6 μL β-mercaptoethanol, and 500 μL recombinant LIF.

2.1.1.2. GENERAL COMMENTS AND REQUIRED EQUIPMENT FOR TISSUE CULTURING

All tissue culture protocols are performed under sterile conditions. The tissue culture facility for cell culture requires the following:

1. Tissue culture hood.
2. 37°C water bath.
3. Coulter cell counter.
4. Humidified incubator at 37°C and 10% CO_2.
5. Inverted microscope with a range of phase contrast objectives.
6. Dissecting microscope.
7. Liquid nitrogen storage tank.
8. Refrigerator (4°C) and freezers (−20°C, −70°C).
9. Tabletop centrifuge.

2.1.2. Derivation of 129/Sv Aprt$^{-/-}$ Mouse and 129SvXC3H/HeJAprt$^{+/-}$ Mouse ES Cells

1. 5.6-kb Aprt-containing genomic DNA fragment isolated from a single λ-phage clone isolated from a mouse DNA library (strain 129/Sv, kindly provided by M. Shull, University of Cincinnati, College of Medicine) *(12)*.
2. Plasmid pMC1NeoPoly A (Stratagene, La Jolla, CA; cat. no. 213201).
3. Restriction enzymes XhoI, SalI, BamHI, and BsmI (New England Biolabs, Beverly, MA; cat. no. R0146, R0138, R0136, R0580).
4. G418 (Invitrogen, Carlsbad, CA; cat. no. 118111-023).
5. 0.5 M EDTA pH 8.0: dissolve 18.61 g EDTA to 80 mL with water. Adjust to pH 8.0 with NaOH (approx 2 g NaOH). Bring the volume to 100 mL with water.
6. 1 M Tris-HCl: dissolve 121.14 g Tris-HCl in 800 mL H_2O. Adjust to pH 8.0 with HCl. Bring the volume to 1 L with water.
7. Genomic DNA isolation buffer: 10 mM Tris-HCl pH 8.0, 100 mM NaCl, 25 mM EDTA pH 8.0, 0.5% sodium dodecyl sulfate (SDS), and 0.1% mg/mL proteinase K. For 100 mL solution, add 1 mL Tris-HCl pH 8.0, 0.58 g NaCl, 5 mL 0.5 M EDTA pH 8.0, 0.1 g proteinase K, and 94 L H_2O.
8. Saturated phenol at pH 6.6 (Fisher, cat. no. BP1750-400).
9. Chloroform (Fisher, cat. no. C298-1).
10. Isopropanol (Fisher, cat. no. A4161).
11. TaqComplete (GeneChoice, Frederick, MD; cat. no. 62-6128-58).

2.1.3. Aprt Selection

1. 0.1 M adenine: 0.135 g adenine (Sigma, cat. no. A8626) dissolved in 1 mL 0.5 M HCl.
2. 2 mg/mL alanosine: 2 mg alanosine (Sigma, cat. no. 9251) dissolved in 1 mL 0.1 M NaOH.
3. 2 mg/mL 2-fluoroadenine (FA): 2 mg 2-FA (adenine analog, Research Products International Corp., Mt. Prospect, IL; cat. no. FA01) dissolved in 1 mL 0.1 M NaOH.
4. 1% crystal violet: 1g crystal violet (Sigma, cat. no. C-0775) dissolved in 10% methanol (Sigma, cat. no C-0775).
5. Aprt$^{-/-}$ mouse embryonic fibroblast (MEF) feeder cells (*see* **Subheading 3.1.2.1.**).

2.1.4. Analysis of the Spectrum of the Mutation

1. The sequence of mouse chromosome 8 microsatellite markers D8Mit155, D8Mit4, D8Mit242, D8Mit321, and D8Mit13 are from Mouse Genome Database, Jackson Laboratory (http://www.informatics.jax.org/searchs/polymorphism_form.shtml).
2. Integrated Technologies Incorporated, Coralville, Iowa, synthesizes the primers of D8Mit155, D8Mit4, D8Mit242, D8Mit321, D8Mit13, and D8Mit56.
3. 0.5X TBE: 5.4 g Tris-HCl, 2.75 g boric acid, and 10 mL 0.5 M EDTA pH 8.0 brought to 1 L with water.
4. 4% high-resolution agarose-1000 (Gibco BRL, cat. no. 10975-019) in 0.5X TBE.

2.2. Cell Cycle Regulation in Response to DNA Damage

2.2.1. Cell Cycle Analysis

1. −20°C absolute ethanol (AAPER Alcohol and Chemical Co., Shelbyville, KY; cat. no. 04A21UB).
2. Deoxyribonuclease-free ribonuclease A (Sigma, cat. no. R4875).
3. 1 mg/mL propidium iodide (Molecular Probes, Eugene, OR; cat. no. P-1304).

2.2.2. Western Blot Analysis of Cell Cycle Proteins

1. 30% acrylamide/*bis*-acrylamide (Sigma, cat. no. A3574).
2. TEMED (N,N,N',N'-tetramethylenediamine; Invitrogen, cat. no. 15524-010).
3. 1.5 M Tris-HCl at pH 6.8: 181.21 g Tris-HCl in 800 mL H_2O, adjust pH to 6.8 with HCl, and then bring the volume to 1 L with water.
4. 1.5 M Tris-HCl at pH 8.8: 181.21 g Tris-HCl in 800 mL H_2O, adjust pH to 8.8 with HCl, and then bring the volume to 1 L with water.
5. Radioimmunoprecipitation (RIPA) buffer: for 100 mL solution, add 5 mL Tris-HCl pH 8.0, 1.15 g NaCl, 1 g NP-40, 0.05 g deoxycholic acid, and 1 g SDS.
6. Bio-Rad protein assay (Bio-Rad Laboratories, Hercules, CA; cat. no. 500-0006).
7. Running buffer: 14.4 g glycine, 3 g Tris-HCl, and 1% SDS in 1 L H_2O.
8. Transfer buffer: 14.4 g glycine, 3 g Tris-HCl, 200 mL methanol, 800 mL H_2O.
9. TBST (Tris-buffered saline plus Tween-20) buffer: 8 g NaCl, 0.2 g KCl, 3 g Tris-HCl, and 1% Tween-20; adjust to pH 7.4 with HCl and then bring the volume to 1 L with water.
10. 2X Laemmli loading sample buffer: 4 mL 10% SDS, 2 mL glycerol, 1.2 mL Tris-HCl pH 6.8, 2.8 mL distilled water, 0.01% bromophenol blue, and 1% β-mercaptoethanol.
11. Immobilon-P transfer membrane (Millipore, Bedford, MA; cat. no. IPVH00010).

2.2.3. Immunofluorescence

1. Formalde-Fresh fixation solution (4% formaldehyde, 1% methanol; (Fisher Scientific, cat. no. FL 12-0298).

ES Cell DNA Damage Response and Mutagenesis

2. TBS (Tris-buffered saline) solution: 8 g NaCl, 0.2 g KCl, and 3 g Tris-HCl in 1 L H_2O at pH 7.4.
3. 1% NP-40 (Sigma, cat. no. I-3021).
4. Blocking buffer: 10% goat serum (Invitrogen, cat. no. 16210-0720).
5. Cover slip (Fisher, cat. no.12-545-80).
6. Primary antibodies to Chk2 and Cdc25A (Santa Cruz, Santa Cruz, CA; cat. no. SC-5278, SC-7389).
7. Secondary antibody coupled to Alexa Fluor 488, Alexa Fluor 594 (Molecular Probes, cat. no. CA-11008, CA-11005).
8. Gel/mount (Biomedia Corp., Foster City, CA; cat. no. M01).

2.3. Measurement of Apoptosis in ES Cells

2.3.1. ES Cell Transfection

For ES cell transfection, use FuGENE 6 (Roche, Indianapolis, IN; cat. no. 1814443).

2.3.2. Apoptosis Analysis

1. Cy5-conjugated annexin V (BD Pharmingen, San Diego, CA; cat. no. 559933).
2. 10X binding buffer (BD Pharmingen, cat. no. 556454).
3. 25 µg/mL propidium iodide (Molecular Probes, cat. no. P-1304).

3. Methods

3.1. Measurement of the Spontaneous and DNA Damage-Induced Mutation Frequency and Spectrum in ES Cells

The use of *Aprt* as a reporter of mutation in mammalian cells is based on its selectable characteristics. The *Aprt* gene encodes for the ubiquitously expressed adenine phosphoribosyltransferase (APRT) enzyme (EC 2.4.4.7), which catalyzes the conversion of adenine to adenosine 5′-monophosphate (AMP) via the following reaction: Adenine + $PRPP^{Mg2+}$ AMP + pyrophosphate (PPi) where PRPP is 5-phosphoribosyl-1-pyrophosphate. APRT is a purine salvage enzyme that allows free adenine to be converted to an utilizable nucleotide. Cells lacking *Aprt* can be selected in the presence of adenine analogs such as FA, which has metabolic products that are cytotoxic to cells with APRT activity. Resistance to the adenine analogs is a consequence of loss of APRT activity. Unlike mutations to the X-linked hypoxanthine phosphoribosyltransferase (*Hprt*) locus, which only allows detection of basepair changes, frame shifts, and small deletions/insertions, *Aprt* maps to an autosome, mouse chromosome 8, that allows the detection of a greater spectrum of mutagenic events, such as mitotic recombination, gene conversion, and chromosome loss.

3.1.1. Derivation of 129/Sv Aprt$^{-/-}$ Mouse

3.1.1.1. CONSTRUCTION OF THE TARGETING VECTOR

1. Isolate a 5.6-kb *Aprt*-containing genomic DNA fragment from a λ-phage mouse genomic library (strain 129/Sv).
2. Introduce an XhoI-Sal I fragment containing the neomycin resistance cassette from pMC1NeoPoly A into a unique BspEI site in exon 3 in the same orientation as *Aprt*.
3. Introduce a herpes simplex virus thymidine kinase gene cassette driven by a kinase promoter into the BsmI site. The vector was named MBSF18 **(Fig. 1)** *(12)*.

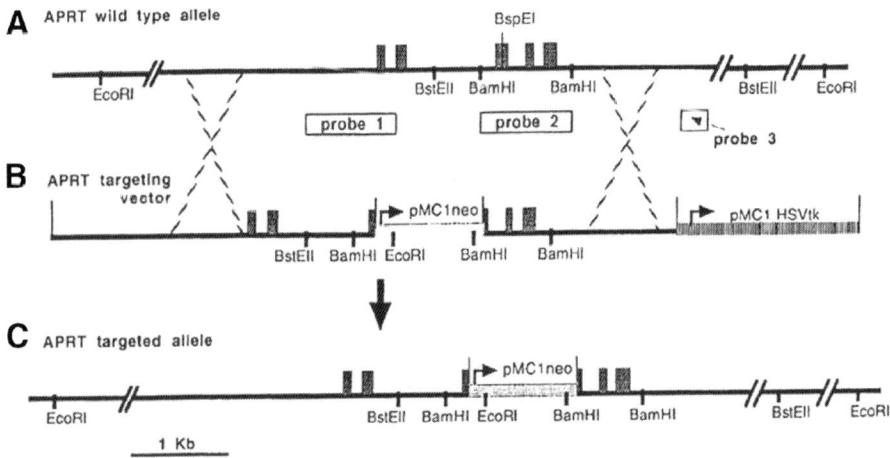

Fig. 1. Schematic for targeting the *Aprt* gene by targeted homologous recombination. (**A**) The wild-type *Aprt* locus showing the 5 exons as shaded boxes and the key restriction enzymes used. Probes used to establish proper targeting are so designated. (**B**) Depiction of the targeting vector with pMC1neo inserted into the BspE1 site within exon 3. Arrows designate the transcriptional orientation of pMC1neo and pMC1HSVtk. (**C**) Schematic of the targeted, null allele.

3.1.1.2. GENE TARGETING

1. Resuspend 5×10^6 ES cells derived from strain 129/Sv mice in 1 mL ice-cold 1X PBS.
2. Add 10 µg targeting vector (MBSF18).
3. Electroporate ES cells in a Cell-porator at 200 V, 1180 µF, at room temperature.
4. Incubate the cells for 10 min at room temperature and seed on mitomycin C-treated MEF feeder layer with ES medium.
5. Add 200 U/mL G418 to medium 48 h after electroporation.
6. Pick the colonies and expand in 24-well dishes after 7–10 d selection.
7. Freeze one-half of the cell cultures and expand the other half for DNA isolation.

3.1.1.3. SOUTHERN BLOT ANALYSIS

1. Isolate the genomic DNA of the G418-resistant clones (*see* **Subheading 3.1.5.1.**).
2. Confirm *Aprt* heterozygosity in which one allele is wild type and the other is correctly targeted by Southern blotting using DNA fragments corresponding to *Aprt* fragment and the *neo* sequence as probes.
3. Introduce correctly targeted ES cells into 3.5-d blastocysts.
4. Introduce manipulated blastocyst into pseudopregnant mice to produce chimerical mice.
5. Produce homozygous knockout mice by sib-mated *Aprt* heterozygosity mice.

3.1.2. Derivation of 129/Sv × C3H/HeJ Aprt$^{+/-}$, Aprt$^{-/-}$ MEFs

1. Produce 129/Sv × C3H/HejAprt$^{+/-}$ mice by mating 129/Sv Aprt$^{-/-}$ mice with C3H/Hej Aprt$^{+/+}$ mice.
2. Sacrifice pregnant mice at 13–15 d postcoitum (dpc).
3. Remove embryos and rinse extensively with 1X PBS.

ES Cell DNA Damage Response and Mutagenesis

4. Dissect away heads, liver, and heart. Cut the remaining portion of the embryos into small piece with forceps and scissors. Save a small piece of tissue for genotyping.
5. Transfer the tissue piece to a sterile Falcon tube and add 4 mL ice-cold trypsin.
6. Incubate overnight (16–18 h) at 4°C.
7. Remove trypsin solution and transfer the tissue to a 37°C water bath for 30 min.
8. Add 4 mL DMEM with 10% FBS.
9. Pipet up and down 20 times and plate the cells onto four 100-mm tissue culture dishes containing 9 mL DMEM with 10% FBS.
10. When the cells become confluent (2–3 d), trypsinize, collect all $Aprt^{+/-}$ and $Aprt^{-/-}$ MEFs and freeze in DMEM with 10% FBS and 10% dimethyl sulfoxide; store in liquid nitrogen.

3.1.2.1. MITOMYCIN C TREATMENT OF FEEDER CELLS

1. Thaw a vial of MEFs (for Aprt selection assay, only $Aprt^{-/-}$ MEF cells were used; *see* **Note 1**) at 37°C.
2. Transfer the cells into a 5-mL Falcon tube and collect the cells by centrifuging at $1000g$ for 3 min.
3. Remove the medium and gently resuspend the cells with fresh medium.
4. Culture the cells in 10 mL DMEM medium with 10% FBS in a 37°C incubator until confluent.
5. Remove the medium and add 5 mL fresh medium and 50 μL 1 mg/mL mitomycin C for 90 min in the incubator.
6. Wash the cells with 1X PBS three times.
7. Trypsinize the cells and seed equally into three 100-mm tissue culture plates.

3.1.3. Derivation of 129/Sv × C3H/HeJAprt$^{+/-}$ Mouse ES Cells

1. 129/Sv $Aprt^{-/-}$ mice were mated to C3H/Hej $Aprt^{+/+}$ mice to produce 129/Sv × C3H/HejAprt$^{+/-}$.
2. Isolate 129/Sv × C3H/Hej $Aprt^{+/-}$ F1 blastocysts at 3.5 dpc.
3. Maintain blastocysts on mitomycin C-treated MEFs in ES medium for approx 5 d.
4. Trypsinize the inner cell mass and transfer to fresh feeder wells.
5. Subculture the possible ES cells to produce a permanent cell line.

3.1.4. Measurement of the ES Cell Mutation Frequency

1. Grow $Aprt^{+/-}$ ES cells in medium containing 0.1 mM adenine and 2 μg/mL alanosine to eliminate preexisting APRT-deficient cells for four population-doubling times (32 h).
2. Treat ES cells with 300 μg/mL and 600 μg/mL alkylating agent ethylmethanesulfonate for 5 h or leave untreated as a control.
3. Culture the cells in nonselective medium (ES medium) for an additional 24 h.
4. Seed 5×10^5 cells (treated or untreated) onto a 100-mm plate with mitomycin C-treated APRT-deficient feeder layer; add 10 mL medium containing 2-FA (2 μg/mL) (*see* **Note 2**).
5. Remove 2-FA after 16 h incubation by three washes with DMEM medium.
6. Stain the plate with 10% crystal violet and count the colonies at d 10–12.
7. Calculate mutation frequency (*see* **Note 3**) (**Fig. 2**).

3.1.4.1. ESTIMATION OF COLONY-FORMING EFFICIENCY

1. Plate 500 ES cells per 100-mm plate on mitomycin C-treated *Aprt*-deficient MEFs.
2. Fix the plate 10 d after plating in selection-free ES medium, stain with crystal violet, and score for the number of colonies (*see* **Note 4**).

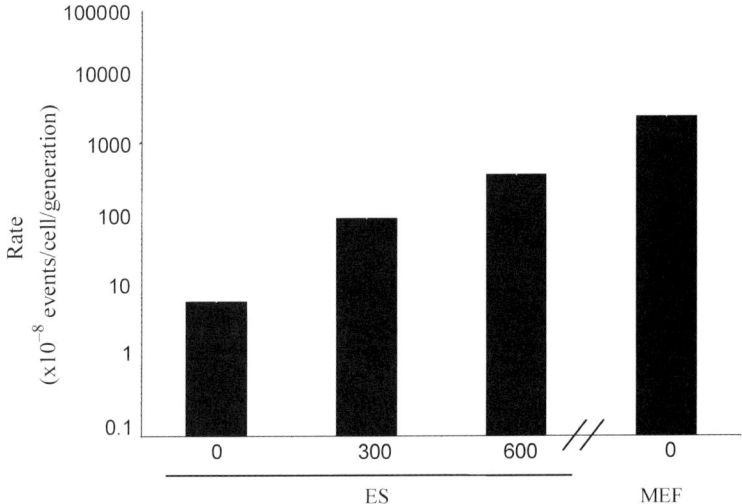

Fig. 2. Spontaneous and induced mutation rates in embryonic stem (ES) cells compared with spontaneous mutation rate in mouse embryonic fibroblasts (MEFs). The solid bars indicate mutation rate. The 0 designation indicates spontaneous mutation rate for ES cells and MEFs. The 300 and 600 designation indicates treatment with 300 μg/mL and 600 μg/mL of the alkylating agent ethylmethanesulfonate.

3.1.5. Determination of Mutation Spectrum

3.1.5.1. ISOLATION OF DNA FROM 2-FA-RESISTANT COLONIES

1. Pick ES cell colonies under dissecting microscopy, trypsinize, and transfer to gelatinized 24-well plates.
2. Add 2 mL ES medium. When cells are confluent, collect the cells by trypsinization and centrifugation at 500g for 5 min.
3. Resuspend the pellets in 100 mL lysis buffer (*see* **Subheading 2.1.2., item 7**) and incubate overnight at 55°C.
4. Extract with phenol/chloroform and precipitate genomic DNA with isopropanol.
5. Wash the pellet with 95% ethanol and resuspend in Tris-EDTA buffer.

3.1.5.2. ALLELE-SPECIFIC POLYMERASE CHAIN REACTION OF *APRT*

1. Isolate genomic DNA from 2-FA-resistant colonies.
2. To 45 μL TaqComplete polymerase chain reaction (PCR) buffer, add 100 ng genomic DNA and 2 μL 100 ng primers derived from exon 3 Aprt 17-2 (5'-CCACAACCTCCCTCCT-TAC-3') and *Aprt* 18-2 (5'-CCACCAAGCAGTTCCTAGTG-3') and a primer derived from neo, which was used to disrupt exon 3, neo2-2 (5'-GAGAACCTGCGTGCAATCCAT CTTG-3').
3. Denature the DNA at 94°C for 2 min and follow by 30 cycles at 94°C (1 min), 55°C (1 min), and 72°C (1 min) with a final extension at 72°C (7 min) (*see* **Note 5**).
4. Group colonies that have lost the untargeted Aprt allele (Aprt$^+$) as class I, and group the other colonies that retained untargeted allele as class II (**Fig. 3A**).

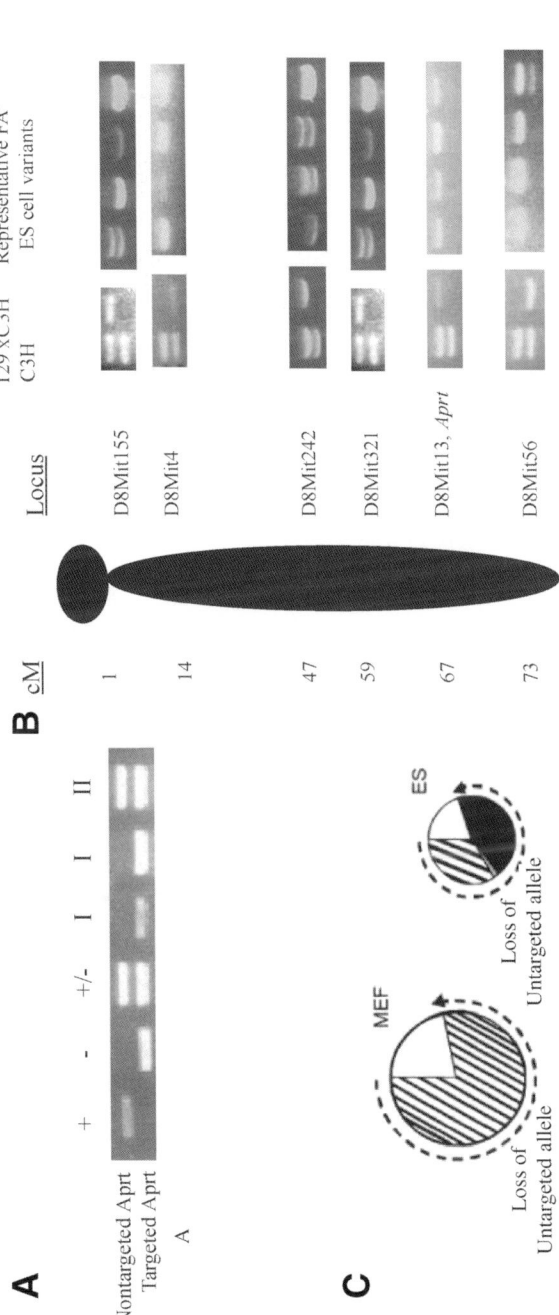

Fig. 3. Different mutational spectra for embryonic stem cells and mouse embryonic fibroblasts determined by polymerase chain reaction (PCR) of polymorphic markers. (**A**) Discrimination between wild-type and targeted *Aprt* by allele specific PCR. (+) Designates DNA from wild-type mice; (−) represents DNA from homozygous null mice with two targeted alleles; and (+/−) designates DNA from mice heterozygous for the *Aprt* allele. The designation I indicates DNA from *Aprt*-negative cells that had lost heterozygosity at *Aprt*; the II designation reflects DNA from *Aprt*-negative cells that had retained the untargeted allele. (**B**) The schematic shows the position in centimorgans of polymorphic markers along the length of chromosome 8, the chromosome that houses *Aprt*. The gel patterns to the right of the schematic chromosome show results from PCR analysis of polymorphic marker D8Mit 155 from DNA of four *Aprt*-negative colonies. Each of the four lanes shows retention of heterozygosity (two bands) or loss of heterozygosity (one band) for each of the markers D8Mit 4, D8Mit 242, D8Mit 321, D8Mit 13, D8Mit 56. (**C**) The dashed line indicates the proportion of events that results in loss of heterozygosity (LOH). The diagonally marked sectors represent events involving LOH because of mitotic recombination. The solid sector represents the proportion of events resulting in LOH as a consequence of nondysjunction. The lightly shaded area represents chromosome loss, and the unshaded sector represents the proportion of events involving point mutation, intragenic deletion, or epigenetic silencing.

321

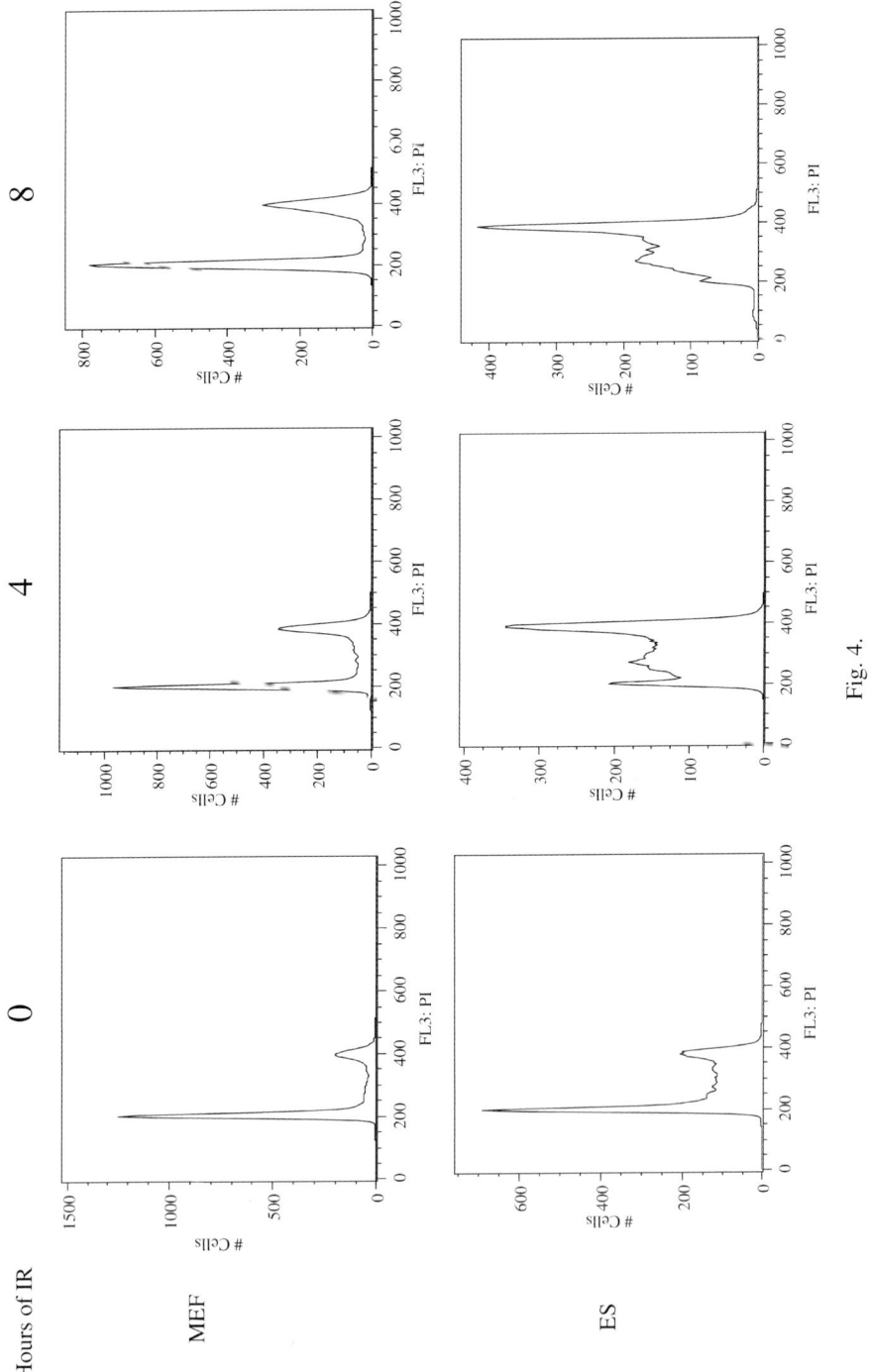

Fig. 4.

3.1.5.3. Sequencing of the Untargeted *Aprt* Allele of Class II Colonies

1. Isolate the class II FA^r variants that retained *Aprt* genomic DNA.
2. Use primers EX1-2U (5'-CGAGGAGGGCACATTCT-3') and EX1-2L (5'GGAGCCAAG-TAGACAGCAGT-3'); EX-5U(5'-TGCTGTCTACTTGGCTCCAT-3') and EX3-5L (5'-CCGGTAGCTCACAAAGGTCA-3') to amplify the *Aprt* exon 1 and 2 (from nucleotide positions 268 to 1710), respectively.
3. Perform PCR at an initial denaturation of 94°C for 3 min followed by 30 cycles at 94°C (1 min), 60°C (1 min), and 72°C (1 min 30 s).
4. Purify and sequence the 1.4-kb and 1.1-kb products.

3.1.5.4. Analysis of Polymorphic Markers in DNA From Colonies That Had Lost Untargeted Allele (Class I)

1. Isolate genomic DNA.
2. Use microsatellite markers from mouse chromosome 8 (**Fig. 3B**).
3. Perform PCR at 94°C for 3 min followed by 30 cycles at 90°C (15 s), 60°C (2 min), and 72°C (1 min 30 s), with a final extension at 72°C (7 min).
4. Fractionate PCR products on a high-resolution 4% agarose in 1X TBE for analysis (**Fig. 3B**).

3.2. Cell Cycle Regulation in Response to DNA Damage

3.2.1. Cell Cycle Analysis

1. Irradiate J11 ES cells with 10-Gy ionizing radiation or leave untreated.
2. Trypsinize and resuspend the cells in 1 mL PBS after 4 or 8 h of irradiation.
3. Slowly add the cells to 4 mL cold 100% ethanol with gentle vortexing.
4. Pellet and stain the fixed cells with 1 mg/mL propidium iodide containing 0.1 mg/mL RNase A.
5. Analyze cell cycle profile by flow cytometry (**Fig. 4**).

3.2.2. Western Blot Analysis of Cell Cycle Checkpoint Proteins

1. Irradiate J11 ES cells with 10-Gy ionizing radiation or leave untreated.
2. Harvest the cells at 30 and 60 min after treatment.
3. Lyse the cells in RIPA buffer and measure the protein concentration with the Bio-Rad protein assay kit according to the manufacturer's instructions.
4. Load equal amounts of whole cell extracts and Laemmli loading sample buffer to an 8% acrylamide gel.
5. Transfer the protein bands to an immobilon-P transfer membrane following the standard protocol *(13)*.
6. Probe the membrane with cell cycle checkpoint protein antibodies such as Chk2 and Cdc25A (**Fig. 5**).

Fig. 4. Embryonic stem (ES) cells do not arrest at a G1/S checkpoint in response to IR treatment. Mouse embryonic fibroblasts (MEFs) and ES cells were subjected to 10 Gy ionizing irradiation or left untreated. At 4 or 8 h after the irradiation, cells were stained with propidium iodide and analyzed by flow cytometry. When MEFs were exposed to IR, they manifested a G1 arrest. In contrast to somatic cells, ES cells had no apparent G1 checkpoint in response to IR irradiation. A large number of cells were arrested in G2, consistent with an active G2 checkpoint *(8)* and a majority of cells arrested in S phase.

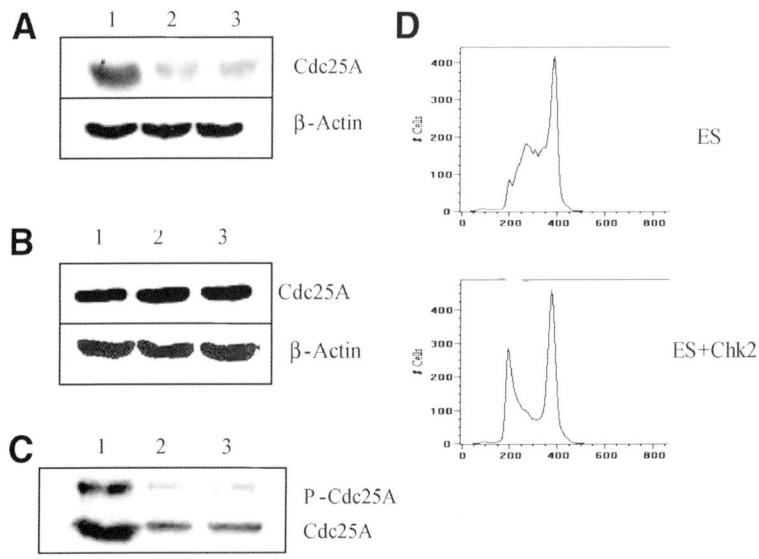

Fig. 5. Cdc25A phosphatase is degraded in MEFs but not in ES cells in response to IR. Ectopic expression of Chk2 induces Cdc25A degradation and restores G1/S checkpoint in stem cells. (**A**) Mouse embryonic fibroblasts (MEFs) were irradiated with 10 Gy X-ray, lysed with ristocetin-induced platelet aggregation buffer, and immunoblotted with Cdc25A monoclonal antibody (F-6, Santa Cruz Biotechnology). **Lane 1** is an untreated sample. **Lanes 2** and **3** are samples 30 and 60 min after IR treatment, respectively. (**B**) ES cells were irradiated as the MEFs and subjected to Western blots with Cdc25A monoclonal antibody at the same time intervals as the MEFs. (**C**) ES cells were transfected with GFP-Chk2 fusion construct. After 24 h, cells were irradiated with 10 Gy and harvested in 30-min intervals. Cell lysates were fractionated by SDS polyacrylamide gel electrophoresis and probed with Cdc25A antibody. **Lane 1** was the unirradiated sample. **Lanes 2** and **3** were samples obtained 30 and 60 min after irradiation, respectively. (**D**) ES cells were transfected with Chk2 (right panel) and untransfected ES cells (left panel) were irradiated with 10 Gy ionizing irradiation. At 8 h after irradiation, the cell cycle distribution was analyzed by flow cytometry.

3.2.3. Immunofluorescence

1. Grow ES cells on gelatinized cover slips overnight in ES medium.
2. Irradiate the cells with 10-Gy ionizing radiation or leave untreated.
3. Fix the cells with Formalde-Fresh for 15 min.
4. Permeabilize the cells in PBS containing 1% NP40 and block with 10% rabbit serum for 1 h.
5. Incubate the cells with monoclonal antibodies to Chk2 and γ-tubulin for 1 h.
6. Wash the cells with 1X PBS and stain the cells with secondary antibody (1:700 dilution) coupled with Alexa Fluor 594 for Chk2, Alexa Fluor 488 for γ-tubulin, and DRAQ5™ (1:700 dilution) for ES cell DNA for 30 min.
7. Wash three times with PBS and mount with antifading gel.
8. Analyze the images by confocal microscopy.

3.3. Measurement of Apoptosis

3.3.1. ES Cell Transfection
1. Culture ES cells at 37°C until confluent.
2. Trypsinize the cells and place in six-well plates for 3 h.
3. Dilute FuGENE by adding 3 µL FuGENE to 100 µL serum-free medium.
4. Add 1 µg Chk2-GFP DNA construct to the diluted FuGENE.
5. After 30-min incubation, add the mix FuGENE solution to the ES cells.
6. Harvest the cells 24 h after transfection.

3.3.2. Apoptosis Analysis
1. Irradiate the wild-type or transfected ES cells with 10-Gy ionizing radiation (IR) or leave untreated.
2. Harvest the cells 16 h after irradiation and resuspend the cells in 1X binding buffer.
3. Stain the cells with 5 µL annexin-V-Cy5 and 2.5 g/mL propidium iodide.
4. Incubate the cell suspension for 15 min at room temperature and analyze by flow cytometry.

4. Notes
1. We use APRT-deficient MEF cells as feeder because the adenine analog 2-FA will kill the cells with APRT activity.
2. Typically, the spontaneous mutation frequency in the ES cell is about 10^{-6}. To get enough mutated colonies to characterize the mutation spectrum, usually we do 20 plates per experiment.
3. Mutation frequency = number of colonies after selection/number of cells seeded × colony-forming efficiency.
4. Colony-forming efficiency = colony number on the plates/500.
5. Primers APRT 17-2 and APRT 18-2 amplified a 700-bp fragment characteristic of wild-type allele; primers APRT 17-2 and neo2-2 amplified a 300-bp fragment characteristic of the targeted allele.

References
1. Hoeijmakers, J. H. (2001) Genome maintenance mechanisms for preventing cancer. *Nature* **411**, 366–374.
2. Hartwell, L. H. and Weinert, T. A. (1989) Checkpoints controls that ensure the order of cell cycle events. *Science* **246**, 629–634.
3. Elledge, S. J. (1996) Cell cycle checkpoints: preventing and identity crisis. *Science* **274**, 1664–1672.
4. Bartek, J. and Lukas, J. (2001) Pathways governing G1/S transition and their response to DNA damage. *FEBS* **490**, 117–122.
5. Odorico, J. S., Kaufman, D. S., and Thomson, J. A. (2001) Multilineage differentiation from human embryonic stem cells lines. *Stem Cells* **19**, 193–204.
6. Stambrook, P. J., Shao, C., Stockelman, M., Boivin, G., Engle, S. J., and Tischfield, J. A. (1996) APRT: a versatile in vivo resident reporter of local mutation and loss of heterozygosity. *Environ. Mol. Mutagen.* **28**, 471–482.
7. Shao, C., Deng, L., Hehegariu, O., Liang, L., Stambrook, P. J., and Tischfield, J. A. (2000) Chromosome instability contributes to loss of hererozygosity in mice lacking p53. *Proc. Natl. Acad. Sci. USA* **97**, 7405–7410.
8. Cervantes, R. B., Stringer, J. R., Shao, C., Tischfield, J. A., and Stambrook, P. J. (2002) Embryonic stem cells and somatic cells differ in mutation frequency and type. *Proc. Natl. Acad. Sci. USA* **99**, 3586–3590.

9. Aladjem, M., Spike, B. T., Rodewald, L. W., et al. (1998) ES cells do not activate p53-dependent stress responses and undergo p53-independent apoptosis in response to DNA damage. *Curr. Biol.* **8,** 145–155.
10. Hiroa, A., Kong, Y-Y., Matsuoka, S., et al. (2000) DNA damage-induced activation of p53 by checkpoint kinase Chk2. *Science* **289,** 1824–1827.
11. de Waard, H., de Wit, J., Gorgels, T. G., et al. (2003) Cell type-specific hypersensitivity to oxidative damage in CSB and XPA mice. *DNA Repair* **2,** 13–25.
12. Engle, S. J., Stockelman, M. G., Chen, J., et al. (1996) Adenine phosphoribosyltransferase-deficient mice develop 2,8-dihydroxyadenine nephrolithiasis. *Proc. Natl. Acad. Sci. USA* **93,** 5307–5312.
13. Sambrook, J., Fritsch, E. F., and Maniatis, T. (1989) *Molecular Cloning,* Cold Spring Harbor Laboratory Press, Cold Spring Harbor, NY.

23

Ultraviolet-Induced Apoptosis in Embryonic Stem Cells In Vitro

Dakang Xu, Trevor J. Wilson, and Paul J. Hertzog

Summary

Embryonic stem (ES) cells, and the inner cell mass from which they are derived, are hypersensitive to DNA damage and appear to have specific cellular defense systems for DNA repair and the elimination of damaged cells. These mechanisms differ from somatic cells and are vital to minimize developmental defects that would potentially result from the continued proliferation and differentiation of abnormal cells into adult cell lineages. Although the DNA damage-induced signaling cascades activated in these cells are known to include p38 and c-Jun N-terminal protein kinase mitogen-activated protein kinase pathways and activation of a variety of transcription factors, including p53, nuclear factor-κB, and activator protein-1, the nature of the specific mechanisms unique to these cells remains to be elucidated. Here, we describe the use of homozygous knockout ES cells to investigate the role of Ets1 in the response to DNA damage in these cells. These studies demonstrate that Ets1 is required for optimal p53 function in this response and further demonstrate the potential for knockout ES cells to elucidate the role of specific genes in early embryonic cell responses.

Key Words: Apoptosis; ES cell; p53; transactivation; UV.

1. Introduction

The inner cell mass of the mouse blastocyst contains between 20 and 40 pluripotent cells that differentiate into all the different cell types that constitute the embryo proper. The small size of this population of cells, which are the basis of embryonic development, suggests that these cells should be equipped with efficient cellular defense mechanisms to cope with DNA damage to avoid the production of mutated daughter cells that will have detrimental effects on embryogenesis and development. Indeed, preimplantation embryos are hypersensitive to DNA damage and can rapidly undergo apoptosis without cell cycle arrest (*1*). Furthermore, apoptosis is observed in the inner cell mass of up to 75% of preimplantation embryos (*2*).

Embryonic stem (ES) cells are derived from the inner cell mass of the blastocyst and are characteristically pluripotent. These cells have been extensively exploited to generate genetically altered mice *(3)* but are also an important resource to investigate these early embryonic checkpoints. Similar to the inner cell mass from which they are derived, ES cells have specific cellular defense systems that include DNA repair mechanisms and the elimination of damaged ES cells via apoptosis *(4)*. The proapoptotic protein p53 plays a critical role in the cellular stress response and appears to be involved in these early embryonic checkpoint controls. Significantly, the absence of p53 resulted in a high rate of embryonic malformations *(1,5)*. This suggests that mechanisms exist to avoid the expansion of potentially abnormal embryonic cells, which could contribute to developmental abnormalities.

DNA damage can occur as a result of replication/recombination errors, oxidative stress, irradiation, and environmental toxins. A variety of agents have been used to experimentally induce DNA damage and induce apoptosis, many inducing oxidative events that result in lipid peroxidation and DNA breaks. Short-wavelength ultraviolet (UV) radiation (UVC; 190–290 nm) is efficiently absorbed by DNA, predominantly causing DNA damage in the form of pyrimidine dimers and (6-4) photoproducts *(6)*.

These DNA lesions halt RNA polymerase II elongation and activate several signaling cascades. In somatic cells, one signaling cascade involves p53-dependent cell cycle arrest via Chk1/ATR and enzymatic repair of the DNA damage by nucleotide excision repair mechanisms. If successfully repaired, then the cells reenter the cell cycle; however, this mechanism could allow accumulation of disease-predisposing mutations if cells with incomplete repair persisted. Other DNA damage-induced signaling cascades include p38 and c-Jun N-terminal protein kinase mitogen-activated protein (JNK MAP) kinase pathways and activation of a variety of transcription factors, including p53, nuclear factor (NF)-κB, and activator protein (AP)-1, which can have both pro- and antiapoptotic properties *(7–9)*. In preimplantation embryos and ES cells, however, cell cycle arrest does not occur following DNA damage. Thus, the mechanisms that control the regulation of DNA repair, cell cycle progression, and apoptosis differ from those in somatic cells and remain to be elucidated in detail.

Gene targeting is one of the most powerful tools to define the function of signaling molecules and their potential role in development and disease etiology. By using these techniques, more than 2000 mutant mouse lines have been produced. In addition to the generation of knockout mice, gene-targeted ES cells can be a useful resource for examining the function of the gene. Homozygous knockout ES cells have been used to investigate the role of genes in the mediation of various cellular activities, such as proliferation, differentiation, and apoptosis. Here, we describe exposure of Ets1 knockout ES cells to DNA-damaging agents, such as UVC irradiation, to elucidate the role of Ets1 in the induction of apoptosis in ES cells in vitro.

2. Materials

2.1. Tissue Culture

1. ES cell lines (J1, D3, W9.5, etc.): experiments were performed using the mouse ES cell lines derived from the 129Sv strain (a kind gift from Whitehead Institute for Biomedical Research, Cambridge, MA) passage 14-30. Cells were cultured as described previously *(10,11)*.

2. ES cell medium: Dulbecco's modified Eagle's medium (DMEM; Gibco, Carlsbad, CA; cat. no. 12430-054) supplemented with 15% fetal calf serum (FCS; CSL, Melbourne, Australia; cat. no. 12003-500M; lot no. 1L0403); 10 mM nonessential amino acids solution (Gibco, cat. no. 11140-050); 90 µM 2β-mercaptoethanol (Gibco, cat. no. 21985-023); penicillin/streptomycin (0.5%) (Gibco; cat. no. 15140-122); 10^3 U/mL leukemia inhibitory factor (LIF; Chemicon International, Temecula, CA; cat. no. ESG 1107). For 600 mL ES cell medium, combine 500 mL DMEM, 90 mL FCS (15%), 6 mL 1 M nonessential amino acids (10 mM), 3 mL penicillin/streptomycin (0.5%), 1 mL 50 mM 2β-mercaptoethanol (90 µM), and 60 µL 10^7 U/mL LIF (10^3 U/mL).
3. 0.25% (w/v) trypsin-EDTA (ethylenediaminetetraacetic acid): 0.25% trypsin-1 mM EDTA, 0.02 M HEPES (Gibco, cat. no. 25200-056).
4. 0.2% Gelatin (Sigma, St. Louis, MO; cat. no. G-1890): 0.2 g gelatin dissolved in 100 mL phosphate-buffered saline (PBS).
5. 100 µg/mL propidium iodide (Sigma, cat. no. P4170) (100X stock solution).
6. Sterile Petri dishes (Falcon).

2.2. Immunofluorescence Analysis

1. PB buffer: 100 mM PIPES at pH 6.8, 2 mM MgCl$_2$, 1 mM EGTA for a stock solution of 1 M PIPES, pH 6.8, dissolve 151.2 g PIPES (free acid) in 400 mL purified water; adjust pH to 6.8 by adding NaOH and adjust to 500 mL with purified water. For 1L PB buffer, combine 100 mL PIPES, pH 6.8, 4.1 g MgCl$_2$·6H$_2$O, and 0.38 g EGTA-NaOH; make up to 1 L with purified water.
2. 3.7% paraformaldehyde in PB buffer: dissolve 0.37 g paraformaldehyde in 10 mL PB buffer by heating at 60°C in a fume hood and stirring until it dissolves. Cool before use.
3. 0.5% NP-40. Prepare a stock of 5% NP-40 in PBS. For 10 mL 0.5% NP-40, add 1 mL 5% NP-40 to 9 mL PB buffer.
4. 0.01% PBS-Tween-20. In 100 mL PBS buffer, add 10 µL Tween-20 (Sigma, cat. no. P7949).
5. Bovine serum albumin (BSA)-PBS-Tween-20: in 10 mL 0.01% PBS-Tween-20, add 0.1 g BSA.
6. Mouse monoclonal antibody against stage-specific embryonic antigen (SSEA) 1 (Chemicon, cat. no. MAB4301).
7. Mouse immunoglobulin G1 isotype control (Dako, Golstrup, Denmark; cat. no. X0931).
8. Fluorescein isothiocyanate-conjugated antimouse antibodies (Molecular Probes, Eugene, OR; cat. no. A-11029).
9. Antifade medium (Bio-Rad, Hercules, CA; cat. no. 170-3140).
10. Fluorescence microscope (Leica, Nussloch, Germany).

2.3. Flow Cytometry

1. Staining buffer: PBS, pH 7.4–7.6, 2% heat-inactivated FCS, and 0.2% sodium azide. To prepare 1 L 1X PBS, combine: 8.233 g Na$_2$HPO$_4$, 2.345 g Na$_2$HPO$_4$·H$_2$O, and 4 g NaCl. Adjust to pH 7.4. In 100 mL PBS, add 2 mL FCS and 0.2 g sodium azide.
2. Perm buffer: staining buffer plus 0.5% saponin (Sigma, cat. no. S-2149). To 100 mL staining buffer, add 0.5 g saponin.
3. Superperm buffer: 3 mL perm buffer plus 1 mL FCS (filtered).
4. Fixation buffer (3.7% paraformaldehyde) (*see* **Subheading 2.2., item 2**).

2.4. Alkaline Phosphatase Activity

To monitor alkaline phosphatase activity, use Vector red alkaline phosphatase substrate kit I (Vector Laboratories, Burlingame, CA; cat. no. SK-5100).

2.5. Apoptosis Analysis

1. Stratalinker 2400 UV crosslinker (Stratagene, La Jolla, CA).
2. Propidium iodide solution: 50 mg/mL propidium iodide (Sigma, cat. no. P4170) in PBS.
3. 80% ethanol (cold).
4. Ribonuclease (RNase) A (Sigma, cat. no. R4875).
5. Flow cytometer (Beckman Coulter, Fullerton, CA; model no. EPICS 752).
6. Lysis buffer for DNA extraction: 100 mM Tris-HCl, pH 8.0, 25 mM EDTA, and 0.5% Triton X-100. Prepare the following stock solutions: for 1 M Tris-HCl, pH 8.0, dissolve 121.14 g Tris-base in 800 mL purified water; titrate to pH 8.0 with concentrated HCl, then adjust to 1 L. For 0.25 M EDTA, add 23.27 g Na$_2$EDTA to 200 mL purified water; stir vigorously on a magnetic stirrer, adjust to pH 8.0 with NaOH, and make up to 250 mL with purified water. For 10 mL lysis buffer, combine 1 mL 1 M Tris-HCl, pH 8.0, and 1 mL 0.25 M EDTA and add 50 µL Triton X-100 plus purified water to a final volume of 10 mL.

2.6. Induction of p53 and Target Gene Analysis

1. 2X sodium dodecyl sulfate (SDS) polyacrylamide gel electrophoresis sample buffer: 0.5 M Tris-HCl, 4% SDS, 10% sucrose, 10% 2-mercaptoethanol, and 0.004% bromophenol blue. To prepare stock solution of 2 M Tris-HCl, pH 6.8, dissolve 242.28 g Tris-base in 800 mL purified water, titrate to pH 6.8 with concentrated HCl, then adjust to 1 L. For 50 mL 2X sample buffer, combine 12 mL 2 M Tris-HCl, pH 6.8, 2 g SDS, 5 g sucrose, 5 mL 2-mercaptoethanol, and 2 mg bromophenol blue and purified water to 50 mL.
2. Immobilon membrane (Millipore, Bedford, MA; cat. no. IPVH00010).
3. Tris-buffered saline plus Tween-20 (TBST): 20 mM Tris-HCl, 135 mM NaCl, pH 7.6, 0.1% Tween-20. To prepare stock solution of 2 M Tris-HCl, pH 7.6, dissolve 242.28 g Tris-base in 800 mL purified water, titrate to pH 7.6 with concentrated HCl, then adjust to 1 L. For 1X TBST (500 mL), combine 5 mL 2 M Tris-HCl, pH 7.6, 4.383 g NaCl, 0.5 mL Tween-20, and 494 mL purified water.
4. 10% fat-free milk powder in 1X TBST.
5. Anti-p53 (Santa Cruz, Santa Cruz, CA; cat. no. sc-6243).
6. β-Tubulin control antibody (Chemicon, Temecula, CA; cat. no. MAB3408).
7. Horseradish peroxidase (HRP)-conjugated secondary antibodies (Dako, cat. no. P0161 for antimouse, cat. no. P0448 for antirabbit).
8. Super-Signal (Pierce, Rockford, IL; cat. no. 34080).
9. BioMax autoradiographic film (Kodak, Rochester, NY; cat. no. X-OMAT 480 RA).
10. 0.2X SSC, 0.1% SDS. In 1 L H$_2$O, combine 1.74 g NaCl, 8.82 g sodium citrate, and 1 g SDS.
11. Fuji BAS 1000 PhosphorImage analyzer and *MacBAS* version 2.5 software (Fuji, Melbourne, Australia).
12. 1X TBE: in 1 L H$_2$O, combine 10.8 g Tris-base, 5.5 g boric acid, and 0.93 g EDTA-2Na.
13. 2X DNA binding buffer: 1 mM EDTA, 10 mM Tris-HCl at pH 8.0, 50 mM NaCl, 5 mM dithiothreitol, 100 µg/mL BSA, and 100 µg/mL poly(dI-dC). For 500 µL, 2X DNA binding buffer, combine 2 µL 250 mM EDTA, 5 µL 2 M Tris-Cl, pH 8.0, 10 µL 5 M NaCl, 5 µL 1 M dithiothreitol, 5 µL 10 mg/mL BSA, 50 µL 1 µg/µL poly(dI-dC).

3. Methods

3.1. Maintenance of ES Cells

ES cells are maintained on an irradiated embryonic fibroblast feeder layer in the presence of LIF with routine media changes and subculturing using trypsin-EDTA *(12)*.

Because the responses of undifferentiated ES cells and differentiated cultures to UV irradiation will differ, it is essential to examine the morphology and differentiation status of each ES cell clone. We have validated the undifferentiated status of ES cell clones by ensuring that the clones each expressed SSEA-1, Oct4 (by immunofluorescence and flow cytometry), and alkaline phosphatase (by histochemical demonstration of enzyme activity). These markers could also be examined by Northern blot or quantitative reverse transcriptase polymerase chain reaction if required.

3.1.1. Immunofluorescence Analysis

1. ES cells were plated onto gelatin-coated cover slips. Cover slips are placed in 12-well plates and covered with warm 0.2% gelatin solution (37°C). After 30 min at room temperature, the gelatin is removed by aspiration, and 5×10^5 ES cells are added in 2 mL culture media.
2. At 24 h after plating, wash cells once with PB.
3. Fix in 3.7% formalin in PB for 20 min.
4. Extract with 0.5% NP-40 in PB for 10 min.
5. Rinse the fixed cells three times with 0.01% PBS-Tween-20.
6. Block nonspecific binding by incubation in 1% BSA-PBS-Tween-20 for 1 h.
7. Incubate the clones on cover slips for 1 h (at room temperature) with SSEA-1 or a mouse-immunoglobulin G1 isotype control at a concentration of 0.35 mg/mL.
8. Rinse four times with 0.01% PBS-Tween-20.
9. Add fluorescein isothiocyanate-conjugated antimouse antibodies at 1:100 dilution for another 1 h at room temperature.
10. Wash cells with PBS-Tween-20 for 15 min, followed by a final wash in PBS.
11. Mount slides in antifade medium (Bio-Rad).
12. Analyze by fluorescence microscopy (Leica).

3.1.2. Intracellular Staining for Flow Cytometry (see **Note 1**)

1. Add 10^6 cells per tube.
2. Wash cells by resuspending in 3 mL PBS, centrifugation at 1500g for 5 min, and aspirating supernatant.
3. Resuspend cells in 0.5 mL PBS and add 0.5 mL fixation buffer. Gently vortex and incubate at room temperature for 20 min.
4. Wash as in **step 1** with PBS (*see* **Note 2**).
5. Wash as in **step 1** with perm buffer and then wash once with superperm buffer.
6. Add primary antibody (SSEA-1 or Oct4) for intracellular staining and incubate at room temperature for 30 min (*see* **Note 3**).
7. Wash twice with perm buffer, add secondary antibody, then incubate at room temperature for 30 min.
8. Wash twice with perm buffer, then wash once with staining buffer.
9. Resuspend in staining buffer and analyze by fluorescent-activated cell sorting (**Fig. 1**).

3.1.3. Alkaline Phosphatase Activity

Endogenous alkaline phosphatase expression by undifferentiated ES cells can be readily detected using the Vector red alkaline phosphatase substrate kit according to the manufacturer's procedures.

1. Fix ES cells on cover slips with cold 95% ethanol for 10 min. Dehydrate in cold 100% ethanol for 10 min and rinse with PBS.

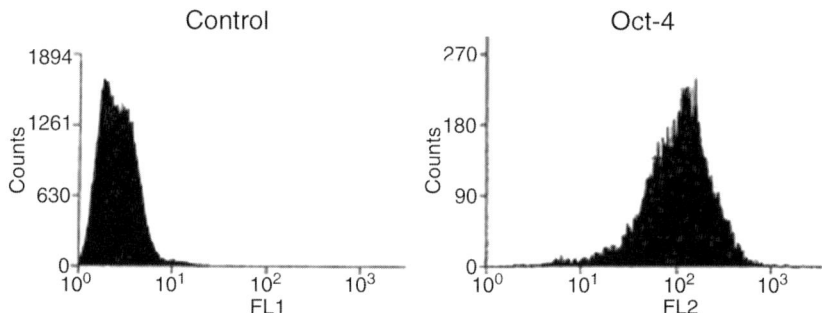

Fig. 1. Embryonic stem cells stained for Oct4. Flow cytometry histogram demonstrating that more than 95% of cells were positive, indicating that this culture contained very few differentiated cells.

2. Prepare the Vector red substrate working solution immediately before use by adding two drops reagent 1 to 5 mL 100 mM Tris-HCl, pH 8.2–8.5 buffer and mixing well; add two drops reagent 2 and mix well; finally, add two drops reagent 3 and mix well.
3. Add substrate solution to fixed ES cells and incubate at room temperature for 20–30 min.
4. Wash in 100 mM Tris-HCl at pH 8.2–8.5 buffer for 5 min and rinse in water.
5. Observe the stained cells with an inverted fluorescence microscope.
6. Determine the proportion of undifferentiated cells by scoring at least 300 randomly chosen ES cell colonies. Colonies consisting entirely of densely packed alkaline phosphatase+ cells are scored as "undifferentiated," whereas colonies consisting of a mixture of unstained and stained cells or entirely of unstained cells with flattened irregular morphology are considered "differentiated."

3.2. Induction of Apoptosis by UV Irradiation

Prior to analysis, ES cells are depleted of feeder cells by differential adherence.

1. Detach cells using trypsin-EDTA.
2. Plate feeder cells and allow to readhere on new gelatin-coated culture dishes for 15 min.
3. Grow nonadherent cells (enriched for ES cells) in monolayers on gelatinized 6- to 10-cm Petri dishes.
4. Repeat this subculture procedure twice before use in analyses.
5. Immediately before assays, stain samples with propidium iodide (*see* **Subheading 3.3.1.**) and analyze by flow cytometry to ensure similar cell cycle status of samples.
6. Remove culture media from exponentially growing cultures and immediately expose to 10 (low) to 50 (high) J/m^2 UVC (254 nm) using a Stratalinker 2400 UV crosslinker.
7. Add fresh culture media and incubate for the indicated periods of time (9–24 h).

3.3. Apoptosis and Cell Cycle Analysis

Apoptosis can be examined by a variety of techniques, including propidium iodide or Hoechst staining *(13)*, detection of activated Caspase3, and DNA fragmentation. It is generally accepted that more than one of these approaches are required to validate that the cell death observed is truly apoptosis *(14)*.

3.3.1. Propidium Iodide Staining

Cell cycle analysis and quantitation of apoptosis were performed as described by Gong et al. *(15)*.

1. Fix more than 10^5 cells (both adherent and floating cells) with 80% cold ethanol (*see* **Note 4**).
2. Wash the fixed cells twice in PBS.
3. Resuspend in 500 µL PBS containing RNase A (0.1 mg/mL final concentration).
4. Incubate for 30 min at room temperature.
5. Add 2 µL of a 50 mg/mL solution of propidium iodide (10 µg/mL final concentration) and analyze cells on a flow cytometer.
6. Gate out cell debris, doublets, and fixation artifacts and record G0–G1, S, G2–M, and apoptotic populations on a logarithmic scale. Apoptotic cells are identified as the lower fluorescence peak (sub-G1 peak on DNA frequency histograms) because of their reduced DNA content. An illustration of the apoptosis effect of UV treatment is presented in **Fig. 2A**.

3.3.2. DNA Fragmentation Assay

1. After stimuli with UV radiation, collect both adherent and nonadherent ES cells for each sample (2×10^6 cells). First, gently resuspend nonadherent ES cells with a pipet and collect into tubes. Then, rinse the plates twice with PBS pooling into the tubes as above. Add 0.5 mL trypsin-EDTA to plates and incubate at 37°C for 3 min. Collect these adherent ES cells and wash twice with PBS pooling as above.
2. Pellet by centrifugation at 1500*g* for 5 min.
3. Remove supernatant and lyse the cells in 0.33 mL of a buffer consisting of 100 mM Tris-HCl, pH 8.0, 25 mM EDTA, and 0.5% Triton X-100.
4. Pellet the insoluble fraction by centrifugation at 27,000*g* for 20 min at 4°C.
5. Transfer the supernatant to a fresh tube and purify the DNA by phenol extraction and ethanol precipitation.
6. Suspend the precipitated DNA in 20 µL H_2O.
7. Treat the samples with 50 µg/mL RNase for 30 min at 37°C.
8. Separate 10 µL of each by electrophoresis in 2% agarose gels and visualize DNA by a UV illuminator (**Fig. 2B**).

3.4. Induction of p53 and Target Gene Analysis

Because UV-induced ES cell apoptosis has been shown to be p53 dependent *(5,13)*, we examined the expression of a number of p53-dependent and p53-independent apoptosis-regulating genes, including perp, mdm2, cyclin G, bax (p53 dependent), Bcl-2, and Bcl-XL (p53 independent; **Fig. 3** and data not shown).

3.4.1. Western Blot Analysis

1. Rinse cells with PBS.
2. Lyse in SDS polyacrylamide gel electrophoresis sample buffer.
3. Boil 0.1–0.2 mg cell lysates for 15 min, then electrophorese the protein 8–12% SDS–polyacrylamide gels.
4. Electroblot the separated proteins to Immobilon membrane using the semidry transfer method (Bio-Rad).
5. Block the membranes with 10% fat-free milk powder in 1X TBST for 2 h at room temperature.

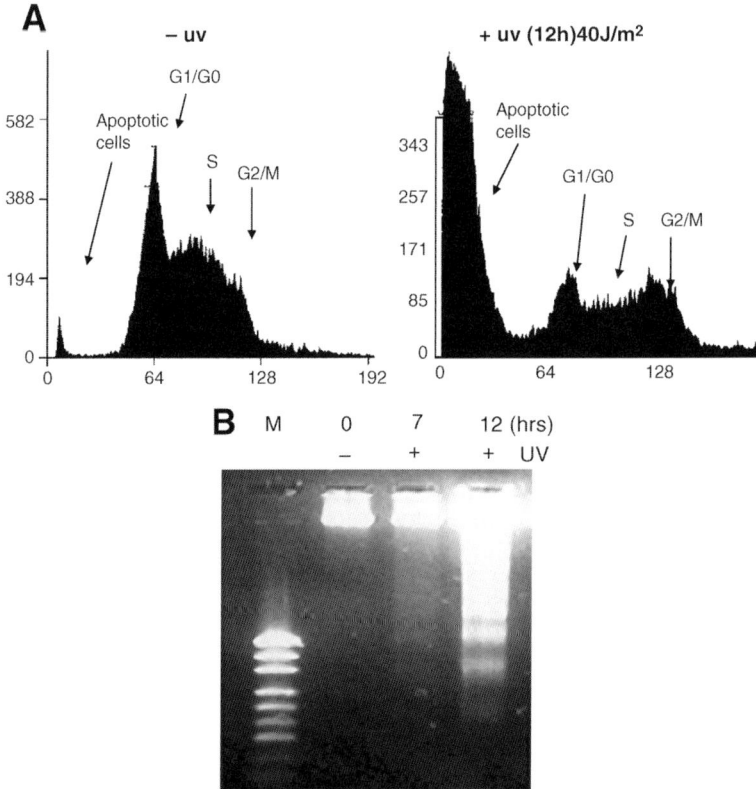

Fig. 2. (**A**) Flow cytometric analysis of embryonic stem (ES) cells stained with propidium iodide 12 h after ultraviolet (UV) irradiation (40 J/m^2). After UV treatment, the proportion of cells that have a hypodiploid (sub-2N) DNA content indicative of apoptosis increases. Cells in G$_0$ and G2/M (by DNA content) are indicated. (**B**) Ethidium bromide staining of DNA from ES cells after 40 J/m^2 UV irradiation analyzed by agarose gel electrophoresis and demonstrating bands of DNA fragmentation characteristic of apoptosis; **lane 1** shows pUC19/HpaII marker.

6. Incubate membranes with appropriate dilutions of specific antibodies in 1X TBST and 10% fat-free milk powder (e.g., anti-p53 or β-tubulin) overnight at 4°C on a rotating wheel.
7. Wash the membranes three times for 10 min with 1X TBST/0.1% Tween-20.
8. Incubate for 1 h at room temperature with appropriate HRP-conjugated secondary antibodies at a 1:1000 dilution in 1X TBST and 10% fat-free milk powder.
9. Wash the membranes another three times for 10 min with 1X TBST/0.1% Tween-20.
10. Use Super-Signal chemiluminescent detection of HRP to visualize the proteins by exposing the membrane to BioMax autoradiographic film. β-tubulin protein expression levels were used to normalize protein loading for each sample (**Fig. 3A**).

3.4.2. Northern Blot Analysis

We have used Northern blot analysis to demonstrate that a variety of genes known to be regulated by p53 are activated following UV irradiation (**Fig. 3**; *[11]*). These data

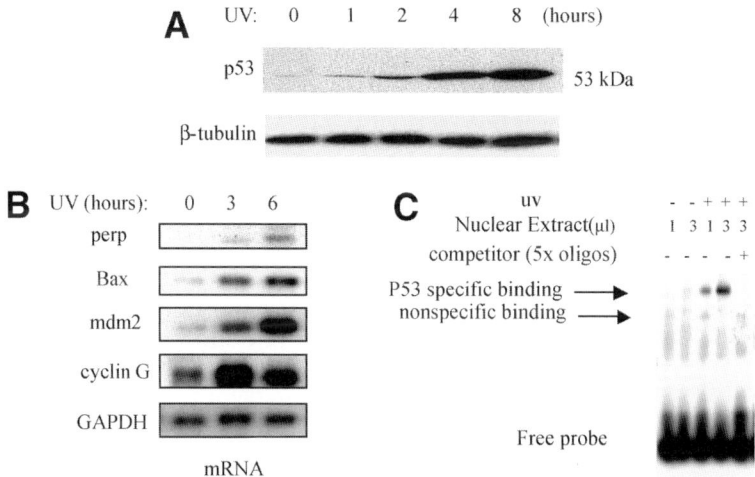

Fig. 3. (**A**) Expression of p53 protein in embryonic stem (ES) cells by Western blot. Cells were isolated at the indicated time points (0, 1, 2, 4, and 8 h) post-ultraviolet (UV) irradiation (40 J/m^2). β-Tubulin loading control is also shown. (**B**) Northern blot analysis of messenger RNA isolated from ES cells at times indicated after UV treatment. Probes specific for mouse perp, mdm2, cyclin G, and bax were used to detect their respective gene levels, and glyceraldehyde-3-phosphate was used as a loading control. (**C**) Nuclear extracts were prepared from untreated or UV-irradiated ES cells (40 J/m^2) 5 h postirradiation. The indicated volumes of extracts were incubated with [^{32}P]dATP-labeled p53 consensus oligonucleotides with or without a fivefold excess of unlabeled oligonucleotide competitor and separated on acrylamide gels. The position of the specific p53–DNA complex is indicated.

demonstrate that these genes are activated, but additional analyses are required to more directly link p53 to the activation of these genes after UV irradiation.

For example, we have previously shown a role for p53 by using promoter/reporter assays (using luciferase coupled to defined promoter elements) with wild-type promoter constructs compared with those with mutations in the p53REI site of the mdm2 promoter or p53REII site of cyclin G promoter, which have been shown to remove p53 responsiveness *(13,16,17)*. Furthermore, we also showed that chromatin immunoprecipitation assays (ChIP) can demonstrate that p53 binds and activates the mdm2 promoter after UV irradiation under physiological conditions by crosslinking chromatin, immunoprecipitating with p53 antibodies, and testing for the mdm2 promoter by polymerase chain reaction and Southern blotting *(13)*.

1. Prepare poly(A)+ messenger RNA from ES cells after various times and doses of UV treatment.
2. Prepare Northern blots as described in **ref. 18**.
3. Hybridize blots with ^{32}P-labeled target gene complementary DNA as described and then wash using high-stringency conditions (0.2X SSC, 0.1% SDS at 65°C) (*see* **Note 5**).
4. Expose Northern blots to a phosphoimager screen and perform a quantitative analysis using a Fuji BAS 1000 PhosphorImage analyzer and *MacBAS* version 2.5 software (**Fig. 3B**).

3.4.3. Electrophoretic Mobility Shift Assays

We have used EMSA to demonstrate that p53 binds to DNA in ES cells following UV irradiation. These data demonstrate that p53-binding activity was induced in ES cells at 5 h after UV treatment (**Fig. 3C**).

1. Treat ES cells (5×10^5 cells/mL) with or without UV irradiation 5 h prior to harvest and prepare nuclear lysates *(19)*.
2. Label 1 ng purified double-stranded p53 consensus oligonucleotides (5'-AGCTTAGACAT-GCCTAGACATGCCTA-3') with [^{32}P]dATP using T4 polynucleotide kinase.
3. Incubate labeled oligonucleotides for 30 min with nuclear lysates in DNA binding buffer. To some reactions, 5- to 100-fold unlabeled oligonucleotides are added to compete with specific oligonucleotide–protein complexes, or 0.5 µg of specific antibodies can be added to the reaction to "supershift" the specific protein–DNA binding complex *(13)* (*see* **Note 6**).
4. Run reactions on a nondenaturing 6% w/v polyacrylamide gel in 0.5X TBE until free oligonucleotides reach the bottom of the gel to maximize band separation.
5. Dry the gels and expose to Biomax film with an intensifying screen.

4. Notes

1. For intracellular staining, all steps prior to fixation are done on ice.
2. All intracellular staining steps should be in saponin-containing buffers.
3. To stain ES cell with primary antibody, use twice the normal amount that would be used for surface staining.
4. After fixation, the cells can be stored for up to 1 wk at 4°C.
5. Also, probe the blots with a ^{32}P-labeled 1.2-kb PstI fragment of glyceraldehyde-3-phosphate. To enable valid comparisons across separate filters, sets of RNA samples were simultaneously electrophoresed, transferred to membranes, and probed with the same preparation of ^{32}P-labeled probe.
6. To incubate labeled oligonucleotides with nuclear lysates in DNA binding buffer, final volume is kept constant at 15 µL.
7. Some genes such as p53 have been reported to be downregulated after ES cell differentiation, so it is important to ensure that the morphology and differentiation status of the ES cell clones are identical, ideally undifferentiated. Thus, normal morphology and staining is verified by examining the percentage of cells expressing SSEA-1 or Oct4 (immunohistology or flow cytometry) or alkaline phosphatase activity (histology). Alternatively, Oct4 gene expression level can be determined. Each of these markers is only expressed in undifferentiated ES cells and not in differentiated derivatives *(20)*.
8. Our data indicate that the proportion of cells undergoing apoptosis in ES cells was UV irradiation dose dependent, with 60% of ES cells apoptosing after 40 J/m^2 UV irradiation *(11)*. Time-course analysis after 40 J/m^2 UV irradiation revealed a background of 3–5% apoptotic cells, which increased to 20–30% apoptotic cells 4 h postirradiation. The percentage of apoptotic cells steadily increased for up to 12 h postirradiation.
9. Undifferentiated ES cell cultures in the log phase of growth are used for UV treatment to minimize differences between cell clones. We also used feeder cell-depleted ES cells to remove any influence of feeder cells on the assays.
10. DNA damage in mammalian cells stabilizes the p53 protein, which then functions as a cell cycle checkpoint, leading to cell cycle arrest or apoptosis *(21)*. This effect is mediated by p53 transcriptional activity in which p53 regulates the expression of many

genes, including mdm2, cyclin G, perp, bax, gadd45, and p21. Significantly, ES cells preferentially undergo apoptosis without cell cycle arrest *(1)*, and a subset of p53 target genes involved in cell cycle arrest, including p21 and gadd45, do not respond to UV irradiation (data not shown *[13,19]*). Thus, it is possible that, in the cellular context of ES cells, other factors such as Ets1 influence the precise subset of p53 target genes that are regulated.

References

1. Heyer, B. S., MacAuley, A., Behrendtsen, O., and Werb, Z. (2000) Hypersensitivity to DNA damage leads to increased apoptosis during early mouse development. *Genes Dev.* **14,** 2072–2084.
2. Hardy, K., Handyside, A. H., and Winston, R. M. (1989) The human blastocyst: cell number, death and allocation during late preimplantation development in vitro. *Development* **107,** 597–604.
3. Thomas, K. R. and Capecchi, M. R. (1987) Site-directed mutagenesis by gene targeting in mouse embryo-derived stem cells. *Cell* **51,** 503–512.
4. Evan, G. and Littlewood, T. (1998) A matter of life and cell death. *Science* **281,** 1317–1322.
5. Corbet, S. W., Clarke, A. R., Gledhill, S., and Wyllie, A. H. (1999) P53-dependent and -independent links between DNA-damage, apoptosis and mutation frequency in ES cells. *Oncogene* **18,** 1537–1544.
6. de Gruijl, F. R., van Kranen, H. J., and Mullenders, L. H. (2001) UV-induced DNA damage, repair, mutations and oncogenic pathways in skin cancer. *J. Photochem. Photobiol. B* **63,** 19–27.
7. Hanawalt, P. C. (2002) Subpathways of nucleotide excision repair and their regulation. *Oncogene* **21,** 8949–8956.
8. Zhou, B. B. and Elledge, S. J. (2000) The DNA damage response: putting checkpoints in perspective. *Nature* **408,** 433–439.
9. Bender, K., Blattner, C., Knebel, A., Iordanov, M., Herrlich, P. and Rahmsdorf, H. J. (1997) UV-induced signal transduction. *J. Photochem. Photobiol. B* **37,** 1–17.
10. Hwang, S. Y., Hertzog, P. J., Holland, K. A., et al. (1995) A null mutation in the gene encoding a type I interferon receptor component eliminates antiproliferative and antiviral responses to interferons α and β and alters macrophage responses. *Proc. Natl. Acad. Sci. USA* **92,** 11,284–11,288.
11. Lahoud, M. H., Ristevski, S., Venter, D. J., et al. (2001) Gene targeting of Desrt, a novel ARID class DNA-binding protein, causes growth retardation and abnormal development of reproductive organs. *Genome Res.* **11,** 1327–1334.
12. Tessarollo, L. (2001) Manipulating mouse embryonic stem cells. *Methods Mol. Biol.* **158,** 47–63.
13. Xu, D., Wilson, T. J., Chan, D., et al. (2002) Ets1 is required for p53 transcriptional activity in UV-induced apoptosis in embryonic stem cells. *EMBO J.* **21,** 4081–4093.
14. Roberts, K. M., Rosen, A., and Casciola-Rosen, L. A. (2004) Methods for inducing apoptosis. *Methods Mol. Med.* **102,** 115–128.
15. Gong, J., Traganos, F., and Darzynkiewicz, Z. (1994) A selective procedure for DNA extraction from apoptotic cells applicable for gel electrophoresis and flow cytometry. *Anal. Biochem.* **218,** 314–319.
16. Zauberman, A., Lupo, A., and Oren, M. (1995) Identification of p53 target genes through immune selection of genomic DNA: the cyclin G gene contains two distinct p53 binding sites. *Oncogene* **10,** 2361–2366.

17. Zauberman, A., Flusberg, D., Haupt, Y., Barak, Y., and Oren, M. (1995) A functional p53-responsive intronic promoter is contained within the human mdm2 gene. *Nucleic Acids Res.* **23,** 2584–2592.
18. Owczarek, C. M., Hwang, S. Y., Holland, K. A., et al. (1997) Cloning and characterization of soluble and transmembrane isoforms of a novel component of the murine type I interferon receptor, IFNAR 2. *J. Biol. Chem.* **272,** 23,865–23,870.
19. Sabapathy, K., Klemm, M., Jaenisch, R., and Wagner, E. F. (1997) Regulation of ES cell differentiation by functional and conformational modulation of p53. *EMBO J.* **16,** 6217–6229.
20. Wobus, A. M., Holzhausen, H., Jakel, P., and Schoneich, J. (1984) Characterization of a pluripotent stem cell line derived from a mouse embryo. *Exp. Cell Res.* **152,** 212–219.
21. Chao, C., Saito, S., Kang, J., Anderson, C. W., Appella, E., and Xu, Y. (2000) p53 transcriptional activity is essential for p53-dependent apoptosis following DNA damage. *EMBO J.* **19,** 4967–4975.

IV

USE OF EMBRYONIC STEM CELLS IN PHARMACOLOGICAL AND TOXICOLOGICAL SCREENS

24

Use of Differentiating Embryonic Stem Cells in Pharmacological Studies

Brigitte Wdziekonski, Phi Villageois, Cécile Vernochet, Blaine Phillips, and Christian Dani

Summary

Adipocytes and osteoblasts are derived from a common precursor cell. It has been proposed that the bone loss commonly seen during aging or in the pathology of osteoporosis might be partly caused by a deregulation of the normal balance between osteoblast and adipocyte differentiation. In vitro differentiation of mouse embryonic stem cells toward the adipogenic and the osteogenic lineages provides a powerful system for testing effects of compounds on the developmental switch between adipogenesis and osteogenesis and identification of pharmacological targets. In this chapter, an improved protocol to commit mouse embryonic stem cells into both lineages at a correct rate is detailed.

Key Words: Adipocytes; embryonic stem cells; osteoblasts.

1. Introduction

The ongoing global explosion in the incidence of obesity has focused attention on the development of adipose cells. Severe obesity is the result of an increase in fat cell size in combination with increased fat cell number. New fat cells arise from a preexisting pool of adipose stem cells that are present irrespective of age. The nutritional and pharmacological control of the pool of stem cells in adipose tissue requires better knowledge of the characteristics of these cells and the regulation of their differentiation. The development of established preadipocyte cell lines has facilitated the study of different steps leading to terminal differentiation. However, these systems are limited for studying early events of differentiation as they represent cells that are already determined for the adipogenic lineage *(1)*. In vitro differentiation of mouse embryonic stem (ES) cells toward the adipogenic lineage provides an alternative source of adipocytes for study in tissue culture and offers the possibility to investigate regulation of the first steps of adipose cell development *(2)*.

From: *Methods in Molecular Biology, vol. 329: Embryonic Stem Cell Protocols: Second Edition: Volume 1*
Edited by: K. Turksen © Humana Press Inc., Totowa, NJ

Adipocytes and osteoblasts are derived from a common precursor cell. It has been proposed that the bone loss commonly seen during aging or in the pathology of osteoporosis might be partly caused by deregulation of the normal balance between osteoblast and adipocyte differentiation. However, factors that regulate the step when progenitor cells are committed to become an adipocyte or an osteoblast remain elusive. The combination of genetic manipulation of undifferentiated ES cells and in vitro differentiation of ES cells into both adipocytes and osteoblasts provides a powerful system for testing effects of compounds on the developmental switch between osteogenesis and adipogenesis and identification of pharmacological targets. In this chapter, a protocol to commit mouse ES cells into the adipogenic and osteogenic lineages at a high rate is detailed.

2. Materials

2.1. Maintenance of Mouse ES Cells

1. Growth medium (1X, stored at 4°C): to 440 mL Glasgow minimum essential medium (MEM)/BHK21 medium (Gibco BRL, Invitrogen, Cergy Pontoise, France; cat. no. 21710-025), add 5 mL 100X nonessential amino acids (Gibco BRL, cat. no. 11140035); 10 mL 200 mM glutamine (Gibco BRL, cat. no. 25030024); 100 mM sodium pyruvate (Gibco BRL, cat. no. 11360039); 0.5 mL 0.1 M 2-mercaptoethanol; and 50 mL selected fetal calf serum (FCS, S. A. Dutscher, Brumath, France).
2. 0.1 M 2-mercaptoethanol: add 100 µL 2-mercaptoethanol (Sigma, Saint-Quentin Fallavier, France; cat. no. M-7522) to 14 mL sterile water. Store up to 3 wk at 4°C.
3. Leukemia inhibitory factor (LIF; *see* **Note 1**).
4. Phosphate-buffered saline (PBS; Gibco BRL, cat. no. 14190-094), calcium and magnesium free. For 1 L, dissolve 8 g NaCl, 0.2 g KCl, 1.44 g Na_2HPO_4 and 0.2 g KH_2PO_4 into water. Adjust to pH 7.4 and filter-sterilize.
5. Trypsin solution: for a 1X solution, add 1 mL 2.5% trypsin (Gibco BRL, cat. no. 2509008), 1 mL 100 mM ethylenediaminetetraacetic acid (EDTA) and 1 mL chicken serum to 100 mL PBS. Aliquot (10 mL) and store at −20°C. Thawed aliquots are stored at 4°C.
6. 0.1% gelatin: purchase gelatin 2% (Sigma, cat. no. G-1393) and dilute to 0.1% with PBS. Store at 4°C.
7. Tissue culture 25-cm^2 flasks (Greiner, S. A. Dutscher, cat. no. 690175, or Corning, S. A. Dutscher, cat. no. 430168).

2.2. Differentiation Into Adipocytes

1. Retinoic acid: for 10 mM stock solution, dissolve 50 mg all-*trans* retinoic acid (RA; Sigma, cat. no. R-2625) in the dark into 17 mL dimethyl sulfoxide. Aliquot and store at −20°C. Subsequent dilutions of RA are prepared in ethanol and used for one experiment only. After dilution into the culture media, the concentration of ethanol never exceeds 0.1%.
2. Adipogenic differentiation medium: this medium consists of growth medium supplemented with antibiotics, 0.5 µg/mL insulin, 2 nM triiodothyronine and thiazolidinedione (*see* **Note 2**). For 100 mL growth medium, add 100 µL antibiotics, 50 µL insulin 1 mg/mL, 100 µL 2 µM triiodothyronine, and 5 µL 10 mM thiazolidinedione.
3. Insulin (Sigma, cat. no. I-5500) is prepared at 1 mg/mL in cold 0.01 N HCl. Dissolve 10 mg insulin in 10 mL 0.01 N HCl. Mix gently and sterilize by filtration. Aliquot (1 mL) and store at −20°C. Thawed aliquots are stored at 4°C.

Differentiating ES Cells in Pharmacological Studies

4. Triiodothyronine (Sigma, cat. no. T-2877) is prepared at 2 µM in ethanol (stock solution). For a 1 mM solution, dissolve 5 mg triiodothyronine into 7.43 mL ethanol. Dilute in ethanol to get a 2 µM 1000X working solution. Store at −20°C.
5. Antibiotics: 1000X penicillin-streptomycin (5000 IU/mL-5000 µg/mL, Gibco BRL, cat. no. 15070-063). Aliquot (1 mL) and store at −20°C.
6. Bacteriological grade 100- and 60-mm Petri dishes (Greiner, cat. no. 633185 and 628102).
7. Tissue culture 100-mm dishes (Corning, cat. no. 664160, or Greiner, cat. no. 664160).

2.3. Differentiation Into Osteoblasts

1. Retinoic acid (see **Subheading 2.2. item 1**).
2. Osteogenic differentiation medium: this medium consists of αMEM medium with 10% FCS and supplemented with 50 µg/mL L-ascorbic acid-2-phosphate, 10 mM β-glycerophosphate, and 0.1 mM dexamethasone (DEX). For 100 mL αMEM medium with 10% FCS, add 100 µL L-ascorbic acid-2-phosphate 50 mg/mL, 666 µL 1.5 M β-glycerophosphate, and 10 µL 1 mM DEX.
3. L-Ascorbic acid 2-phosphate (Sigma, cat. no. A-960) is prepared at 50 mg/mL in water. Dissolve 500 mg L-ascorbic acid 2-phosphate into 10 mL H_2O. Sterilize on filter, and aliquots are stored at −20°C. Aliquots are thawed only once.
4. β-Glycerophosphate (Sigma, cat. no. G-1150) is prepared at 1.5 M in water. Dissolve 3.24 g into 10 mL H_2O. Filter-sterilize, and aliquots are stored at −20°C. Thawed aliquots are then stored at 4°C.
5. DEX (Sigma, cat. no. D-4903) is prepared at 1 mM in 100% ethanol. Dissolve 3.9 mg DEX into 10 mL ethanol. Store at −20°C.
6. Antibiotics (see **Subheading 2.2., item 5**).

2.4. Oil-Red O, Alizarin, and Von Kossa Staining

1. Fix buffer: 0.25% glutaraldehyde in PBS. Dissolve 1 mL 2.5% glutaraldehyde (Sigma, cat. no. G-5882) into 100 mL PBS. Store at 4°C.
2. Wash buffer: Milli-Q water.
3. Oil-Red O stock solution: dissolve 0.5 g Oil-Red O (Sigma, cat. no. O-0625) into 100 mL isopropanol. Working solution: mix 6 vol of stock solution with 4 vol water. Mix and filter. Store at room temperature.
4. Oil-Red O elution buffer: 4 M guanidinium thiocyanate dissolved in 25 mM sodium citrate, pH 7.0, 0.5% sarcosyl, and 20% isopropanol. Dissolve 250 g guanidinium thiocyanate and 1.5 g sarcosyl into 230 mL H_2O. Add 10 mL 750 mM sodium citrate, pH 7.0 (dissolve 22 g sodium citrate into 100 mL H_2O) and 60 mL isopropanol.
5. Alizarin Red S (Sigma, cat. no. A-5533) working solution: 1% in H_2O. Store at room temperature.
6. Alizarin red elution buffer: 4 M guanidinium thiocyanate dissolved in 25 mM sodium citrate, pH 7.0, 0.5% sarcosyl. Prepare as described in **Subheading 2.4., item 4**, except the absence of isopropanol.
7. Von Kossa working solution: 2% silver nitrate solution (Sigma, cat. no. A-913-9).
8. Store solution: 70% glycerol in water (v/v).

2.5. RNA Preparation From Differentiating Embryoid Body Outgrowths

1. Guanidinium lysis buffer: 4 M guanidinium thiocyanate dissolved in 25 mM sodium citrate, pH 7.0, 0.5% sarcosyl (as described in **Subheading 2.2.4.**). This is the stock solution that can be stored 3 mo at room temperature. Just before use, add 0.36 mL 2-mercaptoethanol (Sigma, cat. no. M-7522) in 50 mL stock solution.

2. Sterile 2.2-mL Eppendorf tubes.
3. Phenol-saturated solution, pH 4.0, containing 0.1% 8-hydroxyquinoline (*see* **Note 3**).
4. Chloroform.
5. Isopropanol.
6. 100% ethanol.
7. 5 *M* NaCl; sterilize.
8. TES buffer: 10 m*M* Tris-HCl, pH 7.4, 0.1 m*M* EDTA, 0.1% sodium dodecyl sulfate. Store at room temperature.

2.5. Reverse Transcriptase Polymerase Chain Reaction

1. Reverse transcriptase polymerase chain reaction (PCR) kit (available from several companies).
2. Pairs of primers used for detecting: adipocyte-fatty acid binding protein (a-FABP, an adipocyte-specific gene): 5′ GATGCCTTTGTGGGAACCTGG 3′ and 5′TTCATCGAATTCCACGCCCAG3′; osteocalcin (an osteoblast-specific gene): 5′TCTGCTGACCCTGGCTGC3′ and 5′GGAGCTGCTGTGACATCC3′; and hypoxanthine phosphoribosyl transferase (HPRT; as a standard to balance the amount of RNA and ccDNA used, except for RNA prepared from E14TG2a ES cells that are HPRT deficient): 5′GCTGGTGAAAAGGACCTCT3′ and 5′CACAGGACTAGAACACCTGC3′.

3. Methods

3.1. Coating Flasks With Gelatin

Coating is performed by covering the surface of a 25-cm^2 flask with 5 mL 0.1% gelatin for 15 min at room temperature, followed by careful aspiration of the gelatin solution.

3.2. Maintenance of ES Cells

The conditions outlined are applicable for the maintenance of feeder layer-independent ES cell lines. Cells can be grown on gelatinized coated tissue culture flasks (*see* **Note 4**) and maintained in a multipotent undifferentiated state providing they are exposed to LIF *(3)*.

1. For a 25-cm^2 flask, aspirate medium and wash twice with 5 mL PBS. Aspirate PBS and add 1 mL trypsin solution. Ensure the trypsin covers the cell monolayer and incubate at 37°C, 5% CO_2, for 2–3 min. Check under an inverted microscope that cells are correctly dissociated (*see* **Note 5**).
2. Add 5 mL growth medium to stop trypsinization and suspend the cells by vigorous pipetting. Transfer the cells to a sterile tube and centrifuge at 250*g* for 5 min at room temperature.
3. Aspirate the medium and resuspend the cell pellet with 5 mL growth medium by pipetting up and down two or three times. Count cells.
4. Add 10^6 ES cells into 10 mL prewarmed growth medium containing LIF, then transfer to a freshly gelatinized 25-cm^2 flask (as described in **Subheading 3.3., step 1**).
5. Change medium every day.
6. Trypsinize the cultures 2 d later as in **step 1**. Cultures should be subcultured before cells have reached confluence.

3.3. Differentiation of Embryoid Bodies Into Adipocytes

The first morphological observation of adipocytelike cells derived from ES cells was reported by Field et al. *(4)*. However, the number of ES cell-derived adipocytes was low

Differentiating ES Cells in Pharmacological Studies

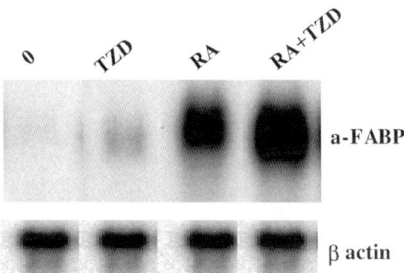

Fig. 1. Effect of retinoic acid (RA) and thiazolidinedione (TZD) on the differentiation of ES cells into adipocytes. ES cells were stimulated from d 2 to 5 with either 0.1% dimethyl sulfoxide (0) or TZD (BRL49653 at 0.5 μM) or RA 10^{-7} M, and a-FABP expression was determined after 20 d. As shown, addition of TZD did not commit ES cells into adipocytes. In contrast, stimulation of EBs by RA from d 2 to 5 plus the addition of TZD from d 7 to 15 (RA + TZD) dramatically increased adipogenesis.

because spontaneous differentiation of ES cells into adipocytes is a rare event. A prerequisite for the commitment of ES cells into the adipogenic lineage at a high rate is to treat ES cell-derived embryoid bodies (EBs) with all-*trans* RA, the biologically active form of vitamin A, for a short period of time *(2,5–7)*.

Two distinct phases can be distinguished in the development of the adipocyte program from ES cells. The first phase, between d 2 and 5 after EB formation, corresponds to the permissive period for the adipogenic commitment of ES cells, which requires RA. The second phase, which corresponds to the period of terminal differentiation, is influenced by adipogenic factors and requires peroxisome proliferators-activated receptors (PPAR)γ. Thus, RA treatment followed by stimulation of PPARγ leads to 60–80% of EB outgrowths containing adipose cells compared to 2–5% in the absence of treatments. Importantly, for commitment of ES cells toward the adipogenic lineage, RA cannot be omitted or substituted. Adipocytes derived from ES cells display both lipogenic and lipolytic activities in response to insulin and to β-adrenergic agonists, respectively, indicating that mature and functional adipocytes are formed.

PPARs are ligand-activated transcription factors belonging to the nuclear hormone receptor family. In vivo, activation of PPARγ improves insulin sensitivity; therefore, thiazolidinediones, which are PPARγ agonists, constitute a new class of pharmacological agents used for the treatment of type 2 diabetes. The effects of thiazolidinediones and the role of PPARγ on the differentiation of ES cells have been investigated *(8,9)*. As shown in **Fig. 1**, activation of PPARγ by thiazolidinedione is not required for the commitment of ES cells but dramatically increases terminal differentiation of RA-treated EBs into adipocytes.

Adipocyte differentiation of ES cells is initiated by aggregation. The aggregates form structures known as embryoid bodies (EBs). To induce adipocyte lineage, the hanging drop method for the formation of EBs is routinely used (*see* **ref. 6** and **Note 6**).

1. Change media on ES cells with complete growth medium supplemented with LIF 2 h before subculture.

2. Aspirate medium and wash twice with 5 mL PBS. Aspirate the PBS and add 1 mL trypsin solution. Incubate at 37°C, 5% CO_2, for 2–3 min.
3. Add 5 mL growth medium to stop trypsinization and suspend the cells by vigorous pipetting. Transfer the cells to a sterile tube and centrifuge at 250g for 5 min at room temperature.
4. Aspirate the medium off and resuspend the cell pellet into 10 mL growth medium without LIF and supplemented with antibiotics (*see* **Subheading 2.2.**, **item 5**). After cell counting, adjust the suspension to a concentration of 5×10^4 cells/mL.
5. Place 20-μL aliquots of this suspension onto the lid of bacteriological-grade dishes (*see* **Note 7**). This is defined as d 0 of EB formation.
6. Invert the lid and place it over the bottom of a bacteriological Petri dish filled with 8 mL PBS containing a few drops of FCS to decrease the surface tension of the liquid. When the lid is inverted, each drop hangs, and the cells fall to the bottom of the drop, where they aggregate into a single clump (EBs).
7. Remove the lid 2 d later, invert it, and collect drops containing EBs in a conical sterile tube. Let it stand for 5 min at room temperature to allow the aggregates to sediment. Aspirate the supernatant and resuspend the pellet in 4 mL growth medium supplemented with $10^{-7} M$ RA. Transfer the suspension into 60-mm bacteriological-grade Petri dishes (*see* **Note 8**).
8. Incubate for 3 d in the presence of RA, changing the medium every day.
9. At d 5 after EB formation, change the medium without the addition of RA and plate two to four EBs per square centimeter in gelatinized tissue culture dishes (*see* **Note 9**).
10. The day after plating, change the growth medium to differentiation medium. Change medium every other day. After 10–20 d in the differentiation medium, at least 50–70% of EB outgrowths should contain adipocyte colonies (*see* **Note 10**).

3.4. Differentiation of EBs Into Osteoblasts

It has been demonstrated that statins, members of the HMG CoA-reductase inhibitory drug family used in therapy for lowering cholesterol, have the unexpected capacity to promote bone mineral density and osteoblast differentiation *(10)*. The importance of statins in osteoblast differentiation of ES cells has been demonstrated *(11)*. Altogether, these results make statins potential drugs for treatment of bone loss. However, we have observed that statins also promote adipocyte differentiation in ES cells *(12)*. If this observation is confirmed in vivo, then it would highlight a potential problem or side effect in the use of statins as antiosteoporotic agents. Because of the interest in using statins in pro-osteogenic therapy, ES cell differentiation would make a useful system to screen the effects of a panel of statins to identify compounds that stimulate osteogenesis but not adipogenesis.

1. Osteoblast differentiation of ES cells, as for adipogenesis, is initiated by formation of EBs by the hanging drop method. RA treatment is required to commit ES cells toward both the adipogenic and the osteogenic lineages (*see* **Fig. 2**). Therefore, to differentiate ES cells into osteoblasts, proceed as indicated from **Subheading 3.3.**, **steps 1–9**.
2. The day after plating, change the complete growth medium to osteoblast differentiation medium (*see* **Note 11**).

3.5. Visualization of Adipose Cells in Outgrowth Cultures
3.5.1. Oil-Red O Staining

Adipose cells with lipid droplets are easily visualized microscopically, especially under bright field illumination. Nonadipose cells containing structures resembling droplets are often detectable in untreated and RA-treated cultures. Therefore, it is essential to identify

Differentiating ES Cells in Pharmacological Studies

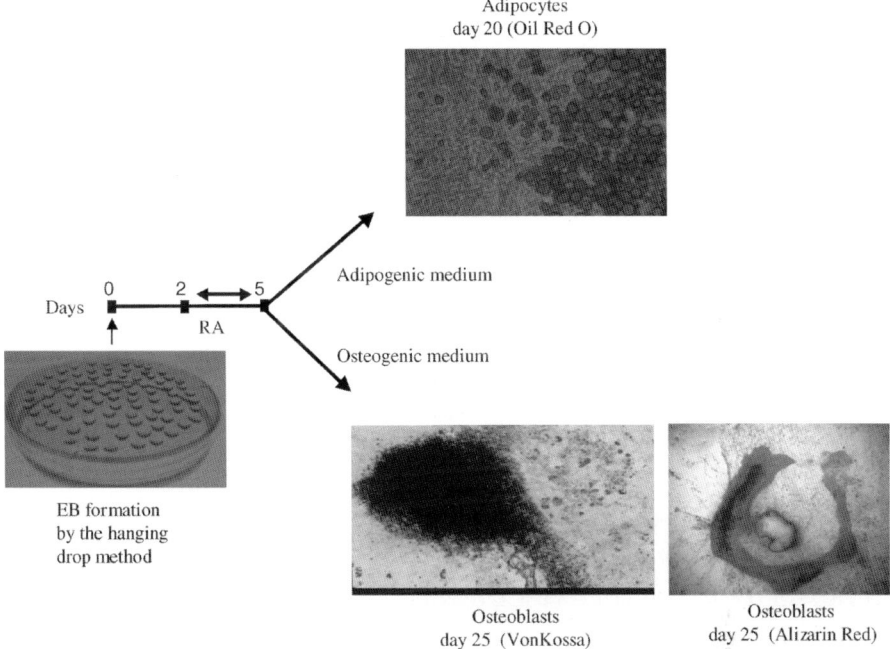

Fig. 2. Experimental protocol used for the commitment of embryonic stem cells into the adipocyte and osteoblast lineages. (Please *see* the companion CD for the color version of this figure.)

dropletlike structures as triglyceride droplets. Staining of cultures with Oil-Red O, a specific stain for triglycerides, gives a good indication of adipocyte differentiation.

1. Aspirate medium and wash cells once with PBS.
2. Fix cells for 15 min with fix buffer at room temperature.
3. Wash twice for 10 min with water.
4. Stain with Oil-Red O solution for 15 min.
5. Wash twice with water. Cover cells with a film of storage solution (70% glycerol in water) or elute the bound stain.
6. Bound stain can be eluted to quantify the formation of adipocytes. Incubate the dish with 4 M guanidine thiocyanate supplemented with 20% isopropanol for 30 min. After elution of the bound stain, the amount of bound Oil-Red O is measured spectrophotometrically at 490 nm.

3.6. Visualization of Mineralized Nodules in Outgrowth Cultures

Bone nodules can be identified either with alizarin red staining, which turns mineralized nodules red, or with von Kossa staining, which reveals nodules in black (*see* **Fig. 2**). Alizarin red can be eluted and quantified.

3.6.1. Alizarin Red Staining

1. Aspirate medium and wash cells once with PBS.
2. Fix cells for 15 min with fix buffer at room temperature.
3. Wash twice for 10 min with Milli-Q water.

4. Stain with alizarin red solution for 10 min.
5. Wash with Milli-Q water. Cover with the store solution or elute the bound stain.
6. Elution of bound alizarin red stain can be performed to quantify the formation of bone nodules. For that purpose, stained cells are washed 10 min with 60% isopropanol in Milli-Q water and wash with Milli-Q water.
7. Cover the dish with alizarin red elution buffer prewarmed at 37°C.
8. After 30 min, elution of the bound stain should be complete. The amount of bound alizarin red is measured spectrophotometrically at 490 nm.

3.6.2. von Kossa Staining

1. Aspirate medium and wash cells once with PBS.
2. Fix cells for 15 min with fix buffer at room temperature.
3. Wash twice for 10 min with Milli-Q water.
4. Aspirate water and serially dehydrate fixed cells by washing them 5 min twice with 70, 90, and 100% ethanol. Air-dry.
5. Serially rehydrate cells by washing them 5 min twice with 90 and 70% ethanol, then with Milli-Q water.
6. Aspirate water and add von Kossa working solution prepared fresh.
7. Expose to light for 1 h.
8. Rinse with Milli-Q water.

3.7. RNA Preparation From EB Outgrowths

The procedure described is an adaptation of the single-step method previously published by Chomczynski and Sacchi *(13)*.

1. Wash cells with PBS.
2. Add 1.6 mL guanidinium lysis buffer per 100-mm tissue culture dish containing EB outgrowths. Split cell lysate into two sterile 2.2-mL Eppendorf tubes. Vortex vigorously.
3. Add 0.8 mL pH 4.0 phenol-saturated solution per tube and mix by inversion. Then, add 0.2 mL chloroform. Shake vigorously for 10 s and keep on ice for 15 min.
4. Centrifuge at 10,000*g* for 20 min at 4°C.
5. Carefully transfer the aqueous phase containing RNA (usually the upper phase; *see* **Note 3**) to a fresh 2.2-mL tube. Take care to avoid any traces of interface material.
6. Repeat **steps 3–5**.
7. Transfer the aqueous phase to a fresh 2.2-mL tube. Add 0.8 mL isopropanol, mix, and place at −20°C for at least 12 h.
8. Centrifuge at 10,000*g* for 20 min at 4°C.
9. Aspirate off the supernatant and collect the RNA pellets from both tubes in 0.3 mL guanidinium lysis buffer.
10. Precipitate with 0.6 mL ethanol at −20°C for 12 h.
11. Centrifuge at 10,000*g* for 20 min.
12. Carefully pour off the supernatant and dissolve the RNA pellet in 0.5 mL TES (10 m*M* Tris-HCl, pH 7.4, 0.1 m*M* EDTA, 0.1% sodium dodecyl sulfate).
13. Determine the RNA concentration (*see* **Note 12**). Store RNA at −20°C.

3.8. Analysis of Adipocyte and Osteoblast Gene Expression

Expression of a-FABP (adipocyte specific) and of osteocalcin (osteoblast specific) in 20-d-old EB outgrowths can be detected by Northern blotting using 20 µg total RNA. However, detection of gene expression in early differentiating outgrowths requires a

Table 1
Detection of Adipocyte- and Osteoblast-Specific Genes by PCR

Gene	Basepairs		Annealing temperature (°C)
	cDNA	Genomic	
a-FABP	213	2400	56
Osteocalcin	860	1400	55
HPRT	249	1100	60

more sensitive method, such as reverse transcriptase PCR. **Table 1** gives the temperatures of annealing for the PCR reaction and the size of expected cDNAs as well as the size of genomic DNA-derived contaminated bands. Gene expression can be detected either by a diagnostic ethidium bromide band or after blotting and hybridization with appropriate cDNA probes.

4. Notes

1. LIF is required to maintain pluripotent ES cells and is omitted to induce the commitment of ES cells toward the adipogenic lineage. LIF is commercially available (named ESGRO by Chemicon, Hampshire, UK; cat. no. ESG 1106) or can be homemade (for details, *see* **ref. 3**).
2. The addition of thiazolidinediones, which are PPARγ activators *(14)*, such as 0.5 µM BRL49653 (SmithKline Beecham Pharmaceuticals, Herts, UK; not commercially available) or 2 µM Ciglitazone (BioMol Research Laboratories, Tebu, Le Perray en Yvelines, France; cat. no. 034GR-205) in the differentiation medium dramatically stimulates the terminal differentiation of RA-treated EBs into adipocytes *(12)*. The regimen used to promote differentiation of 3T3-L1 preadipose cells has been tested on RA-treated EBs *(8)*. With this regimen, 17-d-old EB outgrowths are treated with DEX (400 ng/mL) and methylisobutylxantine (500 nM) for 2 d. Then, DEX and methylisobutylxantine are removed from the culture media, and outgrowths are maintained in differentiation medium. As expected, hormonal treatments that have been previously proved to promote terminal differentiation of preadipocytes into adipocytes are also efficient to induce terminal differentiation of RA-treated EBs. It is important to note that the treatment of EBs with RA is a prerequisite. RA is light sensitive.
3. 8-Hydroxyquinoline stains phenol yellow, which allows the unambiguous identification of the phenol phase and the aqueous phase.
4. Attachment of ES cells to the substratum is susceptible to change according to the tissue culture material. We use Corning or Greiner tissue culture flasks and dishes.
5. It is critical to produce a single-cell suspension for subcultures. This is achieved by knocking the flask several times to ensure complete dissociation during the trypsin treatment.
6. The formation of EBs in mass culture (by maintaining pluripotent ES cells in suspension at a high density, i.e., 5×10^5 cells/mL in a bacteriological-grade Petri dish) leads subsequently to a low number of outgrowths containing adipocyte colonies.
7. We use a multipipeter with a sterile combitip dispensor. Approximately 80 drops can be fitted on the lid of a 100-mm Petri dish. It is essential to cover the bottom of the dish with the liquid to prevent the evaporation of the hanging drops.

8. Bacterial-grade Petri dishes are used to prevent cell attachment to the substrate. EBs have a tendency to attach to the bottom of the plastic dish. EBs that are firmly attached to the dish should be eliminated as these EBs seem to have no adipogenic capacity. Attachment of EBs on Petri dishes can be eliminated by coating the dish with poly(2-hydroxyethylmethylacrylate) supplied as Cellform polymer (ICN Biomedicals, Orsay, France; cat. no. 150207). Dishes are filled with a film of Cellform working solution (0.5 g Cellform dissolved in 42 mL 100% ethanol) and left in a culture hood, with the cover of the dish removed, to evaporate ethanol. Dishes are rinsed with PBS before use.
9. A higher density of EBs can lead to a decrease in the number of EB-containing adipocyte colonies. Two-day-old EBs from four lids of 100-mm Petri dishes are pooled into one 60-mm bacteriological-grade Petri dish, then after RA treatment are plated into one 100-mm tissue culture dish.
10. A wide variety of differentiated derivates, such as neuronelike cells, fibroblastlike cells, and unidentified cell types appear over this period. Spontaneously beating cardiomyocytes should appear 1–5 d after plating the control culture (untreated with RA). At least 40% of EBs should contain beating cardiomyocytes. In contrast, few EBs should contain beating cells from RA-treated cultures.
11. RA is required to differentiate ES cells into osteoblasts at a high rate. Adipocytes appear also in osteogenic conditions.
12. We routinely obtain 100–200 µg RNA from one 100-mm tissue culture plate containing 20-d-old EB outgrowths.

Acknowledgments

The work reported from the authors' laboratory was supported by the Centre National de la Recherche Scientifique (grant UMR6543) and the Association pour la Recherche Contre le Cancer (grant 4625). The authors' laboratory is a member of the FunGenES European Consortium and is granted by the Sixth Framework Programme of the European Union.

References

1. Gregoire, F. M., Smas, C. M., and Sul, H. S. (1998) Understanding adipocyte differentiation. *Physiol. Rev.* **78,** 783–809.
2. Dani, C., Smith, A., Dessolin, S., et al. (1997) Differentiation of embryonic stem cells into adipocytes in vitro. *J. Cell Sci.* **110,** 1279–1285.
3. Smith, A. G. (1992) Mouse embryo stem cells: their identification, propagation and manipulation. *Semin. Cell Biol.* **3,** 385–399.
4. Field, S. J., Johnson, R. S., Mortensen, R. M., Papaioannou, V. E., Spiegelman, B. M., and Greenberg, M. E. (1992) Growth and differentiation of embryonic stem cells that lack an intact c-fos gene. *Proc. Natl. Acad. Sci. USA* **89,** 9306–9310.
5. Dani, C. (1999) Embryonic stem cell-derived adipogenesis. *Cells Tissues Organs* **165,** 173–180.
6. Dani, C. (2002) Differentiation of embryonic stem cells as a model to study gene function during the development of adipose cells. *Methods Mol. Biol.* **185,** 107–116.
7. Wdziekonski, B., Villageois, P., and Dani, C. (2003) Development of adipocytes from differentiated ES cells. *Methods Enzymol.* **365,** 268–277.
8. Rosen, E. D., Sarraf, P., Troy, A. E., et al. (1999) PPAR gamma is required for the differentiation of adipose tissue in vivo and in vitro. *Mol. Cell* **4,** 611–617.

9. Vernochet, C., Milstone, D. S., Iehle, C., et al. (2002) PPARγ-dependent and PPARγ-independent effects on the development of adipose cells from embryonic stem cells. *FEBS Lett.* **510,** 94–98.
10. Mundy, G., Garrett, R., Harris, S., et al. (1999) Stimulation of bone formation in vitro and in rodents by statins. *Science* **286,** 1946–1949.
11. Phillips, B. W., Belmonte, N., Vernochet, C., Ailhaud, G., and Dani, C. (2001) Compactin enhances osteogenesis in murine embryonic stem cells. *Biochem. Biophys. Res. Commun.* **284,** 478–484.
12. Phillips, B. W., Vernochet, C., and Dani, C. (2003) Differentiation of embryonic stem cells for pharmacological studies on adipose cells. *Pharmacol. Res.* **47,** 263–268.
13. Chomczynski, P. and Sacchi, N. (1987) Single-step method of RNA isolation by acid guanidinium thiocyanate-phenol-chloroform extraction. *Anal. Biochem.* **162,** 156–159.
14. Lehmann, J. M., Moore, L. B., Smith, O. T., Wilkison, W. O., Willson, T. M., and Kliewer, S. A. (1995) An antidiabetic thiazolidinedione is a high affinity ligand for peroxisome proliferator-activated receptor gamma (PPARγ). *J. Biol. Chem.* **270,** 12,953–12,956.

25

Embryonic Stem Cells as a Source of Differentiated Neural Cells for Pharmacological Screens

Patrick J. Mee, Carmel M. O'Brien, Hazel Thomson, Sjaak van der Sar, Viktor Lakics, and Timothy E. Allsopp

Summary

The process of bringing a new pharmacologically active drug to market is laborious, time consuming, and costly. From drug discovery to safety assessment, new methods are constantly sought to develop faster and more efficient procedures to eliminate drugs from further investigation because of their limited effectiveness or high toxicity. Because in vitro cell assays are an important arm of this discovery process, it is therefore somewhat unsurprising that there is an emerging contribution of embryonic stem (ES) cell technology to this area. This technology utilizes the in vitro differentiation of ES cells into somatic cell target populations that, when coupled to the use of "lineage selection" protocols, allows for the production of infinite numbers of pure populations of the desired cells for both bioactivity and toxicological screens. Unlike the use of transformed cell lines, ES-derived cells remain karyotypically normal and therefore better reflect the potential responses of cells in vivo, and when selected are more homogeneous than those obtained using primary cultures. In this chapter we discuss the use of ES cell-derived somatic cells in pharmacological screens, with particular emphasis on neural cells, and describe the methods and protocols associated with the development of ES cell-derived neural cell assays.

Key Words: Bioreactor; cellular assay; commercialization; embryoid body; ES cell; β-geo; high-throughput assay; lineage selection; monolayer differentiation; neural differentiation; neural stem cell; neurogenesis; scale-up; serum-free media; sox-1; stem cell; stem cell sciences.

1. Introduction

1.1. Advantages of Embryonic Stem Cell-Derived Cells in Pharmacological Screens

The advent of in vitro cellular assays has simplified discovery and toxicology studies by allowing for dismissal of drugs prior to more costly and time-consuming small animal assays. However, the cellular reagents available for this purpose have been defective in that either transformed cell lines or primary cultures have been used. Transformed cell lines have mutations and aberrant gene expression patterns and so do not resemble the cells found normally in vivo, therefore the relevance of their responses is questionable. Primary cultures on the other hand, are directly cultured from the target tissue and are not transformed. The disadvantage of these cells is that within a single primary culture, there is considerable cellular heterogeneity, which makes interpretation of results difficult. Moreover, there are also difficulties in obtaining consistency between different primary cultures, leading to a lack of confidence in cross comparison.

The biopharmaceutical sector therefore has expressed considerable interest in the use of embryonic stem (ES) cells as a source of more appropriate, disease-orientated cellular screens for therapeutic target validation, optimization of target-acting compounds, and toxicology studies *(1–4)*. ES cells retain their karyotypically normal phenotype even after prolonged culture in vitro. Coupled to their capacity to generate cells of all mesoderm, endoderm, and ectoderm lineages by simple cytokine treatments and selection procedures, this makes the technology particularly attractive *(1,5,6)*.

To date, the commercial use of this technology has been predominantly restricted to the use of mouse ES cells. There is a greater level of understanding because of the large amount of research conducted on these cells. The advent of human ES (hES) cell lines has yet to make an impact because of the relative novelty of hES cells compared to the well-established mouse ES cells, problems with access caused by restrictive licensing practices, as well as relative complexities of growth and maintenance of hES compared to mouse ES cells. However, as this technology develops, a bright future is predicted for the use of hES cells in these assays. Here we concentrate on the use of well-established mouse ES cell technology, but if appropriate point to the advances made in translating this technology to hES cells.

1.2. ES Cell Differentiation Into Neural Cells

The majority of ES cell differentiation protocols to generate mature fully differentiated progeny cells, including neural cell types, utilize an initial embryoid body (EB) aggregation in hanging drop cultures, and adding retinoic acid (RA) as a nonspecific differentiation-inducing agent *(7,8)*. EBs formed in this way are subsequently dissociated and plated on an adhesive substrate for further differentiation. ES cell differentiation in EB cultures resembles processes of early embryonic development because similar gene expression patterns are adopted during the course of differentiation. Outer cells of the EB acquire extraembryonic endodermlike features and express early endodermal genes *(9)*; the inner cells display primitive ectodermlike gene expression patterns. In addition, early mesoderm differentiation markers such as *Brachyury* are expressed in a similar pattern to the mouse embryo, in which they are transiently expressed during gastrulation, such that *Brachyury* expression is similarly downregulated in later stages of EB differentiation *(7)*. Subsequently, progenitor cells are formed in the EB, which can be induced to differentiate into cell types representative of all three embryonic germ layers.

Once the EBs are disaggregated, they are plated in medium containing a cocktail of specific growth and differentiation factors (sometimes applied sequentially) to promote generation of the particular cell type of interest. The generation of neural cells, particularly neurons, from EBs cultured in this manner is widely studied, and several different protocols have been used that include plating onto laminin *(10)*, gelatin *(11)*, or tissue culture plastic *(12)*. Neurons generated in this way have an obvious neuronal morphology and express neurotransmitter receptors, ionic channels, kinases, and proteases representative of many of the major neuronal classes. They also respond to neurotransmitters and depolarizing currents, suggesting that at least to some degree, they mirror the development of these cells in vivo *(10–13)*. Of particular interest for the pharmaceutical industry is the finding that the neural cells express the functional secretase components required to catabolize human amyloid precursor protein into β-amyloid peptide, forming the basis of an ES cell-based Alzheimer's disease screening assay. EB differentiation generates not only neurons, but also astrocytes and oligodendrocytes *(14)*.

A variant of this EB differentiation protocol has been developed that involves the formation of EBs without exposure to RA *(15)*. This protocol depends on the subsequent culture of cells in a serum-free medium that selectively eliminates nonneural cells. Thus, although there is a dramatic decrease in the number of cells that survive, there is an enrichment in nestin-positive neural precursors *(16)*. The majority of neurons generated from ES cells via EB differentiation are GABAergic (γ-aminobutyric acid [GABA]) with a small proportion of glutamatergic neurons. However, techniques have been developed to alter the differentiation process to favor the differentiation of other neuronal types. Thus, to generate dopaminergic midbrain neurons (important because of the prospect of transplants to treat Parkinson's disease; *17,18*) as well as serotonergic neurons, the protocol of plating disaggregated EBs into minimal medium has been further adapted. The neural precursors are exposed to basic fibroblast growth factor (FGF), sonic hedgehog (Shh), fibroblast growth factor (FGF) 8 (molecules with known ventralizing activity in the developing brain *[19]*), as well as ascorbic acid *(20)*. Further modification of this dopaminergic differentiation protocol by overexpressing the midbrain differentiation-inducing transcription factor Nurr1, a member of the orphan nuclear receptor gene family, in ES cells more than doubles the percentage of dopaminergic neurons in culture *(21)*. Conversely, motor neuron identity can be generated from ES cell-derived EB differentiated cells by the use of combined posteriorizing and ventralizing signals *(22)* (RA and Shh agonist, respectively).

Although the protocols described pertain to the use of EB formation as a prerequisite for neural cell differentiation, there are now evolutions of this technology that forgo the requirement for EBs. Generation of dopaminergic neurons from ES cells can be accomplished via the co-culture of ES cells with the bone marrow-derived stromal cell line PA-6 in serum-free conditions. This removes the need for EB formation, RA treatment, or the exogenous addition of cytokines such as Shh and FGF-8 and suggests a direct induction of dopaminergic neuronal cell fate, albeit the frequency of these cells is still low *(23)*.

One of the most intriguing observations of ES cell differentiation has been the result of experiments with defined conditions that allow for neural differentiation without the requirement for any exogenous signals (whether these are via added cytokines or feeder cells) or the formation of EBs. Instead, ES cells are simply plated in monolayer culture into a defined serum-free medium in which they spontaneously differentiate into neural precursors that have the capacity to terminally differentiate into neural populations *(24,25)*. This observation suggests that differentiation of ES cells into neural cells is a default mechanism *(26)* such that, once free of inhibitory signals, ES cells will choose this route of differentiation because of endogenous FGF signaling *(24,27)*. In addition to the obvious interest in the biology of this lineage commitment, from a commercial prospective this gives the opportunity to obtain ES cell-derived neural cells in a simple and low-cost manner as no technically difficult steps exist, and no expensive cytokines are used. This form of differentiation protocol provides consistency of neural differentiation in a cost-effective manner.

All the examples of ES cell neural differentiation are modeled using mouse ES cells. The mouse has proved to be an excellent model organism for drug discovery and will undoubtedly remain at the forefront of the drug-screening process *(1)*. The ultimate goal, however, is to utilize hES cells to provide a source of differentiated cells for drug screens. In a similar manner to murine ES cells, hES cells can be induced to differentiate into all of the three germinal layers *(28–32)*. EB differentiation protocols similar to those used in the mouse have been used to induce differentiation into neural precursors and to promote their development into mature neural cell types, including dopaminergic, GABAergic, glutamatergic, glycinergic, cholinergic, and purinergic cells *(33)*. Once human lines are widely and freely available and the technology surrounding their handling has improved, it is envisaged that a major drive will be made for the commercialization of these cells in drug-screening bioassays utilizing similar technology to that described here for mouse ES cells.

1.3. Lineage Marking and Selection

A significant hurdle to the use of ES cell-derived neural cells as models for drug screening and toxicology studies is that, even with the best available protocols, differentiation is not uniquely neural. Thus, without methods to purify or select for neural cells, cultures will contain contaminating cells of both undifferentiated ES cells and differentiated cells of nonneural lineages. Although the directed differentiation protocols described make it possible to enrich for populations of specific cell phenotypes *(1)*, lineage marking and lineage selection strategies that make use of a "knock-in" targeting approach allow for the generation of highly purified populations of a desired cell type.

The targeted introduction of a reporter gene such as β-galactosidase or green fluorescent protein (GFP) to one allele of a developmentally restricted gene of interest makes it possible to monitor the endogenous gene expression during ES cell in vitro differentiation and in the latter case to purify a specific population of GFP-expressing differentiated progeny by fluorescence-activated cell sorting (FACS). The targeted insertion of a drug resistance gene (e.g., neomycin resistance from the aminoglycosidase phosphotransferase gene) to a gene that is expressed in a specific differentiated precursor cell type enables the purification of a population of these cells by positive selection with drug treatment *(34,35)*.

Of use to these lineage-restricted targeting strategies is the inclusion of an internal ribosome entry site (IRES) element from the 5′-nontranslated regions of picornaviral and human immunoglobulin heavy chain binding protein mRNA, *Drosophila melanogaster* Antennapedia protein, human FGF-2 protein, human insulin growth factor protein, and the human eIF-4G protein *(36–40)*. IRES elements act as a ribosome "landing pad" and allow for cap-independent initiation of translation of bicistronic transcripts, including selectable markers, in mammalian cells *(36,41–46)*. The IRES element from the encephalomyocarditis virus has been widely used for the construction of bicistronic vectors in mammalian transgenesis and has been applied successfully for the expression of a *lacZ* reporter throughout whole mouse embryos and tissues following transfection of ES cells *(47,48)*.

The coupling of IRES transgenes into endogenous genes for lineage marking and selection was initially demonstrated for the introduction of the fusion gene β-*geo* to the *Oct4* gene, a gene only expressed in undifferentiated ES cells and to the *LIF* gene, which is not expressed in ES cells but in the cells of the surrounding differentiated cells *(48)*.

An example for the use of an IRES transgene in isolating a cell population of interest following directed lineage differentiation has been studied by the integration of a bicistronic selectable marker incorporating an improved GFP marker into the *Sox-1* gene locus, enabling the identification and purification of neurectoderm progeny generated by in vitro differentiation of *Sox-1*-targeted ES cells *(24)*. *Sox-1*-mediated GFP expression in ES cell-derived neural populations is validated in this study by the *Sox-1*-generated GFP expression demonstrated in the developing neurectoderm layer of a 10.5-dpc (days postcoitum) mouse embryo following ES cell reintroduction to a host blastocyst *(24,35)*.

Similarly, a dopaminergic neuronal population of cells has been identified and isolated following both directed in vitro differentiation and germline transmission of an ES cell line harboring a targeted GFP integration at the *Pitx3* locus *(49)*. Purified ES cell progeny generated in this way can then be expanded in culture systems supplemented with appropriate mitogens or growth factors and used in screening assays for drug discovery (**Fig. 1**).

A potent negative selection strategy *(6)* can be further applied to such purification strategies to allow for the isolation of desired precursor cells from more primitive cell populations that continue to express the lineage-restricted gene during in vitro differentiation. For example, the introduction of the thymidine kinase gene to the *Oct4* locus in *Sox-2*-targeted ES cells *(50)* enables elimination of residual *Sox-2*-expressing pluripotent ES cells in differentiating neuronal cultures by exposure to the drug gancyclovir. Conversely, the expression of a reporter or selection gene under the constitutive control of the ES cell-restricted genes *Oct4* *(51)* and *Rex-1* *(52)* has been used to purify populations of mouse and human pluripotent cells.

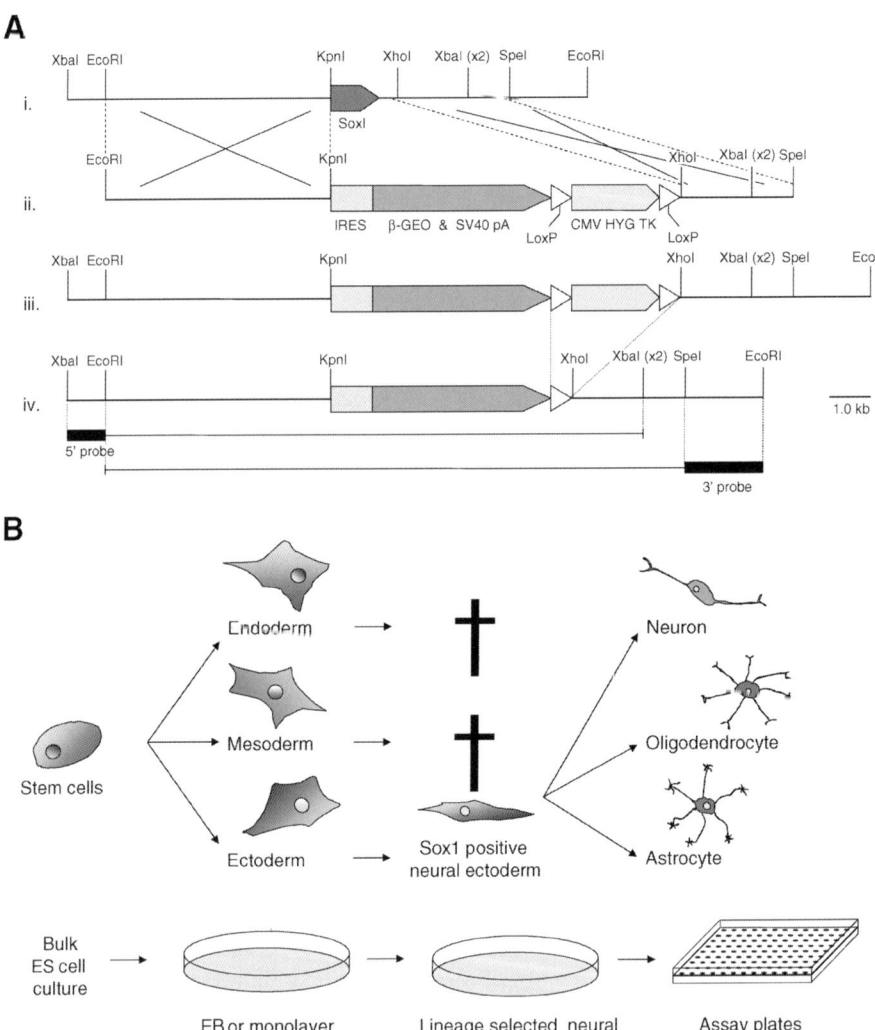

Fig. 1. (**A**) Schematic of *Sox1* targeting: (i) *Sox1* locus; (ii) targeting construct; (iii) integrated structure; (iv) final modified locus after Cre excision of selection cassette. 5' and 3' probes used for confirming integrity are indicated. (**B**) Schematic of rational for production of neural plates for screening.

1.4. Pharmacological Screening

There have been a number of important innovations in screening methodology (automation, reproducibly measuring low-intensity calcium signals in high-density 96-, 384-, or 1536-well plate formats), resulting in the introduction of cell-based assays (e.g., genetically engineered tumor cell lines expressing human targets) very early in the

Pharmacological Screens

screening process. Cell-based systems offer a more complete understanding of compound effects on whole living cells, including cellular toxicity and drug penetration, and they obviate protein purification and expression steps. Unfortunately, the introduction of more complex cellular systems, like differentiated neuroblastomas or primary cultures, parallels an increase in cellular heterogeneity and a decrease in the reproducibility of these assays. The availability of relevant cells (e.g., specific populations of primary neurons from tissue) is also restricted. Therefore, until recently, these model systems were usually employed in secondary screens or proof-of-principle-type experiments. The development of technologies to reliably generate differentiated ES cells at scale that closely resemble primary neurons offers a unique opportunity to solve these problems.

A model system of differentiated mouse ES cell-derived neuronal cells generated by a *Sox-1*-based selection procedure is described here in comparison with a relevant primary neuronal culture. To ascertain that ES cell-derived neural cells can provide similar information to primary cultures, these cells were directly compared to rat cortical neurons and found to be almost identical in terms of expression of neuronal antigens (neurofilament-70 and -200), presence of action potential firing and spontaneous neurotransmitter release (glutamate and GABA, as shown in patch clamp experiments), and the existence of functional neurotransmitter receptors (glutamate and muscarinic acetylcholine receptors) *(53)*. ES cell-derived neuronal cells were also tested in a typical medium-throughput 96-well assay format, measuring drug-induced changes in intracellular calcium concentrations to demonstrate the use of this system for generating concentration-response curves of glutamate receptor agonists. ES cell-derived neural cells provided an easy-to-use, reliable cell-based assay system, and the potencies/efficacies of glutamate receptor agonists correlated well with those measured for primary cortical neurons (**Fig. 2A,B**). Moreover, the presence of the intracellular machinery for glutamate receptor-mediated cell death was also demonstrated: ES cell-derived neuronal cells were vulnerable to excitotoxic cell death, similar to primary cortical neurons *(53)*.

Taken together, there is great potential for differentiated ES cells to be used in pharmacological screens; they can be produced at scale and provide a robust system for medium-throughput screening assays, offering an additional layer of information for lead-optimization decisions compared to tumor cell lines expressing engineered targets.

2. Materials
2.1. Plasmid Vector DNA

1. pGT1.8 (OPT) IRES-βgeo plasmid (provided by P. Mountford, SCS Ltd., Melbourne, Australia; Peter.Mountford@stemcellsciences.com) (*see* **Note 1**).
2. Sox1KO2 plasmid (provided by M. Stavridis, Institute for Stem Cell Research [ISCR], Edinburgh, UK; contact via Professor Austin Smith, austin.smith@ed.ac.uk). Plasmid containing a 5.5-Kb 5' and a 2.5-Kb 3' of homology arms spanning the *Sox-1* monoexon structure isolated from a 129Ola phage library using the Sox-1 complementary DNA.
3. Sox1TV1-targeting plasmid (previously generated by SCS Ltd., Melbourne, Australia).
4. pCAGCreIP Cre-recombinase-expressing plasmid (provided by I. Chambers, ISCR; I.Chambers@ed.ac.uk).

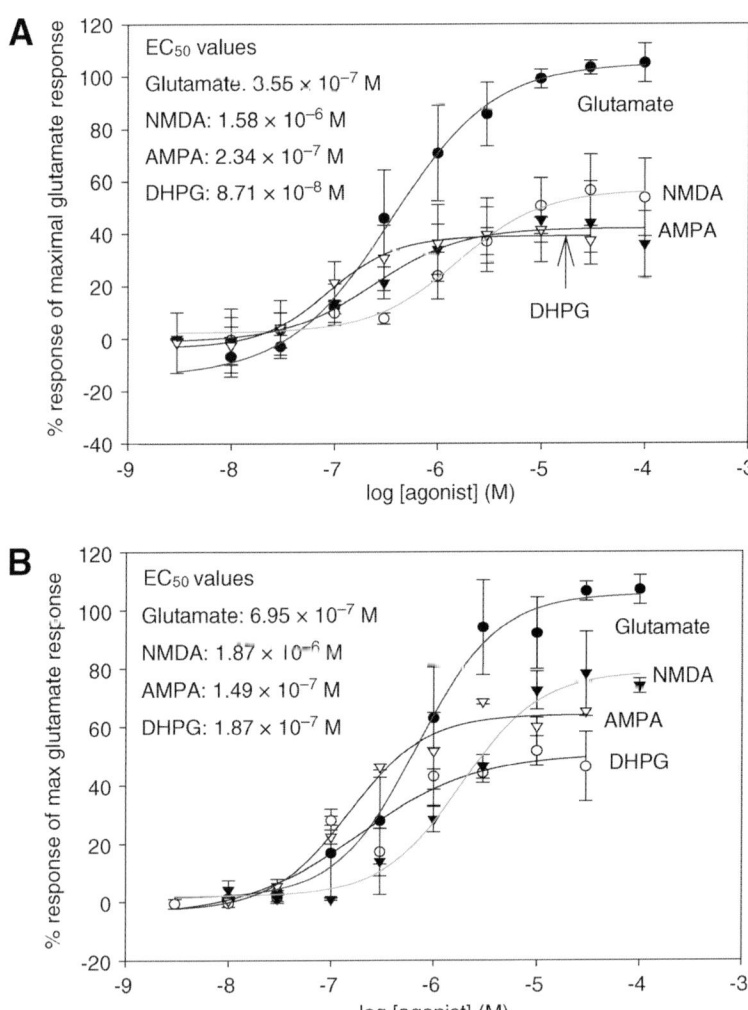

Fig. 2. Concentration-response curves for glutamate receptor agonists in cultures of differentiated embryonic stem cells and primary cortical neurons. Drug-mediated changes in intracellular Ca^{2+} concentration were measured following agonist additions for (**A**) cortical cells 9 d in vitro (average of four experiments, $n = 2$ or 3 for each); (**B**) ES-derived neural cells 12 d after plating ($n = 2$). AMPA, D,L-α-amino-3-hyroxy-5-methyl-4-isoxazolepropionic acid; DHPG, (RS)-3,5-dihydroxyphenylglycine; NMDA, N-methyl-D-aspartate. (Please *see* the companion CD for the color version of this figure.)

2.2. Cell Culture Media and Supplements

1. ES cell culture medium (GMEM-ES): Glasgow minimum essential medium (GMEM) (Sigma, St. Louis, MO; cat. no. G5154) supplemented with 7.5% sodium bicarbonate; 1 mM sodium pyruvate (Gibco-Invitrogen, Paisley, Scotland; cat. no. 25080-060, 11360-039); 10 mM 1X MEM nonessential amino acids (Gibco-Invitrogen, cat. no. 11130-036);

Pharmacological Screens

2 mM L-glutamine (Gibco-Invitrogen, cat. no.25030-024); 10% fetal calf serum (FCS; JRH Biosciences, Lenexa, KS; cat. no. 12003); 0.1 mM β-mercaptoethanol (Sigma, cat. no. M7522); and ESGRO leukemia inhibitory factor (LIF; Chemicon International, Temucula, CA; cat. no. LIF2010) at 10^3 U/mL. To make 500 mL ES cell culture medium, remove 66 mL from a 500-mL bottle of GMEM medium and add 5 mL 100X stock MEM nonessential amino acids, 500 µL of a 10^6 U/mL ESGRO stock solution, 50 mL of FCS, 500 µL 100 mM β-mercaptoethanol stock solution, 10 mL of a stock solution of glutamate and pyruvate made by making a 1:1 mix of 100 mM (100X) sodium pyruvate and 200 mM (100X) L-glutamine.
2. ES cell selection medium: GMEM–ES supplemented with hygromycin at 150 µg/mL (stock solution is 50 mg/mL; Roche, Basel, Switzerland; cat. no. 843 555). To make a 500-mL bottle of selection medium, add 1.5 mL stock hygromycin to 500 mL GMEM-ES.
3. EB medium: GMEM, 10% FCS, glutamine, pyruvate (Gibco-Invitrogen, cat. no. 11710-035, 16140-071). To make 500 mL EB medium, remove 60 mL medium from a 500-mL bottle of GMEM and add 50 mL FCS and 10 mL of a stock solution of glutamate and pyruvate made by making a 1:1 mix of 100 mM (100X) sodium pyruvate and 200 mM (100X) L-glutamine.
4. 1X trypsin: 0.025% trypsin, 1.3 mM ethylenediaminetetraacetic acid, and 0.1% chicken serum. To make 500 mL trypsin, add 0.186 g ethylenediaminetetraacetic acid to 500 mL PBS, then add 5 mL chicken serum (Sigma, cat. no. C-5405) and 5 mL 2.5% trypsin (Gibco-Invitrogen, cat. no.15090-046).

2.3. Pharmacological Screening

1. Tyrode's buffer: 137 mM NaCl, 2.5 mM $CaCl_2$, 2.7 mM KCl, 1 mM $MgCl_2$, 0.2 mM $NaHPO_4$, 12 mM $NaHCO_3$, and 5.5 mM glucose (Sigma, cat. no. T2145).
2. Fluo-3 (Molecular Probes, Eugene, OR; cat. no. F1241).
3. Glutamate receptor agonists: glutamate (Tocris, Ellisville, MO; cat. no. 0218), N-methyl-D-aspartate (Tocris, cat. no. 0114), D,L-alpha-amino-3-hyroxy-5-methyl-4-isoxazolepropionic acid (Tocris, cat. no. 0169), and (RS)-3,5-dihydroxyphenylglycine (Tocris, cat. no. 0805).

3. Methods
3.1. Gene Modification in ES Cells for Selection/Marking of Neural Cells
3.1.1. ES Cell Growth

The general growth and management of ES cell cultures follows standard protocols for feeder-independent ES cell lines grown in suitable containment hood (*see* **Note 2**).

3.1.2. Sox1 Targeting in ES Cells

1. 60 µg of the Sox1TV1 targeting vector plasmid DNA is prepared for each electroporation experiment. The DNA is linearized with SalI digestion at 37°C overnight, with complete digestion confirmed by gel electrophoresis (*see* **Fig. 1** and **Note 3**).
2. Digested DNA is precipitated, pelleted, washed, and resuspended in 100 µL PBS (*see* **Note 4**).
3. The linearized Sox1TV1 vector is introduced into mouse E14Tg2a ES cells by electroporation (*see* **Note 5**).
4. Cells are plated at a density of 2–2.5 × 10^6 cells in 10 mL routine ES cell culture medium per 10-cm culture plate and incubated overnight.
5. After 18–24 h, hygromycin antibiotic is added to both the electroporated culture plates and the nonelectroporated controls at 150 µg/mL.

6. This selection culture is refreshed daily for 3–4 d until cells are dying off and small resistant colonies have started to appear. The selection medium is then changed each 48 h until the cells on the control plate die and resistant colonies are well established on experimental plates.
7. Hygromycin-resistant ES cell colonies (150–200) are picked using pipeter tips and pre-warmed trypsin aliquots and expanded in 24-well plates.
8. Expanding colonies are grown to subconfluency and further expanded into 2 wells of a 24-well plate, 1 well for freezing and 1 well for genomic DNA preparation.

3.1.3. Analysis of Selected ES Cell Clones

1. Genomic DNA is prepared from colony cultures (see **Note 6**).
2. Southern blot hybridization techniques are used to determine those clones with correct targeting of the *Sox-1* gene at the 3′ and 5′ homology regions using DNA probes that are external to these regions (see **Note 7**).
3. Targeted clones shown to be correctly integrated into the endogenous Sox-1 locus at both 5′ and 3′ ends of the insertion are further expanded for Cre deletion and freezing of stocks.

3.1.4. Cre Deletion of a Targeted Sox1TV1 Clone

1. Sox-1-targeted ES cell clones are chosen on the basis of good ES cell morphology and robust maintenance in culture.
2. 15×10^6 ES cells for a chosen clone are transiently transfected with 150 µg of the pCAGCrelP vector, and gancyclovir-resistant colonies (see **Note 8**) are picked, expanded, and frozen in a similar manner to that described in the steps of **Subheading 3.1.3**.
3. The removal of the lox-P-CMV-hygro-tk-SvpA-lox-P cassette is confirmed by Southern blot analysis of gancyclovir-resistant colonies.
4. Cre deletion is further confirmed by the addition of hygromycin (100 µg/mL) to one of a duplicate culture for a Cre-deleted ES cell line displaying a loss of resistance to hygromycin.

3.2. EB Differentiation of ES Cells for Neural Differentiation

1. Passage cells 24 h before establishing the EB cultures such that the flasks are approx 70% confluent for setting up the EB cultures (see **Note 9**).
2. Trypsinize the flasks (see **Note 10**) in the incubator for approx 1 min. Dislodge gently by tapping, gently add EB medium (no LIF) to quench the trypsin, and gently resuspend using a wide-bore pipet. Transfer the cell suspension to a tube and centrifuge using a swing-out rotor at 114*g* for 4 min.
3. Remove the wash medium by aspiration and resuspend in a further 10 mL of EB medium. Count the cells using a hemocytometer and resuspend at 5×10^5 cells per milliliter. Plate 10 mL of the cell suspension into a 10-cm bacterial dish (or 20 mL into a 20-cm dish) and place in a humid 7% CO_2-gassed 37°C incubator.
4. Culture the cells for 2 d, then change the suspension medium using a wide-bore plastic pipet to remove all cells and medium into a fresh tube (taking care not to disrupt the EBs). Allow the EBs to settle to the bottom of the tube and by aspiration gently remove the medium without disturbing the EBs. Replace the medium with fresh EB medium, gently resuspend as before, and return the EBs to a fresh Petri dish.
5. Repeat the medium change as described in **step 4** 2 d later (i.e., d 4 postplating). This time, however, add 10^{-6} *M* RA to the Petri dish after the EBs have been returned (see **Note 11**).

Pharmacological Screens

6. Another 2 d later (i.e., d 6 postplating), following the procedure in **step 4**, change the medium from EB medium to neurodifferentiation medium (*see* **Note 12**). At this stage, antibiotic should also be added for lineage selection; in the case of the Sox-1 β-geo, 200 µg/mL G418 is added.
7. After culturing for 2 d (i.e., d 8 postplating), the EBs are collected and allowed to settle in a tube. The medium is removed as in **step 4** but is replaced with warm sterile PBS and gently mixed. The EBs are again left to settle, and the first wash of PBS is removed and replaced with fresh PBS.
8. Add 0.5 mL 4X trypsin to the tube and place in a 37°C water bath for 2–3 min. Add 9.5 mL EB medium and pipet up and down six to eight times to further dissociate the EBs. Allow any undissociated cells or debris to settle and remove the supernatant containing cells in suspension into a fresh tube. Centrifuge for 3 min at 217g. Wash again by repeating the aspiration, resuspension, and centrifugation steps using additional fresh EB medium.
9. Centrifuge the washed cells and resuspend in 10 mL neurodifferentiation medium (*see* **Note 12**). Count the cells using a hemocytometer (*see* **Note 13**).
10. Centrifuge the cells and resuspend in neurodifferentiation medium containing FGF-2 at 0.5 µg/mL (*see* **Note 14**). If selection is required, then add G418 at 200 µg/mL at this stage.
11. Plate cells on preprepared poly-D-lysine-laminin-coated plates (*see* **Note 15**).
12. The following day after replating, refresh the medium by aspirating and replacing with fresh medium as described in **step 11**. If required, perform counter-selection against undifferentiated ES cells by adding gancyclovir to a final concentration of 2.5 µM (from 10 mM stock).
13. The next day (i.e., 2 d postreplating), change the medium to neurodifferentiation medium without any additives (i.e., no FGF-2 or selection). Culture for as long as desired to observe neuronal differentiation.

3.3. Monolayer Differentiation

1. Passage the cells 24 h in advance to establish the monolayer cultures such that the flasks are approx 70% confluent for setting up the monolayer differentiation cultures (*see* **Note 9**).
2. Trypsinize the flasks (*see* **Note 10**) in the incubator for approx 1 min. Dislodge by tapping gently, carefully add EB medium (no LIF) to quench the trypsin, and gently resuspend with a wide-bore pipet. Transfer the cell suspension to a tube and centrifuge at 144g for 4 min.
3. Remove the wash medium by aspiration and resuspend in a further 10 mL of monolayer differentiation medium. Count the cells using a hemocytometer and plate onto 0.1% gelatin-coated tissue culture plates at a density of 0.5–1.5 × 10^4 cells/cm^2 and place in a humid 7% CO_2, 37°C incubator (*see* **Note 16**).
4. Change medium by aspiration followed by replacement with fresh monolayer differentiation medium every 2 d. Neural cells are visible in the culture 4–6 d after plating, and mature neurons with processes can be seen shortly after this.
5. To further expand the number of neurons, the cultures can be further passaged. After 8 d from the initial plating, the medium is removed by aspiration, and the cells are washed in PBS. The PBS is removed by aspiration, and the cultures are trypsinized in the incubator at 37°C. The cells are then dissociated by gently tapping the side of the flask. The cells are removed using a pipet and transferred to a tube and centrifuged for 3 min at 217g; the cells are resuspended in monolayer differentiation medium. Cells are counted using a hemocytometer and plated on PD-laminated plates (*see* **Note 17**). Addition of FGF-2 at 10 ng/mL improves the viability of the cells on plating but is not necessary for neuronal development (*see* **Note 18**). Culture for as long as desired to observe neuronal differentiation.

3.4. Monitoring Neurogenesis

3.4.1. X-gal Staining

The use of IRES technology as described here to trap the expression patterns of a gene of interest means that, during any stage of the culture, the expression of the marker gene (here Sox-1-IRES-β-geo) can be monitored by staining with X-gal (5-bromo-4-chloro-3-indolyl-β-D-galactoside).

1. Remove the medium from the culture by aspiration and gently wash with PBS.
2. Remove PBS by aspiration and add X-gal, fix for approx 2 min (maximum 10 min), and wash with PBS.
3. Prepare fresh X-gal stain and add to the cultures (*see* **Note 19**).
4. Place the cells at 37°C in an incubator. If the trapped promoter is strongly expressed, then blue staining is visible in as little as 3 h; more moderately expressed traps can take up to 24 h to stain.

3.4.2. Immunostaining

1. Thaw one aliquot of 4% PFA at 37°C. Wash the plates requiring staining using PBS. In the fume hood, place sufficient PFA on the cells to adequately cover the plates and leave for 20 min. Wash twice with PBS.
2. Prepare a solution of primary antibody, in this case 1/100 TuJ (antimouse), in antibody dilution reagent (0.1% Triton X-100, 1% bovine serum albumin, 2% goat serum made up in PBS). Aspirate off the PBS and add antibody for 1–2 h at room temperature.
3. Wash three times in PBS. Prepare a solution of secondary antibody, in this case 1/200 Cy3 (goat antimouse), add to the cells, and incubate for a further 1–2 h.
4. Rinse three times in fresh PBS. At this stage, we often add a Hoechst stain for 20 min to counterstain for nuclei and wash again three times in PBS. Add a drop of Immunofluore mounting solution, mount a glass cover slip, and observe microscopically.

3.5. Plate Screening

A benchtop scanning fluorometer (FlexStation, Molecular Devices, Sunnyvale, CA) with automated liquid handling capabilities was used to measure receptor-mediated calcium responses. In this type of assay, cells are preloaded with a fluorescent calcium-binding dye (Fluo-3, excitation and emission maximums are 490 and 528 nm, respectively, in a Ca^{2+} complex), which emits practically no fluorescence if Ca^{2+} is not present. On Ca^{2+} binding and excitation, the dye emits light proportional to the intracellular calcium concentration. By monitoring this emission, the effect of the addition of various drugs on intracellular calcium concentrations can be quantitatively determined.

1. Aspirate the medium from the wells of the 96-well plate of differentiated ES-cell-derived neuronal cells (10–14 d in vitro).
2. Load the cells with 10 μM Fluo-3 dye in Tyrode's buffer using 50 μL/well dye-buffer solution.
3. Incubate the cells for 1 h at room temperature, wrapping the plate in aluminum foil (Fluo-3 is light sensitive).
4. Replace the dye-buffer solution with 50 μL Tyrode's buffer and wash the cells once.

5. Concentrated stock solutions of drugs should be prepared in 96-well plates for drug additions (20 μL stock solutions will be added automatically by the FlexStation to obtain the desired final concentrations of drugs).
6. Load drug plate, assay the plate with the Fluo-3-loaded cells into the FlexStation, and measure the emitted fluorescence at the emission wavelength of Fluo-3, after drug additions.

4. Notes

1. Previous IRES β-geo constructs *(48,54,55)* have made use of an IRES-*lacZ* fusion sequence in which the 594-bp IRES sequence was modified by mutagenesis of the native 11th translation initiation codon to create a convenient cloning site and force initiation of translation to occur 9 bp downstream at the 12th AUG codon *(44)*. In the pGT1.8 (OPT) IRES β-geo construct, initiation of translation was relocated to the native AUG codon, and the surrounding IRES nucleotide sequence was modified to reflect an optimized Kozak consensus sequence *(56)*.
2. General ES protocols for freezing, passaging, and culture of ES cells can be found in this volume as well as in **ref. 57**.
3. Sal I was added to the reaction to linearize the plasmid and allow for successful gel separation of the 4.6-Kb Xba1 cassette. This cassette contains the IRES and β-geo sequences but does not include the SVpolyA signal.
4. DNA is precipitated by bringing the final salt concentration of the reaction mix to 0.3 M with 3 M sodium acetate, adding 2.5 vol of cold absolute ethanol, and incubating at $-20°C$ for 30 min. The DNA is pelleted and washed twice with 70% (v/v) ethanol, and the ethanol is drained off.
5. For each electroporation, 10^6 ES cells are prepared from trypsin harvesting of confluence cultures and suspended in 0.7 mL PBS. To this, 100 μL linearized plasmid in PBS is added, and the cells are electroporated at 800 V for a pulse length of 0.1 ms in a 0.4-cm cuvet (BTX ECM830 Square Wave Electroporator). Cells are transferred to prewarmed medium within 1 min of electroporation.
6. Genomic DNA is prepared by overnight incubation of clonal cultures in lysis buffer containing 100 μg/mL proteinase K at 37°C and extracted in isopropanol at room temperature for 10–20 min. DNA is pelleted by centrifugation at 17,900g at 4°C for 15 min, washed with 70% ethanol, and resuspended in TE buffer.
7. Genomic DNA is digested in an overnight incubation with the appropriate enzyme for analysis of the 5′ and 3′ end integrations and electrophoresed on 0.8% agarose/TAE gels at 40 V overnight. DNA is depurinated in 0.25 M HCl and transferred to Hybond N+ nylon membrane (Amersham) in 0.4 M NaOH transfer buffer. Probes are labeled using the Rediprime kit (Amersham) and purified using Nick columns (Amersham).
8. Electroporated cells are cultured in flasks for a further 48 h in GMEM-ES culture medium before plating cells at a range of lower densities, such as 2000 and 5000 cells/10-cm plate to minimize bystander killing by gancyclovir-sensitive cells. These low-density plates are cultured for a further 48 h before adding gancyclovir at a concentration of 2.5 μM. Resistant colonies are picked after 4–5 d of gancyclovir culture.
9. For both EB formation and monolayer differentiation, the general health of the culture is enhanced by making a passage 24 h prior to seeding the cells. For EB formation, we also obtain good results when this initial passage is into EB medium containing 10% serum replacement (Invitrogen). We routinely use feeder-independent ES cell lines for EB formation. A 70% confluent medium-size flask (T-75) yields approx 1×10^7 cells; a large flask (T-150) yields approx 5×10^7 cells. Supplemented medium is used on d 0 of EB formation.

10. Cells on the flask should be aspirated of medium, washed with PBS and trypsin (made as described in **Subheading 3.**) should be added (approx 0.9 mL per 25-cm^2, 1.5 mL per 75-cm^2, or 2 mL per 150-cm^2 flasks).
11. All *trans* RA 5×10^{-3} M in dimethyl sulfoxide. Note that RA is extremely light sensitive and a teratogen. Remove a 100-µL aliquot of the 16 mM stock from the freezer and add it to 900 µL EB medium. Mix well and add 6.75 µL to each 10-cm dish (10 mL medium) or 13.5 µL to each 20-cm dish. Swirl the plates gently to mix evenly. Different ES cell lines differ in their response to RA and may require exposure for 4 d instead of the 2-d protocol described here.
12. The serum-free neurodifferentiation medium is a proprietary medium formulation from Stem Cell Sciences Limited; details can be obtained by contacting the authors. Geneticin (G418, Life Technologies) is kept frozen as a 200 mg/mL stock and is diluted 1:1000. When using serum replacements, use 2.5% FCS and 7.5% Knockout™ Serum Replacement (Invitrogen Corp., Groningen, The Netherlands).
13. Counting of cells should be done carefully as the cells are often small, irregular shape, and difficult to distinguish from cell debris. Use trypan blue exclusion or nigrosin staining techniques to help distinguish debris from live cells.
14. Basic FGF (FGF)-2 is obtained from Peprotech. A 50-µg/mL stock is prepared according to the manufacturer's instructions and frozen in 20-µL aliquots. To prepare the neurodifferentiation medium, add 20 µL of this stock to 1 mL of this medium.
15. Typically, for a 4-well plate 2×10^5 cells per well are plated; for a 96-well plate, 1×10^5 cells per well are used. To prepare the poly-D-lysine-laminin plates, first dilute a stock solution of polyD-lysine (1 mg/mL; Sigma) in sterile water to a concentration of 10 µg/mL; coat plates for a minimum of 1 h, aspirate, and wash the plates twice with sterile water. Thaw laminin stock (1 mg/mL) and dilute with sterile PBS to 1 µg/mL. Coat plates at 4°C for 1 h prior to plating the cells. Remove the polyD-lysine-laminin by aspiration when ready to use.
16. Typically, 3×10^5 ES cells in 3 mL monolayer medium are plated into a 6-cm well of a six-well dish. For 10-cm plates, typically 10^6 cells are plated. Different ES cell lines show different survival rates and different kinetics of differentiation in the serum-free medium. Thus, for some ES cell lines the number of cells plated must be reduced to avoid overgrowth of the culture prior to differentiation (this could be up to twice as many cells); other ES lines will require denser initial seeding density.
17. For replating of monolayer differentiated cultures, cells are seeded in plates at $0.5–1.5 \times 10^4$ cells/cm^2; typically, this is 3×10^5 ES cells into a 6-cm well of a 6-well plate, or for a 96-well plate, this is 1×10^3 cells.
18. Many of the neurons generated in this monolayer differentiation method are GABAergic. However, by exposure to cytokines during the course of the culture, other neuronal types can be generated (e.g., addition of Sonic Hedgehog and FGF-8 can generate TH+ dopaminergic neurons).
19. Sufficient X-gal fix and stain should be used to cover the surface of the dishes or plates (e.g., approx 5 mL per 10-cm plate). X-gal fix is 0.2% gluteraldehyde, 2 mM MgCl$_2$, 5 mM EGTA made in 0.1 M sodium phosphate buffer (46 mL 0.5 M NaH$_2$PO$_4$). X-gal is made by the 1/50 dilution of solution A (50 mg/mL X-gal from Sigma in dimethyl formamide; store at $-20°C$ in the dark) in solution B (1.64 mg/mL K$_3$Fe [CN]$_6$ and 2.4 mg/mL K$_4$Fe[CN]$_6$·3H$_2$O made in 0.1 M sodium phosphate buffer 154 mL 0.5 M NaH$_2$PO$_4$ and store at 4°C). X-gal stain should be filtered using a 0.4-µm filter prior to use to remove crystals.

References

1. Gorba, T. and Allsopp, T. E. (2003) Pharmacological potential of embryonic stem cells. *Pharmacol. Res.* **47**, 269–278.
2. Lysaght, M. J. and Hazlehurst, A. L. (2003) Private sector development of stem cell technology and therapeutic cloning. *Tissue Eng.* **9**, 555–561.
3. Lysaght, M. J. and Hazlehurst, A. L. (2004) Tissue engineering: the end of the beginning. *Tissue Eng.* **10**, 309–320.
4. Brower, V. (1999) Human ES cells: can you build a business around them? *Nat. Biotechnol.* **17**, 139–142.
5. O'Shea, K. S. (2001) Neuronal differentiation of mouse embryonic stem cells: lineage selection and forced differentiation paradigms. *Blood Cells Mol. Dis.* **27**, 705–712.
6. Mountford, P. S. and Smith, A. G. (1995) Internal ribosome entry sites and dicistronic RNAs in mammalian transgenesis. *Trends Genet.* **11**, 179–184.
7. Hopfl, G., Gassmann, M., and Desbaillets, I. (2004) Differentiating embryonic stem cells into embryoid bodies. *Methods Mol. Biol.* **254**, 79–98.
8. Ostenfeld, T. and Svendsen, C. N. (2003) Recent advances in stem cell neurobiology. *Adv. Tech. Stand. Neurosurg.* **28**, 3–89.
9. Abe, K., Niwa, H., Iwase, K., et al. (1996) Endoderm-specific gene expression in embryonic stem cells differentiated to embryoid bodies. *Exp. Cell Res.* **229**, 27–34.
10. Bain, G., Kitchens, D., Yao, M., Huettner, J. E., and Gottlieb, D. I. (1995) Embryonic stem cells express neuronal properties in vitro. *Dev. Biol.* **168**, 342–357.
11. Strubing, C., Ahnert-Hilger, G., Shan, J., Wiedenmann, B., Hescheler, J., and Wobus, A. M. (1995) Differentiation of pluripotent embryonic stem cells into the neuronal lineage in vitro gives rise to mature inhibitory and excitatory neurons. *Mech. Dev.* **53**, 275–287.
12. Fraichard, A., Chassande, O., Bilbaut, G., Dehay, C., Savatier, P., and Samarut, J. (1995) In vitro differentiation of embryonic stem cells into glial cells and functional neurons. *J. Cell Sci.* **108**, 3181–3188.
13. Bjorklund, L. M., Sanchez-Pernaute, R., Chung, S., et al. (2002) Embryonic stem cells develop into functional dopaminergic neurons after transplantation in a Parkinson rat model. *Proc. Natl. Acad. Sci. USA* **99**, 2344–2349.
14. Liu, S., Qu, Y., Stewart, T. J., et al. (2000) Embryonic stem cells differentiate into oligodendrocytes and myelinate in culture and after spinal cord transplantation. *Proc. Natl. Acad. Sci. USA* **97**, 6126–6131.
15. Okabe, S., Forsberg-Nilsson, K., Spiro, A. C., Segal, M., and McKay, R. D. (1996) Development of neuronal precursor cells and functional postmitotic neurons from embryonic stem cells in vitro. *Mech. Dev.* **59**, 89–102.
16. Lendahl, U. and McKay, R. D. (1990) The use of cell lines in neurobiology. *Trends Neurosci.* **13**, 132–137.
17. Barberi, T., Klivenyi, P., Calingasan, N. Y., et al. (2003) Neural subtype specification of fertilization and nuclear transfer embryonic stem cells and application in parkinsonian mice. *Nat. Biotechnol.* **21**, 1200–1207.
18. Svendsen, C. N. and Smith, A. G. (1999) New prospects for human stem-cell therapy in the nervous system. *Trends Neurosci.* **22**, 357–364.
19. Ye, W., Shimamura, K., Rubenstein, J. L., Hynes, M. A., and Rosenthal, A. (1998) FGF and Shh signals control dopaminergic and serotonergic cell fate in the anterior neural plate. *Cell* **93**, 755–766.

20. Lee, S. H., Lumelsky, N., Studer, L., Auerbach, J. M., and McKay, R. D. (2000) Efficient generation of midbrain and hindbrain neurons from mouse embryonic stem cells. *Nat. Biotechnol.* **18,** 675–679.
21. Kim, J. H., Auerbach, J. M., Rodriguez-Gomez, J. A., et al. (2002) Dopamine neurons derived from embryonic stem cells function in an animal model of Parkinson's disease. *Nature* **418,** 50–56.
22. Wichterle, H., Lieberam, I., Porter, J. A., and Jessell, T. M. (2002) Directed differentiation of embryonic stem cells into motor neurons. *Cell* **110,** 385–397.
23. Kawasaki, H., Suemori, H., Mizuseki, K., et al. (2002) Generation of dopaminergic neurons and pigmented epithelia from primate ES cells by stromal cell-derived inducing activity. *Proc. Natl. Acad. Sci. USA* **99,** 1580–1585.
24. Ying, Q. L., Stavridis, M., Griffiths, D., Li, M., and Smith, A. (2003) Conversion of embryonic stem cells into neuroectodermal precursors in adherent monoculture. *Nat. Biotechnol.* **21,** 183–186.
25. Ying, Q. L. and Smith, A. G. (2003) Defined conditions for neural commitment and differentiation. *Methods Enzymol.* **365,** 327–341.
26. Hemmati-Brivanlou, A. and Melton, D. (1997) Vertebrate embryonic cells will become nerve cells unless told otherwise. *Cell* **88,** 13–17.
27. Wilson, S. I. and Edlund, T. (2001) Neural induction: toward a unifying mechanism. *Nat. Neurosci.* **4,** 1161–1168.
28. Schuldiner, M., Eiges, R., Eden, A., et al. (2001) Induced neuronal differentiation of human embryonic stem cells. *Brain Res.* **913,** 201–205.
29. Zhang, S. C., Wernig, M., Duncan, I. D., Brustle, O., and Thomson, J. A. (2001) In vitro differentiation of transplantable neural precursors from human embryonic stem cells. *Nat. Biotechnol.* **19,** 1129–1133.
30. Kehat, I., Amit, M., Gepstein, A., Huber, I., Itskovitz-Eldor, J., and Gepstein, L. (2003) Development of cardiomyocytes from human ES cells. *Methods Enzymol.* **365,** 461–473.
31. Odorico, J. S., Kaufman, D. S., and Thomson, J. A. (2001) Multilineage differentiation from human embryonic stem cell lines. *Stem Cells* **19,** 193–204.
32. Carpenter, M. K., Rosler, E., and Rao, M. S. (2003) Characterization and differentiation of human embryonic stem cells. *Cloning Stem Cells* **5,** 79–88.
33. Carpenter, M. K., Inokuma, M. S., Denham, J., Mujtaba, T., Chiu, C. P., and Rao, M. S. (2001) Enrichment of neurons and neural precursors from human embryonic stem cells. *Exp. Neurol.* **172,** 383–397.
34. Li, M., Pevny, L., Lovell-Badge, R., and Smith, A. (1998) Generation of purified neural precursors from embryonic stem cells by lineage selection. *Curr. Biol.* **8,** 971–974.
35. Stavridis, M. P. and Smith, A. G. (2003) Neural differentiation of mouse embryonic stem cells. *Biochem. Soc. Trans.* **31,** 45–49.
36. Pelletier, J. and Sonenberg, N. (1988) Internal initiation of translation of eukaryotic mRNA directed by a sequence derived from poliovirus RNA. *Nature* **334,** 320–325.
37. Macejak, D. G. and Sarnow, P. (1991) Internal initiation of translation mediated by the 5´ leader of a cellular mRNA. *Nature* **353,** 90–94.
38. Vagner, S., Gensac, M. C., Maret, A., et al. (1995) Alternative translation of human fibroblast growth factor 2 mRNA occurs by internal entry of ribosomes. *Mol. Cell Biol.* **15,** 35–44.
39. Teerink, H., Voorma, H. O., and Thomas, A. A. (1995) The human insulin-like growth factor II leader 1 contains an internal ribosomal entry site. *Biochim. Biophys. Acta* **1264,** 403–408.

40. Gan, W. and Rhoads, R. E. (1996) Internal initiation of translation directed by the 5′-untranslated region of the mRNA for eIF4G, a factor involved in the picornavirus-induced switch from cap-dependent to internal initiation. *J. Biol. Chem.* **271,** 623–626.
41. Jang, S. K., Krausslich, H. G., Nicklin, M. J., Duke, G. M., Palmenberg, A. C., and Wimmer, E. (1988) A segment of the 5′ nontranslated region of encephalomyocarditis virus RNA directs internal entry of ribosomes during in vitro translation. *J. Virol.* **62,** 2636–2643.
42. Jang, S. K., Pestova, T. V., Hellen, C. U., Witherell, G. W., and Wimmer, E. (1990) Cap-independent translation of picornavirus RNAs: structure and function of the internal ribosomal entry site. *Enzyme* **44,** 292–309.
43. Jang, S. K. and Wimmer, E. (1990) Cap-independent translation of encephalomyocarditis virus RNA: structural elements of the internal ribosomal entry site and involvement of a cellular 57-kD RNA-binding protein. *Genes Dev.* **4,** 1560–1572.
44. Ghattas, I. R., Sanes, J. R., and Majors, J. E. (1991) The encephalomyocarditis virus internal ribosome entry site allows efficient coexpression of two genes from a recombinant provirus in cultured cells and in embryos. *Mol. Cell Biol.* **11,** 5848–5859.
45. Molla, A., Paul, A. V., Schmid, M., Jang, S. K., and Wimmer, E. (1993) Studies on dicistronic polioviruses implicate viral proteinase 2Apro in RNA replication. *Virology* **196,** 739–747.
46. Yang, Q. and Sarnow, P. (1997) Location of the internal ribosome entry site in the 5′ non-coding region of the immunoglobulin heavy-chain binding protein (BiP) mRNA: evidence for specific RNA-protein interactions. *Nucleic Acids Res.* **25,** 2800–2807.
47. Kim, D. G., Kang, H. M., Jang, S. K., and Shin, H. S. (1992) Construction of a bifunctional mRNA in the mouse by using the internal ribosomal entry site of the encephalomyocarditis virus. *Mol. Cell Biol.* **12,** 3636–3643.
48. Mountford, P., Zevnik, B., Duwel, A., et al. (1994) Dicistronic targeting constructs: reporters and modifiers of mammalian gene expression. *Proc. Natl. Acad. Sci. USA* **91,** 4303–4307.
49. Zhao, S., Maxwell, S., Jimenez-Beristain, A., et al. (2004) Generation of embryonic stem cells and transgenic mice expressing green fluorescence protein in midbrain dopaminergic neurons. *Eur. J. Neurosci.* **19,** 1133–1140.
50. Billon, N., Jolicoeur, C., Ying, Q. L., Smith, A., and Raff, M. (2002) Normal timing of oligodendrocyte development from genetically engineered, lineage-selectable mouse ES cells. *J. Cell Sci.* **115,** 3657–3665.
51. Mountford, P., Nichols, J., Zevnik, B., O'Brien, C., and Smith, A. (1998) Maintenance of pluripotential embryonic stem cells by stem cell selection. *Reprod. Fertil. Dev.* **10,** 527–533.
52. Eiges, R., Schuldiner, M., Drukker, M., Yanuka, O., Itskovitz-Eldor, J., and Benvenisty, N. (2001) Establishment of human embryonic stem cell-transfected clones carrying a marker for undifferentiated cells. *Curr. Biol.* **11,** 514–518.
53. Lakics, V., Allsopp, T., Cluett, T., et al. (2003) Mouse embryonic stem cell derived neuronal progenitors as research tools in drug discovery. Conference abstract, Society for Neuroscience 2003, Prog. 28.3.
54. Chowdhury, K., Bonaldo, P., Torres, M., Stoykova, A., and Gruss, P. (1997) Evidence for the stochastic integration of gene trap vectors into the mouse germline. *Nucleic Acids Res.* **25,** 1531–1536.
55. Bonaldo, P., Chowdhury, K., Stoykova, A., Torres, M., and Gruss, P. (1998) Efficient gene trap screening for novel developmental genes using IRES beta geo vector and in vitro preselection. *Exp. Cell Res.* **244,** 125–136.
56. Kozak, M. (1989) The scanning model for translation: an update. *J. Cell Biol.* **108,** 229–241.
57. Turksen, K. (2002) *Embryonic Stem Cells Methods and Protocols, Methods in Molecular Biology, Vol. 185,* Humana Press, Totowa, NJ.

26

Use of Murine Embryonic Stem Cells in Embryotoxicity Assays

The Embryonic Stem Cell Test

Andrea E. M. Seiler, Roland Buesen, Anke Visan, and Horst Spielmann

Summary

The embryonic stem cell test (EST) takes advantage of the potential of murine embryonic stem (ES) cells to differentiate in culture to test embryotoxicity in vitro. The EST represents a scientifically validated in vitro system for the classification of compounds according to their teratogenic potential based on the morphological analysis of beating cardiomyocytes in embryoid body outgrowths compared to cytotoxic effects on murine ES cells and differentiated 3T3 fibroblasts. Through a number of prevalidation and validation studies, the EST has been demonstrated to be a reliable alternative method for embryotoxicity testing based on the most important mechanisms in embryotoxicity—cytotoxicity and differentiation—as well as on differences in sensitivity between differentiated and embryonic tissues. Improvements of the EST protocol using flow cytometry analysis showed that differential expression of sarcomeric myosin heavy chain and α-actinin proteins quantified under the influence of a test compound is a useful marker for detecting potential teratogenicity. The in vitro embryotoxicity test described in this chapter is rapid, simple, and sensitive and can be usefully employed as a component of the risk/hazard assessment process.

Key Words: Cytotoxicity; developmental toxicity; differentiation; embryonic stem cell test (EST); embryonic stem cells; embryotoxicity; flow cytometry; in vitro; prediction model.

1. Introduction

To obtain information on the toxic effects of chemicals on specific elements of the reproductive cycle, embryotoxicity testing is either performed in vivo using pregnant animals or in vitro on cultured embryos or embryonic cells and tissues from pregnant animals. Both for in vivo and for in vitro testing pregnant animals have to be sacrificed to obtain embryonic cells, tissues, or organs. Taking advantage of the potential of embryonic stem (ES) cells to differentiate in culture into a variety of cell types (reviewed in **ref.** *1*), a new in vitro embryotoxicity test with permanent cell lines from the mouse has been proposed, the embryonic stem cell test (EST) *(2)*.

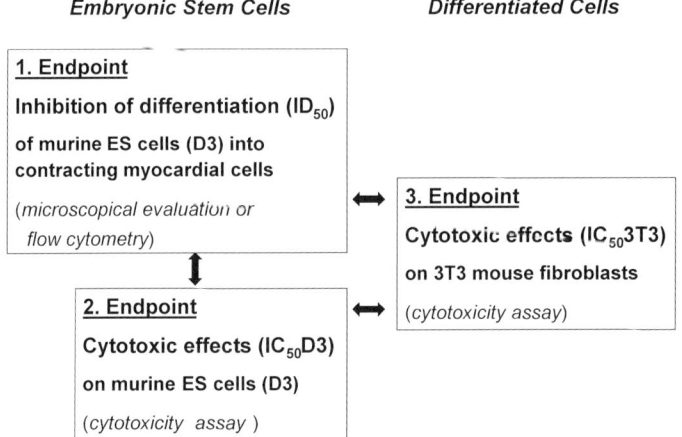

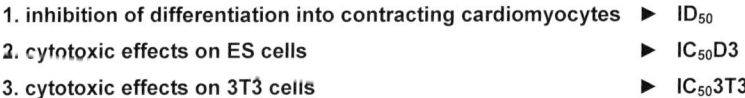

Fig. 1. The embryonic stem cell test (EST). Schematic overview illustrating the principle approach and the end points applied in the test to assess the embryotoxic potential of test compounds using two permanent murine cell lines: 3T3 fibroblasts and D3 ES cells.

In the EST, two permanent mouse cell lines are used: (1) ES cells to represent undifferentiated (embryonic) tissue, and (2) 3T3 fibroblast cells to represent differentiated (adult) tissue. The test was developed after it was found that ES cells could be maintained in vitro in the undifferentiated state in the presence of cytokines (e.g., leukemia inhibitory factor, LIF) (reviewed in **ref. 3**). On withdrawal of LIF, ES cells rapidly lose their undifferentiated phenotype and differentiate. These properties have made ES cells a popular system to study gene function and developmental processes during differentiation in vitro (reviewed in **ref. 4**).

To date, the best-studied model of ES cell differentiation is the formation in suspension culture of multicellular aggregates called embryoid bodies (EBs) (*5*). Within these aggregates, complex interactions between heterologous cell types result in the induction of differentiation of stem cells to derivatives of all three embryonic germ layers (*6*). Plating of the EBs allows further differentiation and EB outgrowth. Cytotoxicity data show that ES cells are more sensitive to toxic agents than adult cells (*7*). The inhibition of differentiation of ES cells into beating cardiomyocytes after 10 d of treatment and the inhibition of growth of ES and 3T3 cells determined by an MTT (3-[4,5-dimethyl-2-thiazolyl]-2,5-diphenyl-2H-tetrazolium bromide) test are selected end points in the EST to predict the embryotoxic potential of chemicals (*see* **Fig. 1**).

To predict the toxic potential of a test compound in vivo, a biostatistically based prediction model (PM) was applied to the results obtained in vitro with the EST. The PM of the EST has been defined previously *(8)* and was scientifically validated in a formal European Centre for the Validation of Alternative Methods (ECVAM) validation study on three in vitro embryotoxicity tests using a set of 20 reference compounds characterized by high-quality in vivo embryotoxicity data assessed in laboratory animals and humans *(9–12)*. The PM is based on linear discriminant analysis, a mathematical procedure that takes three variables into account (50% inhibition of cardiac cell differentiation [ID_{50}], 50% viability of D3 cells [$IC_{50}D3$], and 50% viability of 3T3 cells [$IC_{50}3T3$]) to discriminate among three classes of embryotoxicity (nonembryotoxic, weakly embryotoxic, and strongly embryotoxic).

Results of the validation study showed that, using the EST, chemicals were correctly classified in 78% of the experiments compared to 70% for the micromass test and 80% for the rat postimplantation whole embryo culture test when cytotoxicity data from the EST were included in the analysis. Remarkably, a predictivity of 100% was obtained with strong embryotoxic chemicals in each of the three in vitro tests *(9–12)*. According to the ECVAM Scientific Advisory Committee the three in vitro methods for embryotoxicity testing are scientifically validated and ready for consideration for regulatory acceptance and application *(13)*.

To identify more objective end points of differentiation other than the microscopic evaluation of "beating areas" and to adapt the EST to applications in high-throughput screening systems, we improved and expanded the EST protocol by establishing molecular end points of differentiation *(14–17)*. In the improved EST protocol developed at ZEBET, expression of α-actinin and sarcomeric myosin heavy chain (MHC) genes under the influence of the test chemical are quantified by intracellular flow cytometry *(16)*. As a second approach, quantitative gene expression analyses using TaqMan polymerase chain reaction (PCR) have also been performed by some of our partners in a joint research project *(18,19)*.

For a limited number of test compounds (reference chemicals), it could be demonstrated that the quantitative flow cytometry approach showed the same sensitivity for the classification of substances according to their teratogenic potential as the validated microscopic evaluation of beating cardiomyocytes *(16)*. Furthermore, both measurement protocols yielded almost identical dose-response curves (*see* **Fig. 2**), consistent with a significant correlation between the two end points: the expression of the sarcomere-specific proteins tested in flow cytometry analyses and the development of a functional contractile apparatus that underlies the morphological evaluation. Currently, additional compounds are tested to evaluate the new end point. Preliminary results from a set of 10 chemicals characterized by high-quality in vivo data showed that the quantitative gene expression approach seems to be applicable to testing a diverse group of chemicals with different embryotoxic potentials (A. Seiler, personal communication, 2004).

An important strength of the quantitative flow cytometry approach is its flexibility because it allows several developmental end points to be examined in parallel. For example, additional markers such as channel proteins or components of signal transduction

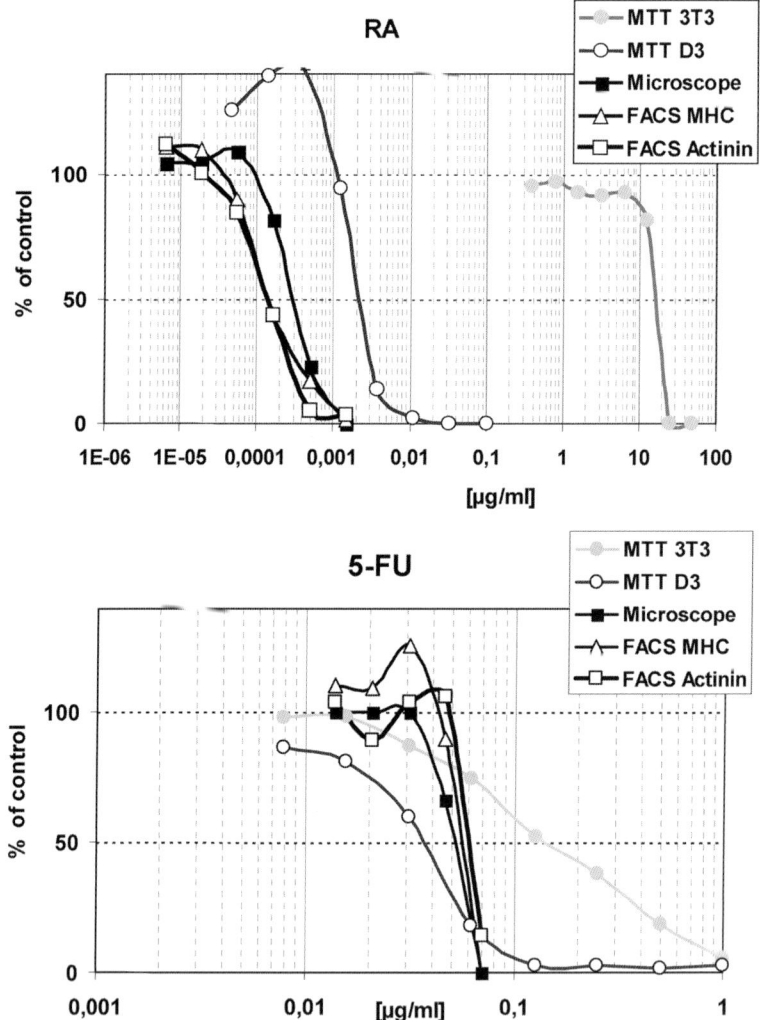

Fig. 2. Representative dose-response curves illustrating the effects of 5-fluorouracil (5-FU) and all-*trans* retinoic acid (RA) on differentiation and survival of mouse ES cells and differentiated fibroblasts. For each chemical, new and validated end points of the EST are presented in the same graph for comparison. End points are (1) quantification of cardiac-specific marker proteins, sarcomeric α-actinin, and myosin heavy chain (MHC) by FACS analysis on d 7 of differentiation (□, FACS actinin; △, FACS MHC); (2) microscopic evaluation of beating cardiomyocytes at d 10 of culture (■, microscope); (3) viability of undifferentiated stem cells (○, MTT D3); and (4) viability of differentiated fibroblasts (●, MTT 3T3) determined by MTT assays at d 10. (Please *see* the companion CD for the color version of this figure.)

pathways such as TGF-β/BMP-2 that are not directly required for cardiomyocyte contractility could be tested along with the sarcomeric proteins used here. In this way, the ability of the test to monitor the cellular response to toxins could be expanded. In this

Embryonic Stem Cell Test

regard, it has been shown that ES cell-derived cardiomyocytes express most, if not all, of the significant pharmacological properties and cardiac-specific signal transduction pathways observed in vivo (reviewed in **refs. 20** and **21**).

The improved EST holds promise as a new predictive screen for risk/hazard assessment with respect to developmental toxicity using stem cell technology and technological advances in the field of gene expression analysis.

2. Materials
2.1. Tissue Culture
2.1.1. Cell Lines

1. Balb/c 3T3 cells, clone A31 (American Type Culture Collection [ATCC], Manassas, VA; cat. no. CCL-163) or ICN-Flow (Eschwege, Germany; cat. no. 03-465-83).
2. ES cells, clone D3 (ATCC, cat. no. CRL-1934; or Prof. Rolf Kemler, Max Planck Institute, Freiburg, Germany).

2.1.2. Reagents

1. Fetal calf serum (FCS), characterized and screened for ES cell growth (500 mL, Perbio Science GmbH, Bonn, Germany; cat. no. CH30160.03) (*see* **Notes 1** and **2**).
2. High-glucose Dulbecco's modified Eagle's medium (DMEM) with L-glutamine (4500 mg/L D-glucose) without sodium pyruvate (500 mL; Gibco, Karlsruhe, Germany; cat. no. 41965-039 [Europe]; or Gibco, cat. no. 11965-092 [USA]).
3. L-Glutamine (100 mL; Gibco, cat. no. 25030-024).
4. 1X trypsin-EDTA (ethylenediaminetetraacetic acid) solution (100 mL; Gibco, cat. no. 25300-054).
5. Penicillin-streptomycin solution: 10,000 U/mL penicillin, 10,000 µg/mL streptomycin (100 mL; Gibco, cat. no. 15140-122).
6. Hybridoma-tested dimethyl sulfoxide (DMSO) (100 mL; Sigma, Taufkirchen, Germany; cat. no. D2650).
7. Nonessential amino acids (NEAA) (100 mL; Gibco, cat. no. 11140-035).
8. Cell culture-tested β-mercaptoethanol (100 mL; Sigma, cat. no. M7522). A 10 mM working solution of β-mercaptoethanol is prepared in phosphate-buffered saline (PBS) and stored at 4°C for 1 wk. For 10 mL, add 7 µL β-mercaptoethanol to a final volume of 10 mL PBS.
9. 10^6 U/mL murine leukemia inhibitory factor (mLIF) (1 mL; Chemicon, Hofheim, Germany; cat. no. ESG 1106). LIF is provided as a solution by the manufacturer and added directly to the culture plates/T-flasks during maintenance of ES cells. Store in aliquots at −20°C. Once thawed, store aliquots at 4°C (stable for up to 1 yr).
10. PBS solution without Ca^{2+} and Mg^{2+} (500 mL; Biochrom, Berlin, Germany; cat. no. L1825).
11. 5-Fluorouracil (1 g; Sigma, cat. no. F-6627).
12. Penicillin G (25×10^6 IU, approx 17 g; Sigma, cat. no. P3032).
13. Trypan blue solution (0.4%) (20 mL; Sigma, cat. no. T8154).
14. Analytical-grade ethanol (EtOH; 2500 mL; Merck, Darmstadt, Germany; cat. no. 100983).
15. Cell culture-tested gelatin (2%) (100 mL; Sigma, cat. no. G1393).
16. Double-distilled water.

2.1.3. Media for ES Cells

Supplemented media (routine culture or assay media) are freshly prepared prior to use without mLIF and used no longer than 1 wk (stored at 4°C). All supplements for media are provided as solutions. They are stored in aliquots at 4°C or −20°C according to the manufacturer's instructions.

1. For routine culture: ES cells are maintained in 1X DMEM supplemented with 15% heat-inactivated FCS, 2 mM glutamine, 50 U/mL penicillin, 50 µg/mL streptomycin, 1% NEAA, 0.1 mM β-mercaptoethanol, and 1000 U/mL mLIF (added directly to the plates). To prepare 50 mL complete medium for ES cell maintenance, mix the following: 40.75 mL 1X DMEM, 7.5 mL heat-inactivated FCS, 0.5 mL glutamine, 0.25 mL penicillin-streptomycin, 0.5 mL NEAA, and 0.5 mL 10 mM β-mercaptoethanol working solution. mLIF is added directly to the plate (5 µL mLIF/5 mL medium) This supplemented DMEM medium with mLIF is called the D3 maintenance medium.
2. For assay: same as for maintenance medium but without mLIF. This medium is called D3 assay medium.
3. For freezing: ES cells are frozen in 1X DMEM supplemented with 40% heat-inactivated FCS, 2 mM glutamine, 50 U/mL penicillin, 50 µg/mL streptomycin, 1% NEAA, 0.1 mM β-mercaptoethanol, and 10% DMSO. To prepare 10 mL D3 freezing medium, add 4 mL heat-inactivated serum, 0.1 mL glutamine, 0.05 mL penicillin-streptomycin, 0.1 mL NEAA, 0.1 mL 10 mM β-mercaptoethanol working solution, and 1 mL DMSO to 4.65 mL 1X DMEM.

2.1.4. Media for Balb/c 3T3 Cells

Supplemented media are freshly prepared prior to use. Complete medium is stored at 4°C for no longer than 1 wk. All supplements for media are provided as solutions. They are stored in aliquots at 4°C or −20°C according to the manufacturer's instructions.

1. For routine culture and assay: 3T3 cells are maintained in 1X DMEM supplemented with 10% heat-inactivated FCS, 4 mM glutamine, and 50 U/mL penicillin, 50 µg/mL streptomycin. To prepare 50 mL complete medium for 3T3 cell maintenance and assay procedure, mix the following: 43.75 mL 1X DMEM, 5 mL heat-inactivated FCS, 1 mL glutamine, and 0.25 mL penicillin-streptomycin. This medium is called 3T3 maintenance/assay medium.
2. For freezing: 3T3 cells are frozen in 1X DMEM supplemented with 20% heat-inactivated FCS, 4 mM glutamine, 50 U/mL penicillin, 50 µg/mL streptomycin, and 10% DMSO. To prepare 10 mL 3T3 freezing medium, add 2 mL heat-inactivated serum, 0.2 mL glutamine, 0.05 mL penicillin-streptomycin, and 1 mL DMSO to 6.75 mL 1X DMEM.

2.1.5. Equipment

1. Incubator (37 ± 1°C), humidified, 5 ± 1% CO_2 in air.
2. Laminar flow cabinet (biological hazard standard).
3. Water bath (37 ± 1°C).
4. Benchtop centrifuge.
5. Refrigerator (4°C) and freezer (−20°C).
6. Liquid nitrogen storage tank.
7. Laboratory burner.
8. Laboratory balance.
9. Cell counter or hemocytometer.
10. Disposable plastic pipets (2, 5, and 10 mL).
11. Eight-channel pipets.
12. Pipetmen (2, 10, 20, 100, 200, and 1000 µL) designated for tissue culture use only.

Embryonic Stem Cell Test

13. Cryotubes (1.8-mL tubes; Nalgene, Rochester, NY; cat. no. 5000-0012).
14. Dilution blocks (Greiner, Frickenhausen, Germany; cat. no. 975 502).
15. Dilution tubes (ICN, Eschwege, Germany; cat. no. 61-226-C2).
16. Safe lock tubes, amber (Eppendorf, Hamburg, Germany; cat. no. 0030 120.191).
17. Safe lock tubes (Eppendorf, cat. no. 0030 120.086).
18. 25-cm^2 T-flasks (Corning Costar Co., Cambridge, MA; cat. no. 430639).
19. 75-cm^2 T-flasks (Corning, cat. no. 430641).
20. Tissue culture dishes (60 × 15 mm; Becton Dickinson, Franklin Lakes, NJ; cat. no. 353004).
21. Tissue culture dishes (100 × 20 mm; Becton Dickinson, cat. no. 353003).
22. Bacterial Petri dishes (60 × 15 mm; Greiner, cat. no. 35628102) (*see* **Note 3**).
23. 96-well flat-bottom tissue culture microtiter plates (Becton Dickinson, cat. no. 353072).
24. 24-well tissue culture plates (Becton Dickinson, cat. no. 353047).
25. 15-mL vials (Becton Dickinson, cat. no. 352095).

2.2. Determination of Differentiation by Morphological Analysis

For determination of differentiation by morphological analysis, use an inverted microscope with phase contrast objectives (×10, ×25, ×100).

2.3. Determination of Differentiation by Flow Cytometry

2.3.1. Reagents

1. Bovine serum albumin (BSA) (100 g; Sigma, cat. no. A-4503).
2. Desoxyribonuclease I (DNase I) from bovine pancreas (one vial of 2000 Kunitz U/mL; Sigma, cat. no. D-4263). Reconstitute one vial with 1 mL cold 0.15 M NaCL to give a standard stock DNase I solution with a concentration of 2000 Kunitz U/mL. Store the stock solution at 2–8°C when not in use; discard after 8 h.
3. EDTA (1%) (100 mL; Biochrom, cat. no. L 2113).
4. 1X trypsin-EDTA solution (100 mL; Gibco, cat. no. 25300-54).
5. PBS without Ca^{2+} and Mg^{2+} (500 mL; Biochrom, cat. no. L1825).
6. Goat serum donor herd (100 mL; Sigma, cat. no. G-6767). Centrifuge and store aliquots at −20°C.
7. Paraformaldehyde (PFA) (500 g; Sigma, cat. no. P-6148). Heat a volume of water equal to slightly less than two-thirds the desired final volume of fixative to 60°C. Weigh out a quantity of PFA that will make a 4% solution and add it with a stir bar to the water. Transfer to fume hood and maintain on heating plate at 60°C with stirring. Add one drop 2 N NaOH with a Pasteur pipet. The solution should become almost clear fairly rapidly, but there will still be some fine particles that will not go away. Be careful not to overheat the solution. Remove from heat and add one-third volume 3X PBS. Bring solution to pH 7.2 with HCl; add water to final volume. Cool to room temperature or to 4°C on ice. Store 1-mL aliquots at −20°C; discard after use.
8. Saponin (25 g; Sigma, cat. no. S-7900).
9. 0.4% trypan blue solution (20 mL; Sigma, cat. no. T8154).
10. Antisarcomeric MHC (anti-MHC; clone MF20, mouse immunoglobulin [Ig] G2b, kappa light chain, supernatant, concentrated, 0.1-mL vial; Hybridoma Bank, Iowa City, IA; cat no. HF20 conc.).
11. Biotin-SP-conjugated AffiniPure goat antimouse IgG FC_y (1.5 mL; Dianova, Hamburg, Germany; cat. no. 115-065-071).
12. R-Phycoerythrin-conjugated streptavidin (1.0 mL; Dianova; cat. no. 016-110-084).

13. Hybridoma-tested DMSO (100 mL; Sigma, cat. no. D2650).
14. Analytical-grade EtOH (1000 ml; Merck, cat. no. 100983).
15. FACSFlow™ (20 L; Becton Dickinson, cat. no. 342 003).
16. Double-distilled water.
17. Dissociation solution: 4 U/mL DNase I in trypsin-EDTA. For 10 mL, add 20 µL DNase I stock to 10 mL trypsin-EDTA solution.
18. Washing solution I: 1% BSA in PBS; prepare fresh daily.
19. Washing solution II: 1% BSA and 0.15% saponin in PBS; prepare fresh daily.
20. Blocking solution: 10% goat serum, 1% BSA, and 0.15% saponin in PBS; prepare fresh daily.

2.3.2. Equipment

1. Fluorescence-activated cell sorting (FACS) cytometer (e.g., FACSCalibur™, Becton Dickinson).
2. Polystyrene round-bottom tubes for FACS cytometer (Becton Dickinson, cat. no. 352052).
3. Refrigerated microcentrifuge.
4. 1.5-mL test tubes (Eppendorf, cat. no. 0030 120086).

2.4. Determination of Cytotoxicity by MTT Assay

2.4.1. Reagents

1. Cell culture-tested MTT (100 mg; Sigma, cat. no. M5655): for 5-mg/mL solution (200 mL), dissolve 1 g MTT in a final volume of 200 mL PBS, filter (0.2 µm), and store 5-mL aliquots at −20°C.
2. Sodium dodecyl sulfate (SDS: Fluka, Taufkirchen, Germany; cat. no. 71727): for 20% stock solution (100 mL), dissolve 20 g SDS in a final volume of 100 mL double-distilled water; store at room temperature.
3. Analytical-grade propan-2-ol (1000 mL; Merck, cat. no. 109634).
4. MTT desorb solution (0.7% SDS in 2-propanol): for 100 mL, add 3.5 mL 20% SDS stock solution to 96.5 mL 2-propanol. Stir at room temperature; prepare fresh prior to use.

2.4.2. Equipment

1. 96-well plate photometer (Immuno reader).
2. Shaker for microtiter plates.
3. CO_2-permeable plastic plate sealers (Dynex, Worthing, West Sussex, UK; cat. no. M30).

2.5. Use of Test Chemicals and Solvents in the ES Cell Differentiation Assay

Test chemicals are dissolved in appropriate solvents. The recommended maximum final solvent concentrations are 1% for PBS, double-distilled water, and DMEM (incomplete medium; see **Note 4**), 0.25% for DMSO, and 0.5% for EtOH. Avoid prolonged exposure to light when testing light-sensitive chemicals (see **Note 5**). The highest test concentration of any chemical is 1000 µg/mL. Chemical solutions have to be weighed and dissolved prior to each experiment, including fluorouracil 5-(FU) for the positive control experiment (see **Subheading 3.2.6.**). The stock preparation made at the beginning of the experiment can be used also on d 3 and 5 (change of medium) if stored in aliquots at −20°C. The final solvent concentration should be noncytotoxic, kept constant throughout the experiment, and should not have any other effects on the cell differentiation at the concentrations indicated. A strategy for pretesting the solubility

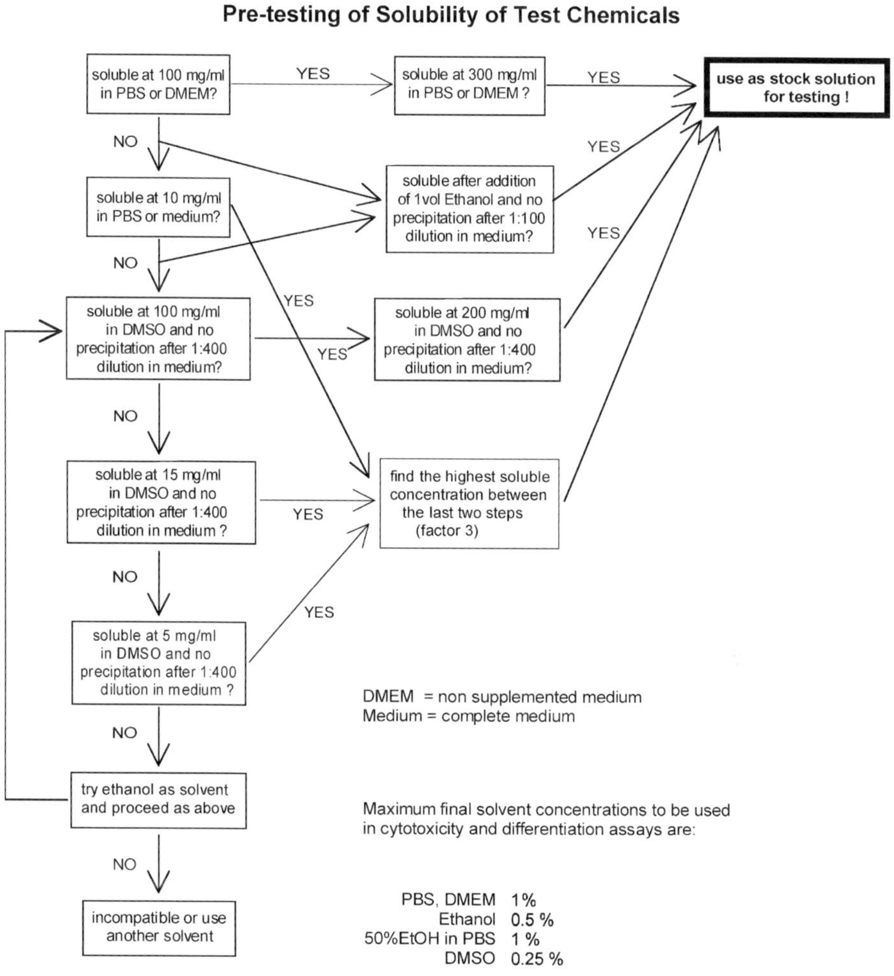

Embryonic Stem Cell Test

Pre-testing of Solubility of Test Chemicals

```
soluble at 100 mg/ml    YES    soluble at 300 mg/ml    YES    use as stock solution
in PBS or DMEM?    ──────►    in PBS or DMEM?    ──────►    for testing!
       │                              │
      NO                             YES
       ▼                              ▼
soluble at 10 mg/ml    YES    soluble after addition
in PBS or medium?    ──────►  of 1vol Ethanol and no
       │                      precipitation after 1:100    YES
      NO                      dilution in medium?
       ▼                              ▲
soluble at 100 mg/ml    YES           │
in DMSO and no       ──────►  soluble at 200 mg/ml
precipitation after 1:400     in DMSO and no
dilution in medium?    YES    precipitation after 1:400
       │                      dilution in medium?
      NO                              │
       ▼                              ▼
soluble at 15 mg/ml           find the highest soluble
in DMSO and no                concentration between
precipitation after 1:400     the last two steps
dilution in medium?    YES    (factor 3)
       │                              ▲
      NO                              │
       ▼                             YES
soluble at 5 mg/ml
in DMSO and no
precipitation after 1:400
dilution in medium?
       │
      NO
       ▼
try ethanol as solvent
and proceed as above
       │
      NO
       ▼
incompatible or use
another solvent
```

DMEM = non supplemented medium
Medium = complete medium

Maximum final solvent concentrations to be used in cytotoxicity and differentiation assays are:

```
PBS, DMEM        1%
Ethanol          0.5 %
50%EtOH in PBS   1 %
DMSO             0.25 %
```

Fig. 3. Schematic overview illustrating the approach of pretesting the solubility of test chemicals.

of test chemicals is presented in **Fig. 3**. Because strong acids and bases may influence the buffer capacity of medium, check the pH of the highest test concentration of a given chemical after dilution in medium by optical inspection. If the medium turns violet or brightly yellow (pH >8.0 or <6.5), then the stock solution of the test chemical should be neutralized with 0.1 N NaOH or 0.1 N HCI (*see* **Note 6**).

2.6. Use of Test Chemicals and Solvents in the Cytotoxicity Analysis

Dissolve test chemical in DMEM or appropriate solvent (*see also* **Subheading 2.5.** and **Fig. 3**). If solvent is used, then the final solvent concentration should be noncytotoxic and kept at a constant concentration. The maximum test concentration of any chemical is

1000 µg/mL (*see also* **Note 5**). Before starting an assay with an unknown chemical, exclude a chemical reaction between MTT, the test chemical, and the medium by measuring the optical density (OD) value at 550–570 nm (add 20 µL MTT solution to 200 µL medium containing the highest test concentration of chemical). After 2 h incubation at 37°C, the OD value should be 0.05 or less (after subtracting blank values). If the OD exceeds this value and if the respective concentration is within the range of the expected IC_{50}, then medium of all wells of the plate (except blanks) is replaced by assay medium (without test chemical) before addition of MTT on d 10 of the assay (*see also* **Note 7**).

For the range finder, use the highest soluble concentration of test chemical and noncytotoxic concentration of solvent as the highest test concentration. Make a dilution series of eight dilutions each with a factor of 1:10.

3. Methods
3.1. Basic Procedures of the EST
3.1.1. Differentiation of ES Cells

1. The mouse ES cell line D3 is cultured permanently in the presence of LIF, a differentiation inhibition factor. In the absence of LIF, ES cells start to differentiate spontaneously.
2. Several concentrations of the test chemical are added to a stem cell suspension.
3. Drops of ES cell suspension in D3 assay medium are placed on the inner side of the lid of a 10-mm Petri dish (hanging drop culture according to **ref. 22**).
4. After cultivation for 3 d, the aggregates (EBs) are transferred into bacterial (nontissue-culture-treated) Petri dishes.
5. 2 d later, EBs are placed into 24-well plates (tissue culture-treated) in which further development of EBs proceeds into different embryonic tissues.
6. Differentiation into contracting myocardial cells is determined by light microscopy after another 5 d of culture. Improvements of the EST protocol showed that quantification of structural proteins of the sarcomere apparatus (α-actinin and MHC) by flow cytometry analysis can be usefully employed as an additional/alternative toxicological end point in the EST (*see* **ref. 16** and **Fig. 2**).

3.1.2. Cytotoxicity Measurement With ES Cells and 3T3 Cells in the MTT Assay

1. Exponentially growing 3T3 cells and ES cells in the absence of LIF are plated into 96-well microtiter plates.
2. 2 h after cell seeding, seven concentrations of the test chemical, dissolved in assay medium or appropriate solvent, are added to each well.
3. After 10 d of culture, the MTT assay *(23)* is performed.
4. The absorbance is read on an enzyme-linked immunosorbent assay reader at 570 nm using a reference wavelength of 630 nm. A spectrum of the formazan product from the MTT substrate is presented in **Fig. 4**.

3.1.3. Historical Data Obtained in the ECVAM Validation Study

ID_{50} and IC_{50} values of 20 chemicals tested in two independent experiments by four laboratories served as a database for evaluating the previously established PM of the EST *(8,9)*. These values are presented in the appendix of **ref. 10** and might be helpful to establish the EST in other laboratories.

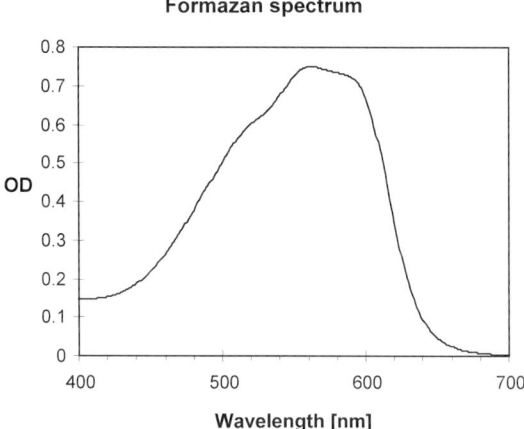

Fig. 4. Absorption spectrum of MTT formazan reagent in 2-propanol.

3.2. ES Cell Differentiation Assay

3.2.1. Differentiation of ES Cells

An outline of the differentiation assay procedure is presented in **Fig. 5**.

3.2.1.1. Day 0

1. Prepare ES cell suspensions (3.75×10^4 cells/mL) from undifferentiated ES cells (60-mm maintenance plates) with:
   a. A concentration range of test chemical (6–8 concentrations) in D3 assay medium (test solutions). Prepare 6–8 concentrations with a two- to fourfold dilution factor covering the relevant range of dose response according to the cytotoxicity range finder experiment (*see* **Subheading 2.6.**).
   b. The solvent in D3 assay medium (solvent control).
   c. D3 assay medium (untreated control) (*see* **Notes 8** and **9**).
2. Use one Petri dish per concentration of the test chemical also for the untreated control (D3 assay medium) and solvent control. Use 60-mm bacterial Petri dishes to prevent adherence of ES cells. Keep the cells in suspension by frequent gentle agitation throughout the plating procedure (*see also* **Note 10**).
3. Check viability by staining an aliquot of the cell suspension with trypan blue. A viability of 90% or higher is acceptable. Do not exceed the highest solvent concentration; keep the solvent at a constant concentration with each concentration of test chemical.
4. Using a pipet, dispense 20 µL of cell suspension (750 cells) containing the appropriate test chemical concentration or solvent or D3 assay medium (without test chemical or solvent) on the inner side of a 100-mm tissue culture Petri dish lid. Pipet 50 drops per lid. Use one Petri dish per concentration of the test chemical as well as for the untreated control (D3 assay medium) and the solvent control.
5. Turn lid carefully into its regular position and put on top of a Petri dish filled with 5 mL PBS.
6. Incubate the hanging drops for 3 d in a humidified atmosphere with 5% CO_2 at 37°C.

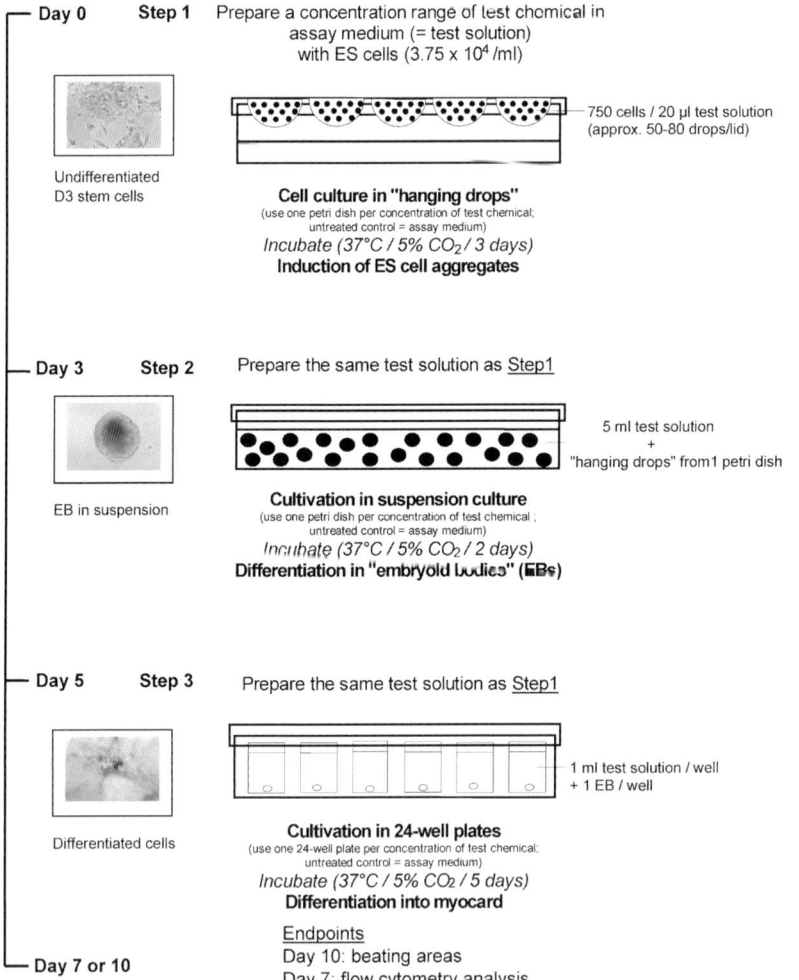

Fig. 5. Outline of the experimental steps of the embryonic stem cell differentiation assay.

3.2.1.2. Day 3

1. Prepare the same test solutions as on d 0.
2. Use one Petri dish per concentration of the test chemical, the untreated control (D3 assay medium), and the solvent control.
3. Pipet 5 mL medium containing the appropriate concentration of test chemical or solvent or D3 assay medium to the lid of the hanging drop culture dish. Hold the lid at approx a 45° angle to rinse the EBs down to the bottom. Using a sterile 5-mL pipet, gently transfer the total suspension to a 60-mm bacterial Petri dish (*see also* **Notes 11** and **12**).
4. Cultivate this EB suspension for 2 d in a humidified atmosphere with 5% CO_2 at 37°C.

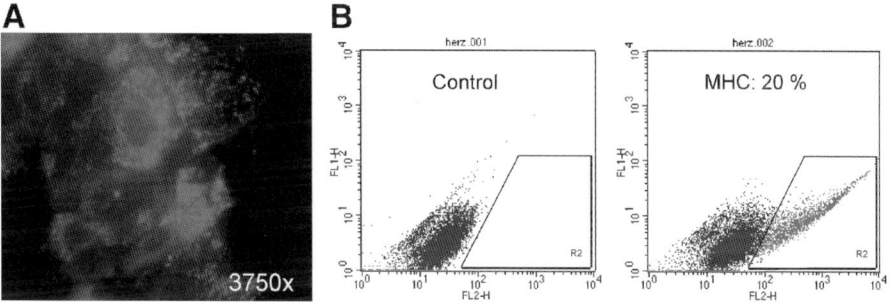

Fig. 6. Quantification of cardiac marker protein expression by intracellular flow cytometry. (**A**) Confirmation of cell-type specificity by indirect immunofluorescence. Microscope images of antisarcomeric myosin heavy chain (MHC)-stained 10-d-old embryoid body (EB) 5 d after attachment. (**B**) Representative dot-blot analyses of antisarcomeric MHC-stained 7-d-old EB 2 d after attachment (murine embryonic stem cell line D3). The x-axis corresponds to the PE fluorescence intensity and the y-axis to the unspecific fluorescence (autofluorescence). (Please *see* the companion CD for the color version of this figure.)

3.2.1.3. DAY 5

1. Prepare the same test solutions as on d 0.
2. Use one 24-well plate per concentration of the test chemical as well as for the untreated control (assay medium) and solvent control.
3. Pipet 1 mL test solution into each well of a 24-well tissue culture plate.
4. Add 1 EB (in a small volume, ≤40 µL, with blue tip or cut yellow tip) per well. Alternatively, for flow cytometry analysis EBs can be plated together in tissue culture dishes (50 EBs per 100-mm dish; *see* **Note 13**).
5. Incubate 100-mm dishes for 2 d (assay end point: flow cytometry) and 24-well plates for 5 d (assay end point: morphological analysis of beating cardiomyocytes) in a humidified atmosphere with 5% CO_2 at 37°C.

3.2.2. Immunofluorescence Staining Method for Flow Cytometry

Alternatively to the morphological analysis at d 10, differentiation can be analyzed at d 7 of the assay by flow cytometry. Indirect staining of cardiomyocytes is performed by an unconjugated primary monoclonal antibody directed against the sarcomeric MHC (clone MF20 *[24]*), followed by an incubation with a secondary biotin-conjugated goat antimouse antibody and finally with the fluorochrome PE-SA (phycoerythrin-conjugated streptavidin). For flow cytometry analysis, the fluorochrome PE-SA is detected at 585 nm in the FL-2 channel from 10,000 viable cells (nonviable cells are prior excluded from flow cytometry analysis by appropriate gating). Untreated cells and cells lacking primary antibody are used as controls. In addition, isotype controls can be used to assess the level of nonspecific antibody binding. All data analyses are carried out using the CellQuest *Pro*™ software. Previous results demonstrated that on d 7 of culture, 10–20% of the untreated viable cell population (gated) were MHC positive (*see* **ref. 16** and **Fig. 6**). If the number of positively stained cells drops below 5%, then the assay has to be repeated.

3.2.2.1. Staining Procedure

Perform antibody staining on ice, use pre-chilled solutions.

1. Use one 100-mm Petri dish of differentiated cells prepared in **Subheading 3.2.1.3., step 5**, per concentration of test chemical as well as for the untreated control (assay medium) and the solvent control.
2. Remove media and rinse with PBS containing 1 mM EDTA.
3. Add 3 mL dissociation solution to each 100-mm dish.
4. Return plate to the incubator for 30 min.
5. Resuspend cells by gently pipetting up and down.
6. Collect all 3 mL into a 15-mL Falcon tube containing 3 mL ice-cold PBS supplemented with 1 mM EDTA and 5% FCS to stop trypsinization.
7. Centrifuge at 170g for 5 min at 4°C.
8. Gently resuspend the cell pellet in 3 mL ice-cold PBS containing 1 mM EDTA and 5% FCS; incubate at 4°C for 30 min.
9. Centrifuge at 170g for 5 min at 4°C.
10. Gently resuspend the cell pellet in 5 mL ice-cold PBS-EDTA.
11. Count a sample of the cell suspension (*see* **Note 14**).
12. Centrifuge the cell suspension (from **step 10**) at 170g for 5 min at 4°C.
13. Gently resuspend the cell pellet in 500 µL washing solution I.
14. Collect all 500 µL to a 1.5-mL Eppendorf tube.
15. To fix the cells, add 500 µL of a 4% PFA solution; incubate on ice for 25 min
16. Centrifuge cell suspension at 1400g for 3 min at 4°C in a refrigerated microcentrifuge.
17. Gently resuspend the cell pellet in 1 mL washing solution I.
18. Wash cells with washing solution I.
19. Gently resuspend the cell pellet in 1 mL blocking solution; incubate on ice for 30 min.
20. For each concentration of the test chemical and the controls, transfer 3×10^5 cells to a new 1.5-mL Eppendorf tube.
21. Centrifuge cell suspension at 1400g for 3 min at 4°C in a refrigerated microcentrifuge.
22. Incubate cells with 100 µL primary antibody in blocking solution (anti-MHC, clone MF20, diluted 1:1600; *see* **Note 15**) for 1 h on ice. Include a sample of untreated cells lacking primary antibody (background control).
23. Rinse with three changes of washing solution II.
24. Incubate with 100 µL appropriate biotin-conjugated second-step antibody in blocking solution (biotin-conjugated goat antimouse IgG, diluted 1:1000) for 30 min on ice.
25. Rinse with three changes of washing solution II.
26. Incubate with 100 µL PE-SA (1:600) in blocking solution for 15 min on ice.
27. Wash extensively with washing solution II.
28. Rinse in washing solution I.
29. Carefully resuspend the cell pellet in 200 µL washing solution I.
30. Add 200 µL FACSFlow and transfer to polystyrene round-bottom tubes.
31. Perform analysis in a flow cytometer equipped with an argon laser.

3.2.3. Assay End Points

3.2.3.1. Morphological Evaluation

On d 10 of the assay, differentiation into contracting myocardial cells is determined under the light microscope. Each well of the 24-well plate is inspected, and

Embryonic Stem Cell Test

the number of wells containing spontaneously contracting cells is recorded. The rate of beating cardiomyocytes derived from treated cells is compared to that obtained with untreated cells (solvent control plate). Use *Excel* spread sheets for data recording.

3.2.3.2. FLOW CYTOMETRY ANALYSIS

On d 7, differentiation is determined by comparing the fluorescence intensity of the treated cells to that of untreated cells obtained from a solvent control plate. Use *Excel* spread sheets for data recording.

3.2.4. Independent Runs

Repeat the assay at least once (two valid experiments). Prepare the medium and the test solution before starting the experiment. Use independent preparations of the reagents in a second experiment.

3.2.5. Quality Control of Cells

At d 10 of differentiation, check the solvent control plates. The assay is acceptable if at least 21 of the 24 EBs have differentiated into spontaneously contracting myocardial cells. According to historical data, a 100% differentiation (24 wells with contracting cells) is obtained in approx 50% of the assays, whereas the acceptable range of 21–24 contracting EBs covers approx 95% of all assays. Compare the data of the untreated control plate (D3 assay medium only) to the data of the solvent control plate to be sure that the solvent has no effect on ES cell differentiation. If the highest allowed solvent concentration is used, then both a solvent control and a medium control have to be performed.

3.2.6. Quality Control of the Assay (Positive Control)

After thawing frozen cells and before testing chemicals of interest, the quality of the assay has to be controlled using 5-FU as a positive reference chemical. ID_{50} values are determined with ES cells (*see* **Subheading 3.4.**). Final concentrations of 0.09, 0.06, 0.04, 0.026, 0.018 µg/mL 5-FU (0.08 µg/mL optional) are tested (prepared from a 2 mg/mL stock solution in PBS). The ID_{50} for 5-FU should be within the range of 0.048–0.06 µg/mL (data obtained in previous assays).

3.2.7. Quality Control of FCS

FCS batches of interest are tested in differentiation assays without chemicals first, which should result in at least 21 of 24 wells containing contracting myocardial cells in at least two independent runs (*see* **Subheading 3.2.5.**). As a next step, 5-FU should be tested as positive control with concentrations according to **Subheading 3.2.6.** at least twice.

3.3. Analysis of Cytotoxicity With ES Cells and 3T3 Cells

3.3.1. Seeding of Monolayers and Assay Procedure

An outline of the MTT assay procedure is presented in **Fig. 7**.

Experimental Steps of the Cytotoxicity Assay (ES and 3T3 cells)

Seed 96-well plate: 500 cells / 50 µl assay medium / well

Incubate (37°C / 5% CO_2 / **2 h**)

↓

Step 1

Add 150 µl test chemical solved in assay medium (7 conc. + 1 pos. control)
(untreated control = assay medium + solvent)

Incubate (37°C / 5% CO_2 / **3 days**)

↓

Remove assay medium

↓

Step 2

Add 200 µl test chemical solved in assay medium
(the same concentrations from step 1)
(untreated control = assay medium + solvent)

Incubate (37°C / 5% CO_2 / **2 days**)

↓

Remove assay medium

↓

Step 3

Add 200 µl test chemical solved in assay medium
(the same concentrations from step 1 and 2)
(untreated control = assay medium + solvent)

Incubate (37°C / 5% CO_2 / **5 days**)

↓

Add 20 µl MTT (5 mg/ml per well)

Incubate (37°C / 5% CO_2 / **2 h**)

↓

Remove MTT medium. Add 130 µl MTT desorb solution
(0.7% SDS, 96.5% propan-2-ol) to each well.

↓

Shake plate for 15 min

↓

Measure OD value at 570 nm with the reference wavelength of 630 nm
to detect formazan absorption

Fig. 7. Outline of the experimental steps of the cytotoxicity assay (embryonic stem and 3T3 cells).

Pipetting Scheme for the Cytotoxicity Assay

	1	2	3	4	5	6	7	8	9	10	11	12
A	b	b	b	b	b	b	b	b	b	b	b	b
B	b	CO	P		test	so	lu	tion			CO	b
C	b	CO	P								CO	b
D	b	CO	P								CO	b
E	b	CO	P	low	con	cen	tra	tion	→	high	CO	b
F	b	CO	P								CO	b
G	b	CO	P								CO	b
H	b	b	b	b	b	b	b	b	b	b	b	b

CO = SOLVENT CONTROL
b = BLANKS (assay medium)
P = POSITIVE or NEGATIVE CONTROL

Fig. 8. Pipetting scheme for the cytotoxicity assay. One test chemical per plate is used (the positive control is placed next to the lowest test concentration).

3.3.1.1. Day 0

1. Prepare a cell suspension in assay medium of 1×10^4 cells/mL; use maintenance plates (D3 and 3T3 cells) from **Subheadings 3.5.1.1., step 8**, and **3.5.1.2., step 6**. One maintenance plate gives enough cells to plate one 96-well plate.
2. Check viability of the cells by staining an aliquot of the cell suspension with trypan blue. A viability of 90% or greater is acceptable.
3. Use a multichannel pipet and dispense 50 µL assay medium (without cells) into the peripheral wells of a 96-well tissue culture microtiter plate (blank).
4. In the remaining wells, dispense 50 µL of the cell suspension of 1×10^4 cells/mL (500 cells/well).
5. Incubate the cells for 2 h in a humidified atmosphere with 5% CO_2 at 37°C. This incubation period allows adherence of the cells.
6. Prepare seven concentrations with a smaller dilution factor covering the relevant range of dose response determined in the range finder and one concentration of the positive control chemical (*see* **Subheading 3.3.4.2.**). The minimum practical dilution factor is 1.5-fold.
7. After a 2-h incubation, add 150 µL assay medium containing the appropriate concentration of test chemical (test solutions) (seven concentrations of test chemical plus one positive control). Be certain that the 150 µL test solution contains 1.333 times the final chemical concentration. Pipet 150 µL assay medium containing solvent into the peripheral wells (blanks = wells without cells, without chemicals) and 150 µL assay medium containing solvent (solvent controls = wells with cells but without chemicals) into the wells of two columns of the plate (columns 2 and 11). A pipetting scheme is presented in **Fig. 8**.
8. Incubate cell cultures in 5% CO_2 and at 37°C for 3 d.

3.3.1.2. Day 3

1. Check cells under the microscope.
2. Remove test solutions (columns 3–10) and solvent controls (columns 2 and 11) using a Pasteur pipet attached to a pump or a multichannel pipet (except peripheral wells) (*see* **Note 16**).
3. Add 200 µL freshly prepared test solution (columns 3–10) and solvent controls (columns 2 and 11) (final concentration/well as on d 0).
4. Incubate cell cultures in 5% CO_2 and at 37°C for 2 d.

3.3.1.3. Day 5

1. Remove test solutions (columns 3–10) and solvent controls (columns 2 and 11) using a Pasteur pipet attached to a pump or a multichannel pipet and add 200 µL fresh test solutions and solvent controls (final concentrations/well as on d 0).
2. Incubate cell cultures in 5% CO_2 and at 37°C for 5 d.

3.3.1.4. Day 10

Determination of cell growth inhibition is performed at d 10 of the assay as in the following sections.

3.3.1.4.1. Microscopic Evaluation

1. Examine cells under a phase contrast microscope.
2. Record changes in morphology caused by cytotoxic effects of the test chemical. This control is performed to exclude experimental errors. Microscopic analysis of cytotoxicity is not used as an end point of the assay.

3.3.1.4.2. Measurement of MTT

1. Add 20 µL MTT solution (5 mg/mL) to all wells of the plate and incubate at 37°C in a humidified atmosphere of 5% CO_2 for 2 h.
2. After 2-h incubation, remove the supernatant using a Pasteur pipet attached to a pump. Place plate upside down on a blotting paper for 1 min.
3. Add exactly 130 µL MTT desorb solution (prewarmed to 37°C) to each well.
4. Shake microtiter plate thoroughly on a microtiter plate shaker for at least 15 min to dissolve blue formazan until the solution is cleared and no more clumps are visible. If aggregates still exist after this incubation, then precipitates can be resuspended by pipetting up and down with a multichannel pipet before measuring the absorption.
5. Measure the absorption of the resulting colored solution at 550–570 nm in a microtiter plate reader using 630 nm as a reference wavelength (*see* **Note 17**).

3.3.2. Independent Runs

1. Repeat the experiments at least once (two valid experiments).
2. Prepare medium and test solution prior to use.
3. Use independent cell culture batches.

3.3.3. Quality Control of Cells

Normal growing behavior of cells is a prerequisite in all cytotoxicity assays based on determination of growth inhibition. Consequently, on d 10 after the MTT assay has been performed, check the absolute OD ($OD_{550-570}$) of solvent control wells (columns 2 and 11 of the 96-well plate; see **Fig. 8**). According to previous data, the following confidence ranges (preliminary calculation) have to be met:

Cell Line	Source	95% Confidence Interval ($OD_{550-570}$)
D3	Kemler	0.15–1.2
	ATCC	0.50–1.6
3T3	CN-Flow	0.50–1.5
	ATCC	0.15–0.6

3.3.4. Quality Control of the Assay (Positive Control)

3.3.4.1. AFTER THAWING FROZEN CELLS AND BEFORE TESTING CHEMICALS OF INTEREST

Quality of the assay is controlled using 5-FU as a positive control and penicillin G as a negative control. The negative control chemical penicillin G can be used from a frozen stock of 100 mg/mL in PBS as well as 5-FU for the fixed concurrent positive control concentration.

Highest test concentrations of 1 µg/mL 5-FU for ES and 3T3 cells (prepared from a 2 mg/mL stock in PBS) are used and diluted in a twofold dilution series. IC_{50} values for 5-FU are determined with both ES cells and 3T3 cells. The IC_{50} values for 5-FU should be in the range of 0.04–0.09 µg/mL with ES cells and 0.12–0.5 µg/mL with 3T3 cells (preliminary calculation). One concentration of penicillin G (1000 µg/mL) is concurrently run in one column of the plate (column 3). Because penicillin G is known to have no adverse effect on reproduction, viability of the cells should be unaffected.

3.3.4.2. CONCURRENTLY IN THE MTT ASSAY

One fixed concentration of the strong embryotoxicant 5-FU (positive control chemical) is included in each cytotoxicity assay (see **Fig. 8**, column 3) concurrently with seven dilutions of the test chemical. The fixed concentration for 5-FU is derived from historical mean IC_{50} values for ES and 3T3 cells. The concentration of 5-FU to be included as the positive control is 0.29 µg/mL for 3T3 cells and 0.06 µg/mL for ES cells. With these concentrations, inhibition should be in the range of 20–80%.

3.4. Evaluation: Prediction of Embryotoxic Potential

3.4.1. General Remarks

1. Evaluation of results is based on three experimental end points determined in two cell lines: ID_{50}, IC_{50} D3, and IC_{50} 3T3. ID_{50} reflects 50% inhibition of ES cell differentiation, and IC_{50} indicates 50% inhibition of cell growth with ES cells ($IC_{50\ D3}$) and 3T3 cells ($IC_{50\ 3T3}$). Half-maximal inhibition concentrations should always be backed by a graded concentration-response curve. It is, therefore, not sufficient to calculate a 50% inhibiting concentration by interpolation between an "all-or-nothing" effect unless the concentration-response curve is extremely steep, and a graded response is not obtainable even with testing narrow concentration steps.

2. To calculate ID_{50}, $IC_{50\ D3}$, and $IC_{50\ 3T3}$ from experimental data, several methods are adequate and may be used: the most simple way is a graphical determination using probability paper with $x = \log$ and $y = $ probit scales, where the test concentrations are assigned to the x-axis and the percentages of the solvent control are assigned to the y-axis. Biostatistical methods modeling the concentration-response curves reveal a more precise calculation of the ID_{50} and IC_{50} and allow calculation of the confidence intervals for these values. For calculation of the ID_{50}, the method of Litchfield and Wilcoxon *(25)* or probit analysis according to Finney *(26)* is recommended. For calculation of IC_{50}, the method of Holzhütter and Quedenau *(27)* is recommended.

3.4.2. Calculation of End Points

3.4.2.1. ES Cell Differentiation Assay

3.4.2.1.1. Microscopic Evaluation

1. On d 10 of differentiation, determine the number of wells with contracting myocardial cells in the 24-well solvent control plate.
2. Set this number as 100%.
3. Determine the number of wells with contracting areas for each of the plates treated with a given concentration of the test chemical.
4. Calculate the inhibition of differentiation as percentage of the solvent control plate.
5. Use an *Excel* spreadsheet for data recording.
6. Use these values for calculating the ID_{50} value (*see* **Subheading 3.4.1., step 2**).

3.4.2.1.2. Flow Cytometry Analysis

1. On d 7, determine the sarc. MHC expression of the solvent control.
2. Set this number as 100%.
3. Determine the MHC expression for each of the samples treated with a given concentration of the test chemical.
4. Calculate the inhibition of differentiation as percentage of the solvent control plate.
5. Use an *Excel* spreadsheet for data recording.
6. Use these values for calculating the ID_{50} value (*see* **Subheading 3.4.1., step 2**).

3.4.2.2. Cytotoxicity Assay With ES and 3T3 Cells

1. Determine the mean $OD_{550-570}$ of the blank wells and subtract this value from all OD values of the 96-well plate (this corrects all values for adherence of the dye to the plastic material of the plates).
2. Determine the mean $OD_{550-570}$ of the untreated solvent controls (columns 2B–G and 11B–G; *see also* **Fig. 8**). Set this value to a cell viability of 100%.
3. Determine the mean $OD_{550-570}$ for each of columns 4–10, each representing a concentration of the test chemical. Express this value as cell viability (percentage of solvent controls).

3.4.3. Classification: The PM of the EST

To predict the toxic potential of a test compound in vivo, a biostatistically based PM is applied to the results obtained in vitro with the EST. The PM of the EST has been defined previously *(8)* and was scientifically validated in a formal ECVAM validation study on three in vitro embryotoxicity tests *(9–12)*. It is based on linear discriminant analysis, a mathematical procedure that takes into account three variables (50% inhibition of cardiac cell differentiation, ID_{50}; 50% viability of D3 cells,

Table 1
Validated Prediction Model of the EST

A		
	Function I	$5.92 \lg(IC_{50}3T3) + 3.50 \lg(IC_{50}D3)$ $- 5.31 \dfrac{IC_{50}3T3 - ID_{50}}{IC_{50}3T3} - 15.7$
	Function II	$3.65 \lg(IC_{50}3T3) + 2.39 \lg(IC_{50}D3)$ $- 2.03 \dfrac{IC_{50}3T3 - ID_{50}}{IC_{50}3T3} - 6.85$
	Function III	$-0.125 \lg(IC_{50}3T3) - 1.92 \lg(IC_{50}D3)$ $+ 1.50 \dfrac{IC_{50}3T3 - ID_{50}}{IC_{50}3T3} - 2.67$
B	Class 1	Nonembryotoxic — If I > II and I > III
	Class 2	Weakly embryotoxic — If II > I and II > III
	Class 3	Strongly embryotoxic — If III > I and III > II

(A) linear discriminant functions I, II, and III and (B) classification criteria *(8,9)*.

$IC_{50}D3$; and 50% viability of 3T3 cells, $IC_{50}3T3$) to discriminate between three classes of embryotoxicity (nonembryotoxic, weakly embryotoxic, and strongly embryotoxic). The linear discriminant functions incorporating the three variables are presented in **Table 1A**. The following procedure is used to classify the chemicals according to the PM: 50% inhibition concentration ($IC_{50}3T3$, $IC_{50}D3$, ID_{50} values) and the relative distance between $IC_{50}3T3$ and ID_{50} are determined and employed in the three linear discriminant functions.

If the result of function I exceeds the results of functions II and III, then the chemical is classified as nonembryotoxic; if the result of function II exceeds the results of function I and III, then the chemical is classified as weakly embryotoxic; finally, if the result of function III exceeds the results of functions I and II, then the chemical is classified as strongly embryotoxic (**Table 1B**). Results from prevalidation and validation demonstrated that all three end points are necessary for the highest rate of correct classification of test compounds *(8,10)*.

3.5. Cell Maintenance and Culture Procedure

3.5.1. Subculture of ES Cells and Balb/c 3T3 Cells

ES cells and Balb/c 3T3 cells are routinely grown as monolayers in tissue culture dishes or culture flasks in a humidified atmosphere in 5% CO_2 at 37°C; *see* **Note 18**. The cells should be examined on a daily basis under a phase contrast microscope. Any changes in morphology or their adhesive properties should be noted. ES cells and Balb/c 3T3 cells are routinely passaged every 2–3 d.

3.5.1.1. Trypsin Treatment of ES Cells

1. Remove the medium and wash the cells twice with 1–3 mL PBS without Ca^{2+} and Mg^{2+}. Wash by gentle agitation to remove any culture medium traces that might inhibit trypsin activity. Discard the washing solution.

2. Add 1–2 mL prewarmed trypsin-EDTA solution to the monolayer for a few seconds up to a maximum of 2 min.
3. Prepare a single-cell suspension by thoroughly pipetting up and down eight times.
4. Transfer the single-cell suspension to a centrifuge tube containing 6 mL D3 assay medium.
5. Centrifuge the cell suspension at 170g for 10 min.
6. Aspirate the medium; gently resuspend the pellet in 3 mL D3 assay medium. Prepare a single-cell suspension by thoroughly pipetting up and down (*see* **Note 19**).
7. Count a sample of the cell suspension using a hemocytometer or cell counter.
8. Plate the cells in D3 maintenance medium for maintenance (*see* **Note 20**) or start a differentiation assay (*see* **Note 8** and **Subheading 3.2.**) or an MTT assay (*see* **Subheading 3.3.**).
9. ES cell cultures must be split every 2–3 d (60–80% confluency). For ES cells, use 25-cm^2 T flasks or tissue culture dishes (60 mm) (*see also* **Notes 21** and **22**).

3.5.1.2. Trypsin Treatment of Balb/c 3T3 Cells

1. Remove the medium and wash the cells twice with 1–3 mL PBS without Ca^{2+} and Mg^{2+}. Wash by gentle agitation to remove any culture medium traces that might inhibit trypsin activity. Discard the washing solution.
2. Add 1–2 mL trypsin-EDTA solution to the monolayer for a few seconds.
3. Remove excess trypsin-EDTA solution and incubate the cells at 37°C.
4. After 2–3 min, lightly tap the culture flask or Petri dish to detach the cells into a single-cell suspension.
5. Count a sample of the cell suspension using a hemocytometer or cell counter.
6. Plate the cells for maintenance (*see* **Note 23**) or start an MTT assay (*see* **Subheading 3.3.**).
7. Balb/c 3T3 cells are routinely passaged every 2–3 d (60–80% confluency); use 75-cm^2 T flasks or dishes (100 mm Ø).

3.5.2. Freezing

Stocks of ES cells and Balb/c 3T3 cells can be stored in sterile cryovials in liquid nitrogen. DMSO is used as a cryoprotective agent.

1. Trypsinize cells in the exponential phase of growth.
2. Pellet cells by centrifugation (170g, 10 min).
3. Discard the supernatant and resuspend the cells in freezing medium (*see* **Subheadings 2.1.3.**, **item 3**, and **2.1.4.**, **item 2**) at a concentration of $1–5 \times 10^6$ cells/mL and fill 1 mL cell suspension per cryovial.
4. Freeze cells at a freezing rate of 1°C/min until −70 to −80°C is reached.
5. Place the frozen vials into liquid nitrogen for storage.

3.5.3. Thawing

1. Remove a cryovial of cells from frozen storage.
2. Immerse vial immediately in warm water (37°C); *see* **Note 24**.
3. Rinse vial with 70% EtOH.
4. Immediately resuspend the cells (1 mL) in 9 mL of the appropriate medium and centrifuge (170g, 5 min) to remove DMSO. Decant supernatant and gently resuspend the cell pellet in fresh medium and transfer to a tissue culture dish (*see* **Note 25**).
5. Incubate in a humidified atmosphere in 5% CO_2 at 37°C.
6. Passage at least two or three times before using the cells in a differentiation or cytotoxicity assay.
7. Cells should not be cultured for more than 25 passages after thawing.

4. Notes

1. The quality of the serum is very important for ES cell growth and differentiation. Batches of serum from different suppliers have to be tested before performing the assays. Suitable batches should then be ordered in large quantities and stored at $-20°C$ for up to 2 yr.
2. Serum is heat inactivated as follows: thaw a bottle of serum overnight at 4°C, then warm the bottle in a water bath to 56°C. Allow the serum to heat inactivate for 30 min at 56°C. Mix the serum approx every 10 min to avoid gelling of serum proteins. When the 30-min heat inactivation period is completed, remove the serum from the water bath and cool rapidly by placing in an ice-water bath. Aliquot into sterile bottles; store in a freezer. Heat inactivation at higher temperatures or for longer periods is not recommended as it may cause gelling.
3. We have found that bacterial (noncell culture treated) Petri dishes from Greiner were reproducibly suitable for the generation of EBs. EBs do not adhere to these plates.
4. Do not use complete (supplemented) media for preparation of stock solutions because serum proteins, test chemicals, or other components may precipitate on repeated freezing and thawing.
5. When testing light-sensitive chemicals, avoid prolonged exposure to light (e.g., under the microscope). Use light-tight tubes (Eppendorf Safe lock amber) or wrap tubes containing chemical solutions in aluminum foil.
6. Prepare highest concentration of the chemical in approx 80% of the solvent volume, measure pH, neutralize, and add solvent to the final volume.
7. Because volatile chemicals tend to evaporate under the conditions of testing, seal each plate with CO_2-permeable plastic film that is impermeable to volatile chemicals.
8. If you want to start a differentiation assay, then the density of the D3 cells should not be lower than 1×10^6 cells/mL at this step.
9. ES cells are trypsinized and added last, after preparation of test solutions, to avoid prolonged storage outside the incubator.
10. Leave cells outside the incubator only for the shortest time period necessary.
11. Use a sterile 5-mL pipet to avoid damage to the EBs.
12. Be certain that the concentrations of test compound of the hanging drops and the Petri dish are identical.
13. Plating of single EBs into the wells of a 24-well plate is only necessary for the morphological analysis of beating cardiomyocytes at d 10 of differentiation. If flow cytometry is performed as an end point, then 50 EBs can be pooled and plated into one 100-mm Petri dish. Use one tissue culture 100-mm Petri dish per concentration of the test chemical as well as for the untreated control (assay medium) and solvent control for the main experiment on d 7.
14. Choose a method that guarantees calculation of the amount of viable cells per pool before fixing the cells. For the untreated control, one 100-mm Petri dish will give $0.7–2.5 \times 10^6$ cells.
15. The optimal amount of antibody that will give strong and specific signals with low background staining should be empirically determined for each lot of antibody.
16. Be certain not to destroy the cell layer at the bottom of the wells.
17. Reference filters may have a tolerance of plus or minus 5%, so reference measurement may still be in the absorption curve of blue formazan (*see* **Fig. 4**). This can significantly reduce the signal. In this case, readings should be performed without reference filter.
18. Prior to use, cell culture medium should be prewarmed to 37°C in a water bath or incubator.
19. It is important to obtain a single-cell suspension for exact counting. Check under microscope if necessary.

20. Use 2.5×10^5 cells/5 mL for each 60-mm dish on Monday and Wednesday and a slightly smaller amount of cells over the weekend ($1.5-1.8 \times 10^5$ cells/5 mL for each 60-mm dish). Do not forget to add 5 µL mLIF directly to the plates.
21. It is important to keep a precise culturing schedule. ES cells are generally subcultured on Monday, Wednesday, and Friday.
22. To maintain pluripotency of ES cells, avoid subculturing up to high passage numbers. Extensive culturing (>25 passages) will result in inconsistent differentiation. This corresponds to 2 mo with three passages per week.
23. Use 1×10^6 cells/10 mL for each 100-mm dish, on Monday and Wednesday and a slightly smaller amount of cells over the weekend (8×10^5 cells/10 mL for each 100-mm dish). Avoid subculturing up to high passage numbers (>25 passages).
24. Thaw cells as quickly as possible to maintain cell integrity. Thawing should be complete within 1–1.5 min.
25. Attachment of ES cells to the plate after thawing is highly improved using gelatin-treated Petri dishes. For routine subculture (after the first passage) non-gelatin-treated dishes are used. Gelatin treatment: use 3 mL 0.1% gelatin in 1X PBS for one 60-mm plate; leave at room temperature for at least 20 min, aspirate gelatin, rinse once with 1X PBS.

References

1. Kirschstein, R. and Skirboll, L. R. (2001) Stem cells: scientific progress and future research directions. Report prepared by the National Institutes of Health. Available at: http://www.nih.gov/news/stemcell/scireport. Sept. 8, 2005.
2. Spielmann, H., Pohl, I., Döring, B., Liebsch, M., and Moldenhauer, F. (1997) The embryonic stem cell test (EST), an in vitro embryotoxicity test using two permanent mouse cell lines. 3T3 fibroblasts and embryonic stem cells. *In Vitro Toxicol.* **10**, 119–127.
3. Smith, A. G. (2001) Embryo-derived stem cells: of mice and men. *Annu. Rev. Cell. Dev. Biol.* **17**, 435–462.
4. Wobus, A. M., Guan, K., Yang, H. T., and Boheler, K. R. (2002) Embryonic stem cells as a model to study cardiac, skeletal muscle, and vascular smooth muscle cell differentiation. *Methods Mol. Biol.* **185**, 127–156.
5. Doetschmann, T., Eistetter, H. R., Schmidt, W., and Kemler, R. (1985) The in vitro development of blastocyst-derived embryonic stem cell lines: formation of visceral yolk sac, blood islands and myocardium. *J. Embryol. Exp. Morphol.* **87**, 7–45.
6. Martin, G. R., Wiley, L. M., and Damjanov, I. (1977) The development of cystic embryoid bodies in vitro from clonal teratocarcinoma stem cells. *Dev. Biol.* **61**, 220–244.
7. Laschinski, G., Vogel, R., and Spielmann, H. (1991) Cytotoxicit test using blastocyst-derived euploid embryonal stem cells: a new approach to in vitro teratogenesis screening. *Reproductive Toxicol.* **5**, 57–64.
8. Scholz, G., Genschow, E., Pohl, I., et al. (1999) Prevalidation of the embryonic stem cell test (EST)—a new in vitro embryotoxicity test. *Toxicol. In Vitro* **13**, 675–681.
9. Genschow, E., Spielmann, H., Scholz, G., et al. (2002) The ECVAM international validation study on in vitro embryotoxicity tests. Results of the definitive phase and evaluation of prediction models. *Altern. Lab. Anim.* **30**, 151–176.
10. Genschow, E., Spielmann, H., Scholz, G., et al. (2004) Validation of the embryonic stem cell test in the international ECVAM validation study on three in vitro embryotoxicity tests. *Altern. Lab. Anim.* **132**, 209–244.
11. Spielmann, H., Genschow, E., Scholz, G., et al. (2004) Validation of the micromass assay (MM) in the ECVAM international validation study on in vitro embryotoxicity. *Altern. Lab. Anim.* **132**, 245–274.

12. Piersma, A. H., Verhoef, A., Spanjersberg, M. Q. I., et al. (2004) Validation of the postimplantation rat whole-embryo culture test in the international ECVAM validation study on three in vitro embryotoxicity tests. *Altern. Lab. Anim.* **32,** 275–307.
13. Balls, M. and Hellsten, E. (2002) Statement of the scientific validity of the embryonic stem cell test (EST)—an in vitro test for embryotoxicity. Statement of the scientific validity of the micromass test—an in vitro test for embryotoxicity. Statement of the scientific validity of the postimplantation rat whole embryo culture assay—an in vitro test for embryotoxicity. *Altern. Lab. Anim.* **30,** 265–273.
14. Spielmann, H., Scholz, G., Pohl, I., Genschow, E., Klemm, M., and Visan, A. (2000) The use of transgenic embryonic stem (ES) cells and molecular markers of differentiation for improving the embryonic stem cell test (EST) *Congenital Anomalies* **40,** 8–18.
15. Seiler, A., Visan, A., Pohl, I., Genschow, E., Buesen, R., and Spielmann, H. (2002) Improving the embryonic stem cell test (EST) by establishing molecular endpoints of tissue specific development using murine embryonic stem cells. *ALTEX* **19,** 55–63.
16. Seiler, A., Visan, A., Buesen, R., Genschow, E., and Spielmann, H. (2004) Improvement of an in vitro stem cell assay for developmental toxicity: the use of molecular endpoints in the embryonic stem cell test. *Reprod. Toxicol.* **18,** 231–240.
17. Buesen, R., Visan, A., Genschow, E., Spielmann, H., and Seiler, A. (2004) Trends in improving the embryonic stem cell test (EST): an overview. *ALTEX* **21,** 15–22.
18. zur Nieden, N. I., Kempka, G., and Ahr, H. J. (2003) In vitro differentiation of embryonic stem cells into mineralized osteoblasts. *Differentiation* **71,** 18–27.
19. zur Nieden, N. I., Kempka, G., and Ahr, H. J. (2004) Molecular multiple endpoint embryonic stem cell test—a possible approach to test for a teratogenic potential of compounds. *Toxicol. Appl. Pharmacol.* **194,** 257–269.
20. Sachinidis, A., Fleischmann, B. K., Kolossov, E., Wartenberg, M., Sauer, H., and Hescheler, J. (2003) Cardiac specific differentiation of mouse embryonic stem cells. *Cardiovasc. Res.* **58,** 278–291.
21. Boheler, K. R., Czyz, J., Tweedie, D., Yang, H. T., Anisimov, S. V., and Wobus, A. M. (2002) Differentiation of pluripotent embryonic stem cells into cardiomyocytes. *Circ. Res.* **91,** 189–201.
22. Wobus, A. M., Wallukat, G., and Hescheler, J. (1991) Pluripotent mouse embryonic stem cells are able to differentiate into cardiomyocytes expressing chronotropic responses to adrenergic and cholinergic agents and Ca^{2+} channel blockers. *Differentiation* **48,** 173–182.
23. Mosmann, T. (1983) Rapid colorimetric assay for cellular growth and survival: application to proliferation and cytotoxicity assays. *J. Immunol. Meth.* **65,** 55–63.
24. Bader, D., Masaki, T., and Fischmann, D. A. (1982) Immunochemical analysis of myosin heavy chain during avian myogenesis in vivo and in vitro. *J. Cell. Biol.* **95,** 763–770.
25. Litchfield, J. T. and Wilcoxon F. (1949) A simplified method for evaluating dose-effect experiments. *J. Pharmacol. Exp. Ther.* **96,** 99–113.
26. Finney, D. G. (1971) *Probit Analysis, 3rd ed.,* Cambridge University Press, Cambridge, UK.
27. Holzhütter, H. G. and Quedenau, J. (1995) Mathematical modelling of cellular responses to external signals. *J. Biol. Syst.* **3,** 127–138.

27

Use of Chemical Mutagenesis in Mouse Embryonic Stem Cells

Sonja Becker, Martin Hrabé de Angelis, and Johannes Beckers

Summary

The chemical mutagenesis of mouse embryonic stem (ES) cells is an approach complementary to chemical mutagenesis of spermatogonia in whole animals. It has great potential to contribute significantly to the generation of a comprehensive collection of multiple alleles for most mammalian genes and to facilitate the progression from gene sequence to gene functional analyses. The general strategy includes the treatment of ES cells with a chemical mutagen and the isolation of individual mutagenized clones that are then cultured in duplicate. Whereas one set of samples is cryopreserved for archiving and subsequent generation of germline chimeras by means of blastocyst injection, the second set is used for genotype- or phenotype-driven screens to identify mutant alleles.

Key Words: Chemical mutagenesis; ES cells; genotype screen; N-ethyl-N-nitrosourea (ENU); methanesulfonic acid ethyl ester (EMS); N-nitroso-N-ethylurea; phenotype screen.

1. Introduction

1.1. Concept of ES Cell Mutagenesis

Mutagenesis in embryonic stem (ES) cells by gene targeting through homologous recombination and gene-trap-based insertional mutagenesis have been performed mostly to generate loss-of-function alleles. With the refinement of recombination techniques in mouse ES cells, it is now possible also to generate conditional and hypomorphic alleles and gain-of-function and regulatory mutants. Despite these tremendous advances in technology, it is still not possible to generate subtle point mutations using the aforementioned methods. Selectable marker genes introduced through homologous recombination were shown to have long-range effects at the targeted locus, and evidence is accumulating that comparable effects may be expected even for short recombination sites, such as *loxP* and flip recombinase target (FRT), which so far cannot be excluded in site-specific recombination techniques. In contrast, chemical mutagenesis using alkylating agents such as N-ethyl-N-nitrosourea (ENU) and methanesulfonic acid ethyl ester

From: *Methods in Molecular Biology, vol. 329: Embryonic Stem Cell Protocols: Second Edition: Volume 1*
Edited by: K. Turksen © Humana Press Inc., Totowa, NJ

(EMS) mainly causes single base exchanges. Thus, in terms of precise gene functional analyses, studies of single protein functional domains and even functional analyses of transcription factor-binding sites of chemical mutagenesis have some major advantages compared to site-specific recombination techniques.

A conceptual difference between ES cell mutagenesis and chemical mutagenesis of spermatogonia lies in the fact that germ cells derived from a particular mutagenized ES cell clone all carry the identical set of mutations. In contrast, the treatment of a single male with a chemical mutagen introduces different sets of mutations in every spermatogonium. As a consequence, in recessive breeding schemes it is required to out-cross the mutagenized male (G_0) with a wild-type female to generate the G_1 founder generation. Offspring of the G_2 generation are then back-crossed to G_1 animals. Subsequently, phenotypic screens for recessive alleles are performed in the G_3 generation. In contrast, when taking the ES cell mutagenesis approach, the chimeras derived from one mutagenized ES cell clone may be out-crossed to generate a G_1 generation that can be back-crossed to the founder chimera to produce the G_2 generation, which may be screened for recessive phenotypes *(1)*.

Thus, shortened breeding schemes are used in phenotype-driven ES cell mutagenesis approaches. With the mutation frequencies tested so far, G_1 sibling intercrosses yielded few or no offspring, suggesting that G_1 animals carry a high load of recessive lethal alleles *(1)*. A further improvement of breeding speed could be achieved if it was possible to differentiate mutagenized ES cells to germ cells in vitro, omitting the blastocyst injection and the generation of chimeras.

Some of the advantages of mutagenesis in ES cells include the possibility to perform ES cell phenotypic screens at high throughput, for example, for mutations affecting basic cellular processes, such as DNA repair mechanisms or metabolic pathways, as well as the feasibility to study cell differentiation in vitro employing high-throughput assays (*see* **Subheading 1.3.**). A major strength of ES cell mutagenesis lies in the fact that it is possible (with affordable efforts) to generate archives of thousands of mutagenized genomes that may be screened in gene-driven approaches for allelic series of a particular gene *(2)*. In such gene-driven mutagenesis approaches, mutations may be detected either from genomic DNA or from reverse-transcribed complementary DNA. This approach is comparable to similar strategies that have successfully been taken also in whole-animal mutagenesis projects in which parallel archives of cryopreserved sperm and DNA were exploited for the generation of mice with allelic series *(3)*. High-throughput techniques for mutation detection, such as single-strand conformation polymorphism (SSCP), denaturing high-performance liquid chromatography (DHPLC), temperature gradient capillary electrophoresis (TGCE), and DNA-chip technology, make the efficient combination of chemical mutagenesis and genotype-driven mutation detection practicable (*see* **Subheading 1.4.**).

1.2. The Mutagens ENU and EMS

The mutagenic effect on ES cells has been assessed for ENU and EMS (both alkylating agents that predominantly induce base substitutions) and ICR191 (mostly causing frameshift mutations in stretches of guanine). At least for ENU and EMS, it was shown that such treated mouse ES cells retain their ability to populate the germline after the generation of chimeras *(1,4)*.

Mutation rates for loss-of-function alleles in the *Hprt* selectable gene were assessed for the mutagens EMS and ICR191 in ES cells *(1)*. Most important, it was also demonstrated that at least EMS-treated ES cells retain their ability to colonize the mouse germline. Munroe et al. *(1)* performed a small-scale recessive screen for visible phenotypes and identified several dysmorphological mutants, mutants with fertility phenotypes in male and female animals, as well as mutants with hearing deficit. Although EMS-induced mutations in the *Hprt* locus predominantly were single base changes, the detailed characterization of two mutants from the recessive phenotype screen provided evidence that EMS-induced chromosomal lesions may at least at low frequency create insertional targets for retrotransposon elements *(5)*.

One mutant line with a recessive syndactyly (*sne*) phenotype was identified as an allele of the *fibrillin 2* gene and designated $Fbn2^{sy-fp-3J}$. The mutation is a deletion of a complete calcium-binding EGF module caused by the insertion of a LINE element. Deletions induced by EMS mutagenesis have also been reported in *Drosophila* and *Caenorhabditis elegans*. At least in the case of *Drosophila* this seems to be related to prolonged sperm storage in the spermathecae and extended periods of exposure to the mutagen *(6,7)*. The second mutant line was phenotypically characterized by dominant polydactyly and recessive lethality and designated $Twist1^{Pde}$. No mutation was identified in the *Twist1* coding sequence. The phenotype may result from an as yet unidentified regulatory mutation that causes a lack of *Twist1* expression *(5)*.

Chen et al. *(4)* explored the feasibility of a strictly genotype-based screen in ES cell chemical mutagenesis. They also assessed the ENU induced mutation frequency and spectrum using the *Hprt* locus. Sequencing of the complete coding region from 1650 mutagenized clones identified 4 loss-of-function mutations and 4 mutagenized cell clones with silent or conservative amino acid exchange sensitive under subsequent 6-TG selection. The overall mutation rate was estimated to be 1 in 200 clones at the *Hprt* locus. The loss-of-function mutation frequencies were estimated to 1 in 1000 and 1 in 1200 for the *Hprt* and HSV-*tk* loci, respectively *(4)*. These loss-of-function frequencies are comparable to the mutation rates observed for EMS in ES cells and for recessive alleles induced by ENU in whole-animal mutagenesis.

A genotype-based screen was also performed for alleles in the intracellular transducers of transforming growth factor-β signaling, *Smad2* and *Smad4* *(8)*. The screening of 2060 ENU mutagenized ES cell clones using reverse transcriptase polymerase chain reaction (PCR) and DHPLC detected 29 ES cell clones with single-nucleotide substitutions: 18 mutations resulted in missense alterations of the proteins, 2 affected splicing, and 9 mutations were silent. The overall mutation rate for nonsilent mutations was approx 1 in 600–700 kb for both genes. ES cell clones of five nonsilent mutations were passed through the germline. The previous generation of complete loss-of-function alleles of *Smad2* and *Smad4* resulted in perigastrulation lethality *(9)*. The phenotypic analysis of hypomorphic alleles of these genes, obtained from ES chemical mutagenesis, that bypass the early embryonic lethality allowed the characterization of yet-unknown *Smad* gene functions and the functional analysis of specific interaction domains of the

SMAD2 and SMAD4 proteins. Such gene functional analyses are generally not possible with complete loss-of-function alleles as they are typically generated in gene-targeting approaches by means of homologous recombination.

1.3. Phenotyping of ES Cells

Although chemical mutagenesis of ES cells has evoked new interest in genotype-based screens in mutagenesis approaches, phenotyping of mutagenized ES cell clones may become an excellent tool for high-throughput screens, particularly for parameters that are rather difficult to assess in whole animals *(1,10)*. The screens that have been performed to identify new alleles in the selectable genes *Hprt* and HSV-*tk* are prototypes of such phenotype-driven screens *(1)*. Whereas genotype-based screens are generally limited to known genes, a major advantage of phenotype-based screens is that they are resourceful tools for the identification of novel genes and unknown gene functions.

It is, thus, conceivable that archives of mutagenized ES cell clones could be efficiently screened, for example, for their response to environmental challenges such as radiation (DNA damage repair), nutritional conditions, chemical compounds, drugs, and the like *(2)*. Expression arrays, as a means to molecularly phenotype ES cells, may represent an excellent tool for the identification of affected molecular pathways and regulatory groups of genes *(11)*. Also, inducing cellular differentiation of mutagenized ES cells in vitro may represent an excellent measure to select mutants with developmental deficit.

1.4. Genotyping in Chemical Mutagenesis Approaches

The resources of thousands of genomes that have been chemically mutagenized in either ES cells as well as whole-animal mutagenesis approaches together with the well-established efficiencies of chemical mutagenesis open the possibility to perform genotype-based screens for novel alleles in known genes *(3,4)*. The method of detection of single-nucleotide mutations would ideally combine the advantages of multiplexing with high sensitivity, relative cost-effectiveness, and moderate logistics.

In addition to direct sequencing, there are several techniques now available that can be used for analysis of single-basepair exchanges such as SSCP, denaturing gradient gel electrophoresis, DHPLC, TGCE, chemical cleavage of mismatches, and enzyme mismatch cleavage. Of these, SSCP and DHPLC are the most commonly used approaches, independent of whether the basis for sequence analysis is transcripts (reverse transcriptase PCR) or genomic DNA.

1.4.1. Single-Strand Conformational Polymorphism

SSCP takes advantage of the fact that single-stranded DNA, when placed in a nondenaturing solution, folds into a specific secondary structure determined by its sequence. DNA strands differing by as little as one base can occupy different conformations, which can be visualized by a difference in mobility of the two strands via electrophoresis. In its original setup, this is a low-cost approach to deliver high sensitivity for fragments up to 200 bp but with limited application for multiplex analysis concerning both reliability of data and requirement for manual inspection of results *(12)*. The combination of this method with capillary electrophoresis might overcome difficulties for high-throughput analysis *(13)*.

1.4.2. Denaturing High-Performance Liquid Chromatography

DHPLC is a column-based technique that uses ion-paired reverse-phase liquid chromatography. Homoduplexes and heteroduplexes of PCR-amplified fragments differentially elute because of an increasing gradient of acetonitrile while the column is maintained at the melting temperature of the PCR fragment. Elution of the PCR fragments at a particular acetonitrile concentration depends on the size and sequence of the DNA duplex. After denaturation and reannealing to allow formation of heteroduplexes, the PCR products are injected onto the column. Different DNA fragments produce characteristic profiles, detected by their absorbance and plotted vs their elution time. The peak of maximum absorbance is the retention time of a DNA sample at a given acetonitrile concentration *(14)*. This technique was successfully used, for example, to isolate allelic series of mutations in the *Smad2* and *Smad4* genes in a genotype-driven screen for ENU-mutagenized mouse ES cells *(8)*. DHPLC features the possibility for both high-throughput and automation, allowing the analysis of several hundred samples in 24 h *(15)*.

1.4.3. Temperature Gradient Capillary Electrophoresis

TGCE combines heteroduplex analysis with capillary electrophoresis. Multiple DNA samples consisting of homoduplexes and heteroduplexes are separated simultaneously by spanning a wide temperature range covering melting temperatures for different regions of a given DNA fragment. Therefore, with this method it is possible to perform simultaneous heteroduplex analyses for various mutations and PCR products with different melting temperatures *(16,17)*. This approach, designed for multiplexing and automation (SpectruMedix), matches the demands of a large-scale mutational analysis, representing a highly sensitive as well as fast method for mutation detection (close to 100 samples may be analyzed in as little as 1 h) *(15,18)*. With these advantageous features, TGCE fits most of the necessary requirements for gene-based mutational screens.

2. Materials

2.1. General Reagents

1. Polystyrene 35-, 60-, 100-, and 150-mm tissue culture dishes (Corning, Corning, NY; cat. no. 430165, 430166, 430167, 430599).
2. Disposable plastic pipets.
3. 15-mL polypropylene Falcon tubes (Becton Dickinson, Franklin Lakes, NJ).
4. Freezing container (Nalgene, Rochester, NY; cat. no. 5100-0001).
5. High-glucose, Dulbecco's modified Eagle's medium (DMEM) with L-glutamine and sodium pyruvate (Gibco, Carlsbad, CA; cat. no. 41966-029).
6. 0.05 M β-mercaptoethanol (Invitrogen, Carlsbad, CA; cat. no. 31350-010).
7. ES-cell qualified fetal bovine serum (Gibco, cat. no. 16141-079).
8. Minimum essential medium-α (Invitrogen, cat. no. 41061-029).
9. Leukemia inhibitory factor (LIF; Chemicon, Temecula, CA; cat. no. ESG1107).
10. 0.05% trypsin/ethylenediaminetetraacetic acid (EDTA; Invitrogen, cat. no. 25300-054).
11. Gelatin (VWR International, Poole, England; cat. no. 440454B): 0.1% in distilled water (dH$_2$O); autoclave and store at 4°C.

12. Gelatinized dishes: in a hood, pipet 0.1% gelatin solution into the dishes, covering the surface of the dishes completely. Leave the dishes for approx 2 min. Remove gelatin solution and allow to dry before using the plates.
13. EMS (Sigma, cat. no. M0880).
14. O^6-benzylguanine (O^6-BG; Sigma, cat. no. B2292).
15. ENU (Serva, Heidelberg, Germany; cat. no. 30800).
16. Mouse embryonic fibroblasts for use as feeders.
17. 2-Amino-6-mercaptopurine or 6 thioguanine (6TG; Sigma, cat. no. A4882) to 10 mg/mL in ES cell medium. Aliquot and store at −20°C.
18. Dimethyl sulfoxide (DMSO; Sigma; cat. no. D2650) for freezing cells and dissolving O^6-benzylguanine.
19. 2 M sodium thiosulfate ($Na_2S_2O_3$; Sigma, cat. no. S1648).
20. Inactivation solution: 10% $Na_2S_2O_3$ (sodium thiosulfate), 1% sodium hydroxide in distilled water (*see* **Note 1**).
21. 10X phosphate-buffered saline (PBS): 1.37 M sodium chloride, 27 mM potassium chloride, 10 mM Na_2HPO_4, 2 mM KH_2PO_4. For 1 L 1X working solution, first dissolve 11.5 g Na_2HPO_4 and 2.0 g KH_2PO_4 in 800 mL distilled water. Then, add 80 g NaCl and 2 g KCl and dissolve. Fill to 1 L with distilled water. Sterilize by autoclaving.
22. Soerensen buffer: 66 mM KH_2PO_4 (Merck, Germany; cat. no. 1.04877; stock solution: 9.078 g dissolved in 1 L dH_2O, autoclaved) and 66 mM $Na_2HPO_4 \cdot 2H_2O$ (Merck, cat. no. 1.06580; stock solution: 11.876 g dissolved in 1 L dH_2O, autoclaved). Adjust pH with 66 mM KH_2PO_4 or 66 mM $Na_2HPO_4 \cdot 2H_2O$ at 4°C to 6.0. For 100 mL, combine 88 mL 66 mM KH_2PO_4 stock solution with 12 mL 66 mM Na_2HPO_4 stock solution.

2.2. Media

1. ES cell medium for EMS mutagenesis: high-glucose DMEM is supplemented with 15% heat-inactivated fetal bovine serum, 0.1 mM β-mercaptoethanol, 2.0 mM L-glutamine, and 1000 U/mL LIF. For 500 mL, to 500 mL DMEM add 75 mL Gibco-ES fetal bovine serum, 1 mL 50 mM β-mercaptoethanol, 5 mL 200 mM L-glutamine, and 50 µL LIF (*see* **Note 2**).
2. ES cell medium for ENU mutagenesis: culture medium consists of MEM-α, 15% fetal calf serum, 0.1 mM β-mercaptoethanol, and 1000 U/mL LIF. For 500 mL, to 500 mL MEM-α add 75 mL Gibco-ES fetal bovine serum, 1 mL 50 mM β-mercaptoethanol, and 50 µL LIF.
3. 2X concentrated freezing medium: 50% fetal calf serum, 30% ES cell medium, and 20% DMSO. For 10 mL, to 5 mL Gibco-ES fetal bovine serum add 3 mL ES cell medium and 2 mL DMSO.

2.3. ENU and EMS Solutions (see Note 3)

2.3.1. ENU Preparation

1. ENU is weighed in a flow hood. For 10 mL 10 mg/mL solution, weigh 100 mg ENU in a 15-mL Falcon tube and pipet 10 mL ice-cold Soerensen buffer into the tube. Because the stability of ENU in solution is pH dependent, the pH of the solvent is critical (*see* **Note 4**).
2. Dissolve ENU completely in ice-cold Soerensen buffer by shaking the solution vigorously. The solution should have a light yellow color. If not, discard.
3. In case of additional O^6-BG treatment (*see* **Subheading 3.3.**), dissolve O^6-BG in DMSO and dilute it to 10 µM in ES medium together with the desired concentration of ENU and filter-sterilize. Use solution immediately.

2.3.2. EMS Preparation

As EMS is an oily liquid, it can be added directly into fresh ES cell medium on each dish under a flow hood at the desired concentration. For example, to one 35-mm dish with 2 mL ES cell medium, pipet 1 µL EMS for a final concentration of 0.6 mg/mL (*see* **Note 5**). Distribute EMS evenly by swaying the dish thoroughly.

3. Methods
3.1. Thawing and Freezing of ES Cells
3.1.1. Thawing

1. Using a 37°C water bath, quickly thaw one vial of frozen ES cells.
2. Transfer cells into a 15-mL Falcon tube containing 10 mL ES cell medium without LIF and collect the cells by centrifugation at 700g for 5 min.
3. Remove supernatant and gently resuspend cell pellet with ES cell medium containing LIF and transfer to a 60-mm culture dish containing a feeder layer.
4. Leave cells to adhere overnight at 37°C and 5% CO_2 in an incubator.
5. Change medium the next day.

3.1.2. Freezing and Storage

1. Remove ES cell medium and wash cells of a 100-mm culture dish (approx 1–2 × 10^7 cells) with 1X PBS.
2. Add 3 mL 0.05% trypsin-EDTA and incubate for 10 min at 37°C in 5% CO_2.
3. Gently pipet the cells up and down to break cell clumps and obtain a single-cell suspension (*see* **Note 6**).
4. Add 10 mL ES cell medium without LIF and centrifuge at 700g for 5 min. Discard supernatant and resuspend cells in 1.5–2.0 mL ES cell medium and keep on ice.
5. Label and precool freezing vials, add an equal amount of cold 2X concentrated freezing medium to the cell suspension, mix gently, and aliquot 1 mL of the suspension into freezing vials.
6. Immediately transfer the vials to a precooled (on ice) freezing container and store it in a –80°C freezer for 24 h.
7. Subsequently, vials can be transferred into liquid nitrogen for long-term storage.

3.2. EMS Mutagenesis (see Note 7)

The EMS mutagenesis protocol is based on the work of Munroe et al. (*1*) with some modifications from on-line protocols of the Jackson Laboratory www.jax.org/index.html).

1. Plate ES cells onto 35-mm dishes with feeders at a concentration of 2 × 10^6 cells/dish and incubate at 37°C in 5% CO_2 overnight.
2. Change medium the next day with 2 mL ES cell medium.
3. Incubate cells at least 4 h at 37°C in 5% CO_2 before adding EMS directly to each dish to a final concentration ranging from 0.4 to 0.8 mg/mL (*see* **Note 8**). The appropriate EMS concentration for reasonable mutation rates has to be determined for the type of ES cells used (*see* **Note 9**).
4. Incubate ES cells 16–20 h at 37°C in 5% CO_2.
5. Subsequently, remove EMS solution (*see* **Note 10**).
6. Wash plates three times with 2 mL 1X PBS.
7. Trypsinize and count the cells. Cells will be divided into three sets for cell death determination, determination of mutation rate, and expansion and freezing (*see* **Subheadings 3.4.–3.6.**).

3.3. ENU Mutagenesis

The ENU mutagenesis protocol is based on Chen et al. *(4)* with some modifications from Soewarto et al. *(19)* and includes optional preincubation of the ES cells with O^6-BG prior to the ENU treatment to inhibit O^6-alkylguanine-alkyltransferase (Agt) enzyme activity. Mammalian Agt is capable of removing various alkyl adducts from the O^6 position of guanine bases in DNA. ES cells as well as primary fibroblasts displayed detectable levels of Agt activity that were inhibited by O^6-BG *(4)* (*see* **Note 11**).

1. Plate ES cells onto 100-mm dishes with feeders and incubate at 37°C and 5% CO_2.
2. Cultivate cells until they reach approx 80% confluence.
3. Exchange ES cell medium with 10 μM O^6-BG diluted in ES cell medium and incubate for 12–16 h at 37°C and 5% CO_2.
4. Remove ES cell medium and wash with 1X PBS. Add 3 mL 0.05% trypsin-EDTA and incubate for 10 min at 37°C in 5% CO_2.
5. Gently pipet the cells up and down to break cell clumps and obtain a single-cell suspension.
6. Add 10 mL ES cell medium without LIF and centrifuge at 700g for 5 min.
7. Discard supernatant and resuspend cells in ES medium (with LIF) at a density of 5×10^5 cells/mL containing 10 μM O^6-BG and the desired concentration of ENU (*see* **Note 12**).
8. Incubate cell suspension with constant rocking at 37°C for 1 h (*see* **Note 13**).
9. Wash and trypsinize again; count cells. Cells will be divided into three sets for cell death determination, determination of mutation rate, and expansion and freezing (*see* **Subheadings 3.4.–3.6.**).

3.4. Calculation of Cell Death

1. Plate 1×10^3 mutagen-treated and untreated (control) cells onto two 60-mm plates with feeders.
2. Count colonies after 1 wk of standard incubation time.
3. Compare colony counts to the untreated plate for plating efficiency and estimation of cell death.
4. Calculate percentage survival by taking the colony number on the plate with mutagen-treated cells divided by the number of colonies from the control plate and multiply by 100 (*see* **Note 14**).

3.5. Determination of Mutation Rate by Hprt Assay

This protocol is based on the work of Munroe et al. *(1)* with some modifications from on-line protocols of the Jackson Laboratory (www.jax.org/index.html).

1. Plate half of the mutagen-treated cells onto a 60-mm gelatinized dish without feeders.
2. Incubate cells for approx 10 d; pass as needed (*see* **Note 15**).
3. Add 1.5 mL 0.05% trypsin-EDTA and incubate for 10 min at 37°C in 5% CO_2.
4. Gently pipet the cells up and down to break cell clumps and obtain a single-cell suspension.
5. After washing with ES cell medium, plate cells onto four 150-mm gelatinized dishes without feeders at a density of 0.5×10^6 cells/plate. Add 6TG to a final concentration of 10 μg/mL ES cell medium. Remove selection after 3 d and count colonies 2 wk after seeding.
6. Also plate 1×10^3 cells onto two 60-mm gelatinized dishes without feeders or 6TG to determine plating efficiency. Feed as needed with ES cell medium and count the colonies

after approx 1 wk. These cells will be used to determine the number of cells that survived plating.
7. Calculate the loss-of-function mutation rate at the *Hprt* locus by multiplying the percentage of cells surviving plating by 2×10^6 cells plated and then divide by the average number of colonies on the Hprt assay plates (*see* **Note 16**).

3.6. Cell Expansion and Freezing (see Note 17)

1. Plate mutagenized cells onto six to eight 100-mm plates and pick isolated colonies into 96-well plates.
2. In the case of O^6-BG-pretreated cells, add 10 μM O^6-BG to the ES cell medium during the first 24 h of postmutagen cultivation to deplete any newly synthesized alkylguanine alkyltransferases.

3.6.1. Option 1: Plate Pool

1. Plate half of the treated cells onto a 100-mm dish with feeders (the other half is used for determination of mutation rate by Hprt assay; *see* **Subheading 3.5.**).
2. Passage as needed. Cells may be expanded further at this point or frozen before expansion. This population of ES cells will be used for injection of blastocysts to create chimeric mice (*see* **Note 18**).

3.6.2. Option 2: Clones

1. Plate half of the cells onto 100-mm dishes with feeders at three different dilutions, one with high plating density for freezing and the other two with low density for picking colonies.
2. Pick colonies onto 96-well plates with feeders. Expand as desired and freeze (*see* **Note 19**).

4. Notes

1. EMS and ENU are mutagens and should be disposed properly. All equipment, pipets, plates, and gloves that come into contact with the mutagen must be rinsed with 10% $Na_2S_2O_3$/1% sodium hydroxide before disposal. All liquids should be brought to a final concentration of 10% $Na_2S_2O_3$/1% sodium hydroxide for at least 10 min before disposal.
2. Make aliquots to save medium. The medium should be used within 2 wk because of instability of L-glutamine and must not be exposed to light for too long because of photoreactivity of riboflavines.
3. Because both substances are genotoxic, persons handling EMS or ENU must wear lab coat, goggles, gloves, and a mask. Inactivation solution has to be prepared in advance. In case of ENU/EMS spills, pour inactivation solution over the spill and wait several minutes before cleaning.
4. In alkaline solutions, ENU decomposes to diazoethane with a half-life of 34 min at pH 7.0 and 37°C *(20)*. At pH 6.0, ENU has a half-life of 31 h.
5. EMS has a density of 1.167 g/mL.
6. It is essential to completely dissociate ES cell colonies into a single-cell suspension; no clumps should be seen on the culture plates.
7. This protocol is well adapted for v6.4 (129/B6) F1 hybrid ES cells *(1)*.
8. Higher EMS concentrations result in increased mutation rates.
9. A mutation frequency at the X-linked *Hprt* locus of 1 in 1200 colonies was observed using CJ7 cells treated with 0.5 mg/mL EMS *(1)*.
10. All solutions containing EMS should be collected and brought to a final concentration of 1.0 M $Na_2S_2O_3$ for at least 10 min before disposal.

11. Preincubation with O^6-BG is only reasonable if high mutation rates are desired. For medium mutational rate, which equals reduced crossing efforts after mutagenesis, O^6-BG treatment is not necessary.
12. The appropriate EMS concentration varies between 0.1 and 0.25 mg/mL and has to be determined for specific ES cell lines.
13. Treating cells in suspension allows homogeneous exposure to ENU and accurate assessment of the mutation frequency *(4)*.
14. A survival range of 10–30% is typically associated with high-efficiency mutagenesis (based on **ref. *1***).
15. Cells must be off feeders prior to 6TG treatment to deplete any feeder-derived HPRT activity and to allow depletion of wild-type HPRT protein.
16. Not all cell lines will tolerate 10 µg/mL 6TG. The appropriate concentration has to be determined for each cell line.
17. This protocol is based on on-line protocols of the Jackson Laboratory (www.jax.org/index.html).
18. Using this approach, each chimera produced is likely to represent a different diploid-mutagenized genome and may represent more than one diploid-mutagenized genome.
19. Remember that lower doses of mutagen result in higher cell survival rates and should be plated at a lower density than higher doses for picking colonies. Each chimeric mouse produced from an isolated ES cell colony represents one diploid-mutagenized genome.

References

1. Munroe, R. J., Bergstrom, R. A., Zheng, Q. Y., et al. (2000) Mouse mutants from chemically mutagenized embryonic stem cells. *Nat. Genet.* **24,** 318–321.
2. Chen, Y., Schimenti, J., and Magnuson, T. (2000) Toward the yeastification of mouse genetics: chemical mutagenesis of embryonic stem cells. *Mamm. Genome* **11,** 598–602.
3. Coghill, E. L., Hugill, A., Parkinson, N., et al. (2002) A gene-driven approach to the identifiation of ENU mutants in the mouse. *Nat. Genet.* **30,** 255–256.
4. Chen, Y., Yee, D., Dains, K., et al. (2000) Genotype-based screen for ENU-induced mutations in mouse embryonic stem cells. *Nat. Genet.* **24,** 314–317.
5. Browning, V. L., Chaudhry, S. S., Planchart, A., Dixon, M. J., and Schimenti, J. C. (2001) Mutations of the mouse Twist and sy (fibrillin 2) genes induced by chemical mutagenesis of ES cells. *Genomics* **73,** 291–298.
6. Ashburner, M. (1989) *Drosophila: A Laboratory Manual,* Cold Spring Harbor Laboratory Press, Cold Spring Harbor, NY.
7. Liu, L. X., Spoerke, J. M., Mulligan, E. L., et al. (1999) High-throughput isolation of Caenorhabditis elegans deletion mutants. *Genome Res.* **9,** 859–867.
8. Vivian, J. L., Chen, Y., Yee, D., Schneider, E., and Magnuson, T. (2002) An allelic series of mutations in Smad2 and Smad4 identified in a genotype-based screen of N-ethyl-N-nitrosourea-mutagenized mouse embryonic stem cells. *Proc. Natl. Acad. Sci. USA* **99,** 15,542–15,547.
9. Weinstein, M., Yang, X., and Deng, C. (2000) Functions of mammalian Smad genes as revealed by targeted gene disruption in mice. *Cytokine Growth Factor Rev.* **11,** 49–58.
10. Littlefield, J. W. (1989) Compaction-defective embryonal carcinoma cell variants. *Dev. Genet.* **10,** 292–297.
11. Beckers, J., Hoheisel, J. D., Mewes, H. W., Vingron, M., and Hrabe de Angelis, M. (2002) Molecular phenotyping of mouse mutant resources by RNA expression profiling. *Curr. Genomics* **3,** 121–129.

12. Beier, D. R. (2000) Sequence-based analysis of mutagenized mice. *Mamm. Genome* **11,** 594–597.
13. Andersen, P. S., Jespersgaard, C., Vuust, J., Christiansen, M., and Larsen, L. A. (2003) Capillary electrophoresis-based single strand DNA conformation analysis in high-throughput mutation screening. *Hum. Mutat.* **21,** 455–465.
14. Lilleberg, S. L. (2003) In-depth mutation and SNP discovery using DHPLC gene scanning. *Curr. Opin. Drug Discov. Dev.* **6,** 237–252.
15. Chen, Y., Vivian, J. L., and Magnuson, T. (2003) Gene-based chemical mutagenesis in mouse embryonic stem cells. *Methods Enzymol.* **365,** 406–415.
16. Li, Q., Liu, Z., Monroe, H., and Culiat, C. T. (2002) Integrated platform for detection of DNA sequence variants using capillary array electrophoresis. *Electrophoresis* **23,** 1499–1511.
17. Zhu, L., Lee, H. K., Lin, B., and Yeung, E. S. (2001) Spatial temperature gradient capillary electrophoresis for DNA mutation detection. *Electrophoresis* **22,** 3683–3687.
18. Murphy, K., Hafez, M., Philips, J., Yarnell, K., Gutshall, K., and Berg, K. (2003) Evaluation of temperature gradient capillary electrophoresis for detection of the Factor V Leiden mutation: coincident identification of a novel polymorphism in Factor V. *Mol. Diagn.* **7,** 35–40.
19. Soewarto, D., Blanquet, V., and Hrabe de Angelis, M. (2003) Random ENU mutagenesis. *Methods Mol. Biol.* **209,** 249–266.
20. Shibuya, T. and Morimoto, K. (1993) A review of the genotoxicity of 1-ethyl-1-nitrosourea. *Mutat. Res.* **297,** 3–38.

V

Epigenetic Analysis of Embryonic Stem Cells

28

Nuclear Reprogramming of Somatic Nucleus Hybridized With Embryonic Stem Cells by Electrofusion

Masako Tada and Takashi Tada

Summary

Cell fusion is a powerful tool for understanding the molecular mechanisms of epigenetic reprogramming. In hybrid cells of somatic cells and pluripotential stem cells, including embryonic stem (ES) and embryonic germ cells, somatic nuclei acquire pluripotential competence. ES and embryonic germ cells retain intrinsic *trans* activity to induce epigenetic reprogramming. For generating hybrid cells, we have used the technique of electrofusion. Electrofusion is a highly effective, reproducible, and biomedically safe in vitro system. For successful cell fusion, two sequential steps of electric pulse stimulation are required for the alignment (pearl chain formation) of two different types of cells between electrodes in response to alternating current stimulation and for the fusion of cytoplasmic membranes by direct current stimulation. Optimal conditions for electrofusion with a pulse generator are introduced for ES and somatic cell fusion. Topics in the field of stem cell research include the successful production of cloned animals via the epigenetic reprogramming of somatic cells and contribution of spontaneous cell fusion to generating intrinsic plasticity of tissue stem cells. Cell fusion technology may make important contributions to the fields of epigenetic reprogramming and regenerative medicine.

Key Words: Cell alignment; cell fusion; cloning; EG cell; electrofusion; epigenetics; ES cell; hybrid cell; reprogramming; stem cell.

1. Introduction

Cell fusion is a phenomenon combining genetic and epigenetic information between two different types of cells, resulting in the creation of a new cellular phenotype and function. Fused cells can grow as polynuclear cells, called heterokaryons. In other cases, one combined nucleus is formed through cell division after cytoplasmic membrane fusion. The phenotype of hybrid cells varies depending on the combination of parental cells for fusion. Specialized gene products from one of the parental cells are maintained; others are often extinguished via a tissue-specific repressive mechanism, a phenomenon called *extinction (1)*. Thus, it has been believed that epigenetic information derived from parental cells coordinately regulates the new phenotypes and functions of somatic hybrid cells.

One well-known cell fusion phenomenon occurring in vivo is fertilization. The union of two pronuclei from an MII oocyte and sperm creates the next generation with genetic diversity. Soon after fertilization, sperm chromatin is reprogrammed by nucleoplasmin because of the exchange of sperm-specific basic proteins to histones *(2)*, almost resulting in the equalization of paternal and maternal chromatin through subsequent cell divisions. Cell fusion is also involved in the technique of animal cloning. The somatic donor nucleus is introduced into a recipient MII oocyte by cell fusion or direct injection. In the cloned embryos, the somatic type of cell phenotype and function is reprogrammed to the pluripotential type as a result of genomewide epigenetic modifications via the activity of *trans*-acting factors preserved in unfertilized oocytes. This phenomenon has been called *nuclear reprogramming*.

Using electrofusion between ES cells and adult somatic cells, it has been demonstrated that ES cells have an intrinsic capacity for the nuclear reprogramming of specialized somatic genomes *(3,4)*. In hybrid cells of ES cells and thymocytes, epigenetic modifications specific to somatic cells are fully reprogrammed to the ES epigenotype, as shown by (1) the successful contribution of the ES hybrid cells to the normal embryogenesis of chimeric embryos, (2) reactivation of the inactivated X chromosome derived from female thymocytes, (3) reactivation of pluripotential cell-specific genes (*Oct4* and *Tsix*) derived from somatic cells, (4) tissue-specific gene expression from the reprogrammed somatic genomes after in vivo and in vitro differentiation, and (5) de-condensed chromatin formation in the reprogrammed somatic nuclei as marked by histone-tail modifications of H3 and H4 hyperacetylation and H3 lysine 4 hypermethylation *(3,5,6)*. Moreover, embryonic germ (EG) cells derived from gonadal primordial germ cells of mouse E11.5 and 12.5 embryos, in which the majority of parental imprints are erased *(7)*, possess additional activity for inducing reprogramming of parental imprints on somatic genomes accompanied by DNA demethylation in the EG hybrid cells *(8)*. Therefore, electrofusion between pluripotential stem cells and somatic cells will contribute to our understanding of the mechanisms of nuclear reprogramming involved in epigenetic modifications on DNA and chromatin.

Cell fusion was first induced by the Sendai virus (HVJ) among Ehrlich's ascites tumor cells, resulting in the formation of giant polynuclear cells *(9)*. Since Harris's group reported cell fusion between two different species, mouse and human, in 1965 *(10)*, the technique has proven to be a powerful approach to analyzing biological interactions between differentiated cell types. Various techniques lead to cell fusion mediated by other viruses such as Paramyxoviruses, including Oncornavirus, Coronavirus, Herpesvirus, and Poxvirus and by treatment with various chemical agents such as calcium ions, lysolecithin, and polyethylene glycol (PEG). However, virus- and chemical-mediated cell fusion has difficulties with respect to the efficiency of inducing cell fusion and reproducibility and biomedical safety for clinical applications. Since the 1980s, efforts have been made to optimize many parameters of electrofusion, resulting in improvements in fusion efficiency, which is now significantly better than that of cell

Electrofusion-Mediated Reprogramming

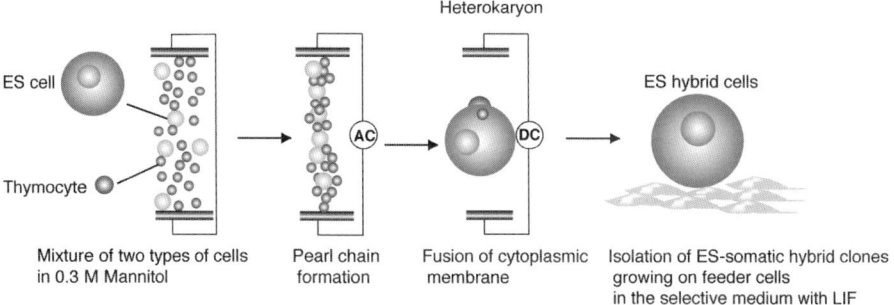

Fig. 1. Scheme of electrofusion system. The cell mixture of embryonic stem (ES) cells and somatic cells suspended in nonelectrolyte 0.3 M mannitol is applied into the 1-mm gap between two electrodes fixed on a Microslide. AC application initiates cell alignment and sequential DC electroporation pulse induces reversible breakage of cytoplasmic membranes, leading to the fusion of cell membranes between adjacent cells. After cell fusion treatment, only ES-somatic hybrid cells survive in the selection medium.

fusion with polyethylene glycol *(11)*. The electrofusion parameters, however, were optimized for making hybridomas for the stable production of antibodies.

In this chapter, we introduce an electrofusion procedure to create hybrid cells from pluripotential stem cells and adult lymphocytes. The overall procedure is summarized (*see* **Fig. 1**). The electrofusion procedure is subdivided into several parts: (1) cell culture of ES and fused cells; (2) pretreatment of somatic cells; (3) setup of the Electro Cell Manipulator (ECM) 2001 (BTX), AC (alternating current)/DC (direct current) pulse generator with a Microslide chamber; (4) operation of the electrofusion generator; (5) selection of postfusion cells; and (6) isolation and cloning of fused cells. In the process of electrofusion, AC pulses induce the alignment and compression of the cells, and DC pulses transiently make revertible pores in cytoplasmic membranes to initiate the process of fusion among cells. The alignment voltage, pulse length, and electroporation voltage and the number of DC pulses should be precisely controlled.

After cell fusion, ES hybrid cells with somatic cells should be isolated with selection medium. Thus, genetically marked ES cells or somatic cells are needed for cell fusion experiments. For example, normal ES cells were hybridized with thymocytes derived from ROSA26 transgenic mice, which carry the ubiquitously expressed *neo/lacZ* transgene *(12)*. Consequently, ES cells hybridized with the thymocytes can survive and grow only in medium supplemented with a protein synthesis inhibitor, G418. The ES hybrid cells and their derivatives are visualized as cells positive for X-gal (5-bromo-4-chloro-3-indolyl-β-D-galactoside) staining, which allows one to observe their contribution in chimeric embryos and tissues *(3,5,8)*.

In another selection system, male ES cells deficient in the *Hprt* gene on the X chromosome are used for isolating ES hybrid cells with wild-type somatic cells. The somatic cell-derived *Hprt* gene rescues the HPRT activity in the hybrid cells. In DNA synthesis, the purine nucleotide can be synthesized by the *de novo* pathway and recycled by the salvage pathway. When electrofusion-treated cells are cultured in medium with HAT (hypoxantine, aminopterin, and thymidine), the *de novo* pathway is interfered with, and only the salvage pathway functions. As a consequence, the *Hprt*-deficient ES cells die; ES hybrid cells are able to survive and grow *(5,6)*.

To succeed in obtaining ES hybrid cells, we recommend that (1) genetic markers highly expressed in the undifferentiated state are used for isolating hybrid clones, (2) the origin of somatic cells is carefully chosen for creating undifferentiated hybrid cells, and (3) the culture condition is optimized for stably maintaining the pluripotential state of ES cells and ES hybrid cells.

2. Materials

2.1. Cells and Animals

1. Normal or *Hprt*-deficient ES cells.
2. Adult mice (6–8 wk old).
3. Neo^r primary embryonic fibroblasts (PEFs) produced from E12.5 embryos of ROSA26 transgenic mice carrying the ubiquitously expressed *neo/lacZ* gene.

2.2. Instruments

2.2.1. Cell Fusion

1. Electro Cell Manipulator ECM 2001 (BTX, Holliston, MA).
2. Microslide chambers with a 1-mm electrode gap (BTX, cat. no. P/N450-1).
3. Micrograbber cables (BTX, cat. no. 464).
4. 100-mm bacterial dishes.
5. Inverted microscope with ×10 and ×20 objectives.

2.2.2. Cell Culture

1. Humidified incubator at 37°C with 5% CO_2 and 95% air.
2. 60-mm plastic tissue culture dishes.
3. 10- and 30-mm well plastic tissue culture plates.
4. 15-mL conical tubes.
5. 0.2-µm microfilter.
6. 200- and 1000-µL disposable pipets with autoclaved tips.

2.2.3. Single-Cell Isolation of Adult Thymic Lymphocytes

1. Syringe (2.5-mL).
2. 18-gage needle.
3. 15- and 50-mL conical tubes.
4. 60-mm bacterial dishes.

2.3. Solutions

1. PEF medium: prepare the culture medium for PEFs in a 500-mL Dulbecco's modified Eagle's medium (DMEM) bottle and store it at 4°C. To 500 mL DMEM (Sigma, St. Louis, MO; cat. no. D5796), add 50 mL fetal bovine serum (FBS); 5 mL 200 m*M* glutamine (Sigma, cat. no. G7513); and 5 mL 10,000 IU/mL penicillin and 10 mg/mL streptomycin (100X penicillin-streptomycin) (Invitrogen, Carlsbad, CA; cat. no. 10378-016).
2. ES medium: prepare the culture medium for ES cells in a 500-mL DMEM/nutrient mixture F12 HAM (F12) bottle and store it at 4°C. To 500 mL DMEM/F12 (Sigma, cat. no. D6421),

Electrofusion-Mediated Reprogramming 415

  add 75 mL FBS; 5 mL 200 mM glutamine; 5 mL 100X penicillin-streptomycin; 5 mL 100 mM sodium pyruvate (Sigma, cat. no. S8636); 8 mL 7.5% sodium bicarbonate (Sigma, cat. no. S8761); 4 µL 10^{-4} M 2-mercaptoethanol (Sigma, cat. no. M7522); and 50 µL 10^7 U/mL recombinant leukemia inhibitory factor (LIF; final 1000 U/mL) (Chemicon, Temecula, CA; cat. no. ESG1107).
  3. 0.25% trypsin/1 mM ethylenediaminetetraacetic acid·4Na. Dispense 0.25% trypsin/1 mM ethylenediaminetetraacetic acid·4Na (Invitrogen, cat. no. 15090-046) into aliquots and store them at $-20°C$.
  4. Ca^{2+}/Mg^{2+} free phosphate-buffered saline (PBS; Sigma; cat. no. D8537). Store at 4°C.
  5. Mitomycin C (MMC; Sigma, cat. no. M4287). Prepare a 0.2-mg/mL MMC solution in PBS (Sigma, cat. no. D8537). Dispense aliquots to 1.5-mL tubes. Store at $-20°C$.
  6. 0.1% gelatin: dissolve 0.1 g gelatin from porcine skin, type A (Sigma, cat. no. G1890) in 100 mL distilled water. Sterilize by autoclaving. Store at 4°C.
  7. Fresh nonelectrolyte solution (0.3 M mannitol): dissolve 2.74 g D-mannitol (Sigma, cat. no. M9546) in 50 mL distilled water. Filter through a 0.2-µm filter. Keep at 4°C.
  8. ES medium with G418: dissolve antibiotic G418 (Geneticin; Sigma, cat. no. A1720) in distilled water at 50 mg/mL. Sterilize through a 0.2-µm filter. Store at 4°C or $-20°C$. Add 50 µL of the G418 solution to 10 mL ES medium to obtain a final concentration of 250 µg/mL.
  9. ES medium with HAT: dissolve HAT media supplement (Sigma, cat. no. H0262) supplied in a vial with 10 mL DMEM (50X stock solution) (*see* **Note 1**). Store at $-20°C$. Add 200 µL stock solution in 10 mL ES medium.

3. Methods
3.1. Electrofusion
3.1.1. Feeder Cells for ES Cells

Inactivated neo^r PEFs are routinely used as feeder cells for culture of ES and hybrid cells and for selection of hybrid cell colonies with G418. Use 4×10^5 feeder cells/30-mm culture dish, 1×10^6 feeder cells/60-mm culture dish, and 2.5×10^6 feeder cells/100-mm culture dish.

  1. Coat 60-mm culture dishes with 0.1% gelatin for at least 30 min at 37°C.
  2. Incubate neo^r PEFs with 10 µg/mL MMC for 2 h at 37°C in the CO_2 incubator.
  3. Prepare the frozen stock of MMC-treated feeder cells at a cell concentration of 5×10^6 cells/mL for each cryotube.
  4. Store in liquid nitrogen.

3.1.2. PEFs for Fused Cells

  1. One day before cell fusion, coat 30-mm culture wells (6-well culture plate) with 0.1% gelatin for at least 30 min at 37°C.
  2. Prepare inactivated PEFs (4×10^5 cells/30-mm well) in 3 mL ES medium.

3.1.3. ES Cells

One of the most important factors for cell fusion experiments is that the culture conditions be optimized for maintaining the pluripotential competence and full set of chromosomes (80 chromosomes) derived from mouse ES cells and somatic cells through numerous cell divisions (*see* **Note 2**).

1. Prepare exponentially growing ES cells cultured on the inactivated PEFs with changes of culture medium once or twice a day.
2. Ascertain that complete sets of chromosomes are maintained in the ES cells before cell fusion.

3.1.4. Somatic Cells

1. Sterilize all dissection instruments (scissors and forceps) by immersion in 70% ethanol, followed by flaming.
2. Sacrifice a 6- to 8-wk-old adult mouse humanely and dissect out the thymus in a clean room if a clean bench is not available.
3. Wash the tissues with sterilized PBS twice in 60-mm Petri dishes.
4. Place the half lobe of thymus in the barrel of a sterile 2.5-mL syringe with a sterile 18-gage needle.
5. Expel and draw up the thymus gently through the needle via the tip of the needle several times in a 50-mL conical tube with 2 mL DMEM to dissociate the thymus into a single-cell suspension.
6. Keep for several min at room temperature.
7. Transfer the supernatant excluding cell clumps to a 15-mL conical tube and add 10 mL DMEM.
8. Spin down the thymocytes in 15-mL conical tubes at $300g$ for 5 min.
9. Resuspend the thymocytes in 10 mL DMEM.

3.1.5. Purification of ES Cells

1. Coat 60-mm culture dishes with 0.1% gelatin for at least 30 min at 37°C.
2. Trypsinize the ES cells and remove excess trypsin quickly.
3. Add 3 mL ES medium to inactivate the trypsin and dissociate the cells into a single-cell suspension by gentle pipetting.
4. Plate the cells on a fresh gelatin-coated 60-mm culture dish.
5. Incubate the ES cells in the CO_2 incubator for 30 min to separate feeder cells from ES cells.
6. Collect unattached ES cells and harvest them by centrifugation at $300g$ for 5 min.
7. Resuspend the cell pellet in 10 mL DMEM and transfer the cell suspension into a 15-mL conical tube.

3.1.6. ES-Somatic Cell Mixture in 0.3 M Mannitol

1. Spin down the ES cells and the thymocytes in 15-mL conical tubes at $300g$ for 5 min, separately.
2. Wash the cells with 10 mL DMEM and spin them down at $300g$ for 5 min and repeat to remove FBS completely.
3. Add 5–10 mL DMEM and adjust the density of ES cells and thymocytes to 1×10^6 cells/mL.
4. Pellet a 1:5 mixture of ES cells and thymocytes (1 mL ES cell suspension and 5 mL thymocyte suspension). Keep the remaining cells for control experiments.
5. Spin down and resuspend the cell pellet in 0.3 M mannitol to the appropriate density of 6×10^6 cells/mL (*see* **Note 3**). Use them for electrofusion immediately.

3.2. Setup of Electro Cell Manipulator ECM 2001

3.2.1. Setup of AC/DC Pulse Generator, ECM 2001, and Accessories

1. Place the ECM 2001 beside the inverted microscope (*see* **Fig. 2A**).
2. Connect the Micrograbber cable to the output jacks on the back of the ECM 2001.
3. Switch power on at the back.

Electrofusion-Mediated Reprogramming

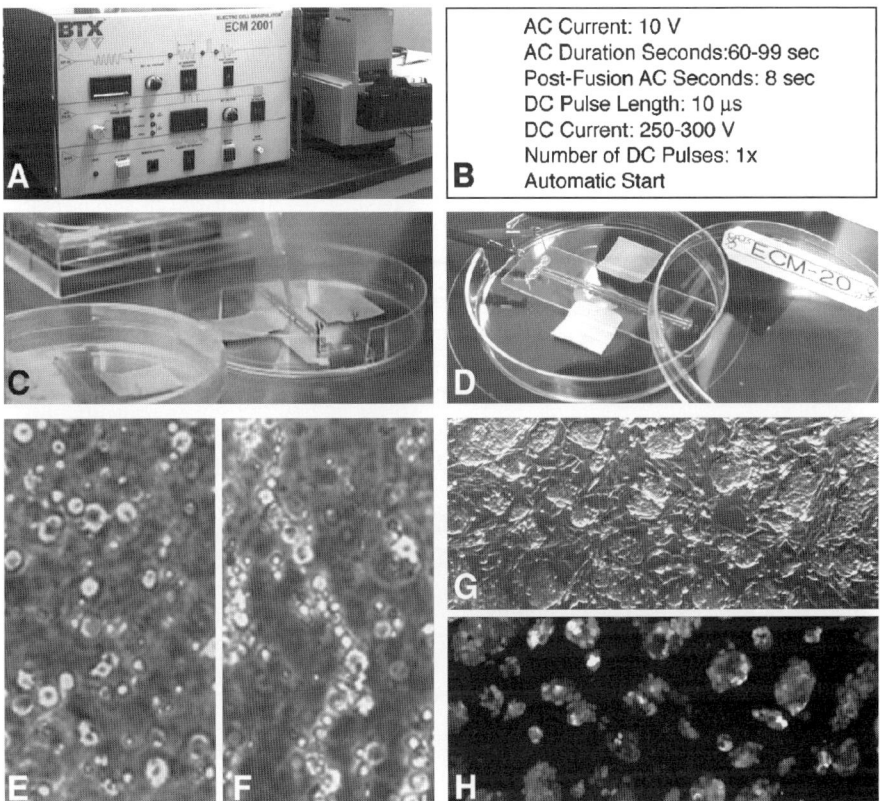

Fig. 2. (**A**) Setup of AC/DC pulse generator Electro Cell Manipulator 2001 (BTX). (**B**) The fusion parameters recommended. (**C**) Cell mixture of embryonic stem (ES) cells and thymocytes applied between electrodes using a micropipet. (**D**) Microslide chamber in the 100-mm plastic dish made with the bacterial dish and Micrograbber cable. Cell mixture (**E**) before AC application and (**F**) during AC pulse stimulation. (**G**) ES hybrid cells cultured on inactivated PEFs expressing the *Oct4-GFP* transgene derived from thymocytes (**H**).

3.2.2. Automatic Operating Parameters

Set the optimized electrical parameters to fuse ES cells and thymocytes (*see* **Fig. 2A,B**).

1. 10 V AC.
2. 60- to 90-s AC duration.
3. 250–300 V (280 V) DC. Adjust DC voltage according to the gap distance between electrodes. Appropriate electric field strength is 2.5–3.0 kV/cm.
4. 10-µs DC pulse length.
5. Number of DC pulses: one.
6. AC postfusion for 8 s.

3.2.3. Setup of Microslide Chambers

1. Sterilize the Microslides by immersion in 70% ethanol followed by flaming.
2. Set a Microslide in a 100-mm plastic dish chamber made from a bacterial dish.
3. Apply 40 µL cell mixture into the 1-mm electrode gap on the Microslide at room temperature (*see* **Fig. 2C**).
4. Connect the Microslide with the Micrograbber cable to the ECM 2001 (*see* **Fig. 2D**).
5. Place the chamber on an inverted microscope to allow observation of cell alignment (*see* **Note 4**).

3.3. Operation of ECM 2001 With Microslides

The automatic operation switch is used to initiate AC followed by DC. AC is utilized to induce a nonhomogeneous electric field, resulting in cell alignment and pearl chain formation (*see* **Fig. 2E,F**) (*see* **Note 5**). DC is utilized to produce reversible temporary pores in the cytoplasmic membranes. When juxtaposed pores in the physically associated cells reseal, cells have a chance to be hybridized by cytoplasmic membrane fusion. AC application after the DC pulse induces the compression of cells, which helps the process of fusion between the cell membranes.

1. Press the automatic operation switch of the ECM 2001.
2. Add 40 µL DMEM to the fusion mixture between electrodes to induce recovery of the membrane reformation immediately after electroporation.
3. Leave the cell mixture for 10 min at room temperature.
4. Transfer the cell mixture directly to a 30-mm culture dish containing inactivated PEFs with 3 mL ES medium with LIF.
5. Repeat the cell fusion procedure sequentially using several Microslides. Usually, the cell suspension recovered from three Microslides (80 µL × 3) is plated into one 30-mm culture dish. As a control, plate the untreated cell mixture and culture it under the same conditions.
6. Incubate the cells at 37°C in the CO_2 incubator for 24 h.
7. Change the medium to ES medium with the proper supplement to select ES hybrid cells 24 h after cell fusion.
8. Change the selection medium once a day.

As a result of the 7-d treatment, nonfused ES cells and hybrid cells derived from ES cells are killed, and thymocytes are nonadherent. Thus, the hybrid cells of ES cells and somatic cells survive and form colonies. Several colonies of hybrid cells per 10^4 host ES cells appear using the previously mentioned procedures for electrofusion.

3.4. Isolation and Cloning of Hybrid Cells

3.4.1. Selection With G418

1. Perform the automatic operation of electrofusion of the mixture of normal ES cells and thymocytes collected from the 6- to 8-wk-old ROSA26 transgenic mice carrying the ubiquitously expressed *neo/lacZ* transgene.
2. Culture the electrofusion-treated cells in ES medium for 24 h.
3. Change the medium to the ES medium supplemented with G418. ES hybrid cell colonies can be detected at 7–10 d.

4. Prepare a 24-well culture plate containing 1×10^5 inactivated neo^r PEFs per 10-mm well and 0.8 mL ES medium with G418 supplement for selection.
5. Pick up the colonies with a micropipet and transfer each colony into the 10-mm well of the 24-well culture plate on inactivated PEFs.
6. Subculture the colonies every 2 or 3 d and gradually expand the number of cells in 30- and 60-mm culture dishes with ES medium on inactivated PEFs (*see* **Fig. 2G,H**) (*see* **Note 6**).
7. Analyze the karyotypes of the hybrid cells before conducting further studies.

3.4.2. Selection With HAT

1. Use ES cells deficient in *Hprt* and normal thymocytes.
2. Culture the electrofusion-treated cells in ES medium for 24 h.
3. Change the medium to the ES medium with HAT supplement. ES hybrid cell colonies can be detected at 7–10 d.

4. Notes

1. Each vial of HAT media supplement contains 5×10^{-3} M hypoxanthine, 2×10^{-5} M aminopterin, and 8×10^{-4} M thymidine.
2. It is strongly recommended that one identify a satisfactory production lot of FBS that can supply supplements to support more effective cell growth without inducing differentiation.
3. Usually, a 1 mL mixture of ES cells and thymocytes is sufficient for the fusion experiment.
4. It is important to determine optimal fusion conditions under the microscope each time.
5. The successful formation of pearl chains during the AC pulse stimulation is extremely important for the efficiency of cell fusion. The pearl chain formation is mainly influenced by cell density and contamination from cell debris, serum, or salts in mannitol buffer. Do the following to improve the conditions:
   a. Pellet the mixed cells by centrifugation and resuspend the cells in a suitable amount of fresh 0.3 M mannitol.
   b. Increase the cell density if pearl chains are poorly formed.
   c. Decrease the cell density if cell movement is disturbed.
   d. Total cell volumes in 0.3 M mannitol have to be carefully controlled to obtain a smooth electric current. When other somatic cells larger than thymocytes and similar in size to ES cells are used as a fusion partner, prepare a 1:1 cell mixture at 2×10^6 cells/mL.
6. When cells become nearly confluent in the 60-mm culture dish, we determine that a hybrid cell line is established as the first passage.

References

1. Weiss, M. C. (1992) Extinction by indirect means. *Nature* **355,** 22–23.
2. Philpott, A. and Leno, G. H. (1992) Nucleoplasmin remodels sperm chromatin in *Xenopus* egg extracts. *Cell* **69,** 759–767.
3. Tada, M., Takahama, Y., Abe, K., Nakatsuji, N., and Tada, T. (2001) Nuclear reprogramming of somatic cells by in vitro hybridization with ES cells. *Curr. Biol.* **11,** 1553–1558.
4. Tada, T. and Tada, M. (2001) Toti-/pluripotential stem cells and epigenetic modifications. *Cell Struct. Func.* **26,** 149–160.
5. Tada, M., Morizane, A., Kimura, H., et al. (2003) Pluripotency of reprogrammed somatic genomes in embryonic stem hybrid cells. *Dev. Dyn.* **227,** 504–510.
6. Kimura, H., Tada, M., Nakatsuji, N., and Tada, T. (2004) Histone code modifications on pluripotential nuclei of reprogrammed somatic cells. *Mol. Cell Biol.* **24,** 5710–5720.

7. Tada, T., Tada, M., Hilton, K., et al. (1998) Epigenotype switching of imprintable loci in embryonic germ cells. *Dev. Gene Evol.* **207,** 551–561.
8. Tada, M., Tada, T., Lefebvre, L., Barton, S. C., and Surani, M. A. (1997) Embryonic germ cells induce epigenetic reprogramming of somatic nucleus in hybrid cells. *EMBO J.* **16,** 6510–6520.
9. Okada, Y. (1962) Analysis of giant polynuclear cell formation caused by HVJ virus from Ehrlich's ascites tumor cells. III. Relationship between cell condition and fusion reaction or cell degeneration reaction. *Exp. Cell Res.* **26,** 119–128.
10. Harris, H., Watkins, J. F., Campbell, G. L., Evans, E. P., and Ford, C. E. (1965) Mitosis in hybrid cells derived from mouse and man. *Nature* **207,** 606–608.
11. Neil, G. A. and Zimmermann, U. (1993) Electrofusion. *Methods Enzymol.* **220,** 174–196.
12. Friedrich, G. and Soriano, P. (1991) Promoter traps in embryonic stem cells: a genetic screen to identify and mutate developmental genes in mice. *Genes Dev.* **5,** 1513–1523.

29

Methylation in Embryonic Stem Cells In Vitro

Koichiro Nishino, Jun Ohgane, Masako Suzuki, Naka Hattori, and Kunio Shiota

Summary

Stem cells raise the possibility of regenerating failing body parts with new tissue. Before stem cells can safely fulfill their promise, many technical problems, including understanding the stem cell phenotype, must be overcome. DNA methylation, which is responsible for gene silencing and is associated with chromatin remodeling, is an epigenetic system that determines the specific characteristic of a variety of cells, including stem cells. Each cell type has a unique DNA methylation profile produced by varied loci-specific methylation. Investigation of such DNA methylation profiles provides a way of identifying pluripotent stem cells. Further, it is likely that analysis of the epigenetic status of stem cells may provide novel information regarding "stemness" within these populations.

Key Words: Cell-type-specific methylation pattern; DNA methylation; epigenetics.

1. Introduction

The inner cell mass of the blastocyst can generate all fetal somatic cells and germ cells. Pluripotent embryonic stem (ES) cell lines can be derived by explant and subsequent in vitro culture of the inner cell mass or epiblast. When placed back into donor blastocysts, ES cells can contribute to all of the germ lineages in chimeric embryos *(1,2)*. ES cells are an indispensable research tool in various biology fields. Human ES cells and nuclear transfer cloning technology have been established *(3–5)*. It is envisioned that these developments could transfer ES cells from use as basic laboratory research tools to important aspects of many therapeutic medical treatments, such as therapeutic applications. Therefore, ES cell lines will have to be thoroughly characterized.

Current characterization techniques rely on the assessment of morphology or molecular markers or the ability of ES cells to form teratomas or contribute to chimeric mice. Global methylation studies in ES cells are now indicating that a more precise assessment of cell state can be achieved by analyzing, in parallel, a large number of loci that are differentially methylated across a range of cell fates.

Although each ES cell from a particular line possesses the same genomic sequence, these cells can differentiate to multiple lineages, with exact gene expression profiles dependent on the final cell fate. We previously found that there are numerous tissue-dependent DNA methylated regions (T-DMR) in several tissues and stem cell lines, indicating that DNA methylation patterns are specific to cell type *(6–12)*. The DNA methylation pattern specific to cells is considered to function as a mechanism for memorizing the set of genes inherent in individual types of cells. Accordingly, this chapter focuses on methods to analyze ES cell DNA methylation status: (1) genomewide analysis of DNA methylation by restriction landmark genomic scanning (RLGS), (2) genomewide analysis of DNA methylation by virtual image RLGS (Vi-RLGS), (3) Southern blotting with methylation-sensitive restriction enzyme, (4) methylation-sensitive quantitative real-time polymerase chain reaction (PCR), (5) sodium bisulfite restriction mapping and sequencing, (6) DNA methylation analysis of teratoma tissues using laser-manipulated microdissection (LMM), and (7) methylation promoter assay.

2. Materials
2.1. Cell Culture
2.1.1. Reagents

1. 1X phosphate buffered saline (PBS): to prepare 1X PBS (1 L) combine the following: 0.45 g sodium phosphate monobasic ($NaH_2PO_4 \cdot 2H_2O$), 3.227 g sodium phosphate dibasic ($Na_2HPO_4 \cdot 12H_2O$), and 8.0 g sodium chloride (NaCl) in distilled water and adjust to pH 7.4. Adjust the volume to 1 L and autoclave.
2. Dulbecco's modified Eagle's medium (DMEM) (Gibco BRL, Rockville, MD; cat. no. 12800-017).
3. Fetal bovine serum (FBS), batch tested and screened for ES cell growth (BioWest, Miami, FL; cat. no. S1820).
4. 100X minimum essential medium (MEM) nonessential amino acids (NEAA) 10 mM solution (Gibco, cat. no. 12383-014).
5. 100X 100 mM MEM sodium pyruvate solution (Gibco, cat. no. 11360-070).
6. 100X 10 mM β-mercaptoethanol (Nacalai Tesque, Kyoto, Japan; cat. no. 214-1725G).
7. 100X 200 mM L-glutamine solution (Gibco, cat. no. 25030-081).
8. 100X penicillin-streptomycin solution (Gibco, cat. no. 1507-063).
9. 0.05% trypsin, 1 mM ethylenediaminetetraacetic acid (EDTA)·4Na in PBS: 0.5% trypsin-EDTA (Gibco, cat. no. 25200-072).
10. Murine leukemia inhibitory factor (LIF), 10^6 U/1 mL (ESGRO™, Chemicon, Temecula, CA; cat. no. ESG1106).
11. Mitomycin C (Sigma-Aldrich, St. Louis, MO; cat. no. M-0503).
12. 2 mM stock of 5-aza-2′-deoxycytidine (5-aza-dC; DNA methylation inhibitor) in PBS (Sigma-Aldrich, cat. no. A-3656). Stored at −20°C.
13. 2 mM stock of trichostatin A (TSA; inhibitor of histone deacetylase 1) dissolved in dimethyl sulfoxide (DMSO; Wako Pure Chemical, Osaka, Japan; cat. no. 047-29353). Stored at −20°C.
14. DMSO (Wako Pure Chemical, cat. no. 046-21981).
15. Polystyrene 100- and 60-mm tissue culture dishes (Nalge Nunc, Rochester, NY).
16. DB Falcon™ bacteriological Petri dishes (BD Biosciences, San Jose, CA; cat. no. 351029).

17. 50-mL plastic centrifuge tube (Corning, Acton, MA; cat. no. 430829).
18. 0.2-μm syringe filters (Pall, East Hills, NY; cat. no. 4612).
19. 4% paraformaldehyde (Wako Pure Chemical, cat. no. 162-16065) in PBS.
20. Tissue-Tek OTC compound (Sakura Finetek, Tokyo, Japan; cat. no. 4583).

2.1.2. Media

1. Embryonic fibroblast (EMFI) medium: EMFI cells are cultured in DMEM supplemented with 10% heat-inactivated FBS, 0.1 mM β-mercaptoethanol, and 1% penicillin-streptomycin. For 1 L, mix 880 mL DMEM, 100 mL FBS, 10 mL β-mercaptoethanol, and 10 mL penicillin-streptomycin.
2. ES medium: MS12 ES cells are maintained in DMEM supplemented with 15% heat-inactivated FBS, 0.1 mM MEM-NEAA, 1 mM MEM sodium pyruvate, 0.1 mM β-mercaptoethanol, 2 mM L-glutamine, 1000 U/mL LIF, and 1% penicillin-streptomycin. For 1 L, mix 800 mL DMEM, 150 mL FBS, 10 mL MEM-NEAA, 10 mL MEM sodium pyruvate, 10 mL β-mercaptoethanol, 10 mL L-glutamine, 1 mL LIF, and 10 mL penicillin-streptomycin.
3. Aza-ES medium: supplement ES medium with 1, 5, or 10 μM 5-aza-dC. For 100 mL, mix 100 mL ES medium and 50, 100, or 500 μL 5-aza-dC, respectively. For control medium, use ES media supplemented only with PBS instead of 5-aza-dC.
4. TSA-ES medium: ES medium is supplemented with 100, 200, or 400 nM TSA. For 100 mL, mix 100 mL ES medium and 5, 10, or 20 μL TSA, respectively. For control medium, use ES medium supplemented only with DMSO instead of TSA.
5. AT-ES medium: a combination of 5-aza-dC and TSA.
6. EB medium: EBs are grown in cultured medium containing DMEM, 10% heat-inactivated FBS, 1 mM L-glutamine, and 1% penicillin-streptomycin. For 1 L, mix 880 mL DMEM, 100 mL FBS, 10 mL L-glutamine, and 10 mL penicillin-streptomycin.

2.1.3. General Comments and Required Equipment for Cell Culturing

As a general rule, all cell culture protocols must be performed using sterile techniques, with great attention given to using clean and detergent-free glassware. Prefiltered media and solutions should be warmed to 37°C before use. Cell culture requires the following:

1. 37°C water bath.
2. Humidified incubator at 37°C and 5% CO_2.
3. Laminar flow cabinet.
4. Sterile glass pipets.
5. Pipetmen designated for cell culture use only.
6. Inverted microscope with a range of phase contrast objectives.
7. Cell counter.
8. Centrifuge.

2.2. Genomic DNA Extraction

1. Proteinase K (Merck, Darmstadt, Germany; cat. no. 70663-4). Prepare to 10 mg/mL in sterilized distilled water; aliquot to 1 mL and store at −20°C for up to several months.
2. Ribonuclease (deoxyribonuclease-free) (Roche, Basel, Switzerland; cat. no. 1 119 915).
3. Lysis buffer: 10 mM Tris-HCl at pH 8.0, 150 mM EDTA, 1% sodium dodecyl sulfate (SDS). Adjust final pH to 8.0.

4. 2-Mercaptoethanol (Nacalai Tesque, cat. no. 214-17).
5. Phenol/chloroform/isoamylalcohol (PCI) 50:49:1: melt 500 g crystal phenol (Nacalai Tesque, cat. no. 26728-45) at 65°C and add 8-hydroxyquinoline (Wako Pure Chemical, cat. no. 085-01212) to a final concentration of 0.1%. Saturate the phenol with 500 mL 0.5 M Tris-HCl at pH 8.0 once, followed by 500 mL 0.1 M Tris-HCl at pH 8.0 three times. Add 490 mL chloroform (Wako Pure Chemical, cat. no. 038-02606) and 10 mL isoamylalcohol (Wako Pure Chemical, cat. no. 135-12015) to the saturated phenol. Store at 4°C in a light-tight bottle up to 2 mo.
6. Ethanol (EtOH; stored at −20°C).
7. TE solution: 10 mM Tris-HCl at pH 8.0 and 1 mM EDTA at pH 8.0.

2.3. Restriction Landmark Genomic Scanning

2.3.1. Reagents

1. Klenow fragment (TaKaRa, Kyoto, Japan; cat. no. 2140A).
2. Sequenase Ver. 2 (USB, Cleveland, OH; cat. no. 70775Y).
3. *Not* I, *Pvu* II, and *Pst* I (Nippongene, Toyama, Japan; cat. no. 310-01454, 311-00281, and 318-01771).
4. dCTPαS and dGTPαS (NEN, Boston, MA; cat. no. NLP-011 and NLP-012).
5. ddATP, ddTTP, ddCTP, and ddGTP (TaKaRa, cat. no. 4031, 4034, 4033, and 4032).
6. [α-^{32}P]dCTP (6000 Ci/mmol) and [α-^{32}P]dGTP (6000 Ci/mmol) (NEN, cat. no. NEN513Z and NEN514Z).
7. *Sca* I (TaKaRa, cat. no. 1084A).
8. pBluescript II SK$^-$ (Stratagene, San Diego, CA, cat. no. 212206).
9. Bovine serum albumin (BSA) and 0.1% Triton X-100.
10. 10X H buffer: 500 mM Tris-HCl at pH 7.5, 100 mM MgCl$_2$, 1 M NaCl, 10 mM dithiothreitol (DTT). Store at room temperature for up to several months.
11. Masking buffer: 4X H buffer, 40 mM DTT, 1.6 µM dGTPαS, 0.8 µM dCTPαS, 1.6 µM ddATP, and 1.6 µM ddTTP; store at −20°C for up to 6 mo.
12. 2.5X SHB: 375 mM NaCl, 0.025% BSA, and 0.025% Triton X-100; store at −20°C for up to 1 yr.
13. Second digestion buffer: 132 µM ddGTP, 132 µM ddCTP, and 15.8 mM MgCl$_2$; store at −20°C up to 6 mo.
14. 10X first-dimension buffer: dissolve 242 g Tris-HCl, 109 g sodium acetate trihydrate, 42 g NaCl, and 23.4 g EDTA-2Na in 1.8 L distilled water. Adjust to pH 8.15 with acetic acid. Adjust volume to 2 L with distilled water and autoclave. Store at room temperature up to several months.
15. SeaKem GTG agarose (FMC Inc., Philadelphia, PA; cat. no. 50070).
16. 6X dye solution for first-dimension electrophoresis: 0.1% bromophenol blue (BPB), 0.1% xylene cyanol (XC), 30% glycerol, and 150 mM EDTA at pH 8.0.
17. 5X TBE: dissolve 270 g Tris-HCl, 138 g boric acid, and 23.3 g EDTA-2Na in 4 L distilled water and adjust the volume to 5 L.
18. Acrylamide and methylenebisacrylamide (Wako Pure Chemicals, cat. no. 019-110025 and 130-08172).
19. Ammonium persulfate and N'-tetramethyl ethylene diamine (Amersham Pharmacia Biotech, Uppsala, Sweden; cat. no. 17-1311-01 and 17-1312-9).
20. Connecting gel: 0.8% GTG agarose and 5% sucrose in 1X TBE buffer.
21. Second-dimension dye solution: 0.25% BPB and 0.25% XC in TE buffer (without glycerol).
22. T4 DNA ligase (Promega, Madison, WI; cat. no. M1801).

23. TaKaRa LA Taq DNA polymerase with GC-rich buffer (TaKaRa, cat. no. RR02BG).
24. MicroSpin™ S-400 column (Amersham Pharmacia Biotech, cat. no. 27-5140-01).
25. *Not* I adaptor (5′-ACGCCAGGGTTTTCCCAGTCACGACGC-3′ and 5′-*p*GGCCGCGTC-GTGACTGGGAAAACCCTGGCGT-3′). Incubate the two oligomers (2 nmol each) in 200 μL TE for 10 min at 70°C and cool to room temperature (*see* **Note 1**). The annealed adaptor generates 5′-protruding end of *Not* I (final adaptor concentration is 10 pmol/μL).
26. *Pst* I adaptor (5′-*p*GTGTACTGCACCAGCAAATCC-3′ and 5′-GGATTTGCTGGT-GCAGTACACTGCA-3′). Annealing reaction is performed under the same conditions as for *Not* I adaptor. The annealed *Pst* I adaptor generates 3′-protruding end of *Pst* I (10 pmol/μL).
27. Primers: Ad-NI (5′-AGGGTTTTCCCAGTCACGACGCGG-3′) and Ad-PI (5′-TTGCTG-GTGCAGTACACTGCAG-3′).

2.3.2. Equipment

1. Dialysis tube: molecular cutoff 14,000 kDa; 100 ft (Sanko Jun-yaku, Tokyo, Japan; cat. no. UC27-32-100): boil or autoclave 15-cm long pieces of the tube in 2% sodium bicarbonate twice, rinse the tubes in sterilized distilled water, and autoclave in distilled water. Store at 4°C up to several months until use.
2. Scotch 3M tape: 483 Labo-sealing tape, 25.4 mm × 32.9 m (Sumitomo 3M, Tokyo, Japan; cat. no. JT-1400-0802-3).
3. Teflon tubing for first-dimension electrophoresis of RLGS: 2.4-mm inner diameter, 3.0-mm outer diameter (Sanplatec Corp., Osaka, Japan; cat. no. 5265E).
4. Teflon tubing for in-gel digestion of RLGS: 3.0-mm inner diameter, 3.7-mm outer diameter (Sanplatec Corp., Osaka, Japan; cat. no. 5267E).
5. RLGS electrophoretic apparatuses (Bio Craft, Tokyo, Japan).
   a. First-dimension apparatus.
   b. Glass tube for first-dimension electrophoresis (the top of the glass tube is beveled and narrowed for fixation of Teflon tubing).
   c. Silicone tubing.
   d. Second-dimension apparatus.
   e. Glass plate with slant at the top for second-dimension electrophoresis.
   f. Spacer for second-dimension gel plates (1 mm thick).
6. Prepare a gel holder of disk gel for first-dimension electrophoresis of RLGS.
   a. Cut Teflon tubing (2.4-mm inner diameter) 70 cm long (*see* **Note 2**).
   b. Pass the Teflon tubing through a glass tube from the broader end and pull up the top about 1 cm with pliers.
   c. Cut off the top of the Teflon tubing, leaving about 2 mm from the top of the glass tube.
   d. Push the top of the Teflon tubing onto the heated top of the flanging tool and push the heated top of the tubing onto a cool, flat surface.
   e. Cut off the bottom of the Teflon tubing, leaving 2 cm below the glass tube.
   f. Seal the gap between the Teflon tubing and glass holder at the bottom with Teflon tape.
7. Biodyne A nylon membrane (PALL, East Hills, NY; cat. no. 60106).
8. Electroeluter (Bio Craft, cat. no. BE-883).

2.4. Southern Blotting to Evaluate Methylation Status

1. Methylation-sensitive restriction enzyme (*see* **Note 3**).
2. pBluescript II SK⁻ (Stratagene, cat. no. 212206).

3. SeaKem GTG agarose (FMC Inc., cat. no. 50070).
4. 1X TAE buffer: 40 mM Tris-acetate and 1 mM EDTA.
5. Gel loading dye: 0.25% BPB, 0 25% XC, and 30% glycerol in water.
6. 0.25 N HCl.
7. Denaturation buffer: 0.5 N NaOH and 1.5 M NaCl.
8. Neutralization buffer: 0.5 M Tris-HCl at pH 7.6 and 1.5 M NaCl.
9. 20X SSC: 3 M NaCl and 0.3 M sodium citrate.
10. Hybridization buffer: 0.4 M phosphate buffer at pH 7.2, 7% SDS, 1 mM EDTA, and 5X Denhardt's solution.
11. 50X Denhardt's solution: 1% each Ficoll, polyvinylpyrrolidone, and BSA in water (stored at $-20°C$ up to 6 mo).
12. 10 mg/mL salmon sperm DNA: dissolve appropriate amount of salmon sperm DNA in TE buffer and sonicate the DNA for several minutes to reduce average molecular weight. Extract the DNA with PCI and precipitate with EtOH. Dissolve the DNA in TE and adjust the concentration to 10 mg/mL with TE.
13. Probe: probes should be located at either side of methylation-sensitive enzyme recognition site (not containing the enzyme site within the probe). This allows for the detection of one methylated band and one unmethylated band in the autoradiogram, with the intensities of both bands proportional to copy number. Thus, the methylation degree of the enzyme recognition site can be calculated by the intensities of methylated and unmethylated bands.
14. DIG DNA labeling kit (Roche, cat. no. 1 175 033), anti-digoxygenin-AP Fab fragment (Roche, cat. no. 1 093 274), and CDP-star (Roche, cat. no. 1 685 627).
15. Maleate buffer: 0.1 M maleic acid and 0.15 M NaCl (adjust to pH 7.5 with NaOH and store at room temperature up to 2 mo).
16. 10% blocking buffer: add 10 g blocking reagent (Roche, cat. no. 1 096 176) to 100 mL maleate buffer, autoclave to dissolve the reagent, and store at 4°C up to 2 mo. For working solution (1%), dilute 10% blocking buffer to 1% with maleate buffer just before use.
17. TBST (Tris-buffered saline plus 0.1% Tween-20): dilute 10X TBS (0.2 M Tris-HCl at pH 7.6, 1.4 M NaCl) and 20% Tween-20 to final concentrations of 1X TBS and 0.1% Tween-20, respectively, with distilled water.
18. Buffer 3: 0.1 M Tris-HCl at pH 9.5 and 0.1 M NaCl.

2.5. Methylation-Sensitive Quantitative Real-Time PCR

1. ABI Prism 7000 Sequence Detection System (Applied Biosystems, Foster City, CA).
2. 96-well PCR plate (Nippon Genetics, Tokyo, Japan; cat. no. 35801) and adhesive cover (Applied Biosystems, cat. no. 4311971).
3. TaqMan Universal PCR Master Mix (Applied Biosystems, cat. no. 4304437).
4. Each 4.5 µM sense and antisense primer mixture: design a primer pair and TaqMan probe using primer design software, *Beacon Designer* 2.13 (Primer Biosoft International, Palo Alto, CA).
5. 2 µM TaqMan probe: probes should be labeled 5'-FAM/3'-TAMRA or 5'-JOE/3'-TAMRA for specific detection of PCR products.
6. TaqMan Rodent GAPDH Control Reagents VIC Probe (Applied Biosystems, cat. no. 4308313).

2.6. Sodium Bisulfite Restriction Mapping and Sequencing

1. *Eco* RI (Takara, cat. no. 1040A).
2. *Hpy* CH4 IV (New England Biolabs, Beverly, MA; cat. no. R0619S).
3. *Taq$^{\alpha}$*-I (New England Biolabs, cat. no. R0149S).

4. 10 *M* NaOH (Wako Pure Chemicals, cat. no. 197-02125).
5. Sodium metabisulfite (Wako Pure Chemicals, cat. no. 197-02365).
6. Hydroquinone (Wako Pure Chemicals, cat. no. 085-01212).
7. Wizard® DNA Clean-Up System (Promega, cat. no. A7280).
8. 6 *M* $CH_3COO-NH_4$ (pH 7.0) (Wako Pure Chemicals, cat. no. 019-02835).
9. TE solution: 10 m*M* Tris-HCl and 1 m*M* EDTA at pH 8.0.
10. AmpliTaq Gold (Applied Biosystems, cat. no. 4311814JP).
11. Qiagen Gel Extraction Kit (Qiagen GmbH, Hilden, Germany; cat. no. 28704).
12. pGEM T Easy Vector System (Promega, cat. no. A1360).
13. Primers: design a primer pair as follows:
    a. Length: 24–26mer each forward and reverse primer.
    b. Fragment size: 200–600 bp.
    c. GC content: 40–60%.
    d. Sequence: do not include CG dinucleotide sequence in primer region before bisulfite reaction. Conform not to include cytosine in forward primer and not to include guanine in reverse primer.

2.7. LMM and Laser Pressure Catapulting

1. 0.17-mm laser pressure catapulting (LPC) membrane-covered slide glass (PALM Microlaser Technologies AG, Bernried, Germany; cat. no. 8150).
2. Clear 2-mm, RIM 500-μL PALM LPC microfuge tubes (PALM Microlaser Technologies AG).
3. PALM Robot-Microbeam System (Robot-Microbeam, PALM Microlaser Technologies AG, an inverted microscope, Carl Zeiz).
4. Mineral oil (Sigma-Aldrich, cat. no. 400-5).
5. 4.0% low melting point agarose/PBS solution: SeaPlaque Agarose (FMC Inc., cat. no. 50101).
6. Lysis buffer: 10 m*M* Tris-HCl at pH 8.0, 10 m*M* EDTA, 1% SDS, and 20 μg/mL proteinase K. Make sure that the final pH is around 8.0 using pH indicator paper.
7. TE solution: 10 m*M* Tris-HCl and 1 m*M* EDTA at pH 7.0, pH 8.0, or pH 9.0.
8. PMSF solution: 40 μg/mL PMSF (Roche, cat. no. 85345621) in TE solution (pH 7.0).
9. *Eco* RI (Takara, cat. no. 1040A).
10. 0.3 *M* and 0.2 *M* NaOH (Wako Pure Chemicals, cat. no. 197-02125).
11. 5 *M* bisulfite solution: 2.5 *M* sodium metabisulfite and 125 μ*M* hydroquinone, pH 5.0.
12. 1 *M* HCl (Wako Pure Chemicals, cat. no. 080-01066).

2.8. Methylation Promoter Assay

1. TaKaRa LA Taq polymerase (TaKaRa, cat. no. RR002A).
2. pGL3 basic vector (Promega, cat. no. E1751).
3. pRL-TK vector (Promega, cat. no. E2241).
4. Bacterial strain SCS110-competent cells (Stratagene, cat. no. 200247). Most *Escherichia coli* hosts contain both DNA adenine methylation (*dam*) and DNA cytosine methylation (*dcm*) genes. These genes code for proteins that methylate specific sequences when DNA is propagated, making subsequent digestion with methylation-sensitive restriction enzymes impossible. SCS110 lacks both *dam* and *dcm* activity.
5. *Sss* I (CpG) methylase (New England BioLabs, cat. no. M0226S).
6. Quantum Prep® plasmid midiprep kit (Bio-Rad, Hercules, CA; cat. no. 732-6120).
7. Effectene transfection reagent (Qiagen GmbH, cat. no. 301425).
8. Dual-Luciferase® Reporter Assay System (Promega, cat. no. E1960).

9. Luminometer (Berthold Technologies GmbH and Co. KG, Bad Wildbad, Germany; model no. Lumat LB 9507).
10. 24-well tissue culture plate (Nalge Nunc, cat. no. 143982).

3. Methods
3.1. ES Cells in Culture
3.1.1. Preparation of EMFI Feeder Layer

ES cells require a feeder layer of mitotically inactivated fibroblast cells or gelatinized tissue culture dishes with the addition of LIF to remain pluripotent in culture. The ES cells that we usually use, the MS12 cell line *(13)*, derived from C57BL/6NCrj strain mice, are maintained on a mitomycin C-treated EMFI feeder layer in standard conditions *(14)*.

1. The day before ES cells are seeded, thaw a frozen vial of EMFI cells, seed onto polystyrene 100-mm tissue culture dishes, and culture in EMFI medium.
2. When confluent, EMFI cells are treated with 10 ng/mL mitomycin C in EMFI medium for 2.5 h.
3. Seed 10^6 mitomycin C-treated EMFI cells per 100-mm dish.
4. Freeze freshly mitomycin C-treated EMFI cells and store in liquid nitrogen for later use. Mitomycin C-treated EMFI cells should be used within 1 wk after thawing.

3.1.2. Maintenance of ES Cells

1. The ES medium must be changed every 2 d.
2. Passage ES cells every 3–5 d by disaggregation of the cells with 0.05% trypsin-EDTA at 37°C for 3–5 min (*see* **Note 4**).
3. For analysis, ES cells are separated from EMFI cells and then collected for DNA extraction (*see* **Subheading 3.1.4.**).

3.1.3. Treatment of ES Cells With 5-aza-dC or TAS

1. Subculture ES cells to a new feeder layer.
2. The next day, gently remove medium and rinse twice with 1X PBS.
3. Add 10 mL Aza-ES medium, TSA-ES medium, or AT-ES medium. Control media also should be used.
4. Culture for 2–3 d.
5. Isolate ES cells from EMFI cells and then collect the cells for DNA extraction (*see* **Subheading 3.1.4.**).

3.1.4. Isolation of ES Cells From EMFI Cells

1. Gently remove medium and rinse twice with 1X PBS.
2. Add 3 mL 0.05% trypsin-EDTA to a 100-mm dish.
3. Return plate to the incubator for 3 min until the cells float with gentle agitation.
4. Add 3 mL ES medium to rinse the dish and collect into a 50-mL conical tube.
5. Add 4 mL ES medium to the dish to rinse again and collect into same tube.
6. Dissociate aggregation of the cells by gently pipetting to single cells.
7. Centrifuge at 1000*g* for 3 min to pellet the cells.
8. Gently remove supernatant, leaving the pellet.
9. Add 10 mL ES medium and gently resuspend the pellet.
10. Plate 10 mL to new 100-mm dish and gently agitate the dish to evenly distribute the cells.

11. Incubate for 30 min to allow EMFI cells to attach to the bottom of the dish.
12. After 30 min, collect medium gently and plate the medium into a new 100-mm dish. Although most EMFI cells are attached to the bottom, almost all ES cells remain floating in medium.
13. Incubate again for 30 min.
14. After incubation, gently collect medium into a 50-mL conical tube.
15. Plate 0.5 mL to new 60-mm dish and culture 1 d; monitor for contamination with EMFI cells.
16. Centrifuge at $1000g$ for 3 min to pellet the cells.
17. Gently remove supernatant. Do not disturb the pellet.
18. Freeze the pellet in liquid nitrogen and store at $-80°C$ until DNA extraction is performed.

3.1.5. Formation of EBs

To induce the formation of EBs, after isolation from EMFI cells, ES cells (10^4 cells) are transferred to a 100-mm bacteriological Petri dish, allowing aggregation of the cells but preventing adherence to the plate. The EB medium should be changed every 2–3 d.

To change EB medium:

1. Gently collect medium, including EBs, into a 50-mL conical tube. Be careful not to break EB aggregates.
2. Leave the tube for 10 min until EBs settle to the bottom.
3. Gently remove the supernatant and add fresh EB medium into the tube.
4. Plate EBs to a new 100-mm bacteriological Petri dish.
5. Culture EBs for 10 d and then collect for DNA extraction.

3.1.6. Formation of Teratomas

1. Graft EBs for 7 wk into the kidney capsule of C57BL/6Ncrj mice *(15)*.
2. After 7 wk, remove teratomas from the recipient mice.
3. For DNA extraction, the teratomas are snap-frozen and kept at $-80°C$.
4. For laser microdissection, the teratomas are fixed with 4% paraformaldehyde and then frozen in Tissue-Tek OTC compound using general methods. Frozen blocks are stored at $-80°C$ until use.

3.2. Genomic DNA Extraction

3.2.1. Purification of High Molecular Weight Genomic DNA From ES Cells and Teratomas

1. Pretreat ES cell samples before PCI extraction as follows:
    a. Collect $1–5 \times 10^7$ cells by centrifugation at $500g$ for 5 min at $4°C$ and discard the supernatant.
    b. Add 5 mL lysis buffer (without SDS) and suspend the pellet.
    c. Add 500 µL 10% SDS to a final concentration of 1% and mix well gently up and down in the tube.
    d. Add 24 µL 10 mg/mL proteinase K, mix gently up and down in the tube, and incubate for 20 min at $55°C$.
2. Pretreat teratoma samples before PCI extraction as follows:
    a. Freeze tissue with 2 mL lysis buffer and break into small pieces with a hammer.
    b. Prechill the mortar, pestle, and 50-mL tubes in liquid nitrogen.

c. Grind down 0.1–0.5 g of tissue into powder using the mortar and pestle.
d. Transfer the tissue powder into a prechilled 50-mL tube and add 10–25 mL lysis buffer containing 150 µL 10 mg/mL proteinase K.
e. Mix gently with a spatula and incubate for 20 min at 55°C.
3. Add an equal volume of PCI, mix by rotating (25 rpm) for 30 min at room temperature, and centrifuge at 3000g for 30 min at room temperature.
4. Transfer the aqueous layer to a new 50-mL tube with a wide-pore cut-off 1000- or 5000-µL tip.
5. Repeat the PCI treatment and centrifugation.
6. Transfer the aqueous layer to a dialysis tube preequilibrated with 10 mM Tris-HCl at pH 8.0.
7. Dialyze three times (2 h, 2 h, and overnight) against 1 L 10 mM Tris-HCl at pH 8.0 per four dialysis tubes.
8. Transfer the dialysate to a 50-mL tube.
9. Add 1:1000 vol of 1 mg/mL ribonuclease A and incubate for 2 h at 37°C.
10. Divide the sample into several 50-mL tubes (less than 10 mL sample in each tube).
11. Slowly add 40 mL ice-cold EtOH retaining the interface between DNA solution and EtOH and rotate slowly to mix the DNA solution and EtOH at 15 rpm for 30 min.
12. Pick up the DNA pellet with a 200-µL yellow tip and transfer to a 1.5-mL tube.
13. Carefully remove EtOH in the tube with a 200-µL yellow tip and dry the pellet briefly for several min (*see* **Note 5**).
14. Dissolve in TE buffer more than several days at 4°C before methylation analyses (*see* **Note 6**).

3.3. Restriction Landmark Genomic Scanning

This section describes the protocol to study genomewide methylation status of gene loci (CpG islands) by RLGS based on two-dimension electrophoresis of genomic DNA. RLGS was originally developed by Hayashizaki et al. *(16)*. A manual also exists with a troubleshooting guide and illustrations *(17)*; the RLGS protocol we describe is a modified version of this *(18)*.

3.3.1. Genomic DNA Treatment Before Electrophoresis

1. Mask nicks in genomic DNA to avoid nonspecific labeling before digestion by restriction enzymes. For a 10-µL reaction, add 7 µL sample genomic DNA (0.5–1.0 µg/µL), 2.5 µL masking buffer, and 0.5 µL Klenow fragment.
2. Mix well by gentle pipetting, incubate for 20 min at 37°C, and inactivate Klenow fragment for 30 min at 65°C.
3. Perform *Not* I digestion. For a 20-µL reaction, add 10 µL masked sample, 8 µL 2.5X SHB, and 2 µL *Not* I (10 U/µL).
4. Transfer 2 µL of the mixture to a 1.5-mL tube containing *Sca* I-linearized pBluescript II SK$^-$ vector (0.1 µg in 8 µL 1X H buffer) for digestion confirmation.
5. Incubate both tubes for 4 h at 37°C.
6. Confirm complete digestion of genomic DNA with electrophoresis (*see* **Note 7**).
7. Labeling: mix 18 µL *Not* I-digested sample, 1 µL [α-^{32}P] dCTP, 1 µL [α-^{32}P] dGTP, 0.3 µL 1 M DTT, and 0.3 µL Sequenase Ver. 2 (13 U/µL); final volume is 20.6 µL.
8. Incubate for 1 h at 37°C.
9. *Pvu* II digestion: for 30.2-µL reaction, mix 20.6 µL labeled sample, 7.6 µL second digestion buffer, and 2 µL *Pvu* II (10 U/µL).
10. Incubate overnight at 37°C (*see* **Note 8**).

3.3.2. First-Dimension Gel Electrophoresis and In-Gel DNA Size Fragmentation With the Third Restriction Enzyme (Pst I)

1. Prepare first-dimension gel: mix 10 mL 10X first-dimension buffer, 0.9 g GTG agarose (0.9% final concentration), and 5 g sucrose (5% final concentration). Bring to 90 mL with distilled water and melt agarose in a microwave oven. Filter through 0.22-µm filter, adjust to 100 mL with prewarmed distilled water, and warm the gel solution in a 55–60°C water bath.
2. Connect the top of the gel holder with 2- to 3-cm silicone tubing at the bottom of a three-way stopcock and connect a 5-mL plastic syringe at the top of the three-way stopcock.
3. Hold the gel holder connected with the syringe on a stand. Suction up the first-dimension gel gradually from the bottom to a height 1 cm below the top of the gel holder and close the stopcock.
4. After the gel solidifies, open the stopcock and remove the silicone tubing with stopcock.
5. Set the gel holders to the first-dimension electrophoresis tank and fill 350 mL 1X first-dimension buffer to the bottom tank and 250 mL first-dimension buffer to the upper tank.
6. Remove air inside the gel holder using a 5-mL syringe with a 19-gage needle (check the electrical current before applying hot samples; 100 V with four gels is usually about 2–3 mA).
7. Apply 12.5 µL sample mixed with 2.5 µL 6X dye (*see* **Note 9**).
8. Run the electrophoresis at 100 V for 1 h and at 230 V overnight until the BPB dye reaches 50 cm from the top of the gel.
9. Remove the buffer from the top tank with an aspirator and bring out the gel holders (*see* **Note 10**).
10. Push out the gel noodle slowly with the in-gel digestion buffer (1X H buffer) using a cut-off 200-µL tip on a 1-mL syringe.
11. Cut the gel noodle at a point about 5 cm lower than the BPB dye (about 500 bp) and discard the gel below the 500-bp position.
12. Put the gel noodle into a 50-mL tube containing 40 mL 1X H buffer and rock the tube gently for 10 min to equilibrate the gel.
13. Change the buffer and rock the tube once more for 10 min.
14. Put the equilibrated gel noodle into a white plastic tray and put a black celluloid sheet (about half the size of the tray) on the bottom of the tray (*see* **Note 11**).
15. Pour the 1X H buffer into the tray.
16. Gently suction up the gel noodle from the BPB side into the Teflon tubing (3-mm inner diameter, 42 cm long) connected to a 5-mL syringe with 1.5-cm silicone tubing.
17. Remove the buffer inside the Teflon tubing into the syringe by pulling slowly on the piston.
18. Slowly suction up 1200 µL 1X H buffer containing 1200 U *Pst* I and 0.01% BSA into the tubing (*see* **Note 12**).
19. Remove the syringe and connect one end of the tube to the other with 1.5-cm silicone tubing to loop the Teflon tubing.
20. Incubate for 2 h at 37°C.

3.3.3. Second-Dimension Electrophoresis

1. Wipe the surface of glass plates for second-dimension apparatus with 70% EtOH and siliconize one side of glass plates (the side without slope).
2. Assemble the second-dimension apparatus and seal the side holes near the bottom of the apparatus with Scotch 3M tape.
3. Prepare 2.5 L 5% acrylamide (120.8 g acrylamide and 4.2 g methylenebisacrylamide; acrylamide:bis 29:1) gel solution in 1X TBE and add 1.7 g ammonium persulfate.

4. Add 677 μL N'-tetramethyl ethylene diamine and pour the acrylamide gel solution into second-dimension electrophoretic tank.
5. Cover the gel surface with water-saturated 2-butanol and wait for about 2 h until the gel is solidified.
6. Remove the tape around the side holes.
7. Wash out 2-butanol and fill the top of the gels with 1X TBE buffer until electrophoresis (see **Note 13**). **Steps 1–7** must be completed before the in-gel digestion is finished.
8. Rinse the top of the second acrylamide gel with 1X TBE and gently wipe the gel surface with Whatman 3MM filter paper.
9. Expel the *Pst* I-treated gel noodle from the Teflon tubing into a 50-mL tube containing 40 mL 1X TBE.
10. Rock the tube gently for 10 min to equilibrate the gel.
11. Discard the buffer and put the equilibrated gel onto a black celluloid sheet.
12. Wipe the surface of the gel with Kimwipe.
13. Transfer the gel noodle onto the top of the second acrylamide gel.
14. Connect the gel noodle and second acrylamide gel with about 2 mL connecting agarose gel using a 5-mL syringe with a 19-gage needle.
15. Cover the connected gel surface with 6X dye solution without glycerol.
16. Run electrophoresis at 100 V for 1 h and at 150 V for about 20 h until BPB dye reaches the side holes at the bottom.

3.3.4. Autoradiography

1. Carefully remove the glass plate over the gel.
2. Place a Whatman 3MM filter paper (34.5 × 42.5 cm; 5 mm each smaller than 35 × 43 cm cassette size) at the center of acrylamide gel and cut the gel along the filter paper with a razor blade.
3. Pick up the excised gel with the filter paper and cover the gel surface with Saran Wrap.
4. Dry the gel for about 20 min at 80°C.
5. Check the radioactivity of the gel at the center and the most radioactive point of the top side of second dimension with a survey meter.
6. Autoradiograph at −80°C for 2–4 wk depending on radioactivity (**Fig. 1**) (see **Note 14**).

3.3.5. Spot DNA Cloning

3.3.5.1. Two-Dimension Electrophoresis for Cloning

1. Label the genomic DNA sample that can visualize spots of interest by the same method as in **Subheading 3.3.1.** for RLGS.
2. Digest the same genomic DNA sample with *Not* I and *Pvu* II (without labeling).
3. Mix the labeled and unlabeled DNA samples in the following order (see **Note 15**): for a 125-μL reaction, mix 30.2 μL labeled DNA, 25 μL 6X dye solution for first-dimension electrophoresis (containing 150 mM EDTA), 39.6 μL TE, and 30.2 μL unlabeled DNA.
4. Apply 12.5 μL each of mixed sample (a total of 10 gels) and run RLGS.
5. Dry the second-dimension gel at 65°C and staple the dried gel with X-ray film at multiple sites (more than 10 per gel) to identify the gel and X-ray film position.
6. Autoradiograph for 1 mo, develop, and restaple the gel and autoradiogram.

3.3.5.2. Adaptor Ligation and PCR Amplification

1. Punch out spots of interest and peel Saran Wrap from dried spot gel.
2. Electroelute the DNA for 20 min at 200 V in 1X TBE buffer.

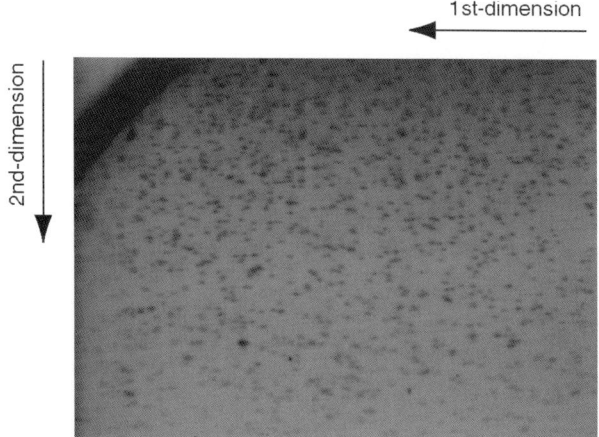

Fig. 1. Restriction landmark genomic scanning profile using the combination of *Not* I-*Pvu* II-*Pst* I digestion. This profile includes approx 1500 spots.

3. Extract the DNA with an equal volume of PCI and centrifuge at 15,000g for 1 min.
4. Precipitate with EtOH in the presence of coprecipitant and dissolve in 3 μL TE.
5. Assemble the following and mix well by pipetting: for a 10-μL reaction, mix 3 μL eluted spot DNA, 0.5 μL *Not* I adaptor (10 pmol/μL), 0.5 μL *Pst* I adaptor (10 pmol/μL), 5 μL 2X ligation buffer, and 1 μL T4 DNA ligase.
6. Incubate overnight at 4°C.
7. Add 30 μL TE and pass through a MicroSpin S-400 column at 700g for 1 min to remove excess adaptors.
8. Mix the following reagents to perform PCR. For a 50-μL reaction, mix 14.5 μL adaptor-ligated spot DNA, 25 μL 2X GC-rich buffer II, 8 μL dNTP (3.2 m*M* each), 1 μL Ad-NI primer (10 pmol/μL), 1 μL Ad-PI primer (10 pmol/μL), and 0.5 μL LA-Taq.
9. Amplify the linker-ligated DNA for 30 cycles at 94°C for 1 min, 65°C for 30 s, and 72°C for 1 min.
10. Electrophorese 5 μL PCR product in an agarose gel. If the resulting band is faint or is not seen, then reamplify the spot DNA using 1 μL first PCR solution as a template.
11. Double digest the PCR product with *Not* I and *Pst* I.
12. Ligate into pBluescript II SK⁻ digested with *Not* I and *Pst* I for 2 h at 16°C.
13. Transform into *E. coli*-competent cells (XL-1Blue MRF') and select positive colonies containing plasmid with *Not* I site.

3.3.6. Further Analyses

It has been reported that approx 90% of *Not* I sites are in CpG islands, which are the dense CpG regions. Genomic regions, including *Not* I sites as restriction landmarks, which are revealed to be differentially methylated among cells, can be further analyzed by other methods, including Vi-RLGS, DNA methylation-sensitive PCR, and sodium bisulfite restriction mapping and sequencing (described in this chapter).

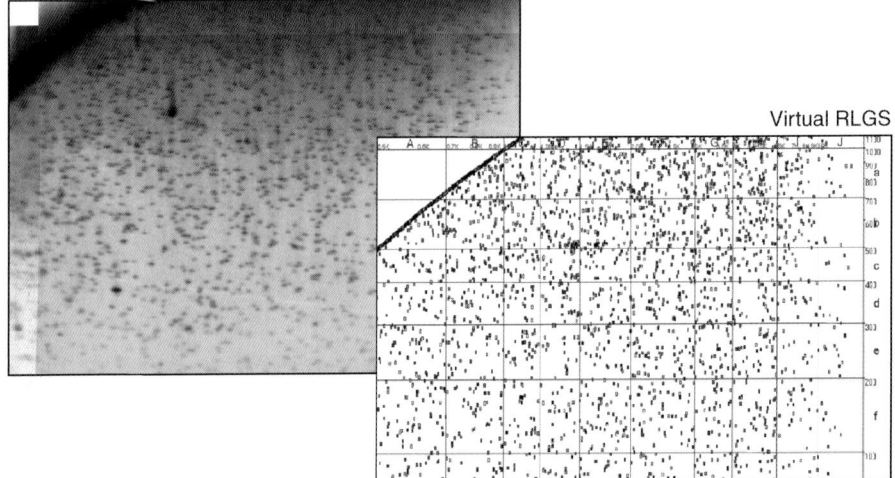

Fig. 2. Comparison between "real" restriction landmark genomic scanning (RLGS) and virtual-image RLGS. By clicking the candidate spot, the corresponding sequence data are retrieved. (Please *see* the companion CD for the color version of this figure.)

3.4. Virtual Image RLGS

Vi-RLGS software, which processes any sequence data in GenBank or FASTA format and simulates two-dimension electrophoresis using the resulting fragment, was developed and described elsewhere *(19,20)*.

1. Download mouse draft genome sequence (MGCSv3_release3 or later version) in masked FASTA format from the GenBank ftp site (ftp://ftp.ncbi.nih.gov/genomes/M_musculus/).
2. Process with the combination of Not I-Pvu II-Pst I recognition sequences.
3. By matching the Vi-RLGS and "real" RLGS profiles to identify candidate spots (**Fig. 2**), corresponding sequences are retrieved by clicking the spot on the virtual image and are used as queries for sequence analysis in BLAST (http://www.ncbi.nlm.nih.gov/genomes/seq/MmBlast.html) and ensemble (http://www.ensembl.org/Mus_musculus/) to obtain the surrounding sequence information, chromosomal position, and gene names.
4. Using the sequence information, the primers and probes were designated for Southern blotting with methylation-sensitive restriction enzyme and methylation-sensitive quantitative real-time PCR (*see* **Subheadings 3.5. and 3.6.**).
5. Identification of repetitive elements in spot data is performed using RepeatMasker (http://ftp.genome.washington.edu/cgi-bin/RepeatMasker).

3.5. Southern Blotting With Methylation-Sensitive Restriction Enzyme

This section describes the procedure for Southern blotting to evaluate the methylation status of genomic DNA by methylation-sensitive restriction enzymes using *Not* I as an example. To simplify the results, genomic DNA is first fragmented with *Pst* I that does not contain methylatable CpGs within its recognition sequence.

3.5.1. Restriction Enzyme Digestion and Electrophoresis (see **Note 16**)

1. Digest genomic DNA with *Pst* I. For a 50-µL reaction, mix genomic DNA (20 µg), 5 µL 10X H buffer, and 3 µL *Pst* I (10 U/µL). Bring to 50 µL with sterilized distilled water and mix well.
2. Transfer 2 µL of the mixture to a 1.5-mL tube containing *Sca* I-linearized pBluescript II SK⁻ vector (0.1 µg in 8 µL 1X H buffer) to ensure the complete digestion of the genomic DNA.
3. Incubate both tubes overnight at 37°C.
4. Run a sample of the plasmid on a 1.5% agarose gel. The completely digested plasmid results in 1.9- and 1.1-kb bands in addition to the genomic band. If a 3.0-kb undigested plasmid band can still be seen, then the sample should be further digested with *Pst* I until complete digestion is confirmed (*see* **Note 7**).
5. Precipitate *Pst* I-digested DNA with EtOH and dissolve in 30 µL TE.
6. Digest half of the *Pst* I-digested genomic DNA. For a 30-µL reaction, mix 15 µL *Pst* I-digested DNA, 3 µL 10X H buffer, 3 µL 0.1% BSA, 3 µL 0.1% Triton X-100, 2 µL *Not* I (10 U/µL), and 4 µL sterilized distilled water.
7. Transfer 2 µL of the mixture to a 1.5-mL tube containing *Sca* I-linearized pBluescript II SK⁻ vector (0.1 µg in 8 µL 1X H buffer) for digestion confirmation.
8. Incubate both tubes overnight at 37°C.
9. Ensure complete digestion as for **step 4**.
10. Precipitate the double-digested DNA with EtOH and dissolve in 16 µL TE.
11. Measure the DNA concentration and adjust to 5 µg/15 µL by TE.
12. Apply 15 µL of the DNA (5 µg) with gel-loading buffer and run on a 1.4% GTG agarose gel in 1X TAE buffer.
13. Stain the DNA in 0.5 µg/mL ethidium bromide solution.
14. Photograph the gel under ultraviolet light, placing a transparent ruler alongside the gel.

3.5.2. Capillary Transfer of DNA to Nylon Membrane

1. Gently rock the gel in distilled water to remove excess ethidium bromide.
2. Soak upper third of the gel in 0.25 *N* HCl by slanting the plastic tray containing the gel and gently rock the gel for 10 min for partial degradation of high molecular weight DNA.
3. Denature the DNA by soaking the gel twice for 15 min in denaturation buffer with gentle rocking.
4. Neutralize the gel by soaking the gel twice for 15 min in neutralization buffer with gentle rocking.
5. While the gel is soaking in neutralization buffer, cut the Biodyne A nylon membrane to the same size as the gel and equilibrate the membrane with distilled water followed by 20X SSC (before equilibration, record information such as date and sample names with a pencil to identify the membrane).
6. Capillary transfer the neutralized DNA from the gel to the membrane with 20X SSC (*see* **Note 17**).
7. Mark the well positions with a ballpoint pen and peel the gel.
8. Air-dry the membrane and ultraviolet crosslink at 120,000 µJ three times.
9. Store the membrane in a heat-sealable plastic bag at room temperature until use.

3.5.3. Probe Labeling

1. Prepare DNA fragment of the interested region of genomic DNA.
2. Denature 300 ng purified DNA by boiling for 10 min followed by chilling on ice for 5 min.

3. Assemble the following (DIG DNA labeling kit): for a 20-μL reaction, mix 15 μL heat-denatured DNA, 2 μL hexanucleotide mixture, 2 μL dNTP mixture, and 1 μL Klenow fragment. Incubate overnight at 37°C.
4. Stop the reaction by adding 1 μL 0.5 M EDTA and precipitate the labeled DNA with EtOH in the presence of coprecipitant.
5. Dissolve the DNA in 50 μL TE. Using 300 ng template DNA, 500 ng labeled probe is generated by an overnight reaction. Thus, the final probe concentration is about 10 ng/μL.
6. Store the labeled probe at −20°C up to 6 mo.

3.5.4. Hybridization and Detection

1. Boil 150 μL 10 mg/mL salmon sperm DNA for 10 min followed by chilling on ice for 3 min.
2. Add the denatured salmon sperm DNA to the membrane with 15 mL hybridization buffer in a plastic tray.
3. Incubate for 1 h at 60°C for prehybridization.
4. Boil 160 ng of DIG-labeled probe in 100 μL TE for 10 min followed by chilling on ice for 3 min.
5. Add the denatured probe (final probe concentration is 20 ng/mL) to the membrane with hybridization buffer (8 mL for a 6 × 11 cm membrane) in a plastic bag and mix well before sealing.
6. Remove air bubbles and seal the plastic bag.
7. Incubate overnight at 60°C for hybridization.
8. Open the bag and discard the probe-containing hybridization buffer.
9. Wash the membrane three times in a plastic tray containing 2X SSC/0.1% SDS for 5 min each at 68°C.
10. Wash the membrane three times in 0.1X SSC/0.1% SDS for 5 min at 68°C.
11. Rinse the membrane in TBS-T to remove SDS for 3 min with gentle rocking.
12. Briefly rinse the membrane with maleate buffer followed by 1% blocking buffer for 1 min each.
13. Place the membrane in 15 mL 1% blocking buffer and rock the membrane gently for 1.5 h at room temperature for blocking.
14. Add 1.5 μL of anti-Digoxigenin-AP Fab fragment to the membrane in blocking buffer and incubate for 1 h with gentle rocking (antibody is diluted to 1:10,000).
15. Wash the membrane three times in a plastic tray containing TBS-T for 10 min at room temperature.
16. Equilibrate the membrane in buffer 3 for 10 min at room temperature.
17. Dilute 4 μL CDP-star to 400 μL with buffer 3 and spot on the membrane as small drops.
18. Place the membrane in the plastic bag and seal the plastic bag after removal of air bubbles.
19. Expose to an X-ray film for 15 min and develop the film.
20. Re-expose and develop X-ray films depending on the hybridization signal intensity.

3.6. Methylation-Sensitive Quantitative Real-Time PCR

This section describes the procedure for methylation-sensitive quantitative real-time PCR using ABI Prism 7000 Sequence Detection System. Troubleshooting and illustrations of real-time PCR are described in the manufacturer's instruction. Methylation status at specific loci detected by RLGS is evaluated using the combination of the methylation-sensitive restriction digestion and quantitative real-time PCR *(20,21)*. Genomic DNA is digested by methylation-sensitive restriction enzymes using *Not* I as an example.

3.6.1. Quantitative Real-Time PCR

1. Digest genomic DNA using restriction enzymes *Pst* I and *Not* I as described in **Subheading 3.5.1., steps 1–9**.
2. Precipitate the double-digested DNA with EtOH and dissolve in TE.
3. Measure the DNA concentration and adjust to 10 ng/μL by TE.
4. Assemble the following for PCR: for a 20-μL reaction, mix 4 μL enzyme-digested DNA (10 ng/μL), 10 μL TaqMan Universal PCR Master Mix, 4 μL 4.5 μM each primers mixture, and 2 μL 2 μM TaqMan probe.
5. Amplify the enzyme-digested DNA for 40 cycles of 95°C for 15 s and 60°C for 1 min with an initial cycle of 95°C for 10 min using the ABI Prism 7000 Sequence Detection System.

3.6.2. Analyzing Methylation Rate

1. Determine cycle number when a reaction reaches threshold, the cycle threshold (CT).
2. Generate a standard curve from a dilution series of the *Not* I-untreated control for each primer set.
3. Determine PCR efficiency from the slope of the standard curve generated for each detection approach using the following equation: PCR efficiency (E) = $10^{-1/slope} - 1$.
4. Estimate the amount of undigested DNA in both *Not* I-treated and untreated genomic DNA using the following equation: $DNA_0 = 1/(1 + E)^{CT}$.
5. Normalize the initial DNA amount in the reaction mix with TaqMan Rodent GAPDH Control Reagents VIC Probe.
6. Determine DNA methylation rate using the following equation, where DNA_C is *Not* I-treated DNA_0, and DNA_{UC} is *Not* I-untreated DNA_0. Therefore, Percentage methylated DNA = (DNA_C/DNA_{UC}) × 100. All quantitative data are analyzed using *ABI Prism 7000 SDS* software. For all samples, at least three independent PCRs performed in duplicate are repeated.

3.7. Sodium Bisulfite Restriction Mapping and Sequencing

The bisulfite reaction, in which cytosine is converted to uracil and 5′-methylcytosine remains nonreactive *(22)*, is carried out as previously described *(9–12)*. Using the combination of the sodium bisulfite PCR and restriction enzyme digestion, methylation status at specific loci is evaluated *(12,23)*. The sodium bisulfite sequence provides further information of methylation status of individual CpG sites.

3.7.1. Bisulfite Reaction and PCR

1. Digest genomic DNA with *Eco* RI. For a 50-μL reaction, mix genomic DNA (10 μg), 5 μL 10X H buffer, and 2 μL *Eco* RI (10 U/μL); bring to 50 μL with sterilized distilled water.
2. Incubate both tubes overnight at 37°C.
3. Precipitate *Eco* RI-digested DNA with EtOH and dissolve in 32 μL sterilized distilled water.
4. Add 1 μL 10 M NaOH and incubate at 37°C for 15 min.
5. Assemble the following for bisulfite reaction: for a 250-μL reaction, mix 33 μL denatured DNA, 12.5 μL 10 mM hydroquinone, 200 μL 2.5 M sodium metabisulfite (pH 5.0), and 4.5 μL sterilized distilled water.
6. Add mineral oil on top of the aqueous layer to prevent evaporation.
7. Incubate in the dark at 55°C for 16 h.
8. Purify the genomic DNA with Wizard DNA Purification Resin and elute in 100 μL sterilized distilled water.

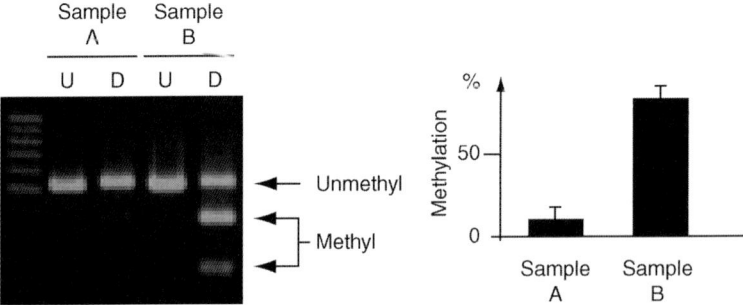

Fig. 3. Bisulfite restriction mapping. (**Left**) Electrophoresis image after bisulfite polymerase chain reaction and restriction enzyme digestion. Sample B was cleaved more than sample A, indicating that sample B was more hypermethylated than sample A. (**Right**) Methylation ratio was estimated based on the intensity of each band on the right panel. U, uncut; D, digested with Hpy CH4 IV.

9. Mix 96 µL modified DNA and 3 µL 10 M NaOH and incubate at 37°C for 15 min.
10. Add 99 µL of 6 M CH_3COO-NH_4 and 500 µL of cold EtOH.
11. Place at −20°C for 30 min and centrifuge at 15,000g for 15 min at 4°C.
12. Discard the supernatant and rinse with cold 70% EtOH.
13. After centrifugation, dry for 3 min and dissolve in 50 µL TE.
14. Assemble the following for PCR: for a 20-µL reaction, mix 2 µL bisulfite-treated genomic DNA, 2 µL 2 µM forward primer, 2 µL 2 µM reverse primer, 2 µL 10X AmpliTaq Gold buffer, 2 µL 2 mM dNTP mixture, 2 µL 25 mM $MgCl_2$, 0.2 µL AmpliTaq polymerase, and 7.8 µL sterilized distilled water.
15. Amplify the enzyme-digested DNA for 43 cycles of 94°C for 30 s, 55°C for 30 s, and 72°C for 1 min with an initial cycle of 95°C for 10 min.
16. Store PCR products at −20°C until use.

3.7.2. Restriction Mapping

For restriction mapping, restriction enzyme, *Hpy* CH4 IV or *Taq* I, is routinely used. *Hpy* CH4 IV and *Taq* I recognize 5′-ACGT-3′ and 5′-TCGA-3′ sequences, respectively. Therefore, after the bisulfite reaction, unmethylated DNA remains intact following Hpy CH4 IV or Taq I digestion, whereas the methylated DNA is cleaved (**Fig. 3**). Methylation rate is estimated from the intensity of cleaved bands based on electrophoresis image.

1. Digest half of the PCR products with restriction enzyme *Hpy* CH4 IV or *Taq* I: for a 20-µL reaction, mix 10 µL PCR products, 2 µL 10X enzyme buffer, 0.5 µL *Hpy* CH4 IV or *Taq* I (10 U/µL), and 7.8 µL sterilized distilled water.
2. Incubate for 3 h at 37°C for *Hpy* CH4 IV or at 65°C for *Taq* I.
3. Run PCR products on an agarose gel and stain with ethidium bromide. The other half of the PCR products are also run as undigested controls.
4. Determine the intensity of each band using an image analyzing program (*NIH Image*, version 1.6.1; ftp://rsbweb.nih.gov/pub/nih-image v1.6.1).
5. Determine DNA methylation rate plated into the following equation: percentage methylated DNA = $I^{Me}/(I^U + I^{Me}) \times 100$, where I^U is the intensity of the undigested band, and I^{Me} is the intensity of cleaved band. For all samples, at least three independent PCR reactions are repeated.

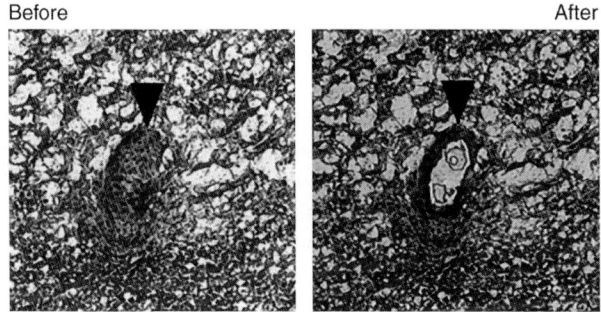

Fig. 4. Representative photographs of a teratoma before and after laser-manipulated microdissection (LMM) and laser pressure catapulting (LPC) (magnification ×100). (**Left**) A 5-μm frozen section of a cartilagelike tissue (arrow) in a teratoma 7 wk after transplantation stained by hematoxylin and eosin before LMM and LPC. (**Right**) The same specimen after LMM and LPC. The cartilage microdissected and catapulted. (Please *see* the companion CD for the color version of this figure.)

3.7.3. Sequencing

1. After **step 3** in **Subheading 3.7.2.**, the PCR bands are recovered from the agarose gel using the Qiagen gel extraction kit.
2. Elute the PCR fragments in 50 μL sterilized distilled water.
3. Ligate into pGEM T Easy vector and subclone.
4. Analyze the sequences of at least five clones/PCR reaction. For all samples, at least three independent PCR reactions are repeated. Then, determine the methylation status of individual CpG sites. The unmethylated cytosine is changed to thymine, whereas the methylated cytosine remains as cytosine in sequencing data. As a control, plasmid DNA, which is unmethylated, should also be subjected to the same analysis in parallel to confirm the completion of the bisulfite reaction.

3.8. DNA Methylation Analysis of Teratoma Tissues Using LMM

LMM is a method to cut out single cells or a limited tiny region from a specimen under microscopic observation by a laser beam. LPC is a method to push up and collect samples that were microdissected using a strong laser *(24–26)*. This section describes the protocol to analyze DNA methylation status of teratoma tissues using LMM. Collected samples are treated with a bisulfite reaction using agarose beads *(27,28)*, and then restriction mapping is performed.

3.8.1. LMM and LPC

1. Cut the teratoma using a cryostat.
2. Mount 5-μm thick sections of the teratoma onto an LPC membrane-coated slide glass.
3. Stain specimens with hematoxylin and eosin by conventional methods. Do not apply a cover glass after air-drying (*see* **Note 18**).
4. Set the specimen on a microscope stage of PALM Robot Microbeam System and observe with a charge-coupled device camera.
5. Set the cap of the LPC PCR tube, which is coated on the inside with mineral oil, on the micromanipulator.
6. Focus the laser on the specimen and dissect the sample as well as the thin membrane by laser beam (**Fig. 4**).

7. Collect pieces of the sample using laser pressure of LPC into the cap of the LPC PCR tube.
8. Reset the cap to LPC PCR tube and spin down the pieces and mineral oil to the bottom.

3.8.2. Genome Extraction From the Sample Pieces

1. Dissolve 4% agarose/PBS by heating. Keep the solution warmed to 40°C.
2. Place 300 µL mineral oil in a 2-mL tube on ice.
3. Add 5 µL TE to the sample pieces and spin down.
4. Add 5 µL 4% agarose/PBS solution to the sample.
5. Pipet 10 µL agarose/sample mixture into the cold mineral oil. The agarose/sample drop immediately solidifies in the oil.
6. Add 400 µL lysis buffer and spin down for several seconds.
7. Incubate at 50°C overnight.
8. Remove mineral oil and lysis buffer.
9. Add 1 mL TE (pH 9.0).
10. Incubate at room temperature for 15 min, and then remove the solution.
11. Repeat **steps 9** and **10**.
12. Add 400 µL PMSF solution.
13. Incubate at room temperature for 45 min, and then remove the solution.
14. Repeat **steps 12** and **13**.
15. Repeat **steps 9** and **10** twice and then remove the solution.
16. Add 100 µL 1X *Eco* RI enzyme buffer.
17. Incubate at room temperature for 15 min, and then remove the solution.
18. Repeat **steps 16** and **17**.
19. Add 100 µL *Eco* RI enzyme solution, including 1X *Eco* RI enzyme buffer and 10 U *Eco* RI. Incubate at 37°C for 5 h.
20. Remove the enzyme solution.
21. Add 500 µL 0.3 M NaOH.
22. Incubate at room temperature for 15 min, and then remove the solution.
23. Repeat **steps 21** and **22**.
24. Add 300 µL mineral oil and heat at 80°C until the agarose beads are completely dissolved.
25. During heating, place 300 µL mineral oil into a new 2-mL tube on ice.
26. Resolidify the agarose beads in cold mineral oil.
27. Store at 4°C until use.

3.8.3. Sodium Bisulfite Reaction for Agarose Beads

1. Add 500 µL 5 M bisulfite solution to the agarose beads.
2. Gently mix by inverting several times.
3. Add 300 µL mineral oil and spin down for several seconds.
4. Incubate in the dark at 55°C for 16 h.
5. Remove mineral oil and bisulfite solution.
6. Add 1 mL TE (pH 8.0).
7. Leave at room temperature for 10 min, then remove the solution.
8. Repeat **steps 6** and **7** six times.
9. Add 500 µL 0.2 M NaOH.
10. Leave at room temperature for 15 min.
11. Repeat **steps 9** and **10** twice.
12. Add 100 µL 1 M HCl and gently mix by inverting several times.

Methylation in ES Cells

13. Remove solution and add 1 mL TE (pH 8.0).
14. Leave at room temperature for 10 min.
15. Repeat **steps 6** and **7** three times and then remove TE completely.
16. Add 200 µL mineral oil.
17. Store at 4°C until PCR reaction.

3.8.4. PCR Reaction

Analysis of DNA methylation using the agarose bead method usually requires a nested PCR step because the amounts of genomic DNA extracted from the sample pieces are very small.

1. Prepare the PCR mixture: for a 50-µL reaction, mix 10 µL agarose beads, 5 µL 20 µM forward primer, 5 µL 20 µM reverse primer, 5 µL 10X AmpliTaq Gold buffer, 5 µL 2 mM dNTP mixture, 5 µL 25 mM MgCl$_2$, 0.5 µL AmpliTaq polymerase, and 14.5 µL sterilized distilled water.
2. Amplify for 43 cycles of 94°C for 30 s, 55°C for 30 s, and 72°C for 1 min with an initial cycle of 95°C for 10 min.
3. Assemble the following for nested PCR (see **Note 19**): for a 20-µL reaction, mix 2 µL first PCR product, 2 µL 2 µM nested forward primer, 2 µL 2 µM nested reverse primer, 2 µL 10X AmpliTaq Gold buffer, 2 µL 2 mM dNTP mixture, 2 µL 25 mM MgCl$_2$, 0.2 µL AmpliTaq polymerase, and 7.8 µL sterilized distilled water.
4. Amplify for 43 cycles of 94°C for 30 s, 55°C for 30 s, and 72°C for 1 min with an initial cycle of 95°C for 10 min.
5. Proceed with bisulfite restriction mapping method (see **Subheading 3.7.2.**).

3.9. Methylation Promoter Assay

This section describes the protocol to examine if DNA methylation affects promoter activity by use of a methylation promoter assay *(11,12)*. This is different from a deletion promoter assay. This protocol permits assessment of whether DNA methylation in a promoter sequence suppresses gene expression.

3.9.1. Establishment of Reporter Constructs

1. Regulatory region including a promoter of a target gene is isolated from genomic DNA by genomic PCR using LA Taq polymerase. The thermocycling program is an initial cycle of 95°C for 1 min followed by 94°C for 30 s, 55–60°C (depending on primer sequences) for 30 s, and 72°C for 1 min/kb for 35 cycles.
2. Subclone the PCR fragment purified into pGL3-Basic Vector.
3. Cloning of constructs is carried out using the bacterial strain SCS110, which is deficient for two methylases found in most strains of *E. coli*, *dam,* and *dcm* methylase, permitting the effects of DNA methylation on gene expression clearly.
4. To methylate CpG sites in a reporter construct, assemble the following: for a 100-µL reaction, mix 10 µg reporter construct, 10 µL 10X *Sss* I buffer, 0.5 µL 32 mM S-adenosylmethionine, 30 U *Sss* I (CpG) methylase, and sterilized distilled water.
5. Incubate at 37°C for 3 h.
6. Extract the DNA with PCI and precipitate with EtOH.
7. Dissolve the DNA in TE and adjust the concentration to 200 ng/µL with TE.
8. Confirm completion of the methylation by resistance to *Hpa* II (methylation-sensitive restriction enzyme) digestion.

3.9.2. Transfection and Promoter Activity

After passage onto fresh EMFI cells, ES cells are transiently transfected with a reporter construct using Effectene transfection reagent. To normalize the luciferase activity of the reporter constructs, an internal control plasmid (pRL-TK vector, 1/10 mol of the reporter construct) expressing *Renilla* luciferase is cotransfected.

For transfection of the reporter construct into ES cells:

1. On the day of transfection, collect harvest ES cells with feeder cells described in **Subheading 3.1.2.**
2. Seed 2.0×10^5 cells/well of 24-well dishes in 150 μL ES medium (*see* **Note 20**).
3. Dilute 0.25 μg DNA, including reporter construct and pRL-TK vector with buffer EC, to a total volume of 67.5 μL.
4. Add 2 μL enhancer and mix by vortex for 1 s.
5. Incubate at room temperature for 2–5 min.
6. Spin down the mixture.
7. Add 5.5 μL Effectene reagent to the DNA enhancer solution.
8. Mix by pipetting up and down five times.
9. Incubate the sample for 5–10 min at room temperature to allow transfection complex formation.
10. Add 150 μL ES medium to the tubes containing the transfection complexes.
11. Mix by pipetting up and down twice.
12. Immediately add the transfection complexes onto the cells in the 24-well dishes.
13. After 6 h, remove medium containing the transfection complexes, then add fresh ES medium.
14. Incubate ES cells under their normal growth conditions (37°C and 5% CO_2) for 48 h (*see* **Note 21**).
15. After 48 h, the activities of both luciferases are determined using the Dual-Luciferase Reporter Assay System. Assays are performed at least three times each, in triplicate.

4. Notes

1. The 1.5-mL annealing reaction tube is floated on water at 70°C in a 500-mL beaker. Leave the tube in the warm water until it reaches room temperature.
2. The top of Teflon tubing is cut sharply.
3. The REBASE homepage (http://rebase.neb.com/rebase/rebase.html) has an excellent list of hundreds of methylation-sensitive restriction enzymes as well as methylases. This list also contains information on enzyme suppliers and detailed position effects of methylated bases on digest ability. Refer to this information to decide which enzyme is appropriate to evaluate the methylation status of the genes of interest.
4. Timing of subculture depends on proliferation of the ES cells.
5. Do not dry DNA pellet too much.
6. The pellet is dissolved in 50–100 μL for cultured cell samples. Purification of high-quality genomic DNA is most important to obtain fine profiles of RLGS. Accordingly, degradation by endogenous DNases and mechanical shearing must be avoided.
7. Restriction enzymes can simply be added to the reaction mixture, ensuring the total enzyme volume does not exceed 10% of the reaction mixture. Purifying genomic DNA by PCI extraction and EtOH precipitation may help to digest DNA in the case of certain contaminants inhibiting the enzyme reaction.
8. The *Pvu* II-digested samples can be stored at 4°C for a few days before electrophoresis.

9. Labeled genomic DNA applied to first-dimension electrophoresis must be less than 1 µg. Running a large amount of DNA results in an unclear spot pattern in the high molecular weight region and high background. Usually, up to a 1/5 dilution of labeled DNA with TE gives the appropriate DNA concentration for first-dimension electrophoresis.
10. The buffer after electrophoresis is contaminated with ^{32}P.
11. Use a white tray because white helps to distinguish the BPB and XC dyes, and black visualizes the gel shape.
12. High glycerol levels remaining from the in-gel digestion can cause high background in autoradiograms. Thus, high-concentration *Pst* I enzyme should be purchased and used for in-gel digestion to minimize glycerol content.
13. Check that the buffer does not leak.
14. Checking the RLGS spot pattern with an image analyzer is simple and quick. However, it is better to use X-ray films for obtaining fine profiles and for comparison of spot patterns.
15. Labeled DNA gives spots in autoradiograms, and unlabeled DNA at the same position with the spots can be cloned. Because addition of 150 m*M* EDTA in 6X dye ensures the complete inactivation of Sequenase, unlabeled DNA must not be added before the 6X dye is added to the labeled DNA.
16. Recognition sites of methylation-sensitive enzymes containing methylatable CpG dinucleotides exist 1/4 to 1/5 less frequently in the mammalian genome than expected. Thus, to separate methylated and unmethylated bands clearly on ordinary agarose gels, genomic DNA must first be fragmented with one or two appropriate restriction enzymes, the recognition sites of which do not contain methylatable CpGs.
17. Remove air bubbles between the gel and membrane completely.
18. Stained specimens with hematoxylin and eosin should be used within 1 wk.
19. Sequential PCR, using a 50- to 100-fold dilution of the product of the previous PCR as a template, can help amplify low-abundance signals.
20. Construct DNA is not transfected into feeder cells because feeder cells do not proliferate.
21. ES cells transfected with a luciferase reporter construct are typically incubated for 48 h posttransfection to obtain the maximal level of gene expression.

References

1. Martin, G. R. (1981) Isolation of a pluripotent cell line from early mouse embryos cultured in medium conditioned by teratocarcinoma stem cells. *Proc. Natl. Acad. Sci. USA* **78**, 7634–7638.
2. Nagy, A., Rossant, J., Nagy, R., Abramow-Newerly, W., and Roder, J. C. (1993) Derivation of completely cell culture-derived mice from early-passage embryonic stem cells. *Proc. Natl. Acad. Sci. USA* **90**, 8424–8428.
3. Thomson, J. A., Itskovitz-Eldor, J., Shapiro, S. S., et al. (1998) Embryonic stem cell lines derived from human blastocysts. *Science* **282**, 1145–1147.
4. Wilmut, I., Schnieke, A. E., McWhir, J., Kind, A. J., and Campbell, K. H. (1997) Viable offspring derived from fetal and adult mammalian cells. *Nature* **385**, 810–813.
5. Wakayama, T., Rodriguez, I., Perry, A. C., Yanagimachi, R., and Mombaerts, P. (1999) Mice cloned from embryonic stem cells. *Proc. Natl. Acad. Sci. USA* **96**, 14,984–14,989.
6. Shiota, K., Kogo, Y., Ohgane, J., et al. (2002) Epigenetic marks by DNA methylation specific to stem, germ and somatic cells in mice. *Genes Cells* **7**, 961–969.
7. Shiota, K. and Yanagimachi, R. (2002) Epigenetics by DNA methylation for development of normal and cloned animals. *Differentiation* **69**, 162–166.

8. Ohgane, J., Wakayama, T., Kogo, Y., et al. (2001) DNA methylation variation in cloned mice. *Genesis* **30,** 45–50.
9. Imamura, T., Ohgane, J., Ito, S., et al. (2001) CpG island of rat sphingosine kinase-1 gene: tissue-dependent DNA methylation status and multiple alternative first exons. *Genomics* **76,** 117–125.
10. Cho, J. H., Kimura, H., Minami, T., et al. (2001) DNA methylation regulates placental lactogen I gene expression. *Endocrinology* **142,** 3389–3396.
11. Hattori, N., Nishino, K., Ko, Y. G., Ohgane, J., Tanaka, S., and Shiota, K. (2004) Epigenetic control of mouse Oct-4 gene expression in embryonic stem cells and trophoblast stem cells. *J. Biol. Chem.* **279,** 17,063–17,069.
12. Nishino, K., Hattori, N., Tanaka, S., and Shiota, K. (2004) DNA methylation-mediated control of Sry gene expression in mouse gonadal development. *J. Biol. Chem.* **279,** 22,306–22,313.
13. Kawase, E., Suemori, H., Takahashi, N., Okazaki, K., Hashimoto, K., and Nakatsuji, N. (1994) Strain difference in establishment of mouse embryonic stem (ES) cell lines. *Int. J. Dev. Biol.* **38,** 385–390.
14. Matise, M., Auerbach, W., and Joyner, A., (2000) Production of targeted embryonic stem cell clones, in *General Targeting: A Practical Approach,* (Joyner, A., ed.), Oxford University Press, New York, pp. 101–132.
15. Kremenskoy, M., Kremenska, Y., Ohgane, J., et al. (2003) Genome-wide analysis of DNA methylation status of CpG islands in embryoid bodies, teratomas, and fetuses. *Biochem. Biophys. Res. Commun.* **311,** 884–890.
16. Hayashizaki, Y., Hirotsune, S., Okazaki, Y., et al. (1993) Restriction landmark genomic scanning method and its various applications. *Electrophoresis* **14,** 251–258.
17. Hayashizaki, Y. and Watanabe, S. (1997) *Restriction Landmark Genomic Scanning (RLGS),* Springer-Verlag, Tokyo.
18. Ohgane, J., Aikawa, J., Ogura, A., Hattori, N., Ogawa, T., and Shiota, K. (1998) Analysis of CpG islands of trophoblast giant cells by restriction landmark genomic scanning. *Dev. Genet.* **22,** 132–140.
19. Matsuyama, T., Kimura, M. T., Koike, K., et al. (2003) Global methylation screening in the Arabidopsis thaliana and Mus musculus genome: applications of virtual image restriction landmark genomic scanning (Vi-RLGS). *Nucleic Acids Res.* **31,** 4490–4496.
20. Hattori, N., Abe, T., Suzuki, M., et al. (2004) Preference of DNA methyltransferases for CpG islands in mouse embyonic stem cells. *Genome Res.* **14,** 1733–1740.
21. Heid, C. A., Stevens, J., Livak, K. J., and Williams, P. M. (1996) Real time quantitative PCR. *Genome Res.* **6,** 986–994.
22. Frommer, M., McDonald, L. E., Millar, D., et al. (1992) A genomic sequencing protocol that yields a positive display of 5-methylcytosine residues in individual DNA strands. *Proc. Natl. Acad. Sci. USA* **89,** 1827–1831.
23. Xiong, Z. and Laird, P. W. (1997) COBRA: a sensitive and quantitative DNA methylation assay. *Nucleic Acids Res.* **25,** 2532–2534.
24. Bonner, R. F., Emmert-Buck, M., Cole, K., et al. (1997) Laser capture microdissection: molecular analysis of tissue. *Science* **278,** 1481–1483.
25. Schutze, K., Posl, H., and Lahr, G. (1998) Laser micromanipulation systems as universal tools in cellular and molecular biology and in medicine. *Cell Mol. Biol. (Noisy-le-grand)* **44,** 735–746.

26. Nagasawa, Y., Takenaka, M., Matsuoka, Y., Imai, E., and Hori, M. (2000) Quantitation of mRNA expression in glomeruli using laser-manipulated microdissection and laser pressure catapulting. *Kidney Int.* **57,** 717–723.
27. Olek, A., Oswald, J., and Walter, J. (1996) A modified and improved method for bisulphite based cytosine methylation analysis. *Nucleic Acids Res.* **24,** 5064–5066.
28. Malik, K., Salpekar, A., Hancock, A., et al. (2000) Identification of differential methylation of the WT1 antisense regulatory region and relaxation of imprinting in Wilms' tumor. *Cancer Res.* **60,** 2356–2360.

VI

TUMOR-LIKE PROPERTIES

30

Identification of Genes Involved in Tumor-Like Properties of Embryonic Stem Cells

Kazutoshi Takahashi, Tomoko Ichisaka, and Shinya Yamanaka

Summary

Embryonic stem (ES) cells are pluripotent stem cells derived from preimplantation stage embryos. ES cells proliferate infinitely while maintaining pluripotency. These properties make them attractive sources for stem cell therapies and regenerative medicine. However, undifferentiated ES cells produce tumors when transplanted, which may preclude their therapeutic usage. It is largely unknown why ES cells can possess tumorigenicity without having chromosomal abnormalities. In this chapter, we introduce the methods to identify genes that play roles in tumor-like properties of ES cells.

Key Words: Embryonic stem cells; ERas; phosphatidylinositol-3-kinase; teratoma; tumorigenicity.

1. Introduction

Embryonic stem (ES) cells were first established from the inner cell mass of mouse blastocysts in 1981 *(1,2)*. These cells can proliferate infinitely while maintaining an undifferentiated state. Pluripotent cells were subsequently generated from human blastocysts in 1998; these are considered promising sources for cell therapy to treat patients with degenerative diseases such as diabetes and Parkinson's disease. However, ES cells have a propensity to produce tumors called teratomas when transplanted, which may preclude their therapeutic usage *(3)*. Transcription factors expressed predominantly in ES cells, such as Oct3/4 *(4,5)* and Nanog *(6,7)*, are essential for the maintenance of pluripotency. Several cytokines, including leukemia inhibitory factor (LIF), bone morphogenetic protein 4, and Wnt could sustain self-renewal of ES cells *(8–11)*. However, it remains elusive why ES cells possess tumor-like properties without chromosomal abnormalities.

Previous studies have shown that PTEN (phosphatase and tensin homolog deleted on chromosome ten)-deficient ES cells form significantly larger tumors than wild-type ES cells *(12)*, and PTEN-null ES cells proliferate much faster in vitro *(13)*. On the other hand, deletion of *Akt1*, also known as protein kinase B-α, by gene targeting in PTEN-null ES cells resulted in suppression of both tumorigenicity and proliferation rate in vitro *(14)*.

In addition, we reported that undifferentiated ES cells specifically express a constitutively active Ras protein, which we designated ERas (ES cell-expressed Ras). ERas-deficient ES cells form significantly smaller teratomas and show reduced proliferation rate in vitro than wild-type ES cells do. We also demonstrated that ERas interacts with phosphatidylinositol-3-kinase (PI3K) catalytic subunit p110 and constitutively activates this pathway *(15)*. The role of PI3K in the cell cycle control of ES cells has been also shown by a study in which the PI3K-specific inhibitor LY294002 markedly increased the proportion of ES cells in the G0/G1 phase *(16)*. However, ERas-null ES cells still produce teratomas, albeit smaller than normal ES cells do, suggesting the existence of other factors activating the PI3K pathway. Factors involved in other signaling pathways might also be involved in tumorigenicity of ES cells.

Identification of these factors is essential for understanding the tumor-like properties of ES cells and for their clinical application. In this chapter, we summarize our protocols to evaluate candidate genes for tumorigenicity of ES cells.

2. Materials

2.1. Cell Culture

1. Dulbecco's modified Eagle's medium (DMEM; Nacalai tesque, Kyoto, Japan; cat. no. 14247-15).
2. Fetal bovine serum (characterized for ES cells; Biowest, Rue de la Caille, France; cat. no. S1820).
3. Calf serum (Sigma, St. Louis, MO; cat. no. C6278).
4. Nonessential amino acids solution (Invitrogen, Carlsbad, CA; cat. no. 11140-050).
5. L-Glutamine (Invitrogen, cat. no. 25030-081).
6. 2-Mercaptoethanol (Invitrogen, cat. no. 21985-023).
7. Penicillin/streptomycin (Invitrogen, cat. no. 15140-122).
8. LIF (we routinely prepare it ourselves); ESGRO, which is recombinant LIF suitable for ES cell culture (Chemicon, Temecula, CA; cat. no. ESG1107).
9. Phosphate-buffered saline (PBS; Nacalai tesque, cat. no. 14249-95).
10. Trypsin/1 mM ethylenediaminetetraacetic acid (EDTA; for ES cells and STO cells; Invitrogen, cat. no. 25200-056).
11. Trypsin/5.3 mM EDTA (10X solution for other cells) (Sigma, cat. no. T4174).
12. Gelatin (Sigma, cat. no. G1890).
13. 100-mm tissue culture dishes (Iwaki, Tokyo, Japan; cat. no. 3020-100).
14. 60-mm tissue culture dishes (Iwaki, cat. no. 3010-060).
15. 24-well tissue culture plates (Iwaki, cat. no. 3820-024).
16. 50-mL polypropylene conical tubes (Iwaki, cat. no. 2345-050).
17. 15-mL polypropylene conical tubes (Iwaki, cat. no. 2325-015).
18. Puromycin (Sigma, cat. no. P8833) dissolved in PBS at 10 mg/mL.
19. Brastcidine S (Funakoshi, Tokyo, Japan; cat. no. KK-400) dissolved in PBS at 100 mg/mL.
20. Coulter counter (Beckman Coulter, Fullerton, CA; model no. Z2).
21. 1-mL disposable plastic pipets (Iwaki, cat. no. 7051-001).
22. 5-mL disposable plastic pipets (Iwaki, cat. no. 7053-005).

23. 10-mL disposable plastic pipets (Iwaki, cat. no. 7054-010).
24. 25-mL disposable plastic pipets (Iwaki, cat. no. 7055-025).
25. 500-mL bottle-top filter unit (Iwaki, cat. no. 8120 500).
26. 0.22-μm pore size syringe filter (Millex GP; Millipore, Billerica, MA; cat. no. SLGP033RS).
27. ES cell medium: ES cells are maintained in DMEM containing 10% fetal bovine serum, $1 \times 10^{-4}\,M$ nonessential amino acids, 2 mM L-glutamine, $1 \times 10^{-4}\,M$ 2-mercaptoethanol, and 50 U penicillin/streptomycin (*see* **Note 1**) *(17)*. To prepare 500 mL ES cell medium, mix 50 mL fetal bovine serum, 5 mL nonessential amino acids, 5 mL L-glutamine, 1 mL 2-mercaptoethanol, and 2.5 mL penicillin/streptomycin and then fill to 500 mL with DMEM.
28. MG1.19 ES cells are grown on gelatin-coated tissue culture plates in the medium containing LIF (*see* **Notes 2** and **3**) *(18)*.
29. Medium for primary mouse embryonic fibroblasts (PMEFs) and PLAT-E packaging cells. PMEFs and PLAT-E cells are maintained in DMEM containing 10% fetal bovine serum and 50 U penicillin/streptomycin. To make 500 mL medium, mix 50 mL fetal bovine serum and 2.5 mL penicillin/streptomycin and then fill to 500 mL with DMEM. In the case of PLAT-E, 1 μg/mL puromycin and 10 μg/mL brastcidine S are supplemented into the medium (*see* **Note 4**) *(19)*.
30. Medium for NIH3T3 cells: NIH3T3 cells are cultured in DMEM containing 10% calf serum. To prepare 500 mL of this medium, mix 50 mL calf serum and 2.5 mL penicillin/streptomycin and then fill to 500 mL with DMEM.

2.2. Transfection and Retroviral Infection

1. Lipofectamine 2000 (Invitrogen, cat. no. 11668-019).
2. Fugene 6 transfection reagent (Roche, Grenzacherstrasse, Switzerland; cat. no. 1 814 443).
3. Polybrene (Nacalai tesque, cat. no. 17736-44).
4. 0.45-μm cellulose acetate filter unit (FP30/0.45 CA-S; Schleicher and Schuell, Relliehausen, Germany; cat. no. 10462100).
5. 10-mL disposable syringes (sterilized; Terumo, Tokyo, Japan; cat. no. SS-10ESZ).
6. Crystal violet (Nacalai tesque, cat. no. 09804-52).

2.3. Teratoma Formation

1. 1-mL disposable syringes (sterilized; Terumo, cat. no. SS-01T).
2. 23-gage needles (sterilized; Terumo, cat. no. NN-2325R).
3. Diethyl ether (Nacalai tesque, cat. no. 15402-35).
4. Nude mice (4- to 6-wk-old female mice).
5. Formaldehyde (Nacalai tesque, cat. no. 16223-55).

3. Methods

3.1. Analyses of Tumor-Like Properties in ES Cells

3.1.1. Forced Expression in ES Cells by Supertransfection

To understand the role of genes of interest in ES cells, the episomal expression system developed by Dr. Austin Smith's laboratory *(19)* is suitable. This method provides rapid and efficient establishment of both polyclonal and monoclonal ES cells expressing genes of interest. MG1.19 ES cells are derived from CCE ES cells and stably express the large T antigen of murine polyoma virus. When plasmids containing the replication

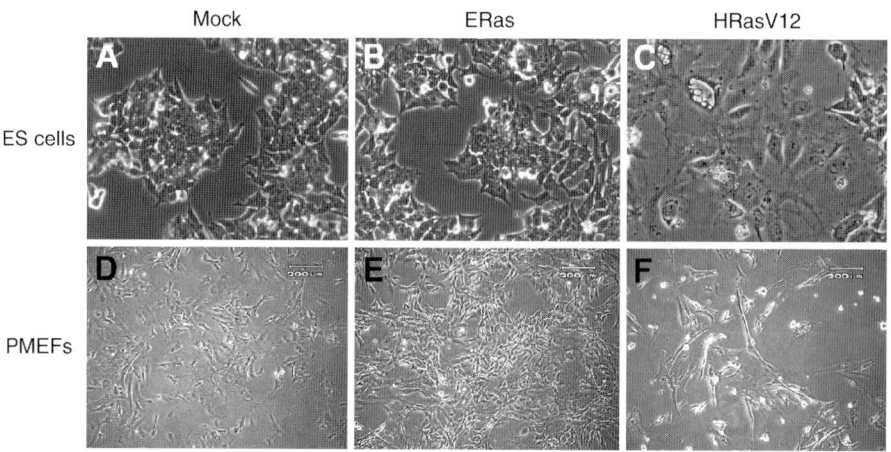

Fig. 1. Morphological changes by embryonic stem (ES) cell-expressed Ras (ERas) or HRas. The morphologies of MG1.19 ES cells and primary mouse embryonic fibroblasts transfected with (**A,D**) mock control vector, (**B,E**) ERas, or (**C,F**) HRasV12, which is a constitutively active mutant of HRas.

origin of polyoma virus (pPyCAG-IP, etc.) are introduced into these cells, they can be replicated without integration into chromosomes *(20)*.

1. The day before transfection, seed MG1.19 cells at 1×10^5 cells per well of gelatin-coated 24-well plates.
2. The next day, introduce pPyCAG-IP encoding the genes of interest into MG1.19 cells with Lipofectamine 2000; dilute 1 µg plasmid DNA and 2 µL Lipofectamine 2000 each in 150 µL DMEM.
3. Combine the diluted DNA/Lipofectamine 2000, mix by vortexing briefly, and incubate for 20 min at room temperature.
4. During incubation, remove the medium and wash cells once with DMEM.
5. After incubation, apply the DNA/Lipofectamine 2000 mixture to ES cells and incubate for 4 h at 37°C with 5% CO_2. The parent plasmid or green fluorescent protein control plasmid should be transfected at the same time as controls.
6. After 4-h incubation, wash the cells once with PBS and treat with 100 µL 0.25% trypsin/ 1 m*M* EDTA for 5 min at room temperature.
7. Add 400 µL of the ES medium, suspend the cells by pipetting, and replate the cells on gelatin-coated 60-mm dishes (*see* **Note 5**).
8. At 24 h after transfection, start selection of transfected cells with 2 µg/mL puromycin.
9. Change medium every day until nontransfected control cells are completely eliminated. After selection, cells can be used as polyclonal cell lines expressing genes of interest for various assays.

3.1.2. Characterization of ES Cells Expressing Genes of Interest

When selection is completed, you should confirm whether the gene of interest induced differentiation and growth retardation.

Genes Involved in ES Cell Tumor-Like Properties

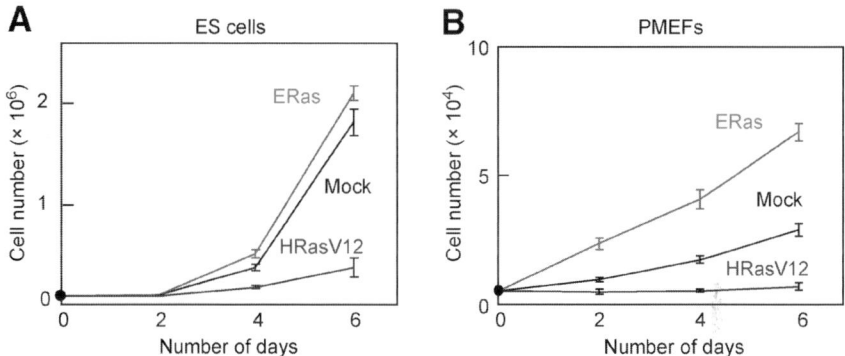

Fig. 2. Effect of embryonic stem (ES) cell-expressed Ras or HRasV12 on proliferation (**A**) on MG1.19 ES cells and (**B**) primary mouse embryonic fibroblasts.

1. Check the morphologies and expression of maker genes such as *Oct3/4*, *Rex1*, *ERas*, or *Nanog* by reverse transcriptase polymerase chain reaction (PCR), Northern blotting, Western blotting, or the like (*see* **Note 6** and **Fig. 1**).
2. To measure the growth rate of these cells, replate cells at 1×10^4 cells per well of gelatin-coated 24-well plates. Count cell number every 2 d using a Coulter counter (*see* **Note 7** and **Fig. 2A**).

3.2. Analyses of Tumor-Like Properties in PMEFs

3.2.1. Forced Expression in PMEFs by Retroviral System

To analyze genes of interest in fibroblasts, we use a retroviral transfection system. This method provides efficient and rapid establishment of stable cell lines. Transfection efficiency greater than 80% is easily obtained.

1. The day before transfection, detach the packaging cell line PLAT-E from a semiconfluent 100-mm dish with 4 mL 0.05% trypsin/0.53 m*M* EDTA.
2. Resuspend in 10 mL of the medium devoid of puromycin and brastcidine S by pipetting, seed 2×10^6 cells per 60-mm dish, and incubate overnight.
3. The next day, introduce pMX-IP retroviral vectors into PLAT-E with Fugene 6 transfection reagent according to the manufacturer's recommendation: dilute 9 μL Fugene 6 transfection reagent in 100 μL DMEM and incubate for 5 min at room temperature.
4. Add 3 μg plasmid DNA (pMX-IP retroviral vector with gene of interest) into the mixture and incubate for 15 min at room temperature.
5. After incubation, add the DNA/Fugene 6 complex drop by drop into PLAT-E plate and incubate overnight at 37°C with 5% CO_2.
6. At 24 h after transfection, replace medium. Prepare PMEFs by seeding at 2×10^5 cells (at subconfluent density) per 60-mm dish (*see* **Note 8**).
7. At 24 h later, filter virus-containing supernatant of PLAT-E dishes through a 0.45-μm cellulose acetate filter. Add 4 μg/mL polybrene and replace medium of PMEFs with the virus- and polybrene-containing supernatant.
8. Incubate the cells for 4 h to overnight.
9. After infection, remove the virus-supernatant from the PMEF plate and add 10 mL fresh medium. Allow the cells to grow and express the introduced gene for a day after infection.
10. At 48 h after infection, select cells with 2 μg/mL puromycin for 4 or 5 d.

3.2.2. Characterization of PMEFs Expressing Genes of Interest

PMEFs generally grow much slower than ES cells and do not contain genetic abnormalities. It should be noted that forced activation of the mitogen-activated protein kinase pathway caused growth arrest because of premature senescence in PMEFs. In contrast, upregulation of the PI3K pathway promotes their proliferation. ES cells resemble PMEFs in that mitogen-activated protein kinase activity enhances differentiation and induces growth retardation, whereas the PI3K pathway facilitates their tumorigenicity and cell growth (**Figs. 1** and **2**). To monitor the effects of the genes of interest on cell proliferation, using PMEFs is much clearer than ES cells.

1. After obtaining cells expressing the genes of interest, replate the cells at 5×10^3 cells per well of 24-well tissue culture plates.
2. Count cell number every 2 d using a Coulter counter (**Fig. 2B**).

3.3. Analyses of Tumor-Like Properties in NIH3T3 Cells

NIH3T3 cells have been historically used for screening of oncogenes. These cells can grow infinitely because of inactivated $p16^{ink4a}$ loci. In undifferentiated ES cells, expression level of p16 is low. Overexpression of p16 in ES cells did not induce growth arrest or changing of cell cycles *(21)*. These data suggest that ES cell proliferation is driven by machinery independent of p16. Therefore, overexpression experiments in NIH3T3 are suitable for screening of genes that promote ES cell growth.

3.3.1. Establishment of NIH3T3 Stably Expressing the Genes

NIH3T3 cells are immortalized but nontransformed cell lines. They are transformed by introduction of active oncogenes, such as HRas. Genes of interest can be effectively introduced into NIH3T3 cells by retroviral vectors as described in **Subheading 3.2.1.** Transformed cells show higher refractivity and spindlelike morphology (**Fig. 3A–C**). They also show a faster growth rate than nontransformed cells. Transformation can be confirmed by the assays described next (*see* **Note 9**).

3.3.1.1. COLONY FORMATION IN SOFT AGAR

1. Autoclave MilliQ water containing 1% agar and then incubate in 42°C water bath.
2. When the 1% agar solution cools, mix with an equal amount of 20% calf serum-containing medium prewarmed to 42°C.
3. Apply the agar-containing medium to 60-mm tissue culture dishes (3 mL per dish). Solidify the mixture (DMEM containing 0.5% agar and 10% calf serum) in the dishes on the bench for 15 min. Keep the remaining mixture in 42°C water bath to avoid solidification.
4. Harvest NIH3T3 cells by trypsinization, suspend with the medium, and count the cell number using a Coulter counter. Dilute the cell suspension with the medium at 7500 cells/mL.
5. Mix the cell suspension with DMEM containing 0.5% agar and 10% calf serum at a 1:2 ratio and seed on the prepared 0.5% agar dishes (3 mL per dish).
6. Incubate the dishes in 37°C CO_2 incubator for 2 or 3 wk and count colony number under a microscope (**Fig. 3D–F**).

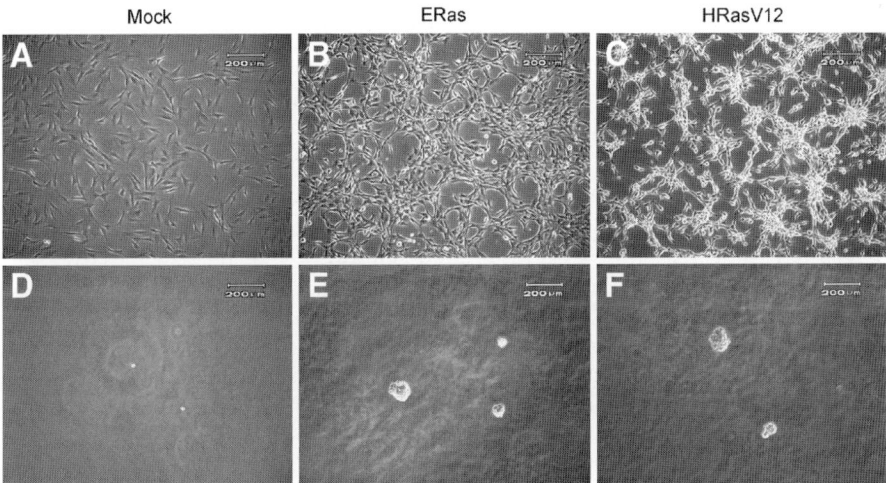

Fig. 3. Transformation of NIH3T3 cells by embryonic stem cell-expressed Ras and HRasV12. Transformation was evaluated with (**A–C**) morphological changes and (**D–F**) anchorage-independent proliferation in soft agar.

3.3.1.2. Transformation Assay

1. Seed the cells on six-well plates at 2×10^5 cells per well and incubate overnight.
2. The next day, transfect cells with expression vectors for genes of interest (e.g., pcDNA3) by lipofection or the calcium phosphate method.
3. Two days later, transfer the cells to 100-mm dishes and maintain for 14 d in DMEM containing 5% calf serum.
4. Stain the cells with 1% crystal violet and count colony numbers.

3.4. Teratoma Formation

Tumorigenicity of ES cells and NIH3T3 cells can be best characterized by subcutaneous injection of cells into nude mice. Genes of interest are overexpressed in either ES cells or NIH3T3 cells or inactivated in ES cells by gene targeting. For the latter purpose, we routinely use the β-geo (β-galactosidase and neomycin-resistant fusion gene) knock-in targeting vector to obtain heterozygous mutant clones *(22)*. Subsequently, a targeting vector containing another selection maker is introduced into heterozygous mutant cells to obtain homozygous mutant cells. Alternatively, antibiotic selection with higher concentration may result in the isolation of the null mutant clones. Clones that remain heterozygous after these treatments should be used as control cells.

1. Trypsinize and resuspend cells at 1×10^7 cells/mL in ES cell medium devoid of antibiotics. Keep the cells on ice until injection.
2. For injection, nude mice are anesthetized with diethylether (*see* **Note 10**). Transfer cell suspension into a 1-mL sterilized syringe, attach a 23-gage needle, and then remove air bubbles. Inject 100 µL cell suspension (1×10^6 cells) subcutaneously to the dorsal flank area of mice.

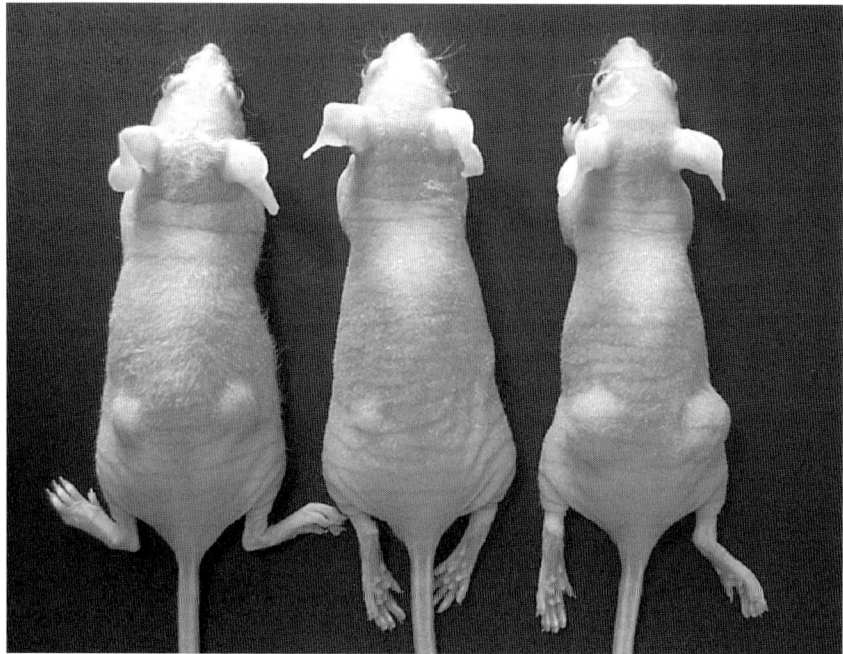

Fig. 4. Tumor formation in nude mice.

3. From 2–8 wk after injection, surgically dissect tumors from mice (*see* **Note 11** and **Fig. 4**). Weigh the tumors and fix them in PBS containing 4% formaldehyde. Store them at 4°C until histological or biochemical analyses.

4. Notes

1. DMEM should contain 4500 mg/L glucose. PBS should not contain calcium or magnesium. To prepare LIF, 293T cells are transfected with pPyCAG-IP encoding human LIF using Fugene 6 transfection reagent as described in **Subheading 3.2.1.** At 24 h after transfection, replace with fresh medium and incubate for 3 d. After incubation, supernatant is collected and filtrated through a 0.22-μm pore size filter. LIF should be stored in small quantity in sterile tubes at −20°C until use.
2. All dishes and plates used for ES cell culture should be gelatinized. Gelatin powder is dissolved in Milli-Q water at 1% w/v and then autoclaved. This solution is stored at 4°C as 10X stock. The 10X solution is melted by microwave, diluted 10-fold with Milli-Q water, and sterilized through a 0.22-μm pore size filter to prepare a 1X solution. Cover the surface of dishes or plates with 1X gelatin solution; incubate at least 30 min at 37°C. Remove the solution and dry before use.
3. MG1.19 cells should be subcultured every 2 or 3 d. When these cells reach subconfluency, wash the cells once with PBS, harvest by trypsinization, then add the culture medium to neutralize and suspend to single cells by pipetting. This suspension is transferred into new gelatin-coated dishes or plates at 1:4 to 1:8 dilution.
4. PLAT-E cells are derived from the 293T cell line and used as packaging cells for retroviral vectors. These cells contain gag-pol-IRES-blas and env-IRES-puro driven by EF1-α

promoter. To obtain high-titer retrovirus, PLAT-E cells should be maintained in medium containing 1 μg/mL puromycin and 10 μg/mL blastcidine S. These cells should be subcultured every 3 d.
5. Normally, transfection efficiency higher than 50% can be achieved in the case of green fluorescent protein expression. To avoid reaching overconfluency, cells should be subcultured to larger dishes after transfection.
6. The PCR program for *Nanog* consists of the initial denaturation at 94°C for 2 min; 25 cycles of 94°C for 30 s, 55°C for 30 s, 72°C for 1 min; and the final extension at 72°C for 5 min. A specific product of approx 350 bp should be obtained. The PCR program for *Oct3/4* consists of the initial denaturation at 94°C for 2 min; 25 cycles of 94°C for 10 s, 62°C for 30 s, and 72°C for 1 min; and the final extension at 72°C for 5 min. The specific product of approx 500 bp should be observed. Primer sequences for reverse transcriptase PCR are as follows: *Nanog* forward 5'-AGG GTC TGC TAC TGA GAT GCT CTG-3' and reverse 5'-CAA CCA CTG GTT TTT CTG CCA CCG-3'; *Oct3/4* forward 5'-CTG AGG GCC AGG CAG GAG CAC GAG-3' and reverse 5'-CTG TAG GGA GGG CTT CGG GCA CTT-3'.
7. Mouse ES cells proliferate so rapidly that further increases in proliferation by genes of interest is not obvious in most cases. Measurements should be repeated as often as possible to obtain reliable and statistically significant results.
8. PMEFs are isolated with standard protocols. It is advisable to make large stocks of passage 1 cells. PMEFs should be used for assays within passage 3 to avoid replicative senescence.
9. NIH3T3 should be cultured at low density because these cells frequently transform spontaneously when left overconfluent.
10. Try not to anesthetize mice too deeply.
11. To obtain reproducible data, make sure to inject cells subcutaneously, not intracutaneously or intraperitoneally.

Acknowledgments

We thank Yoshimi Tokuzawa, Masayoshi Maruyama, Mirei Murakami, and other members in the Yamanaka laboratory for valuable discussion; Yukiko Ikeguchi, Junko Iida, and Masako Shirasaka for technical and administrative support; Dr. Hitoshi Niwa for MG1.19 ES cells and pPyCAG-IP; and Drs. Kenji Kohno and Toshio Kitamura for PLAT-E cells and pMX retroviral vector.

References

1. Evans, M. J. and Kaufman, M. H. (1981) Establishment in culture of pluripotential cells from mouse embryos. *Nature* **292,** 154–156.
2. Martin, G. R. (1981) Isolation of a pluripotent cell line from early mouse embryos cultured in medium conditioned by teratocarcinoma. *Proc. Natl. Acad. Sci. USA* **78,** 7634–7638.
3. Freed, C. R. (2002) Will embryonic stem cells be a useful source of dopamine neurons for transplant into patients with Parkinson's disease? *Proc. Natl. Acad. Sci. USA* **99,** 1755–1757.
4. Nichols, J., Zevnik, B., Anastassiadis, K., et al. (1998) Formation of pluripotent stem cells in the mammalian embryo depends on the POU transcription factor Oct4. *Cell* **95,** 379–391.
5. Niwa, H., Miyazaki, J., and Smith, A. G. (2000) Quantitative expression of Oct-3/4 defines differentiation, dedifferentiation or self-renewal of ES cells. *Nat. Genet.* **24,** 372–376.
6. Chambers, I., Colby, D., Robertson, M., et al. (2003) Functional expression cloning of Nanog, a pluripotency sustaining factor in embryonic stem cells. *Cell* **113,** 643–655.

7. Mitsui, K., Tokuzawa, Y., Itoh, H., et al. (2003) The homeoprotein Nanog is required for maintenance of pluripotency in mouse epiblast and ES cells. *Cell* **113,** 631–642.
8. Smith, A. G., Heath, J. K., Donaldson, D. D., et al. (1988) Inhibition of pluripotential embryonic stem cell differentiation by purified polypeptides. *Nature* **336,** 688–690.
9. Williams, R. L., Hilton, D. J., Pease, S., et al. (1988) Myeloid leukaemia inhibitory factor maintains the developmental potential of embryonic stem cells. *Nature* **336,** 684–687.
10. Ying, Q. L., Nichols, J., Chambers, I., and Smith, A. (2003) BMP induction of Id proteins suppresses differentiation and sustains embryonic stem cell self-renewal in collaboration with STAT3. *Cell* **115,** 281–292.
11. Sato, N., Meijer, L., Skaltsounis, L., Greengard, P., and Brivanlou, A. H. (2004) Maintenance of pluripotency in human and mouse embryonic stem cells through activation of Wnt signaling by a pharmacological GSK-3-specific inhibitor. *Nat. Med.* **10,** 55–63.
12. Di Cristofano, A., Pesce, B., Cordon-Cardo, C., and Pandolfi, P. P. (1998) Pten is essential for embryonic development and tumour suppression. *Nat. Genet.* **19,** 348–355.
13. Sun, H., Lesche, R., Li, D. M., et al. (1999) PTEN modulates cell cycle progression and cell survival by regulating phosphatidylinositol 3,4,5,-trisphosphate and Akt/protein kinase B signaling pathway. *Proc. Natl. Acad. Sci. USA* **96,** 6199–6204.
14. Stiles, B., Gilman, V., Khanzenzon, N., et al. (2002) Essential role of AKT-1/protein kinase B alpha in PTEN-controlled tumorigenesis. *Mol. Cell Biol.* **22,** 3842–3851.
15. Takahashi, K., Mitsui, K., and Yamanaka, S. (2003) Role of ERas in promoting tumour-like properties in mouse embryonic stem cells. *Nature* **423,** 541–545.
16. Jirmanova, L., Afanassieff, M., Gobert-Gosse, S., Markossian, S., and Savatier, P. (2002) Differential contributions of ERK and PI3-kinase to the regulation of cyclin D1 expression and to the control of the G1/S transition in mouse embryonic stem cells. *Oncogene* **21,** 5515–5528.
17. Tokuzawa, Y., Kaiho, E., Maruyama, M., et al. (2003) Fbx15 is a novel target of Oct3/4 but is dispensable for embryonic stem cell self-renewal and mouse development. *Mol. Cell Biol.* **23,** 2699–2708.
18. Gassmann, M., Donoho, G., and Berg, P. (1995) Maintenance of an extrachromosomal plasmid vector in mouse embryonic stem cells. *Proc. Natl. Acad. Sci. USA* **92,** 1292–1296.
19. Morita, S., Kojima, T., and Kitamura, T. (2000) Plat-E: an efficient and stable system for transient packaging of retroviruses. *Gene Ther.* **7,** 1063–1066.
20. Niwa, H., Masui, S., Chambers, I., Smith, A. G., and Miyazaki, J. (2002) Phenotypic complementation establishes requirements for specific POU domain and generic transactivation function of Oct3/4 in embryonic stem cells. *Mol. Cell Biol.* **22,** 1526–1536.
21. Savatier, P., Lapillonne, H., van Grunsven, L. A., Rudkin, B. B., and Samarut, J. (1996) Withdrawal of differentiation inhibitory activity/leukemia inhibitory factor up-regulates D-type cyclins and cyclin-dependent kinase inhibitors in mouse embryonic stem cells. *Oncogene* **12,** 309–322.
22. Mountford, P., Zevnik, B., Duwel, A., et al. (1994) Dicistronic targeting constructs: reporters and modifiers of mammalian gene expression. *Proc. Natl. Acad. Sci. USA* **91,** 4303–4307.

31

In Vivo Tumor Formation From Primate Embryonic Stem Cells

Takayuki Asano, Kyoko Sasaki, Yoshihiro Kitano, Keiji Terao, and Yutaka Hanazono

Summary

To achieve human embryonic stem (ES) cell-based transplantation therapies, allogeneic transplantation models of nonhuman primates would be particularly useful. In this chapter, we describe an example of this model. We prepared cynomolgus ES cells genetically marked with the green fluorescent protein. The cells were transplanted into the allogeneic fetus because the fetus is immunologically premature and does not induce immune responses to transplanted cells. In addition, fetal tissue compartments are rapidly expanding, presumably providing space for engraftment. At 3 mo posttransplantation, a fluorescent teratoma, obviously derived from transplanted ES cells, was found in the fetus. However, transplanted cell progeny were also detected (approx 1%) in multiple fetal tissues. The cells were solitary and indistinguishable from surrounding host cells as assessed by *in situ* polymerase chain reaction. Transplanted cynomolgus ES cells can engraft in allogeneic fetuses. The cells will, however, form a tumor if they "leak" into an improper space, such as the thoracic cavity.

Key Words: Allogeneic transplantation; genetic marking; green fluorescent protein; immunological tolerance; *in situ* PCR; *in utero* transplantation; primate embryonic stem cells; teratoma.

1. Introduction

Because human embryonic stem (ES) cell lines have dual abilities to proliferate indefinitely and differentiate into multiple tissue types *(1,2)*, human ES cell-based transplantation therapies are considered to hold a great potential in the treatment of a variety of diseases and injuries. To address the safety and efficacy of these therapies, allogeneic transplantation models of large animals, especially nonhuman primates, would be useful. However, it has been difficult to transplant primate ES cells or their derivatives into allogeneic hosts. There are two major reasons for this. First, the efficient and stable marking of primate ES cells has been difficult. It is necessary to distinguish transplanted allogeneic ES cell progeny from surrounding host cells. Second, the immune rejection of transplanted cells must be circumvented for a sustained engraftment. The cells would otherwise be cleared by immune responses.

We have previously reported highly efficient gene transfer into cynomolgus ES cells using a lentivirus vector derived from the simian immunodeficiency virus *(3)*. Lentiviral transgene expression in ES cells is stable, with minimal levels of transcriptional silencing *(4,5)*. In addition, cynomolgus ES cell sublines stably expressing green fluorescent protein (GFP) were established after electroporation of a GFP-expressing plasmid *(6)*. By using such cynomolgus ES cells genetically modified to express GFP, it is now possible to distinguish transplanted allogeneic ES cell progeny from surrounding host cells as GFP will serve as a good genetic tag.

The early gestational fetus is a good recipient with which to circumvent immune rejection because the immune system is premature *(7,8)*. Furthermore, in the animal fetus, "space" would be relatively available for engraftment as compared to the adult because of the rapid expansion of fetal tissue compartments. Thus, transplanted cells could engraft without conditioning of recipients, such as by irradiation or immunosuppressive treatment.

In this chapter, we show a method to transplant nonhuman primate (cynomolgus macaque) ES cells *(9)* into xenogeneic immunodeficient mice to form teratoma. In addition, we show methods to transplant nonhuman primate (cynomolgus macaque) ES cells stably expressing GFP *(3,6)* into the allogeneic fetus *in utero* and to examine the in vivo fate of transplanted cells using GFP as a genetic tag. At 3 mo after the allogeneic *in utero* transplantation, a fluorescent tumor, obviously derived from transplanted ES cells, was found in the thoracic or abdominal cavity. Notably, transplanted cell progeny were also detected (approx 1%) in multiple fetal tissues. The cells were solitary and indistinguishable from surrounding host cells as assessed by *in situ* polymerase chain reaction (PCR). Thus, transplanted cynomolgus ES cells can engraft in allogeneic fetuses. However, the cells will form a tumor if they "leak" into an improper space, such as the thoracic and abdominal cavities *(10)*.

2. Materials
2.1. Cells
1. Cynomolgus ES cells stably expressing GFP (*see* Chapters 20 and 21, this volume).
2. Mouse embryonic fibroblasts from CD-1 (also referred to as ICR) (Charles River, Wilmington, MA) or BALB/c mice (Charles River).

2.2. Teratoma Formation in Immunodeficient Mice
1. 6- to 8-wk-old non-obese diabetic/severe combined immunodeficient (NOD/SCID) mice (Jackson Laboratory, Bar Harbor, ME) (*see* **Note 1**).
2. Hanks' balanced salt solution (HBSS; Invitrogen, Carlsbad, CA; cat. no.14025-092).
3. Dulbecco's modified Eagle's medium/nutrient mixture F-12 1:1 mixture (DMEM/F12) (Invitrogen, cat. no. 11330-032).
4. ES cell-qualified fetal bovine serum (Invitrogen, cat. no. 10439-024).
5. 10,000 IU/mL penicillin-10,000 µg/mL streptomycin (100X; Invitrogen, cat. no. 15070-063).
6. 200 mM L-glutamine (100X; Invitrogen, cat. no. 25030-081).
7. 2-Mercaptoethanol (Sigma, St. Louis, MO; cat. no. M3148).

Tumor Formation From Primate ES Cells

8. Culture medium for primate ES cells: DMEM/F12 containing 15% ES cell-qualified fetal bovine serum, 2 m*M* L-glutamine, 100 IU/mL penicillin-100 μg/mL streptomycin, and 0.1 m*M* 2-mercaptoethanol.
9. 0.25% trypsin in HBSS (2.5% trypsin 10X liquid; Invitrogen, cat. no. 15090-046).
10. 1% bovine serum albumin (BSA fraction V; Sigma, cat. no. A4503) in HBSS.

2.3. Teratoma Formation in Allogeneic Fetuses

1. Anesthetic and surgical facilities for primates (including ultrasound and inhalation anesthesia equipment) *(11)*.
2. A time-dated pregnant cynomolgus monkey of 50- to 70-d gestation (*see* **Note 2**) *(12)*.
3. Ketamine hydrochloride (Ketalar® 50; Sankyo, Tokyo, Japan).
4. Isoflurane (Forane®; Dainippon Pharmaceutical, Osaka, Japan).
5. A percutaneous transhepatic cholangiography (PTC) needle (22-gage, Sonoguide PTC needle type B; Hakko Medical, Nagano, Japan; cat. no. 22412210).
6. A 1-mL syringe (Terumo, Tokyo, Japan; cat. no. SS-01T) filled with graft cells (10^5–10^7 cells in 200–500 μL).
7. A 1-mL syringe (Terumo, cat. no. SS-01T) filled with normal saline (for flushing).

2.4. Sample Preparation

1. 4% paraformaldehyde (Wako, Osaka, Japan; cat. no. 169-18432) and 8% sucrose (Wako, cat. no. 192-00012) in phosphate-buffered saline (PBS; Invitrogen, cat. no. 10010-023).
2. OCT compound (Tissue Tek series; Sakura, Zoeterwoude, Netherlands; cat. no. 4583) containing 10% sucrose.

2.5. In Situ PCR

1. A PTC100 Peltier thermal cycler (MJ Research, Waltham, MA).
2. 20 μg/mL proteinase K (Sigma, cat. no. 39450-01-6) in PBS.
3. 0.1% Triton X-100 (Sigma, cat. no. T8787) in PBS.
4. A slide frame for *in situ* PCR (slide seal; Takara, Shiga, Japan; cat. no. 9066 [25 μL] or cat. no. 9067 [65 μL]).
5. 50 μL Digoxigenin dNTP labeling mix (Roche, Basel, Switzerland; cat. no. 1277065).
6. Rabbit anti-Digoxigenin polyclonal antibody, horseradish peroxidase labeled (Dako, Glostrup, Denmark; cat. no. P5104) diluted (1:100) in 2% BSA and 5% horse serum (Invitrogen, cat. no. 16050-130) in PBS.
7. A Vector SG substrate kit (Vector, Burlingame, CA; cat. no. SK-4700).
8. Kernechtrot solution (0.1% Kernechtrot in aluminum sulfate; Muto, Tokyo, Japan; cat. no. 4087).

3. Methods

3.1. Teratoma Formation in Immunodeficient Mice

1. Wash ES cells with HBSS twice and add 0.25% trypsin to the dish at 37°C for 3 min. Neutralize trypsin with ES culture medium and make a suspension of ES cell clumps.
2. Transfer the cell suspension into a 50-mL conical tube, centrifuge it at 140*g* for 4 min, and resuspend the pellet with 20 mL 1% BSA/HBSS.
3. Centrifuge the cell suspension again at 140*g* for 4 min and resuspend the pellet with an appropriate volume of 1% BSA/HBSS (10^6 cells in 150–200 μL per injection site).
4. Aspirate the ES cell suspension into a 1-mL syringe with a 23-gage needle and inject the suspension into NOD/SCID mice subcutaneously (*see* **Note 3**).

5. Resulting tumors will be palpable at 8–13 wk after the injection. Expose, observe, and excise tumors.
6. Fix tumor samples (5 × 5 × 3 mm) at 4°C for 4 h in 4% paraformaldehyde and 8% sucrose in PBS and embed the samples in paraffin for histological examination. To prepare fresh frozen samples, embed samples (5 × 5 × 3 mm) in OCT compound containing 10% sucrose, freeze them in liquid nitrogen, and store them at −80°C.

3.2. Teratoma Formation in Allogeneic Fetuses

3.2.1. Anesthesia

1. Prepare a pregnant monkey around the end of first trimester (50–70 d; full term 165 d) (*see* **Note 2**).
2. Give the monkey 10 mg/kg ketamine hydrochloride intramuscularly. Secure the monkey on a table and monitor maternal heart rate by electrocardiography (*see* **Note 4**).
3. Induce and maintain anesthesia by inhalation of isoflurane (1.5–2%) mixed with 100% oxygen via a mask.

3.2.2. In Utero Transplantation

1. Shave whole abdomen and sterilize the surface with iodine solution (from **Subheading 3.2.1., step 3**).
2. Determine fetal position by transabdominal ultrasound with a 7.5-MHz convex probe (*see* **Note 5**).
3. Let an assistant secure the other side of the uterus while an operator holds the transducer parallel to the intended course of the needle.
4. Select an optimal entry site into the uterine cavity, avoiding the placental tissue.
5. Insert a 23-gage PTC needle through the maternal skin and uterine wall into the amniotic cavity and then into the desired site (e.g., peritoneal cavity, brain, or liver) under continuous ultrasound guidance (*see* **Note 6** and **Fig. 1**). A small push of an injector can visualize a tip of the needle on echocardiography.
6. Let an assistant gently inject the cells (200–500 µL) and flush the needle with 100 µL normal saline. The operator should focus on keeping the tip of the needle in an appropriate position.
7. Confirm adequate heart beats after the procedure (*see* **Note 7**).

3.2.3. Caesarian Section

1. Prepare the pregnant monkey after transplantation as described in **Subheading 3.2.1**. *In utero* transplantation is usually done around the end of the first trimester (50–60 d) (*see* **Subheading 3.2.2.**). The full term is 165 d; therefore, the *in utero* incubation time of transplanted ES cells is about 3 mo.
2. Expose the gravid uterus through a midline incision and deliver the fetus through a low transverse hysterotomy (*see* **Note 8**).
3. Clamp and divide the cord. Remove the placenta and cord. Close the uterus and abdomen with absorbable sutures.
4. Insert a small catheter (24-gage intravenous catheter) into the umbilical vein and irrigate the newborn with normal saline to completely wash out fetal blood for mercy killing. Open the chest and abdomen, observe the whole body, and excise tumors (*see* **Fig. 2A–C**). Collect tissues.
5. Fix tissue samples (5 × 5 × 3 mm) at 4°C for 4 h in 4% paraformaldehyde and 8% sucrose in PBS and embed the samples in paraffin for histological examination. To prepare fresh

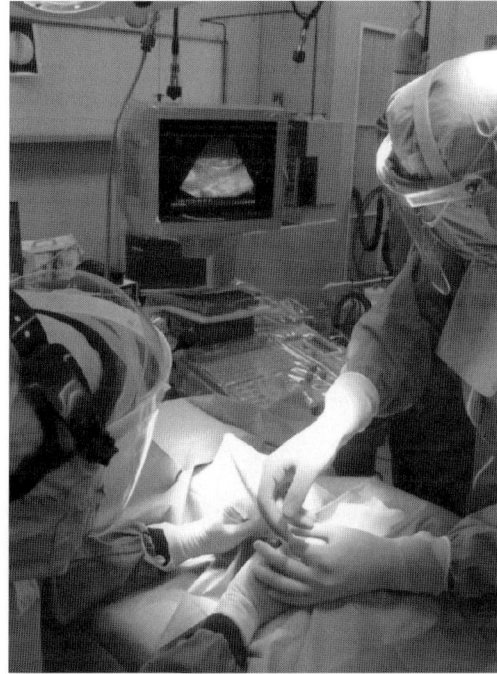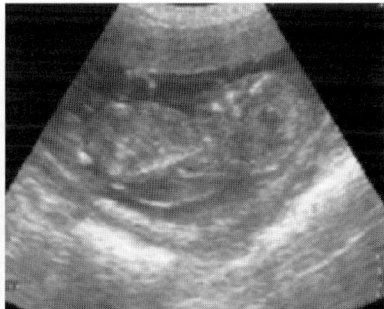

Fig. 1. *In utero* transplantation. Pregnant monkeys were anesthetized by intramuscular administration of ketamine hydrochloride (Ketalar). Cynomolgus ES cells genetically modified to express GFP (10^6 cells/fetus) were injected into the fetal abdominal cavity or liver through a 23-gage needle using an ultrasound-guided technique around the end of the first trimester (**left**). The full term is 165 d. The weight of the fetus at the time of transplantation was estimated at 20 g, which is equivalent to that of an adult mouse (**right**).

frozen samples, embed samples ($5 \times 5 \times 3$ mm) in OCT compound containing 10% sucrose, freeze them in liquid nitrogen, and store them at $-80°C$.

3.3. In Situ *Detection of Transplanted Cell Progeny*

You may examine tissue sections for in vivo fate of transplanted cell progeny by *in situ* PCR, which amplifies marker (GFP) sequences *(10,13)*. It is especially useful when it is difficult to identify cells by staining specific surface markers, when GFP fluorescence is hampered by the high autofluorescence of tissue samples, or when the transgene expression is shut down ("silenced") in vivo.

3.3.1. Cell Wall Permeabilization

1. (Optional) If a tissue section is embedded in paraffin, then dewax it by dipping the slide in xylene three times, each for 10 min, and then in 100% ethanol three times, each for 10 min. Air-dry the slide.
2. Soak the slide in 20 µg/mL proteinase K/PBS and incubate it at 37°C for 10 min (*see* **Note 9**).

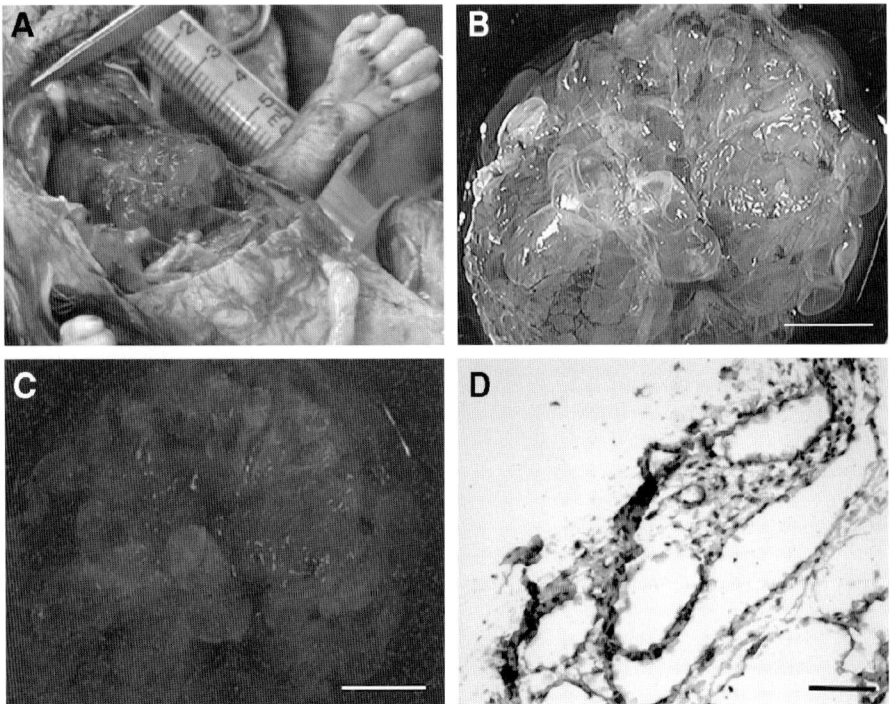

Fig. 2. Teratoma formation in a cynomolgus fetus after transplantation of allogeneic ES cells. (**A**) A tumor (4 × 3 × 2.5 cm) was detected in the thoracic cavity 3 mo after allogeneic transplantation of ES cells expressing GFP. The tumor was observed in (**B**) a bright field and (**C**) a dark field (**C**) under a fluorescence stereomicroscope. GFP was expressed in the tumor, clearly indicating that the tumor was derived from transplanted ES cells. (**D**) The GFP gene was also detected in the tumor cells by *in situ* PCR (stained black). Bar in **B** and **C** = 1 cm. Bar in **D** = 50 µm.

3. Soak the slide in 0.1% Triton X-100/PBS for 5 min and wash it with PBS twice, each for 5 min.
4. Soak the slide in 95% ethanol for 10 min and then in 100% ethanol twice, each for 10 min, to remove proteins and air-dry it.

3.3.2. In Situ PCR

1. Attach a slide frame to the slide (from **Subheading 3.3.1.**, **step 4**) and incubate it at 95°C for 5 min.
2. Apply a master mix of *in situ* PCR to the slide at room temperature (*see* **Note 10** and **Table 1**).
3. Cover the slide with a film (*see* **Note 11**).
4. Place the slide upside down in a PTC100 Peltier thermal cycler and start a cycling program (i.e., 94°C for 1 min and 55°C for 2 min with 15 cycles; *see* **Note 12**).

3.3.3. Detection

1. Remove the slide frame gently after PCR (from **Subheading 3.3.2.**, **step 4**) and soak the slide in two changes of PBS for 5 min each.
2 Dropwise add an horseradish peroxidase-labeled anti-Digoxigenin solution (diluted 1:100 with 2% BSA and 5% horse serum in PBS) onto the slide and incubate it at 37°C for 2 h.

Table 1
In Situ PCR Reaction Mixture

Master mix	Volume per reaction	Final concentration
10X PCR buffer (Mg^{2+} free)	2.5 µL	1X
25 mM $MgCl_2$	4.5 µL	4.5 mM
dNTPs mixture (2.5 mM each)	3 µL	420 µM dATP
Digoxigenin (DIG) DNA labeling mix (Roche)	3 µL	420 µM dCTP
		420 µM dGTP
		378 µM dTTP
		42 µM DIG-dUTP
Forward primer (10 µM)	2 µL	0.8 µM
Reverse primer (10 µM)	2 µL	0.8 µM
Takara Taq polymerase (5 U/µL)	0.8 µL	0.16 U/µL
Water	7.2 µL	NA
Total volume of master mix	25 µL	NA

NA, not applicable.
Primer set for the GFP sequence: 5'-CGT CCA GGA GCG CAC CAT CTT C-3' and 5'-GGT CTT TGC TCA GGG CGG ACT-3'.

3. Soak the slide in two changes of PBS for 5 min each.
4. Dropwise add a Vector SG solution onto the slide, stain it for 3–10 min, and gently wash it with water for 10 min.
5. Dropwise add a Kernechtrot solution and incubate the slide at room temperature for 1–2 min to stain nucleotides and gently wash it with water for 10 min.
6. Mount the slide with glycerol and observe it under a light microscope (see **Note 13** and **Fig. 2D**).

4. Notes

1. Although we use NOD/SCID mice (lack of B and T lymphocytes but presence of natural killer cells), SCID mice are usually used in many other laboratories to form teratomas from ES cells. NOD/SCID mice are more highly immunodeficient than SCID mice; thus, NOD/SCID mice may be better in the setting of xenotransplantation.
2. Cynomolgus or rhesus monkeys are the most appropriate to work with because of ES cell availability and their size. In the monkey fetus, "the window of opportunity" for successful tolerance induction may be earlier and narrower than thought *(14)*. To avoid immune responses, transplantation at earlier days (around 40–50 d) may be better.
3. It is not necessary to disperse ES cell clumps to single cells when transplanting ES cells into mice (or other animals). We transplanted about 1×10^6 ES cells (corresponding to two confluent 60-mm dishes) per site in mice. There is, however, considerable variation among reports: from 10–15 clumps (200 cells) per site *(2)* to 5×10^6 cells per site *(15)*. ES cells are usually transplanted subcutaneously into the hind leg muscle, testis capsule, or abdominal cavity. In our experiments, a teratoma was formed in any site. It is recommended to choose injection sites you can observe easily from the outside and from which you can easily excise tumors.
4. For ultrasound-guided transplantation operations, endotracheal intubation is not necessary.
5. We prefer a small convex transducer rather than a big linear transducer because of the small size of the monkey fetus. Although a needle adapter is available, we prefer the freehand technique.

6. You may puncture transplacentally when the placenta is located anteriorly. Bleeding from the placenta usually stops spontaneously. However, we recommend every effort to avoid this approach by manipulation.
7. The survival rate with this *in utero* transplantation technique is currently 100%, excluding those fetuses that died from massive teratoma formation.
8. Uterine atony requiring oxytocin administration is quite rare in primates.
9. The treatment with proteinase K may need longer time depending on samples.
10. The amount of master mix per slide is 25 µL for Takara cat. no. 9066 and 65 µL for cat. no. 9067.
11. Slides are attached to the Takara slide seal kit. Be careful not to trap air under films.
12. The PCR conditions should be optimized for each *in situ* PCR.
13. The results should be observed within the same day. On the following day, the tissue would peel off, making examination difficult.

Acknowledgments

We thank Dr. Naohide Ageyama for help in monkey operations. We conducted all monkey operations in the Tsukuba Primate Research Center, National Institute of Biomedical Innovation (Ibaraki, Japan).

References

1. Thomson, J. A., Itskovitz-Eldor, J., Shapiro, S. S., et al. (1998) Embryonic stem cell lines derived from human blastocysts. *Science* **282**, 1145–1147.
2. Reubinoff, B. E., Pera, M. F., Fong, C. Y., Trounson, A., and Bongso, A. (2000) Embryonic stem cell lines from human blastocysts: somatic differentiation in vitro. *Nat. Biotechol.* **18**, 399–404.
3. Asano, T., Hanazono, Y., Ueda, Y., et al. (2002) Highly efficient gene transfer into primate embryonic stem cells with a simian lentivirus vector. *Mol. Ther.* **6**, 162–168.
4. Lois, C., Hong, E. J., Pease, S., Brown, E. J., and Baltimore, D. (2002) Germline transmission and tissue-specific expression of transgenes delivered by lentiviral vectors. *Science* **295**, 868–872.
5. Pfeifer, A., Ikawa, M., Dayn, Y., and Verma, I. M. (2002) Transgenesis by lentiviral vectors: lack of gene silencing in mammalian embryonic stem cells and preimplantation embryos. *Proc. Natl. Acad. Sci. USA* **99**, 2140–2145.
6. Takada, T., Suzuki, Y., Kondo, Y., et al. (2002) Monkey embryonic stem cell lines expressing green fluorescent protein. *Cell Transplant.* **11**, 631–635.
7. Harrison, M. R., Slotnick, R. N., Crombleholme, T. M., Golbus, M. S., Tarantal, A. F., and Zanjani, E. D. (1989) *In-utero* transplantation of fetal liver haemopoietic stem cells in monkeys. *Lancet* **2**, 1425–1427.
8. Tarantal, A. F., Goldstein, O., Barley, F., and Cowan, M. J. (2000) Transplantation of human peripheral blood stem cells into fetal rhesus monkeys (*Macaca mulatta*). *Transplantation* **69**, 1818–1823.
9. Suemori, H., Tada, T., Torii, R., et al. (2001) Establishment of embryonic stem cell lines from cynomolgus monkey blastocysts produced by IVF or ICSI. *Dev. Dyn.* **222**, 273–279.
10. Asano, T., Ageyama, N., Takeuchi, K., et al. (2003) Engraftment and tumor formation after allogeneic *in utero* transplantation of primate embryonic stem cells. *Transplantation* **76**, 1061–1067.

11. Honjo, S. (1985) The Japanese Tsukuba Primate Center for Medical Science (TPC): an outline. *J. Med. Primatol.* **14,** 75–89.
12. Honjo, S., Cho, F., and Terao, K. (1984) Establishing the cynomolgus monkey as a laboratory animal. *Adv. Vet. Sci. Comp. Med.* **28,** 51–80.
13. Haase, A. T., Retzel, E. F., and Staskus, K. A. (1990) Amplification and detection of lentiviral DNA inside cells. *Proc. Natl. Acad. Sci. USA* **87,** 4971–4975.
14. Lindton, B., Markling, L., Ringden, O., Kjaeldgaard, A., Gustafson, O., and Westgren, M. (2000) Mixed lymphocyte culture of human fetal liver cells. *Fetal Diagn. Ther.* **15,** 71–78.
15. Xu, C., Inokuma, M. S., Denham, J., et al. (2001) Feeder-free growth of undifferentiated human embryonic stem cells. *Nat. Biotechnol.* **19,** 971–974.

VII

ANIMAL MODELS AND THERAPY

32

Directed Differentiation and Characterization of Genetically Modified Embryonic Stem Cells for Therapy

Adeline A. Lau, Kim M. Hemsley, Adrian Meedeniya, Aaron J. Robinson, and John J. Hopwood

Summary

Lysosomal storage disorders are rare, inherited diseases caused by a deficiency of a specific, lysosomal enzyme. In the case of mucopolysaccharidosis type IIIA, a lack of active sulfamidase enzyme results in heparan sulfate accumulation, severe and progressive neurological deficits, and usually premature death. Embryonic stem cells can be genetically modified to overexpress lysosomal enzymes, providing a renewable reservoir of cells that can be readily expanded in culture. Screening clonal lines of embryonic stem cells for desirable properties such as high levels and maintenance of enzyme activity throughout terminal differentiation to neural phenotypes theoretically provides a reproducible population of cells that can be fully characterized in vitro before implantation within the central nervous system in animal models of lysosomal storage disorders.

Key Words: Clonal; differentiation; embryonic stem cells; in vitro; lysosomal enzyme; lysosomal storage; therapy; transgenic.

1. Introduction

Mucopolysaccharidosis type IIIA (MPS IIIA) is an autosomal-recessive lysosomal storage disorder (LSD) resulting from a deficiency of functional lysosomal sulfamidase enzyme. Patients are usually diagnosed between 2 and 6 yr of age and present with profound and progressive mental deterioration, usually resulting in premature death in the second decade of life. Enzyme replacement therapies have proven effective in other LSDs, such as Gaucher, Fabry, and MPS I *(1–4)*, for which regular intravenous infusion of the deficient enzyme is capable of treating somatic pathology.

However, this strategy is ineffective for the two-thirds of LSDs with neurological symptoms because of the blood-brain barrier. Proof-of-principle studies in a murine model of MPS IIIA have established that if recombinant human sulfamidase (rhNS) can be delivered to the central nervous system (CNS) of affected mice, behavioral learning deficits can be reduced, and histological correction can occur *(5)*. We propose that embryonic stem (ES) cell-derived neural cells, genetically modified to overexpress rhNS, may be useful as a vector for delivering a continuous supply of recombinant enzyme to the CNS of affected individuals.

ES cells are derived from the inner cell mass of the blastocyst of preimplantation embryos *(6)*. The pluripotent nature of ES cells (and their amenability to expansion in culture) theoretically provides an unlimited reservoir for cell-based therapies. ES cells are also able to undergo spontaneous or directed differentiation into progressively more developmentally restricted lineages, including all CNS cell phenotypes, namely, neurons, astrocytes, and oligodendrocytes *(7–10)*. Sustained expression of the marker gene enhanced green fluorescent protein has been observed for more than 4 wk postimplantation when murine ES cells, predifferentiated in vitro to form neural precursor cells, were implanted in the striatum of adult rats *(11)*. ES cell-derived dopaminergic neurons have also been shown to correct motor deficits in a parkinsonian mouse model, with well-integrated grafts seen at 2 mo postimplantation *(12)*. Driving the differentiation of ES cells into a specific neuronal phenotype by overexpression of the transcription factor Nurr1 by stable transfection was demonstrated *(13)*. These cells showed efficacy in an animal model of disease.

The ready expandability and pluripotency of ES cells make them the ideal system for the creation and study of genetically modified cells capable of expressing foreign transgenes while undergoing differentiation into clinically relevant cell types. Extensive in vitro characterization of clonal lines can then be undertaken before implantation and cryopreserved stocks may provide a renewable and reproducible source of cells for future transplants.

Murine ES cells have been genetically modified to express rhNS *(14)* and have been directed to terminally differentiated neural phenotypes using the Med II-conditioned media system *(15)*. The availability of naturally occurring animal models of LSD, such as the MPS IIIA mouse and dog *(16,17)*, will provide the opportunity for analyzing the potential of these cells as implanted, therapeutic grafts in vivo and the resultant effect on cellular pathology and behavior.

2. Materials
2.1. Cell Culture
2.1.1. Cell Lines

1. Human hepatocellular carcinoma cell line, HepG2 (American Type Culture Collection, Manassas, VA; cat. no. HB-8065) *(18)*.
2. D3 mouse ES cells (American Type Culture Collection, cat. no. CRL-1934) *(19)*.

2.1.2. Reagents, Plasticware, and Equipment

1. Dulbecco's modified Eagle's media (DMEM) without HEPES (Invitrogen, Mt. Waverley, Australia; cat. no. 12800).
2. F12 media (Invitrogen, cat. no. 11765).
3. 10X phosphate-buffered saline (PBS): for 1 L 10X solution, dissolve 80 g NaCl, 2 g KCl, 11.5 g Na_2HPO_4, 2 g KH_2PO_4 in 1 L MilliQ water. Adjust to pH 7.4 and autoclave. Dilute to 1X stock before use.

4. Tissue culture ware, including 10-cm plates, 15-cm plates, 6-well trays, 24-well trays, and 96-well trays.
5. Bacterial Petri dishes.
6. Fetal calf serum (FCS; JRH Biosciences, Brooklyn, Australia; cat. no. 12103-500M). Heat inactivate by incubating at 56–60°C for 40 min and then store at −20°C (*see* **Note 1**).
7. Trypsin (JRH Biosciences, cat. no. 59430-100M). Dilute 1:20 with sterile PBS. Store at −20°C.
8. Penicillin-streptomycin (Invitrogen, cat. no. 15140-122): use at a final concentration of 100 U penicillin and 100 µg streptomycin.
9. L-Glutamine (Invitrogen, cat. no. 25030156). Use at a final concentration of 2 mM.
10. Trypan blue (Sigma, Castle Hill, Australia; cat. no. T8154).
11. Leukemia inhibitory factor (LIF): LIF is produced from a COS-1 cell line transfected with an expression vector that encodes the mouse LIF gene as described in **refs.** *15* and *20*. Stock concentration is equivalent to 10^6 U/mL.
12. Insulin-transferrin-sodium selenite supplement (Boehringer Mannheim, Mannheim, Germany; cat. no. 1074547).
13. Human basic fibroblast growth factor 2 (Invitrogen, cat. no. 13256). Store at −20°C in aliquots and use at a final concentration of 10 ng/mL.
14. β-Mercaptoethanol (Sigma, cat. no. M7522). For 100 mM stock, mix 100 µL neat β-mercaptoethanol and 14.1 mL 1X PBS and sterilize through 0.22-µm filter. Stable at 4°C for 1 wk.
15. Hemocytometer.

2.1.3. Media

1. ES cell medium: 10 mL FCS, 0.1 mL β-mercaptoethanol, 100 µL LIF, 1 mL penicillin-streptomycin, and 1 mL L-glutamine made up to 100 mL in DMEM.
2. 50% Med II medium: 50 mL Med II, 5 mL FCS, 0.1 mL 100 mM β-mercaptoethanol, and 1 mL L-glutamine made up to 100 mL with DMEM.
3. Serum-free medium: 50% F12 media, 50% DMEM, 1 mL insulin-transferrin-sodium selenite, 10 ng/mL fibroblast growth factor 2, and 1 mL L-glutamine.
4. Neutralizing medium/HepG2 medium: DMEM with 10% FCS and 1 mL L-glutamine.

2.1.4. Freezing Apparatus

1. Nalgene™ Cryo 1°C freezing container (Nalge Nunc, Rochester, NY; cat. no. 5100-0001).
2. Isopropanol.
3. Cryotubes.
4. Freezing mix: 10% (v/v) dimethyl sulfoxide (DMSO) and 90% (v/v) FCS cooled to 4°C before use.

2.2. Production of Transgenic Clonal Cell Lines

1. Plasmid midikit (Qiagen, Doncaster, Australia; cat. no. 12143).
2. *Pvu*I or *Sca*I restriction endonucleases.
3. 70% (v/v) and 100% ethanol (EtOH).
4. 3 M sodium acetate (NaAc) at pH 5.2.
5. Spectrophotometer.
6. Agarose and gel electrophoresis equipment.
7. Bio-Rad Gene Pulser or other cell electroporator.

8. Cuvets.
9. Puromycin.

2.3. Characterization of Genetically Modified ES Cells

2.3.1. Enzymology

1. 12- to 14-kDa dialysis membrane (Biolabs, Mulgrave, Australia; cat. no. SCT-132676).
2. 2% sodium bicarbonate (w/v) and 1 mM ethylenediaminetetraacetic acid (EDTA) (pH 8.0) made up in MilliQ water: for 100 mL, dissolve 2 g sodium bicarbonate and 0.037 g EDTA disodium salt in MilliQ water. Adjust to pH 8.0 and bring to a final volume of 100 mL. Store at room temperature.
3. 1 mM EDTA (pH 8.0) in MilliQ water.
4. 0.9% (w/v) NaCl.
5. PBS lysis buffer: for 500 mL, mix 5 g sodium deoxycholate, 0.5 g sodium dodecyl sulfate, and 2.5 mL Nonidet P40 in sterile PBS. Store at 4°C.
6. Micro BCA™ protein assay reagent (Pierce, Rockford, IL; cat. no. 23235).
7. Tritiated, heparin-derived tetrasaccharide substrate for sulfamidase *(21)*.
8. 200 mM NaAc at pH 5.0.
9. 100 mM NH$_4$OH.

2.3.2. Pulse/Chase and Immunoprecipitation

1. Cysteine/methionine-free DMEM media (MP Biomedicals, Irvine, CA; cat. no. 1642449).
2. EXPRE^{35}S^{35}S protein-labeling mix (NEN, Boston, MA; cat. no. NEG072).
3. Leupeptin (Sigma, cat. no. L2884).
4. Pepstatin A (Sigma, cat. no. P4265).
5. Pansorbin® cells, *Staphylococcus aureus* (Merck Bioscience, Kilsyth, Australia; cat. no. 507861).
6. Normal rabbit serum (Women's and Children's Hospital, Adelaide, Australia).
7. Immunoprecipitation wash buffer: 250 mM NaCl, 20 mM Tris-HCl, pH 7.0. For 100 mL, dissolve 1.461 g NaCl and 0.242 g Tris-base in MilliQ water. Adjust to pH 7.0 and store at room temperature.
8. Monoclonal mouse anti-rhNS antibody (clone 23B2; Women's and Children's Hospital, Adelaide, Australia) *(22)*.
9. Rocking platform mixer.
10. 4X Laemmli buffer: for 10 mL, dissolve 0.8 g sodium dodecyl sulfate, 0.04 g bromophenol blue, and 0.2422 g Tris-base in 5 mL MilliQ water. Adjust to pH 6.8, add 4 mL neat glycerol, and make up volume to 10 mL with MilliQ water. Store at −20°C. Supplement with 2% (v/v) β-mercaptoethanol immediately before use in a fume hood (i.e., 20 µL β-mercaptoethanol per 1 mL of 4X loading buffer).
11. Precast 10% polyacrylamide gels (Gradipore, French Forest, Australia; cat. no. NG21-010).
12. C^{14} protein molecular weight markers (1 µCi; Amersham, Castle Hill, Australia; cat. no. CFA626).
13. 10X reservoir buffer: for 1 L 10X solution, dissolve 30.28 g Tris-base, 150.14 g glycine, and 10 g sodium dodecyl sulfate in MilliQ water. Dilute to 1X solution in MilliQ water before use.
14. Fixing solution: 25% (v/v) propanol and 65% (v/v) glacial acetic acid in MilliQ water.
15. Amplify solution (Amersham, cat. no. NAMP100).
16. Gel drier.
17. X-ray film and cassette.

2.4. Differentiation of ES Cells to Embryoid Bodies

2.4.1. Cover Slip Coating

1. Poly-L-ornithine (Sigma, cat. no. P4957), final concentration of 20 µg/mL in sterile water.
2. Laminin (Sigma, cat. no. L2020), final concentration of 1 µg/mL in sterile water.
3. Thermanox® cover slips (Nalge Nunc, cat. no. 174950).
4. Sterile forceps.

2.4.2. Cytocentrifugation

1. Paraformaldehyde (Sigma, cat. no. P6148): weigh 4 g paraformaldehyde in fume hood and dissolve in 100 mL PBS at pH 7.4 (*see* **Subheading 2.1.2., item 3**) overnight at room temperature with stirring.
2. Filter cards (Shandon, Pittsburg, PA; cat. no. 5991022).
3. Cytocentrifugation funnels (Shandon, cat. no. 5991040).
4. Shandon-Elliot cytocentrifuge (Shandon, model SCA0030).
5. Sodium azide (Sigma, cat. no. 58032).

2.4.3. Immunofluorescence

1. Pap pen.
2. DMSO (Sigma, cat. no. D8418).
3. Wash buffer: 3 mL Triton X-100 in 1 L PBS.
4. Normal donkey serum (NDS; Jackson Immunoresearch, West Grove, PA; cat. no. 017-000-121): prepare neat stock by reconstituting powder with 10 mL sterile water and aliquot. Prepare working stock by mixing 100 µL neat NDS to 900 µL wash buffer.
5. For primary and secondary antibody details, *see* **Table 1**. All antibodies are diluted 1:1 with neat glycerol (i.e., 100 µL antibody plus 100 µL glycerol), aliquoted, and frozen at −20°C. For a 1:1000 dilution, add 1 µL diluted antibody stock to 999 µL 10% NDS and 90%wash buffer.
6. Vectashield mounting media containing 4′,6-diamidino-2-phenylindole (DAPI; Vector Laboratories, Burlingame, CA; cat. no. H1200).
7. Humidity chamber (Genex, Coulsden, UK; cat. no. 4300M521; *see* **Note 2**).
8. Cover slips.
9. Forceps.
10. Microscope slides.
11. Nail polish.
12. Small wire hook (optional).
13. Fluorescent microscope.

3. Methods

3.1. D3 ES Cell Culture

3.1.1. Thawing

1. Prepare fresh ES cell medium and prewarm to 37°C in a water bath.
2. Remove a cryovial from liquid nitrogen and quickly thaw cells in a 37°C water bath.
3. Ethanol swab the tube before opening and transfer the thawed cells to a 25-mL tube.
4. Add 10 mL neutralization medium dropwise with continual mixing using a swirling motion.
5. Pellet the cells at 25*g* for 2 min at room temperature and resuspend in 1 mL ES cell medium.

Table 1
Commonly Used Antibodies and Suggested Working Dilutions

Antibody	Antibody type	Cell type	Distributor	Dilution of 1:1 stock
Rabbit anti-Ki67	Polyclonal	Proliferative cells	Novocastra, Newcastle upon Tyne, UK; cat. no. NCL-Ki67p	1:2000
Mouse anti-microtubule-associated protein (Map) 2	Monoclonal	Neurons	Chemicon, Temecula, CA; cat. no. MAB3418	1:1000
Mouse anti-nestin	Monoclonal	Neural progenitor cells	Developmental Studies Hybridoma Bank, Iowa City, IA; cat. no. Rat-401	1:500
Mouse antineuronal nuclei (NeuN)	Monoclonal	Neurons	Chemicon, cat. no. MAB377	1:1000
Mouse antineurofilament 160 kDa (NF160)	Monoclonal	Neurons	Novocastra, cat. no. NCL-NF160	1:500
Mouse anti-smooth muscle actin (SMA)	Monoclonal	Smooth muscle actin	Sigma, clone 1A4, cat. no. A2547	1:3000
Cy3-conjugated donkey antimouse immunoglobulin G			Jackson Immunoresearch Laboratories, cat. no. 715-165-150	1:300
FITC-conjugated donkey antirabbit immunoglobulin G			Jackson Immunoresearch Laboratories, cat. no. 711-095-152	1:300

6. Determine the viability of the cells and seed at 2×10^6 cells/10-cm plate (*see* **Note 3**).
7. Incubate overnight at 37°C, 10% CO_2.
8. Replace with 10 mL fresh ES cell medium the following day and then continue passaging as in **Subheading 3.1.2.**

3.1.2. Passaging

1. Remove the expired medium and wash the adherent D3 ES cell layer twice in 5 mL PBS.
2. Detach the cells with 3 mL prewarmed trypsin per 10-cm plate and gently triturate with a wide-bore pipet tip to encourage detachment.
3. Neutralize trypsin with 3 mL neutralizing medium or FCS and pellet at 25*g* for 2 min at room temperature.
4. Discard the supernatant, resuspend the cells in 2 mL ES cell medium, and assess viability using trypan blue dye exclusion and a hemocytometer. Viable cells will appear white (*see* **Note 4**).
5. Evenly seed $1–2 \times 10^6$ viable cells per untreated 10-cm tissue culture dish into warm 10 mL ES cell medium and incubate in a humidified 37°C incubator containing 10% CO_2 (*see* **Note 5**).
6. View cells by light microscopy regularly. Discard ES cells if extensive spontaneous differentiation (i.e., spicule formation) is evident. Cells are fed daily and will generally require passaging every 2–3 d.

3.1.3. Freezing

1. Detach and determine the total number of viable cells as in **Subheading 3.1.2.**
2. Pellet cells at $25g$ for 2 min at room temperature and discard the supernatant.
3. Slowly add freezing mix (precooled to 4°C) dropwise to the cell pellet with continual agitation to a final concentration of 2×10^6 cells/mL.
4. Mix the cell suspension thoroughly and aliquot into cryotubes in 1-mL portions.
5. Transfer tubes into a freezing container and store at −70°C overnight. Cryotubes may then be transferred into liquid nitrogen for long-term storage.

3.2. Production of Transgenic Clonal Cell Lines

3.2.1. Preparation of DNA

1. Obtain high-quality plasmid DNA using a plasmid midikit (*see* **Note 6**).
2. Linearize approx 20 µg DNA by restriction endonuclease digestion with *Pvu*I or *Sca*I in the ampicillin gene sequence.
3. Confirm complete digestion of plasmid DNA using agarose gel electrophoresis.
4. Precipitate DNA by mixing 50 µL digested DNA with 5 µL 3 *M* NaAc at pH 5.2 and 100 µL 100% EtOH. Vortex tubes and microfuge at maximum speed on a benchtop centrifuge (approx $17,900g$) for 30 min at 4°C.
5. Decant supernatant and sterilize the cell pellet in 500 µL 70% EtOH at approx $17,900g$ for 10 min at 4°C.
6. Air-dry plasmid DNA in sterile conditions for 2–3 h and then resuspend in 10 µL sterile water.
7. Determine the concentration of DNA by diluting a sample and using a spectrophotometer at an absorbance of 260 nm.

3.2.2. Electroporation of ES Cells and Generation of Transgenic Clonal Cell Lines

1. Detach the ES cells with trypsin as in **Subheading 3.1.2.** Pellet the cells at $25g$ for 2 min and resuspend in 10 mL ice-cold PBS.
2. Determine the number of viable cells and transfer 5×10^7 cells into a prechilled cuvet in a total volume of 900 µL.
3. Add 5 µg sterile plasmid DNA, cap the cuvets, and electroporate the ES cells at 200 V and a capacitance of 500 µF for approx 10 ms. Include an electroporated control without DNA.
4. Quickly transfer the cells into 5 mL ES cell media without puromycin and dissociate cell clumps with a pipet tip.
5. Plate all of the cells into a 15-cm dish containing an additional 25 mL ES medium and incubate overnight at 37°C, 10% CO_2.
6. Replace medium daily with 15 mL ES medium containing 1.5 µg/mL puromycin to remove cellular debris. Continue to incubate cells at 37°C and 10% CO_2 until there are no viable cells in the control plate without DNA, and small individual colonies of transgenic cells are clearly evident by eye in DNA-transfected plates. Colonies should generally be clearly visible and ready to pick after 7–10 d.
7. Establish clonal lines by gently washing the attached cells once with PBS and aspirating most of the liquid with vacuum suction. Collect well-separated colonies with a fresh pipet tip for each colony and transfer into a 96-well plate containing 100 µL trypsin per well. Dissociate the cells using trituration and then plate in a 24-well tray containing 500 µL/well of ES cell medium supplemented with puromycin.

8. Continue to change ES cell medium daily and upscale the size of the culture vessel from 24-well trays to 6-well trays and then to 10-cm tissue culture dishes by passaging the cells as in **Subheading 3.1.2.**
9. Freeze stocks of early passage clonal lines as soon as practical as in **Subheading 3.1.3.**

3.3. Characterization of Genetically Modified ES Cells

3.3.1. Enzymology

3.3.1.1. CONDITIONED MEDIA

1. Prepare dialysis tubing by boiling in 1 L sodium bicarbonate/1 mM EDTA (pH 8.0) in MilliQ water for 10 min. Rinse tubing in distilled water several times. Boil the tubing a second time in 1 mM EDTA (pH 8.0) for 10 min, rinse thoroughly in distilled water, and then store in MilliQ water at 4°C. Keep the dialysis tubing submerged at all times.
2. Remove the lids and the lips of 1.5-mL Eppendorf tubes using a flat blade and discard the bodies of the tubes.
3. Remove cellular debris from the conditioned medium by microfuging at approx 17,900g for 5 min.
4. Aliquot 200 µL conditioned medium into the lid of the Eppendorf tube, overlay with a small square of dialysis tubing, and lock the lip of the Eppendorf in place to form a "drum."
5. Dialyze the medium in 0.9% NaCl overnight at 4°C with slow stirring.
6. Remove the dialyzed medium by puncturing the membrane with a pipet tip and store at 4°C in a fresh Eppendorf tube.

3.3.1.2. CELL LYSATES

1. Detach cells with trypsin treatment (*see* **Subheading 3.1.2.**) and pellet the cells at 25g for 2 min.
2. Wash the cell pellet twice with 5 mL PBS.
3. Solubilize the cells with 10 mL PBS lysis buffer and allow to equilibrate at 4°C for more than 24 h.
4. Determine the total protein content of cells using the MicroBCA™ protein assay kit according to the manufacturer's instructions.

3.3.1.3. SULFAMIDASE ACTIVITY ASSAY

This is based on the method described in **ref. 21**.

1. Mix 1 µL dialyzed conditioned media or cell lysate sample with 1 µL substrate, 3 µL 200 mM NaAc at pH 5.0, and 7 µL water and incubate at 60°C for 4 h. Quench with 100 µL 100 mM NH$_4$OH.
2. Briefly centrifuge and analyze by high-performance liquid chromatography.
3. Determine the conversion rate of substrate into product and calculate the activity of sulfamidase enzyme per microliter of sample.
4. Normalize the enzyme activity to total protein as determined in **Subheading 3.3.1.2.** and compare to levels observed in untransfected controls.

3.3.2. Pulse/Chase and Immunoprecipitation

This method is adapted from **ref. 22**.

3.3.2.1. METABOLIC LABELING OF ES CELLS

1. Dialyze 50 mL FCS three times in 1 L PBS at 4°C overnight. Sterilize through 0.22-µm filter and store at 4°C.
2. Detach the ES cells as in **Subheading 3.1.2.** and seed 1×10^6 cells per 25 cm^2 flask. Incubate overnight at 37°C and 10% CO$_2$.

Characterization of Genetically Modified ES Cells

3. Wash the cells twice with 5 mL PBS and then deplete cells in 3 mL cysteine/methionine-free media containing 10% dialyzed FCS per flask for 45 min at 37°C.
4. Replace with the same media containing 100 µCi/mL L-[^{35}S] and incubate for exactly 30 min at 37°C (*see* **Notes 7** and **8**).
5. Wash cells twice with 5 mL PBS and harvest cells for time 0 as in **step 8**.
6. Chase the remaining flasks with 4 mL ES cell medium per flask and incubate at 37°C and 10% CO_2 for 0.5, 1, 2, 4, 8, and 24 h.
7. Harvest conditioned medium by collecting into a 10-mL tube and removing debris by centrifugation at 25*g* for 3 min at room temperature. Store at 4°C.
8. Harvest cells as in **Subheading 3.3.1.2.** and resuspend in 1 mL PBS lysis buffer supplemented with 1 µ*M* leupeptin and 1 µ*M* pepstatin A. Incubate overnight at 4°C.

3.3.2.2. Immunoprecipitation of Sulfamidase

1. Equilibrate 1.5-mL aliquots of Pansorbin cells by centrifugation/resuspension four times in PBS lysis buffer at approx 17,900*g* at 4°C with a final resuspension in 1 mL PBS lysis buffer.
2. Treat the Pansorbin cells with an equal volume of normal rabbit serum for 4 h at 4°C.
3. Wash the Pansorbin cells three times with immunoprecipitation wash buffer and resuspend to 1 mL per original 1.5-mL aliquot (*see* **Note 9**).
4. Add 60 µL Pansorbin cells to each metabolically labeled sample and mix for 2 h at 4°C on a rocking table. Pellet the samples at 25*g* for 5 min and transfer the supernatant to a fresh tube.
5. Add 100 µg of monoclonal mouse anti-rhNS and mix overnight at 4°C.
6. Add 50 µL Pansorbin cells to the samples and incubate for a minimum of 2 h at 4°C.
7. Centrifuge at 25*g* for 5 min, discard the supernatant, and wash the cell pellet twice with 1 mL PBS lysis buffer and then once with MilliQ water.
8. Resuspend cells in 10 µL 1X Laemmli buffer containing 2% β-mercaptoethanol (*see* **Note 10**).
9. Boil samples for 5 min and pellet at approx 17,900*g* for 15 min at room temperature.
10. Load samples and C^{14} marker and electrophorese at 40 mA through a 10% polyacrylamide gel until the dye front reaches the bottom of the gel.
11. Fix for at least 30 min in fixing solution and then transfer to amplify solution for 1.5 h.
12. Cover the gel with cellophane and dry the gel under vacuum for approx 1 h at 80°C.
13. Expose the gel to X-ray film in a dark room and store the cassette at −70°C and develop X-ray after approx 1 wk.

3.4. Differentiation of ES Cells to Embryoid Bodies

3.4.1. Med II Conditioned Medium Preparation

This method is described in **ref. 15**.

1. Passage HepG2 cells by washing the cell layer twice in PBS, detaching with trypsin treatment, and neutralizing with DMEM with 10% FCS. Pellet cells at 25*g* for 2 min and assess the viability of cells using trypan blue dye exclusion.
2. Seed cells 1:2 into 175-cm^2 flasks containing 50 mL HepG2 medium and culture at 37°C containing 5% CO_2 for 3–4 d.
3. Collect conditioned media (Med II) and sterilize through a 0.22-µm filter.
4. Pool Med II fractions into one stock and store in 50-mL aliquots at −70°C (*see* **Note 11**).

3.4.2. Poly-L-Ornithine and Laminin Coating of Cover Slips

1. Using sterile forceps, place one sterile cover slip per well in 24-well trays.
2. Add 300 µL/well poly-L-ornithine and incubate overnight at room temperature.

3. Rinse each well three times with 1 mL sterile water.
4. Coat the wells with laminin (300 µL/well) overnight at 37°C (*see* **Note 12**).
5. Rinse the wells three times in sterile water, once in PBS, and once in DMEM (*see* **Note 13**).
6. Aliquot 500 µL/well of the appropriate medium and allow to equilibrate for 10 min before use.

3.4.3. Generation of Embryoid Bodies in Suspension Culture From ES Cells

1. Detach cells with trypsin treatment (as described in **Subheading 3.1.2.**) and seed 1×10^6 ES cells into 10 mL 50% Med II medium in bacterial Petri dishes. Incubate plates at 37°C and 10% CO_2 for 2 d. The day of seeding is designated d 0.
2. On d 2 and 4, split cells 1:2 by swirling embryoid bodies to the center of the plate, removing the expired medium (aspirated from the edge of the plate using vacuum suction) and replacing with 10 mL fresh 50% Med II medium. Collect the embryoid bodies resuspended in the fresh medium and distribute evenly between two new Petri dishes (i.e., 5 mL per plate) before adding an additional 5 mL medium per plate. Continue to incubate at 37°C and 10% CO_2 (*see* **Note 14**).
3. Days 5 and 6: feed embryoid body cell suspensions with 10 mL fresh 50% Med II medium by aspirating medium as described in step 2 and replacing with fresh medium in the same dish.
4. Day 7: feed the embryoid bodies with 10 mL serum-free medium (*see* **Note 15**).

3.4.4. Adhesion Culture and Differentiation of Embryoid Bodies

1. Coat cover slips in 24-well trays with poly-L-ornithine/laminin as in **Subheading 3.4.2.**
2. Day 9: fill the wells in the coated 24-well tray with 500 µL serum-free medium per well and transfer two or three aggregates to each well with a pipet tip (*see* **Note 16**).
3. Day 12: gently remove the expired medium, leaving a residue of approx 100 µL using vacuum suction. Replace with 500 µL fresh serum-free medium, taking care to avoid disruption of delicate cell processes.
4. Seeded embryoid bodies can be harvested from d 15 onward, when extensive, arborizing cell processes should be evident.

3.4.5. Cytocentrifugation of Cell Suspensions

1. Prepare single-cell suspension by treating aggregates with trypsin and perform cell counts as in **Subheading 3.1.2.**
2. Pellet the single-cell suspension at 25*g* for 2 min and then fix in 4% paraformaldehyde at room temperature for 1 h. Mix the cells by inversion at regular intervals to prevent clumping.
3. Centrifuge the fixed cells at 25*g* for 2 min and resuspend in PBS at a concentration of 1×10^6 cells/mL.
4. Cytocentrifuge 200 µL cell suspension onto Superfrost® Plus slides and check cell density using a light microscope.
5. Air-dry the slides for 2 h; store in 0.1% (w/v) sodium azide at 4°C until ready for immunostaining.

3.4.6. Immunofluorescent Detection of Marker Genes in Cytospun Cells

1. Encircle the cell suspension with a wax PapPen to keep the reagents localized (*see* **Note 17**).
2. All incubations are performed at room temperature. Gently rinse the slides in buffer for 5 min and then permeabilize the cells with DMSO for 10 min.

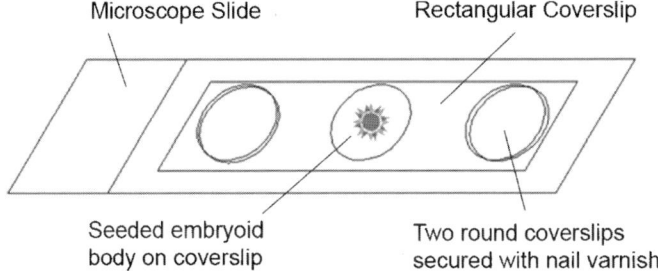

Fig. 1. Schematic diagram of immunostained seeded embryoid bodies grown on cover slips.

3. Wash three times with buffer for 5 min each.
4. Block in 10% NDS (50 μL per cytospin) for 30 min (*see* **Note 18**).
5. Incubate in the primary antibody (diluted in 10% NDS) in a humidified chamber overnight at room temperature (*see* **Note 19**).
6. Wash the cells four times in buffer for 10 min each.
7. Incubate the cells overnight in the humidity chamber using the appropriate secondary antibody (diluted in 10% NDS) conjugated to a fluorophore. Avoid exposure to light from this step onward to minimize fading of fluorescence.
8. Wash the cells four times in buffer for 10 min each.
9. Mount under a cover slip with one drop of mounting media containing the nuclear stain DAPI and store flat at 4°C.
10. View immunostaining results using a fluorescence microscope.

3.4.7. Immunofluorescent Detection of Marker Genes in Seeded Bodies

1. Culture seeded bodies as in **Subheadings 3.4.3.** and **3.4.4.**
2. Remove expired medium from differentiated seeded bodies using gentle vacuum suction and immunostain in the 24-well culture dish as in **Subheading 3.4.6.** (*see* **Note 20**).
3. Prepare microscope slides as in **Fig. 1** using nail polish and round cover slips and allow to dry.
4. After the final wash, remove the remaining liquid with a pipet and gently lift the cover slip containing the cells with a fine wire hook or fine forceps.
5. Place the cover slip containing the embryoid bodies onto a microscope slide with the cells facing the air interface.
6. Add a single drop of DAPI mounting medium to the cells and quickly overlay with a rectangular cover slip (secured to the round cover slips with nail polish), forming a "bridge" structure.
7. Store flat at 4°C.

4. Notes

1. Test each batch of FCS for the ability to support ES cell and differentiated embryoid body growth.
2. A humidity chamber can be constructed from a rectangular, plastic Tupperware container with two to four serological pipets taped in place to raise the microscope slides from the bottom of the chamber. Add tissues moistened with water for humidity and seal tightly during staining.

3. Downscale the size of the culture vessel if cell viability is low.
4. An average yield per 10-cm culture dish following 2 d of culture is 4×10^6 to 8×10^6 cells.
5. Tissue culture plates can be gelatinized to improve attachment of cells if desired. To prepare plates, incubate 10 mL 0.2% gelatin/PBS per 10-cm plate for 2–3 h at room temperature. Wash plates once in PBS before use.
6. The rhNS-pCAGIPuro vector used in our experiments contained the human cytomegalovirus intermediate early enhancer/chicken β-actin promoter, the rhNS gene, an internal ribosomal entry site sequence and a puromycin resistance gene.
7. ^{35}S has a half-life of 87.4 d, and the decay factor of ^{35}S should be considered during calculations.
8. Contents are radioactive and care should be taken when handling cells and waste materials after metabolic labeling. Work areas should be regularly monitored for spills before and after use, and tissue culture hoods, incubators, and waste should be labeled appropriately.
9. Equilibrated Pansorbin cells may be stored at 4°C for up to 7 d.
10. Samples can be stored at −20°C after this step.
11. HepG2 cells should be replenished from frozen stocks every 2–3 mo.
12. The laminin incubation can be shortened to a minimum of 3 h.
13. After the PBS wash, trays can be wrapped in Parafilm and can be stored at 4°C for up to 1 wk in the last PBS wash. Do not let the cover slips dry out at any stage.
14. Embryoid bodies can also be centrifuged gently at $10g$ for 4 min to remove expired medium if necessary.
15. Following this differentiation protocol, cells may be harvested for implantation or further differentiated for characterization in vitro.
16. Do not disturb the trays for at least 24 h to allow the embryoid bodies to adhere to the coated cover slips.
17. Do not let the cells dry out at any stage as this may induce autofluorescence.
18. The blocking agent should be serum of the same species in which the secondary antibody is raised.
19. If double labeling, then ensure that the two primary antibodies are not raised in the same species.
20. The minimum total volume for the primary and secondary antibody incubations in 24-well trays is 150 µL per well. Use 500 µL per well for each of the wash steps.

Acknowledgments

We acknowledge the helpful discussions provided by staff at the Women's and Children's Hospital and BresaGen Pty Limited, in particular Peter Cartwright and Bruce Davidson (both formerly of BresaGen Pty Limited). We also thank Peter Clements and Vivienne Muller (Women's and Children's Hospital) for the production of the tetrasaccharide substrate. This work was supported by the Canadian Sanfilippo Children's Research Foundation and the National Health and Medical Research Council of Australia.

References

1. Barton, N. W., Brady, R. O., Dambrosia, J. M., et al. (1991) Replacement therapy for inherited enzyme deficiency—macrophage-targeted glucocerebrosidase for Gaucher's disease. *N. Engl. J. Med.* **324,** 1464–1470.
2. Eng, C. M., Guffon, N., Wilcox, W. R., et al. (2001) Safety and efficacy of recombinant human α-galactosidase A—replacement therapy in Fabry's disease. *N. Engl. J. Med.* **345,** 9–16.

3. Schiffmann, R., Kopp, J. B., Austin, H. A., III, et al. (2001) Enzyme replacement therapy in Fabry disease: a randomized controlled trial. *JAMA* **285,** 2743–2749.
4. Kakkis, E., Muenzer, J., Tiller, G., et al. (2001) Enzyme-replacement therapy in mucopolysaccharidosis I. *N. Engl. J. Med.* **344,** 182–188.
5. Gliddon, B. L. and Hopwood, J. J. (2004) Enzyme-replacement therapy from birth delays the development of behaviour and learning problems in mucopolysaccharidosis type IIIA mice. *Pediatr. Res.* **56,** 1–8.
6. Evans, M. J. and Kaufman, M. (1981) Establishment in culture of pluripotent cells from mouse embryos. *Nature* **292,** 154–156.
7. Bain, G., Kitchens, D., Yao, M., Huettner, J. E., and Gottlieb, D. I. (1995) Embryonic stem cells express neuronal properties in vitro. *Dev. Biol.* **168,** 342–357.
8. Rathjen, J., Haines, B. P., Hudson, K. M., Nesci, A., Dunn, S., and Rathjen, P. D. (2002) Directed differentiation of pluripotent cells to neural lineages: homogeneous formation and differentiation of a neurectoderm population. *Development* **129,** 2649–2661.
9. Tang, F., Shang, K., Wang, X., and Gu, J. (2002) Differentiation of embryonic stem cells to astrocytes visualised by green fluorescent protein. *Cell Mol. Neurobiol.* **22,** 95–101.
10. Liour, S. S. and Yu, R. K. (2003) Differentiation of radial glia-like cells from embryonic stem cells. *Glia* **42,** 109–117.
11. Andressen, C., Stocker, E., Klinz, F. J., et al. (2001) Nestin-specific green fluorescent protein expression in embryonic stem cell-derived neural precursor cells used for transplantation. *Stem Cells* **19,** 419–424.
12. Barberi, T., Klivenyi, P., Calingasan, N., et al. (2003) Neural subtype specification of fertilization and nuclear transfer embryonic stem cells and application in parkinsonian mice. *Nat. Biotechnol.* **21,** 1200–1207.
13. Kim, J. H., Auerbach, J. M., Rodriguez-Gomez, J. A., et al. (2002) Dopamine neurons derived from embryonic stem cells function in an animal model of Parkinson's disease. *Nature* **418,** 50–56.
14. Lau, A. A., Hemsley, K. M., Meedeniya, A., and Hopwood, J. J. (2004) In vitro characterization of genetically modified embryonic stem cells as a potential therapy for murine mucopolysaccharidosis type IIIA. *Mol. Genet. Metab.* **81,** 86–95.
15. Rathjen, J., Lake, J. A., Bettess, M. D., Washington, J. M., Chapman, G., and Rathjen, P. D. (1999) Formation of a primitive ectoderm like cell population, EPL cells, from ES cells in response to biologically derived factors. *J. Cell Sci.* **112,** 601–612.
16. Bhaumik, M., Muller, V. J., Rozaklis, T., et al. (1999) A mouse model for mucopolysaccharidosis type III A (Sanfilippo syndrome). *Glycobiology* **9,** 1389–1396.
17. Jolly, R. D., Allan, F. J., Collett, M. G., Rozaklis, T., Muller, V. J., and Hopwood, J. J. (2000) Mucopolysaccharidosis IIIA (Sanfilippo syndrome) in a New Zealand Huntaway dog with ataxia. *N. Z. Vet. J.* **48,** 144–148.
18. Knowles, B. B., Howe, C. C., and Aden, D. P. (1980) Human hepatocellular carcinoma cell lines secrete the major plasma proteins and hepatitis B surface antigen. *Science* **209,** 497–499.
19. Doetschman, T. C., Eistetter, H., Katz, M., Schmidt, W., and Kemler, R. (1985) The in vitro development of blastocyst-derived embryonic stem cell lines: formation of visceral yolk sac, blood islands and myocardium. *Embryol. Exp. Morphol.* **87,** 27–45.
20. Smith, A. G. (1991) Culture and differentiation of embryonic stem cells. *J. Tiss. Cult. Meth.* **13,** 89–94.

21. Hopwood, J. J. and Elliott, H. (1982) Diagnosis of Sanfilippo A syndrome by estimation of sulphamidase activity using a radiolabelled tetrasaccharide substrate. *Clin. Chim. Acta* **123,** 241–250.
22. Perkins, K. J., Byers, S., Yogalingam, G., Weber, G., and Hopwood, J. J. (1999) Expression and characterization of wild type and mutant recombinant human sulfamidase. *J. Biol. Chem.* **274,** 37,193–37,199.

33

Use of Differentiating Embryonic Stem Cells in the Parkinsonian Mouse Model

Fumihiko Nishimura, Hayato Toriumi, Shigeaki Ishizaka, Toshisuke Sakaki, and Masahide Yoshikawa

Summary

Progressive loss of dopaminergic neurons in the substantia nigra pars compacta and the following reduction in striatal dopamine cause Parkinson's disease (PD). Transplantation of dopamine-producing cells into the striatum is a proposed treatment modality. In this report, we describe a model experiment assessing the effectiveness of mouse embryonic stem (ES) cell-derived dopaminergic neurons using a mouse model of PD. ES cells were shown to be an attractive and promising source for the generation of dopaminergic neurons, and the mouse PD model was useful to assess the efficacy of transplantation therapy with dopamine-producing cells, including ES cell-derived dopaminergic neurons.

Key Words: Dopaminergic differentiation; embryoid body (EB); embryonic stem (ES) cells; 6-hydroxydopamine (6-OHDA); multistep culture; Parkinson's disease (PD); tyrosine hydroxylase (TH).

1. Introduction

Parkinson's disease (PD) is considered to be one of the most appropriate diseases for cell replacement therapy because its etiology, progressive loss of dopaminergic neurons in the substantia nigra pars compacta leading to a reduction in striatal dopamine, is relatively clear and simple. Clinical trials of transplantation using human fetal nigral tissues have shown symptomatic relief; however, technical and ethical difficulties in obtaining sufficient and appropriate graft tissues have limited the application of this therapy. Embryonic stem (ES) cells are attractive as a potential novel source for new therapeutic strategies in the treatment of PD because of their capacity to renew themselves and differentiate into various kinds of cell types, including dopaminergic neurons.

The differentiation of mouse ES cells and primate ES cells, including those from humans, into dopaminergic neurons has been reported *(1–3)*. Before the era of human

ES cell-based therapy for PD can be realized, the potential usefulness of these cells must first be confirmed in animal models of PD (*4,5*). We describe here our method of successful transplantation of mouse ES cell-derived dopaminergic neurons using a mouse PD model. We believe that this model can be used to assess the usefulness of nonmouse ES cell-based therapy under the support of appropriate immunosuppression and even without any treatment to influence the immune system because of the special characteristics of the brain as an immunoprivileged site.

2. Materials
2.1. Murine ES Cell Lines: EB3 Cells, G4-2 Cells

We utilized a mouse ES cell line, G4-2, derived from EB3 ES cells that carry the enhanced green fluorescent protein (EGFP) gene under the control of the CAG expression unit (*6*). EB3 cells, a subline derived from E14tg2a ES cells (*7*) that carry the blasticidin S-resistant selection marker gene driven by the Oct3/4 promoter (active under undifferentiated status) (*6*), were also used. Undifferentiated G4-2 cells were maintained on gelatin-coated dishes without feeder cells in medium containing 10 μg/mL blasticidin S to eliminate spontaneously differentiated cells.

2.2. Maintenance of Undifferentiated ES Cells

1. Dulbecco's modified Eagle's medium (DMEM; Sigma, St. Louis, MO; cat. no. D6429) stored at 4°C).
2. ES-qualified fetal calf serum (FCS; Gibco/BRL, Grand Island, NY; cat. no. 15141-079, special lot required) (*see* **Note 1**) stored at 4°C after 56°C for 30 min.
3. 100X nonessential amino acids (NEAA; Gibco/BRL, cat. no. 11140-035) stored at 4°C.
4. 100X sodium pyruvate (Sigma, cat. no. S8636).
5. Leukemia inhibitory factor (LIF; 10^7 U; Chemicon, Temecula, CA; cat. no. ESG1107).
6. Trypsin-ethylenediaminetetraacetic acid (EDTA): 0.25% trypsin, 1 mM EDTA (Gibco/BRL; cat. no. 25200-056/072).
7. 2-Mercaptoethanol (2-ME; Sigma, cat. no. M7522) (*see* **Note 2**).
8. 100 mg blasticidin S (Funakoshi, Tokyo, Japan; cat. no. KK-400) (*see* **Note 3**).
9. Gelatin (porcine skin; Sigma, cat. no. G9391 or G2625), gelatin-coated dish (*see* **Note 4**).
10. Petri dish (90 × 15 mm; Bio-Bik Ina Optica, Nagano, Japan; cat. no. I-90).
11. 100X penicillin and streptomycin (pen-strep) solution (Sigma, cat. no. P4333).
12. Maintenance medium for mouse ES cells: DMEM containing 10% FCS, 1X NEAA, 1 mM sodium pyruvate, 1X pen-strep, 10^{-4} 2-ME, and 1400 U/mL LIF. To make 500 mL of the maintenance medium, add 5 mL 10 mM MEM NEAA, 5 mL 100 mM sodium pyruvate, 5 mL pen-strep solution, 500 μL 0.1 M 2-ME, 70 μL 107 U/mL LIF, and 55 mL FCS.

2.3. Induction of Dopaminergic Differentiation

1. 100X insulin-transferrin-selenium (ITS) supplement (Gibco/BRL, cat. no. 51300-044).
2. Fibronectin (Fn; Invitrogen; Carlsbad, CA; cat. no. 33010-018).
3. 100X N2 supplement (Gibco/BRL, cat. no. 17502-048).
4. Laminin (Invitrogen, cat. no. 23017-015).
5. Basic fibroblast growth factor (bFGF, bovine brain derived) (RD Systems, Minneapolis, MN; cat. no. 133-FB/CF).
6. Murine N-terminal fragment from Sonic Hedgehog (Shh; RD Systems, cat. no. 461-SH).

Differentiating ES Cells in Parkinsonian Model 487

7. Murine FGF8 isoform b (RD Systems, cat. no. 423-F8).
8. Ascorbic acid (Nakarai tesque, Kyoto, Japan; cat. no. 034-22).

2.4. PD Model Mice and Materials

1. Adult male 129/SvJ mice (10–12 wk old; CLEA Japan Inc., Tokyo, Japan).
2. Pentobarbital.
3. Stereotactic frame (Narishige, Tokyo, Japan).
4. 6-Hydroxydopamine HCl (6-OHDA; Sigma, cat. no. H-4381).
5. 22-Gage 10-µL microsyringe.
6. Apomorphine (Sigma, cat. no. A-4393).

2.5. Immunochemistry

2.5.1. Immunocytochemistry for Cultured Cells

1. 4% paraformaldehyde (PFA) in PBS (Wako Pure Chemicals, Osaka, Japan; cat. no. 163-20145).
2. Antibodies: mouse anti-TH (tyrosine hydroxylase) monoclonal antibody at 1:1000 (Chemicon, cat. no. AB-152), mouse anti-map 2 monoclonal antibody at 1:200 (Sigma, cat. no. M1406), rabbit anti-glial fibrillary acidic protein (GFAP) polyclonal antibody at 1:200 (Sigma, cat. no. G-9269), mouse antinestin monoclonal antibody at 1:500 (Chemicon, cat. no. MAB353), rhodamine-labeled secondary antibody at 1:200 (goat antimouse immunoglobulin [Ig] G affinity purified; Chemicon, cat. no. AP124R), and fluorescein isothiocyanate (FITC)-labeled secondary antibody at 1:200 (goat antirabbit IgG affinity purified; Chemicon, cat. no. AP307F).
3. 4,6-Diamidino-2-phenyindole dilactate (DAPI; Sigma, cat. no. D9564).
4. Triton X (Wako Pure Chemicals, cat. no. A16046).

2.5.2. Immunohistochemistry for Brain Tissues

1. Sucrose (10, 15, and 20% in PBS).
2. Sakura Tissue-TEK® OCT™ compound.
3. Liquid nitrogen.
4. 3-Aminopropyltriethoxysilane-coated slides.
5. Antibodies: mouse anti-TH monoclonal antibody at 1:1000 (Chemicon, cat. no. AB-152) and rabbit anti-GFP polyclonal antibody at 1:9000 (Affinity BioReagents, Golden, CO; cat. no. PA1-980) diluted in PBS with 0.1% Triton X-100. The following are used as secondary antibodies: rhodamine-labeled goat antimouse IgG antibody (Chemicon; cat. no. AP124R) at 1:200 and FITC-labeled goat antirabbit IgG antibody (Chemicon, cat. no. AP307F) at 1:200.
6. Laser confocal microscope.

3. Methods

3.1. Maintenance of Undifferentiated ES Cells

Undifferentiated ES cells were maintained on gelatin-coated dishes without feeder cells in maintenance medium containing 10 µg/mL blasticidin S (*see* **Notes 5–7**). Undifferentiated ES cells formed colonies from single cells.

1. Wash colonies of undifferentiated ES cells on a 9-cm dish twice with 10 mL PBS.
2. Add 2.0 mL trypsin-EDTA solution and incubate for 2 min at 37°C.

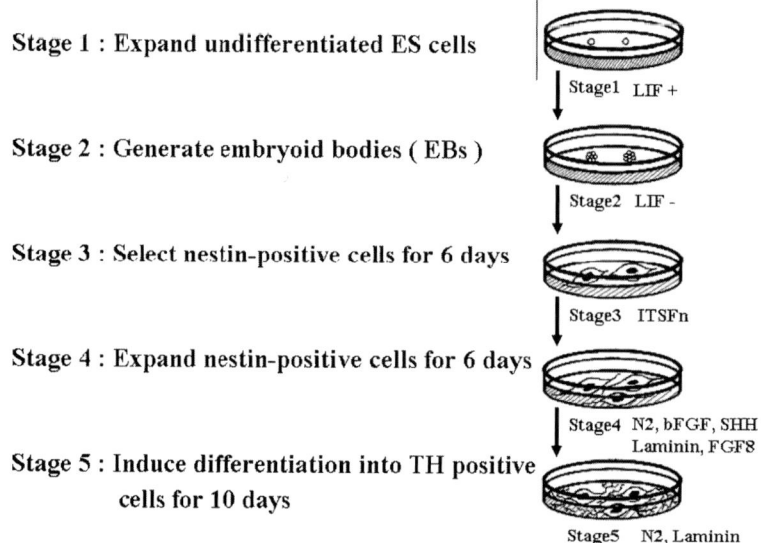

Fig. 1. General scheme of in vitro embryonic stem (ES) cell culture. TH-positive populations were generated from undifferentiated ES cells by an embryoid body-based multistep in vitro differentiation method. ITSFn, insulin-transferrin-selenium-fibronectin; bFGF, basic fibroblast growth factor; SHH, sonic hedgehog; TH, tyrosine hydroxylase.

3. Dislodge the ES cells by gentle pipetting and make a single-cell suspension.
4. Neutralize trypsin by adding 2 mL maintenance medium containing FCS.
5. Spin for 5 min at 500g.
6. Resuspend the pellet in 5 mL PBS and count the number of cells in the solution.
7. Plate an appropriate quantity of solution containing 5×10^5 undifferentiated ES cells onto a gelatin-coated 9-cm dish filled with 10 mL ES maintenance medium (see **Note 8**).
8. Repeat **steps 1–6** every 3 or 4 d.

3.2. Induction of Dopaminergic Differentiation

In vitro differentiation of ES cells into TH-positive neurons was carried out as previously described (1) with some minor modifications (**Fig. 1**). The lack of LIF in the medium promotes the start of differentiation; thus, the basal medium used for induction of dopaminergic differentiation is ES maintenance DMEM lacking LIF.

1. Make a single-cell suspension of undifferentiated ES cells (stage 1) using the procedures outlined in **Subheading 3.1., steps 1–3**.
2. Neutralize trypsin by adding 2 mL basal DMEM medium containing FBS.
3. Spin down the cells at 250g for 5 min and resuspend the pellet with 5 mL basal DMEM medium.
4. Take an appropriate quantity of solution to make a cell suspension at 2.5×10^4 cells/mL in basal medium.
5. Form embryoid bodies (EBs) using the hanging drop method for 4 d (stage 2) (see **Note 9**).

6. Plate the resulting EBs onto plastic 100-mm gelatin-coated dishes and allow them to attach for an outgrowth culture.
7. After 24 h of culture, replace the medium with serum-free insulin-transferrin-selenium-fibronectin (ITSFn) medium to select nestin-positive cells (stage 3) (*see* **Note 10**).
8. After 6–10 d of selection, replace the medium with N2 medium, supplemented with 1 µg/mL laminin and 10 ng/mL bFGF, in the presence of murine Shh (500 ng/mL) and murine FGF8 isoform b (100 ng/mL) (*see* **Note 11**). Nestin-positive cells are then grown for 6 d (stage 4).
9. Remove bFGF and culture for 6–15 d in N2 medium supplemented with laminin (1 µg/mL) and ascorbic acid (200 µM) (stage 5) for the further induction of dopaminergic differentiation (*see* **Note 12**).

3.3. Animal Preparation

Adult male 129/SvJ mice (10–12 wk old) (*see* **Note 13**) were used to make PD model mice by unilateral dopamine denervation with 6-OHDA.

1. Place the mice in a stereotactic frame under pentobarbital anesthesia (60 mg/kg ip).
2. Administrate 6-OHDA to the left midstriatum area (0.4 mm anterior, 1.8 mm lateral, 3.5 mm ventral), as determined from the bregma and surface of the skull, to produce a dopamine-denervated striatum (*see* **Notes 14** and **15**).
3. Assess apomorphine-induced rotational behavior at 2, 4, 6, 8, 10, and 12 wk after the 6-OHDA injection (*see* **Note 16**).

3.4. Preparation and Transplantation of ES-Derived TH-Positive Cells

We used cells at stage 5 of in vitro differentiation as grafts and transplanted them into the mice 4 wk after the 6-OHDA injection.

1. Trypsinize the cultured cells at stage 5 with 0.25% trypsin for 5 min at 37°C.
2. Neutralize trypsin by adding 2 mL PBS containing 10% FCS.
3. After spinning down the cells at 250g for 5 min, resuspend them in 2 mL ice-cold PBS.
4. Prepare cell solutions at appropriate cell densities (*see* **Note 17**).
5. Infuse a 2-µL suspension of the cell solution into the striatum area (0.9 mm anterior, 2.0 mm lateral, 3.0 mm ventral), as determined from the bregma and skull surface (*see* **Note 18**).
6. Hold the needle in place for 4 min before removing.
7. Assess apomorphine-induced rotational behavior at 2-wk intervals after transplantation (*see* **Note 19**).

3.5. Immunochemistry

3.5.1. Immunocytochemistry for Grafted Cells

1. Fix the cultured cells at stage 5 with 4% PFA at room temperature for 10 min.
2. Wash the cells twice with PBS for 5 min each time.
3. Permeabilize the cells with 0.1% Triton X-100 in PBS for 10 min for detection of intracellular antigens.
4. Wash once with PBS.
5. Block with 1% BSA in PBS for 20 min.
6. Incubate the cells with the primary antibody in PBS overnight at 4°C.
7. Wash with PBS three times.

8. Incubate with the secondary antibody solution for 60 min at room temperature.
9. Wash with PBS.
10. Stain with DAPI.
11. Wash with PBS.
12. Mount the sections with aqueous permanent mounting medium.

3.5.2. Immunohistochemistry of Brain Specimens

1. Anesthetize the mice with an overdose of pentobarbital given intraperitoneally at the appropriate time after transplantation (*see* **Note 20**).
2. Remove the brains and fix them for 24 h in 4% PFA in PBS at 4°C, then prepare sections.
3. Equilibrate the sections in 10% sucrose in PBS for 4 h at 4°C, then in 15% sucrose in PBS for 4 h at 4°C, and finally in 20% sucrose in PBS overnight at 4°C.
4. Embed the sections in OCT compound and freeze in liquid nitrogen.
5. Cut the sections into 8-μm thick slices using a cryostat and place on 3-aminopropyltriethoxysilane-coated slides.
6. Rinse the sections in PBS.
7. Incubate overnight at 4°C with a primary antibody in PBS with 0.1% Triton X-100.
8. Rinse in PBS and incubate in medium containing fluorescent-labeled secondary antibodies for 60 min at room temperature.
9. Wash with PBS.
10. Mount the sections with aqueous permanent mounting medium.

4. Notes

1. It is important to choose a serum lot that supports the growth of ES cells and does not generate differentiated cells. Batch-related FCS variations affect the differentiation of ES cells.
2. We prepared a 10^{-1} solution (1000X final) by adding 0.1 mL 2-ME to 14.1 mL PBS. This solution can be stored for up to 4 wk at 4°C after filtration with a 0.2-μm filter.
3. We prepared a 100-mg/mL solution in distilled water and stored it at 4°C.
4. The gelatin solution (0.1% gelatin in PBS) can be stored at 4°C after autoclaving. We added 8.0 mL 0.1% gelatin solution into a 9-cm dish and kept it for 10 min or more to prepare the gelatin-coated dishes.
5. G4-2 and EB3 ES cells do not require MEF feeder cells to remain in an undifferentiated state.
6. We have found it best to add blasticidin S to every maintenance culture to keep the ES cells undifferentiated. However, it can be left out of the maintenance culture medium as long as the shape of the colonies is compact without scattering, showing a reliable undifferentiated state. Further, occasional treatment with blasticidin S, such as intermittently for two or three successive maintenance cultures, was useful.
7. The maintenance culture medium contained 10% FCS. Although we have not encountered any trouble in inducing neural differentiation using the present EB-based multistep differentiation protocol, FCS may contain substances that inhibit neuronal differentiation of ES cells. Instead of FCS, Knockout serum replacement might also be useful. The EB-based multistep differentiation protocol includes serum-free steps. If another protocol is used to induce dopaminergic neural differentiation, including the SDIA method *(8,9)*, then we recommend the use of Knockout serum replacement.
8. Precise cell number counting may not be necessary. The cells can be seeded at a range of 1:10 to 1:20 and harvested after 3 d.
9. Place 20-μL aliquots of the cell suspension (2.5×10^4 cells/mL) onto the dish lid. This is defined as d 0 of EB formation; one hanging drop contains 500 cells.

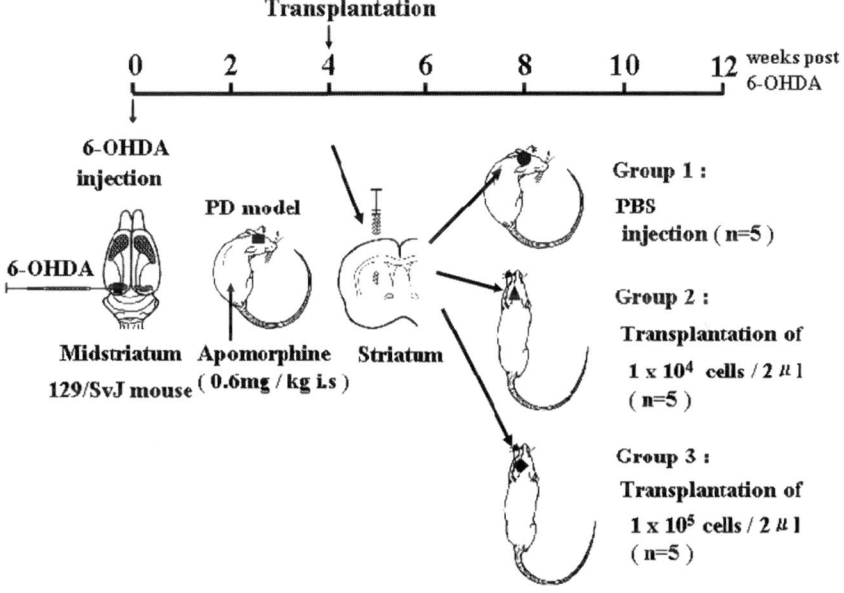

Fig. 2. Experiment outline. We divided 15 hemiparkinsonian mice into three different treatment groups of 5 animals each. At 4 wk after 6-OHDA injection, group 1 received phosphate-buffered saline into the dopamine-denervated striatum. Group 2 received a suspension of 1×10^4 cells, and group 3 received a suspension of 1×10^5 cells. Graft function was determined by reviewing apomorphine-induced rotation at 2-wk intervals for 8 wk after transplantation. Apomorphine-induced rotational behavior was assessed at 2, 4, 6, 8, 10, and 12 wk following 6-OHDA injection.

10. ITSFn: final concentration = 1X ITS, 5 µg/mL fibronectin.
11. In the original description, the cultured cells are made into a single-cell solution by trypsinization (0.05% trypsin, 0.04% EDTA in PBS) at the end of stage 3 and replated at the start of stage 4. However, this process decreases the viability of cells, and we recommend that it be omitted. After 6–10 d of culture in stage 3, it is only necessary to replenish the culture with the medium for stage 4. The addition of Shh and FGF8 led to increased generation of dopaminergic neurons because Shh and FGF8 have been shown to promote ventral midbrain fates.
12. Approximately 32% of the cells at stage 5 were TH positive; MAP2-, GFAP-, and nestin-positive cells comprised approx 80, 10, and 5%, respectively, of the total cells.
13. C57BL/6 mice may be used instead of 129/SvJ mice as the immunological background of both is H-2^b.
14. We infused 2 µL physiological saline containing 4 µg 6-OHDA and 0.02% ascorbic acid to each mouse at a rate of approx 0.5 µL/min using a 22-gage, 10-µL microsyringe and then left the microsyringe in position for an additional 4 min before retraction.
15. We used the position at 0.2 mm anterior, 2.0 mm lateral, and 3.4 mm ventral for C57BL/6 mice.
16. We placed the mice in individual plastic hemispherical bowls and allowed them to habituate for 10 min before injecting them with a subcutaneous dose of apomorphine (0.6 mg/kg). Rotational behavior was monitored for 30 min in a closed room without any environmental

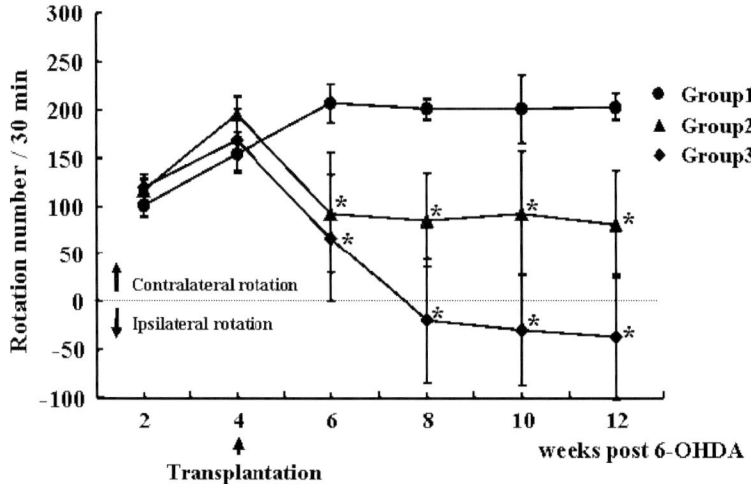

Fig. 3. Apomorphine-induced rotational behavior. Data were presented as mean plus or minus standard deviation. Mice with grafted cells in group 2 (triangle) and group 3 (square) exhibited significantly reduced rotational behavior compared with those in group 1 (circle) after transplantation (*p = 0.05). However, mice in group 3 showed ipsilateral rotation toward the lesioned side from 4 wk after transplantation, which was unexpected. In group 3, one mouse died on d 42 after transplantation.

disturbance. Mice that turned contralateral toward the lesion side at a rate of three or more rotations per minute were selected as PD models.
17. In previous experiments, we used cell concentrations of 5×10^6/mL and 5×10^7/mL, and 2 µL of those gave 1×10^4 and 1×10^5, respectively, graft cells. We divided 15 PD mice into three groups. Group 1 received PBS into the dopamine-denervated striatum at 4 wk after 6-OHDA administration, group 2 received a suspension of 1×10^4 graft cells, and group 3 received a suspension of 1×10^5 graft cells (**Fig. 2**).
18. We used the position at 0.7 mm anterior, 2.2 mm lateral, and 3.0 mm ventral for C57BL/6 mice.
19. Assessment of apomorphine-induced rotation is shown in **Fig. 3**. Occasionally, rotation toward an unexpected direction was seen, as reported previously *(5,10)*.
20. We performed histological observations of the brain 8 wk after transplantation. Our results showed ES-derived TH-positive neurons by double staining with a rhodamine-conjugated secondary antibody directed to the anti-TH antibody and an FITC-conjugated secondary antibody directed to the anti-GFP antibody *(10)*.

References

1. Lee, S. H., Lumelsky, N., Studer, L., Auerbach, J. M., and McKay, R. D. (2000) Efficient generation of midbrain and hindbrain neurons from mouse embryonic stem cells. *Nat. Biotechnol.* **18,** 675–679.
2. Kawasaki, H., Suemori, H., Mizuseki, K., et al. (2002) Generation of dopaminergic neurons and pigmented epithelia from primate ES cells by stromal cell-derived inducing activity. *Proc. Natl. Acad. Sci. USA* **99,** 1580–1585.

3. Park, S., Lee, K. S., Lee, Y. J., et al. (2004) Generation of dopaminergic neurons in vitro from human embryonic stem cells treated with neurotrophic factors. *Neurosci. Lett.* **359,** 99–103.
4. Barberi, T., Klivenyi, P., Calingasan, N. Y., et al. (2003) Neural subtype specification of fertilization and nuclear transfer embryonic stem cells and application in parkinsonian mice. *Nat. Biotechnol.* **21,** 1200–1207.
5. Kim, J. H., Auerbach, J. M., Rodriguez-Gomez, J. A., Velasco, I., Gavin, D., and Lumelsky, N. (2002) Dopamine neurons derived from embryonic stem cells function in an animal model of Parkinson's disease. *Nature* **418,** 50–56.
6. Niwa, H., Miyazaki, J., and Smith, A. G. (2000) Quantitative expression of Oct-3/4 defines differentiation, dedifferentiation or self-renewal of ES cells. *Nat. Genet.* **24,** 372–376.
7. Hooper, M., Hardy, K., Handyside, A., Hunter, S., and Monk, M. (1987) HPRT-deficient (Lesch-Nyhan) mouse embryos derived from germline colonization by cultured cells. *Nature* **326,** 292–295.
8. Kawasaki, H., Mizuseki, K., Nishikawa, S., Kaneko, S., Kuwana, Y., and Nakanishi, S. (2000) Induction of midbrain dopaminergic neurons from ES cells by stromal cell-derived inducing activity. *Neuron* **28,** 31–40.
9. Kawasaki, H., Mizuseki, K., and Sasai, Y. (2002) Selective neural induction from ES cells by stromal cell-derived inducing activity and its potential therapeutic application in Parkinson's disease, in *Methods in Molecular Biology Vol. 185 Embryonic Stem Cells. Methods and Protocols* (Turksen, K., ed), Humana Press, Totowa, NJ, pp. 217–227.
10. Nishimura, F., Yoshikawa, M., Kanda, S., et al. (2003) Potential use of embryonic stem cells for the treatment of mouse parkinsonian models: improved behavior by transplantation of in vitro differentiated dopaminergic neurons from embryonic stem cells. *Stem Cells* **21,** 171–180.

Index

A

Adipocyte,
   embryoid body differentiation culture for pharmacological studies,
      embryoid body culture, 344–346, 349, 350
      gene expression analysis, 348, 349
      Oil-Red O staining, 346, 347
      rat embryonic stem cell differentiation, 54
Alizarin red, osteoblast staining, 347, 348
Alkaline phosphatase (AP),
   chicken blastodermal cell marker detection, 30
   horse embryonic stem cell staining, 62, 68
   medaka embryonic stem cell staining, 10, 11, 15
AP, *see* Alkaline phosphatase
Apoptosis,
   DNA damage assay,
      annexin staining, 325
      materials, 317
      transfection, 325
   p53 role, 328
   TUNEL assay, 145, 146, 148
   ultraviolet-induced apoptosis in Ets knockout mouse embryonic stem cells,
      DNA fragmentation assay, 333
      maintenance culture and marker analysis, 330–332, 336
      materials, 328–330
      overview, 327, 328
      p53 induction and target gene analysis,
         electrophoretic mobility shift assay, 336
         Northern blot, 334–336
         Western blot, 333, 334
      propidium iodide staining, 333
      ultraviolet irradiation, 332

B–C

Bisulfite restriction mapping, *see* DNA methylation
Buffalo rat liver-conditioned media, preparation, 203, 204
Capillary electrophoresis, *see* Temperature gradient capillary electrophoresis
Cardiomyocyte,
   Cripto in embryonic stem cell differentiation, 152
   differentiation induction of embryonic stem cells,
      Cripto expression analysis,
         protein extraction, 160–162, 167, 168
         Western blot, 162, 166
      Cripto induction in knockout cells,
         Cripto addition, 163, 164, 168
         Cripto purification, 163
      embryoid body formation, 159, 160, 167
      immunofluorescence microscopy, 164, 168
      materials, 152–156, 166
      tissue culture,
         feeder layer preparation, 157
         freezing and thawing, 156
         gelatin coating of dishes, 157

maintenance, 159
subculture, 159
timetable, 158
Cell cycle, *see* Apoptosis; DNA damage
Cell fusion,
electrofusion,
cell culture,
embryonic stem cells, 415, 416, 419
somatic cells, 416, 419
cell purification, 416
feeder cell preparation, 415
hybrid cell isolation and cloning, 418, 419
materials, 414, 415, 419
microslide chamber setup, 418
principles, 413
selection systems, 413, 414, 418, 419
setup and operation of pulse generator, 416, 417–419
embryonic stem cell hybrid findings, 412
extinction of gene products, 411, 412
fertilization analogy, 412
virus-mediated fusion, 412, 413
Cell replacement therapy, *see* Parkinson's disease
Chemical mutagenesis,
alkylating agent mechanisms of action, 398, 399
cell death calculation, 404, 406
cell expansion and freezing, 405, 406
N-ethyl-*N*-nitrosourea mutagenesis, 404, 406
freezing and thawing of cells, 403, 405
genotype-based screens, 399, 400
genotyping of mutants,
denaturing high-performance liquid chromatography, 401
single-strand conformational polymorphism, 400, 401
temperature gradient capillary electrophoresis, 401

materials, 401–403, 405
methanesulfonic acid ethyl ester mutagenesis, 403, 405
mutation rate determination, 404, 405
phenotyping of embryonic stem cells, 400
principles, 397, 398
Chicken embryonic stem cells,
culture,
blastodermal cell isolation and maintenance, 28–30, 33
marker detection in undifferentiated blastodermal cells,
alkaline phosphatase, 30
immunofluorescence microscopy, 31
materials, 20, 21
STAT3, 31–33
materials, 19, 20, 33
leukemia inhibitory factor preparation, *see* Leukemia inhibitory factor
Chromosome analysis,
horse embryonic stem cell Giemsa chromosome staining, 62, 67, 68, 77
medaka embryonic stem cells, 10, 11, 15
monkey embryonic stem cell karyotyping, 87
transfected cell karyotyping, 292–294
CMV promoter, *see* Cytomegalovirus promoter
Cripto,
cardiomyocyte differentiation induction of embryonic stem cells,
Cripto addition, 163, 164, 168
Cripto purification, 163
embryonic stem cell differentiation role, 152
neuron differentiation of knockout cells, 164

Cytomegalovirus (CMV) promoter,
  monkey embryonic stem cell expression of green fluorescent protein, 309
  stable transgene expression in mouse embryonic stem cells,
    freezing and thawing of cells, 288, 293
    karyotyping, 292–294
    maintenance culture, 288, 293
    materials, 285–287, 293
    overview, 283, 285
    plasmid propagation and purification, 288, 289, 293
    transfection, 290, 291
    transgene expression assays,
      flow cytometry, 291, 293, 294
      immunofluorescence, 291, 292, 294
    transient transfection of embryonic stem cells, 284, 285, 289

D

DAG, *see* Diacylglycerol
Denaturing high-performance liquid chromatography (DHPLC), genotyping of embryonic stem cell mutants, 401
Desert hedgehog, *see* Hedgehog
DHPLC, *see* Denaturing high-performance liquid chromatography
Diacylglycerol (DAG), phospholipase C signaling, 128, 129
Digital differential display, *see* Oct3/4
DNA damage, *see also* Chemical mutagenesis,
  apoptosis assay,
    annexin staining, 325
    materials, 317
    transfection, 325
  cell cycle regulation response in mouse embryonic stem cells,
    immunofluorescence microscopy, 324
    materials, 316, 317
    Western blot analysis of cell cycle checkpoint proteins, 323
  embryonic stem cell consequences, 313, 314
  etiology, 328
  mutation frequency measurement in mouse embryonic stem cells,
    Aprt knockout mouse generation, gene targeting, 318
    Southern blot, 318
    targeting vector construction, 317
    colony-forming efficiency assay, 319, 325
    feeder cell preparation, 318, 319, 325
    frequency calculation, 319, 325
    materials, 314–316
    mutation spectrum analysis, 320, 323
    principles, 317
DNA methylation,
  embryonic stem cell studies,
    bisulfite restriction mapping and sequencing, 437–439
    cell culture, 428, 429, 442
    genomic DNA extraction, 429, 430, 442
    laser-manipulated microdissection and teratoma analysis, 439–441, 443
    materials, 422–428, 442
    methylation-sensitive quantitative real-time polymerase chain reaction, 436, 437
    overview, 421, 422
    promoter methylation assay, 441, 442
    restriction landmark genomic scanning,
      autoradiography, 432, 443

first-dimension gel
electrophoresis, 431, 443
genomic DNA treatment, 430, 442
second-dimension gel
electrophoresis, 431, 432, 443
spot DNA cloning, 432, 433
virtual image technique, 434
Southern blot,
DNA transfer to membranes,
435, 443
hybridization and detection, 436
probe labeling, 435, 436, 443
restriction enzyme digestion and
electrophoresis, 435, 443
tissue-dependent DNA methylated
genes, 422

E

EB, *see* Embryoid body
Electrofusion, *see* Cell fusion
Electron microscopy,
peri-implantation studies, 123
RNA interference and morphological
analysis, 254, 255, 258
Electrophoretic mobility shift assay
(EMSA),
Oct3/4 binding targets,
binding reaction, 229
cell lysis, 228, 229
dialysis, 229
electrophoresis, 229
plasmid preparation, 228
probe preparation and labeling, 229
p53 induction and target gene
analysis in ultraviolet-induced
apoptosis, 336
Electroporation,
genetically modified embryonic
stem cells, 477, 478
monkey embryonic stem cells, 309–311

Embryoid body (EB),
differentiation markers, *see*
Glyceraldehyde-3-phosphate
dehydrogenase; Hypoxanthine
phosphoribosyltransferase; β-
Tubulin
differentiation, *see specific cell types*
genetically modified embryonic stem
cell differentiation,
adhesion culture, 480, 482
cover slip coating, 479, 480, 482
cytocentrifugation, 480
immunofluorescent detection of
marker genes, 480–482
materials, 475, 481
medium preparation, 479, 482
suspension culture, 480, 482
Hedgehog signaling, *see* Hedgehog
monkey embryonic stem cells and
formation, 88
peri-implantation development
studies, *see* Peri-implantation
development
Embryonic stem cell test (EST),
applications, 375
cytotoxicity assay,
independent runs, 388, 393
materials, 378
MTT assay, 380, 385, 387, 388, 393
quality control, 388, 389
differentiation assay, 380–383, 393
embryotoxic potential prediction,
389–391
flow cytometry,
end points of differentiation, 373, 374
independent runs, 385
materials, 377, 378
quality control, 385
staining, 383, 384, 393
freezing and thawing of cells, 392, 394
materials, 375–380, 393

morphological analysis of
differentiation, 377, 384, 385
prediction model, 373, 390, 391
principles, 371, 372
test chemicals and solvents,
378–380, 393
trypsinization of cells, 391–394
validation, 373
Embryotoxicity assay, *see* Embryonic
stem cell test
EMS, *see* Methanesulfonic acid ethyl
ester
EMSA, *see* Electrophoretic mobility
shift assay
Endothelial cell,
horse embryonic stem cell
differentiation, 68, 69
rat embryonic stem cell
differentiation, 53, 58
ENU, *see* N-Ethyl-N-nitrosourea
EST, *see* Embryonic stem cell test
N-Ethyl-N-nitrosourea (ENU),
embryonic stem cell chemical
mutagenesis, *see* Chemical
mutagenesis
mechanisms of action, 398, 399
solution preparation, 402, 403, 405

F–G

Flow cytometry,
embryonic stem cell test,
end points of differentiation, 373, 374
independent runs, 385
materials, 377, 378
quality control, 385
staining, 383, 384, 393
RNA interference and gene
expression analysis, 247, 248
simian immunodeficiency virus
vector transduction efficiency
evaluation, 299, 301
transgene expression assay, 291,
293, 294

GAPDH, *see* Glyceraldehyde-3-
phosphate dehydrogenase
Gene expression profiling, *see* Serial
analysis of gene expression
Gene therapy, *see* Genetically-modified
embryonic stem cell
Genetically-modified embryonic stem
cell,
characterization,
materials, 474
pulse/chase and
immunoprecipitation, 478, 479
sulfamidase assay, 478
culture and transgenesis,
DNA preparation, 477
electroporation, 477, 478
freezing, 477
materials, 472–474, 481
passaging, 476, 482
thawing, 475, 476, 482
embryoid body differentiation,
adhesion culture, 480, 482
cover slip coating, 479, 480, 482
cytocentrifugation, 480
immunofluorescent detection of
marker genes, 480–482
materials, 475, 481
medium preparation, 479, 482
suspension culture, 480, 482
stable transfection of transgenes in
disease treatment, 472
GFP, *see* Green fluorescent protein
Glial cell, rat embryonic stem cell
differentiation, 52
Glyceraldehyde-3-phosphate
dehydrogenase (GAPDH),
embryonic stem cell
differentiation marker analysis,
experimental design and statistics, 110
materials, 102–104, 110
reverse transcriptase polymerase
chain reaction,
amplification reactions, 108, 111

gel electrophoresis and analysis of
products, 109–111
reverse transcription, 107, 108, 111
RNA isolation and quality
analysis, 106, 107, 111
tissue culture and embryoid body
formation, 104–106, 111
Green fluorescent protein (GFP),
lentivirus-mediated gene transfer
reporter, 277
monkey embryonic stem cell
expression,
applications, 306
colony isolation, 310, 311
culture, 308, 311
electroporation, 309–311
feeder layer preparation, 307,
308, 311
freezing of cells, 310, 311
materials, 306, 307
passage, 308, 309, 311
transplanted cell progeny
detection *in situ*, 463–466
vector preparation, 309
promoter analysis in embryonic
stem cells,
advantages, 188, 192
fluorescence assay, 191, 192

H

Hedgehog (Hh),
receptor, 171
signaling in embryoid bodies,
embryoid body formation, 178, 183
embryonic stem cell culture, 177,
180, 183
hematopoietic cells,
differentiation, 180, 183
staining, 180–183
inhibitors and enhancers, 173, 174
materials for study, 174–177, 182

neural cells,
differentiation, 178
staining, 178–180, 183
overview, 171–173
types, 172
Hematopoietic cells, Hedgehog
signaling studies in embryoid
bodies,
differentiation, 180, 183
staining, 180–183
Hepatocyte,
polymerase chain reaction of
markers, 49, 58
rat embryonic stem cell
differentiation, 55, 56, 58
Herpes simplex virus-1 (HSV-1)
amplicon vectors,
advantages with embryonic stem
cells, 265, 266
features, 266
growth factor transfer to embryonic
stem cells,
embryoid body formation, 269
infection, 268, 269, 271
maintenance culture, 268, 271
materials, 266–268, 271
neural progenitor cells,
detection with nestin
antibodies, 270, 271
infection, 270, 271
neuron differentiation, 271, 272
selection, 270
Hh, *see* Hedgehog
High-performance liquid
chromatography (HPLC), *see
also* Denaturing high-
performance liquid
chromatography,
inositol phosphates, 138–140, 147
Horse embryonic stem cells,
alkaline phosphatase staining, 62, 68

derivation and maintenance,
  embryo thawing, 65
  feeder cell preparation, 64, 65, 76
  freezing and thawing, 65, 76
  isolation and culture, 65, 67,
    76, 77
  materials, 60–62, 76
  overview, 60
differentiation,
  endothelial cells, 68, 69
  immunohistochemical staining,
    62, 63, 69, 72, 77
  materials for study, 62, 63, 76
  neurons, 68
  reverse transcription polymerase
    chain reaction,
    discrimination between DNA
      and RNA, 74, 77
    materials, 63, 64, 76
    RNA denaturation, 75
    RNA extraction, 73, 74, 77
    SuperScript system, 75, 77
gene targeting, 59, 60
Giemsa chromosome staining, 62,
  67, 68, 77
sex determination, 72, 73, 77
HPLC, *see* High-performance liquid
  chromatography
HPRT, *see* Hypoxanthine
  phosphoribosyltransferase
HSV-1 amplicon vectors, *see* Herpes
  simplex virus-1 amplicon
  vectors
Human immunodeficiency virus, *see*
  Lentivirus-mediated gene transfer
Hypoxanthine
  phosphoribosyltransferase
  (HPRT), embryonic stem cell
  differentiation marker analysis,
experimental design and statistics, 110
materials, 102–104, 110

reverse transcriptase polymerase
  chain reaction,
  amplification reactions, 108, 111
  gel electrophoresis and analysis of
    products, 109–111
  reverse transcription, 107, 108, 111
  RNA isolation and quality
    analysis, 106, 107, 111
  tissue culture and embryoid body
    formation, 104–106, 111

I

Immunohistochemistry,
  dopaminergic embryonic stem cells
    in mouse Parkinson's disease
    model, 489, 490, 492
  RNA interference and gene
    expression analysis, 253, 254,
    257, 258
Indian hedgehog, *see* Hedgehog
Inositol phosphates,
  embryonic stem cell production assays,
    cell proliferation assays,
      hemocytometer cell counting,
        141–143, 148
      MTT assay, 143, 148
      overview, 130
      TUNEL apoptosis assay, 145,
        146, 148
      viability staining, 130,
        143–145, 148
    extraction,
      inositol phosphates, 135, 147
      phosphoinositides, 136, 147
    inositol phosphate analysis,
      Dowex resin separation, 137,
        138, 147
      high-performance liquid
        chromatography, 138–140, 147
      principles, 136, 137
    materials, 130–134, 146, 147

overview, 129, 130
phosphoinositide analysis,
　identification and
　　quantification, 141, 148
　separation with thin-layer
　　chromatography, 140, 141,
　　147, 148
　tissue culture, 134
　tritiated inositol labeling, 134,
　　135, 147
inositol recycling, 129–134
signaling system, 128, 129
Internal ribosome entry site (IRES), cell lineage marking and selection, 356, 357
IRES, see Internal ribosome entry site

L

Laser-manipulated microdissection (LMM), DNA methylation analysis in teratomas, 439–441, 443
Lentivirus-mediated gene transfer, see also Simian immunodeficiency virus vector,
　advantages with embryonic stem cells, 273, 274
　mouse embryonic stem cell gene transfer,
　　infection,
　　　CRE-mediated recombination, 277
　　　reporter gene expression, 277
　　maintenance culture, 276, 277, 279
　　materials, 274–276
　　vector preparation, 277, 279
　safety, 296
Leukemia inhibitory factor (LIF),
　chicken protein preparation,
　　affinity chromatography, 26, 27
　　competent cell transformation, 25, 34
　　expression vector construction, 24, 34
　　gel filtration, 28

　　large-scale bacteria culture and sonication, 25, 26
　　materials, 18, 19, 33
　　overview, 18
　　polymerase chain reaction, 23, 24, 34
　　reverse transcription, 23
　　RNA extraction, 21, 23, 34
　embryonic stem cell maintenance culture, 18
LIF, see Leukemia inhibitory factor
LMM, see Laser-manipulated microdissection
Luciferase, see Promoter analysis, embryonic stem cells
Lysosomal storage diseases,
　enzyme replacement therapy, 471
　mucopolysaccharidosis type IIIA, 471, 472
　sulfamidase expression in embryonic stem cells, see Genetically modified embryonic stem cell

M

Medaka embryonic stem cells,
　advantages of study, 3, 4
　cell lines, 4
　characterization,
　　alkaline phosphatase staining, 10, 11, 15
　　chromosome preparation and analysis, 10, 11, 15
　chimera formation, 11–15
　gene targeting, 5, 13
　isolation and differentiation,
　　culture initiation, 8, 9, 14
　　directed differentiation, 14
　　fish embryo extract preparation, 7, 8, 14
　　fish serum preparation, 8
　　freezing, 10
　　materials, 5–7
　　overview, 4, 5
　　subculture, 10, 14
　　thawing, 10, 14

Index

prospects for study, 14
transfection and drug selection, 13
Methanesulfonic acid ethyl ester (EMS),
   embryonic stem cell chemical mutagenesis, see Chemical mutagenesis
   mechanisms of action, 398, 399
   solution preparation, 403, 405
Monkey embryonic stem cells,
   characterization,
      differentiation potency, 87, 88
      karytopying, 87
      marker expression, 87
   clinical significance, 81–83
   embryoid body formation, 88
   establishment,
      expansion and maintenance, 85, 86, 88
      feeder cell layer preparation, 84, 88
      inner cell mass,
         culture, 85, 88
         isolation from blastocysts, 84, 88
      materials, 83, 84, 88
      overview, 82
   gene transfer vectors, see Simian immunodeficiency virus vector
   green fluorescent protein expression,
      applications, 306
      colony isolation, 310, 311
      culture, 308, 311
      electroporation, 309–311
      feeder layer preparation, 307, 308, 311
      freezing of cells, 310, 311
      materials, 306, 307
      passage, 308, 309, 311
      vector preparation, 309
   human embryonic stem cell comparison, 306
   mouse embryonic stem cell comparison, 295

tumor formation,
   allogeneic monkey fetus teratoma formation, 462, 463, 465, 466
   immunodeficient mice and teratoma formation, 461, 462, 465
   materials, 460, 461, 465
   overview, 459, 460
   transplanted cell progeny detection *in situ*, 463–466
Mouse embryonic stem cells,
   cell replacement therapy, see Parkinson's disease
   differentiation culture for pharmacological studies,
      adipocyte differentiation,
         embryoid body culture, 344–346, 349, 350
         gene expression analysis, 348, 349
         Oil-Red O staining, 346, 347
      maintenance culture, 344, 349
      materials, 342–344, 349
      osteoblast differentiation,
         alizarin red staining, 347, 348
         embryoid body culture, 346, 350
         gene expression analysis, 348, 349
         von Kossa staining, 348
      overview, 341, 342
      RNA preparation from embryoid body outgrowths, 348, 350
   embryotoxicity assay, see Embryonic stem cell test
   gene expression profiling, see Serial analysis of gene expression
   historical perspective, 91, 92
   injection into blastocysts, 96, 97
   mutation, see Chemical mutagenesis; DNA damage
   promoter strength studies, see Promoter analysis, embryonic stem cells

serum- and feeder-free culture,
    disaggregation and expansion,
        95–97
    embryo recovery, 94, 96
    freezing, 96, 97
    materials, 92–94
    passaging, 96
    rationale, 92
ultraviolet-induced apoptosis, see
    Apoptosis,
viral vectors, see Cytomegalovirus
    promoter; Herpes simplex
    virus-1 amplicon vectors;
    Lentivirus-mediated gene
    transfer
Mouse trophoblast stem cells,
    applications, 36
    derivation and culture,
        establishment from blastocysts,
            38–41, 43
        feeder cells,
            removal from culture,
                41, 42
            stock preparation, 38, 42, 43
        feeder-free culture, 42
        freezing and thawing, 42, 43
        materials, 36–38
        overview, 36
        passage, 41, 43
    origins, 35, 36
MTT cytotoxicity assay,
    embryonic stem cell test,
        independent runs, 388, 393
        materials, 378
        MTT assay, 380, 385, 387,
            388, 393
        quality control, 388, 389
    embryonic stem cells in inositol
        phosphate detection,
        143, 148
Mutation, see Chemical mutagenesis;
    DNA damage

N

Neuron,
    Cripto knockout embryonic stem cell,
        differentiation, 164
        immunofluorescence microscopy,
            165, 166, 168
    embryonic stem cell neural
        differentiation culture for
        pharmacological screening,
        advantages, 353, 354
        Cre deletion, 362, 365
        embryoid body differentiation
            culture, 354–356
        embryoid body differentiation
            culture, 362, 363, 366
        high-throughput screening,
            358, 359
        lineage marking and selection,
            356, 357
        materials, 359–361, 365
        monolayer differentiation, 363,
            365, 366
        neurogenesis monitoring,
            immunostaining, 364
            X-Gal staining, 364, 366
        plate screening, 364, 365
        selected clone analysis, 362, 365
        Sox-1-based selection, 359, 361,
            362, 365
    Hedgehog signaling studies in
        embryoid bodies,
        differentiation, 178
        staining, 178–180, 183
    herpes simplex virus-1 amplicon
        vector transfer of growth
        factors to neural progenitor
        cells,
        detection with nestin antibodies,
            270, 271
        infection, 270, 271
        neuron differentiation, 271, 272
        selection, 270

horse embryonic stem cell differentiation, 68
rat embryonic stem cell differentiation, 52
Northern blot, p53 induction and target gene analysis in ultraviolet-induced apoptosis, 334–336

O

Oct3/4
  embryonic stem cell expression, 223
  functions, 223, 224
  target gene identification,
    digital differential display,
      binding site identification in candidate genes, 226, 227
      materials, 224
      overview, 224, 225
      software utilization, 225, 226, 230
      Web resources, 225, 230
    electrophoretic mobility shift assay,
      binding reaction, 229
      cell lysis, 228, 229
      dialysis, 229
      electrophoresis, 229
      plasmid preparation, 228
      probe preparation and labeling, 229
    reporter gene analysis of candidate gene regulatory elements,
      luciferase assay, 228
      plasmid construction, 227, 228
      transfection, 228, 230

O–P

Oil-Red O, adipocyte staining, 346, 347
Osteoblast embryoid body differentiation culture for pharmacological studies,
  alizarin red staining, 347, 348
  embryoid body culture, 346, 350
  gene expression analysis, 348, 349
  von Kossa staining, 348
PA6 cells, embryonic stem cell differentiation induction, 152
Parkinson's disease (PD),
  cell replacement therapy rationale, 485
  mouse model studies of differentiating embryonic stem cells,
    animal preparation, 489, 491, 492
    cell maintenance culture, 487, 488, 490
    dopaminergic differentiation induction, 488–491
    immunohistochemistry, 489, 490, 492
    materials, 486, 487, 490
    overview, 485, 486
    transplantation of cells, 489, 492
PCR, see Polymerase chain reaction
PD, see Parkinson's disease
Peri-implantation development,
  analytical applications, 118
  embryoid body differentiation,
    embryonic stem cell cluster preparation, 121, 123
    induction, 121, 123, 124
    preparation for microscopy,
      electron microscopy, 123
      frozen sectioning and immunostaining, 121–124
    sequence, 115, 117, 118
  embryonic stem cell culture,
    feeder cell preparation, 120, 123
    maintenance, 120, 121, 123
    trypsinization, 120
  extracellular matrix assembly, 113, 115
  materials for study, 119, 120
  mutant studies, 118
  overview, 113–115
Pharmacological screening,
  embryonic stem cell neural differentiation culture,
    advantages, 353, 354

Cre deletion, 362, 365
embryoid body differentiation culture, 354–356
embryoid body differentiation culture, 362, 363, 366
high-throughput screening, 358, 359
lineage marking and selection, 356, 357
materials, 359–361, 365
monolayer differentiation, 363, 365, 366
neurogenesis monitoring,
    immunostaining, 364
    X-Gal staining, 364, 366
plate screening, 364, 365
selected clone analysis, 362, 365
Sox-1-based selection, 359, 361, 362, 365
mouse embryonic stem cell differentiation culture for pharmacological studies,
adipocyte differentiation,
    embryoid body culture, 344–346, 349, 350
    gene expression analysis, 348, 349
    Oil-Red O staining, 346, 347
maintenance culture, 344, 349
materials, 342–344, 349
osteoblast differentiation,
    alizarin red staining, 347, 348
    embryoid body culture, 346, 350
    gene expression analysis, 348, 349
    von Kossa staining, 348
overview, 341, 342
RNA preparation from embryoid body outgrowths, 348, 350
Phosphatidylinositol, *see* Inositol phosphates
Phosphoinositides, *see* Inositol phosphates

Phospholipase C,
embryonic stem cell assays, *see* Inositol phosphates
signaling system, 128, 129
Polymerase chain reaction (PCR),
ditag amplification in SAGE, 209, 210
DNA methylation analysis,
    bisulfite restriction mapping and sequencing, 437–439
    laser-manipulated microdissection and teratoma analysis, 439–441, 443
    methylation-sensitive quantitative real-time polymerase chain reaction, 436, 437
embryonic stem cell differentiation marker analysis with reverse transcriptase polymerase chain reaction,
    amplification reactions, 108, 111
    gel electrophoresis and analysis of products, 109–111
    reverse transcription, 107, 108, 111
    RNA isolation and quality analysis, 106, 107, 111
gene expression analysis in differentiated embryoid bodies, 348, 349
hepatocyte markers, 49, 58
horse embryonic stem cell,
    reverse transcription polymerase chain reaction of differentiation markers,
    discrimination between DNA and RNA, 74, 77
    materials, 63, 64, 76
    RNA denaturation, 75
    RNA extraction, 73, 74, 77
    SuperScript system, 75, 77
    sex determination, 72, 73, 77
mutation spectrum analysis in mouse embryonic stem cells, 320, 323

Index 507

RNA interference and gene
  expression analysis with
  reverse transcription
  polymerase chain reaction,
  248–250, 256, 257
simian immunodeficiency virus
  vector transduction efficiency
  evaluation, 299, 301, 302
Promoter analysis, embryonic
  stem cells,
  embryonic stem cell culture,
    freezing and thawing, 190, 192
    gelatin coating of plates, 190, 192
    passaging, 190, 192
    plating, 189, 190
  green fluorescent protein reporter,
    advantages, 188, 192
    fluorescence assay, 191, 192
  luciferase assay, 192
  materials, 188, 189, 192
  overview, 187, 188
  reporter DNA construct preparation,
    191, 192
  transfection, 191

R

Rat embryonic stem cells,
  culture,
    fibroblast feeder layer
      preparation, 50
    freezing, 49, 50, 58
    maintenance, 50, 58
    materials, 46, 47, 57
    Matrigel culture, 53
    overview, 45, 46
    thawing, 50, 58
  differentiation,
    adipocytes, 54
    carboxyfluorescein succinimidyl
      ester labeling, 52, 58
    cellular transplant preparation,
      52, 58
    embryonic bodies, 51
    endothelial cells, 53, 58
    glial cells, 52
    hepatocytes, 55, 56, 58
    immunostaining, 48, 49, 56, 57
    materials, 47, 48
    neurons, 52
    polymerase chain reaction of
      hepatocyte markers, 49, 58
    signs, 51, 53
Restriction landmark genomic scanning
  (RLGS), DNA methylation
  analysis,
  autoradiography, 432, 443
  first-dimension gel electrophoresis,
    431, 443
  genomic DNA treatment, 430, 442
  second-dimension gel
    electrophoresis, 431, 432, 443
  spot DNA cloning, 432, 433
  virtual image technique, 434
Retinoic acid, embryonic stem cell
  differentiation induction, 152
RLGS, *see* Restriction landmark
  genomic scanning
RNA interference,
  embryonic stem cell studies,
    gene expression monitoring,
      flow cytometry, 247, 248
      immunohistochemistry, 253,
        254, 257, 258
      morphological analysis with
        microscopy, 254, 255, 258
      reverse transcription
        polymerase chain reaction,
        248–250, 256, 257
      Western blot, 250–253, 257
    materials, 241–245, 255, 256
    nonspecific interferon-like
      response, 246, 247, 256
    rationale, 233, 234, 255
    transfection, 245, 246, 256

expression plasmids,
  delivery to embryonic stem cells, 240
  inducible RNA interference, 238, 239
  RNA polymerase III promoters, 236–238
  short-hairpin RNA expression constructs, 236–238
mechanisms, 234, 235
targeting constructs,
  development, 235, 236
  selection markers, 239, 240

S

SAGE, *see* Serial analysis of gene expression
Serial analysis of gene expression (SAGE),
  cell extract preparation, 207, 218
  complementary DNA,
    cleavage and binding to magnetic beads, 208
    linker ligation, 208, 209
    synthesis, 207, 208
    tag release and blunt ending, 209, 218
  concatemer cloning and sequencing, 212–215, 218, 219
  data collection and analysis, 215, 216, 219
  ditags,
    large-scale amplification, 210, 211
    ligation and polymerase chain reaction amplification, 209, 210, 212
    purification, 211, 212
  DNA preparation,
    ethanol precipitation, 206, 208
    isopropanol precipitation, 206, 218
    phenol-chloroform extraction, 205, 206

embryonic stem cell culture,
  buffalo rat liver-conditioned media preparation, 203, 204
  overview, 203
  passaging, 203–205, 218
linkers,
  ligation control reaction, 206, 207
  ligation to bound cDNA, 208, 209
  phosphorylation, 206
materials, 198–202, 217, 218
principles, 195–198
reference libraries for mouse embryonic stem cells, 216, 217
RNA preparation, 207, 218
Simian immunodeficiency virus (SIV) vector,
  monkey embryonic stem cell gene transfer,
    materials, 296, 297
    specificity, 296
    transduction, 298, 300
    transduction efficiency evaluation,
      flow cytometry, 299, 301
      overview, 298, 299
      polymerase chain reaction, 299, 301, 302
    vector construction,
      harvesting and titering, 298, 300
      transfection, 297–299
      VSV-G-pseudotyped virus, 297
  safety, 296
Single-strand conformational polymorphism (SSCP),
  genotyping of embryonic stem cell mutants, 400, 401
SIV vector, *see* Simian immunodeficiency virus vector
Small-interfering RNA, *see* RNA interference
Sonic hedgehog, *see* Hedgehog

Southern blot,
　DNA methylation analysis,
　　DNA transfer to membranes, 435, 443
　　hybridization and detection, 436
　　probe labeling, 435, 436, 443
　　restriction enzyme digestion and electrophoresis, 435, 443
　knockout mouse genotyping, 318
SSCP, *see* Single-strand conformational polymorphism
STAT3, Western blot from chicken blastodermal cells, 31–33

T

Temperature gradient capillary electrophoresis (TGCE), genotyping of embryonic stem cell mutants, 401
Teratoma, *see also* Tumorigenicity,
　formation assay, 455–457
　laser-manipulated microdissection and DNA methylation analysis, 439–441, 443
TGCE, *see* Temperature gradient capillary electrophoresis
Thin-layer chromatography (TLC), phosphoinositides, 140, 141, 147, 148
TLC, *see* Thin-layer chromatography
Transcriptome, *see* Serial analysis of gene expression
β-Tubulin, embryonic stem cell differentiation marker analysis,
　experimental design and statistics, 110
　materials, 102–104, 110
　reverse transcriptase polymerase chain reaction,
　　amplification reactions, 108, 111
　　gel electrophoresis and analysis of products, 109–111
　　reverse transcription, 107, 108, 111

　RNA isolation and quality analysis, 106, 107, 111
　tissue culture and embryoid body formation, 104–106, 111
Tumorigenicity, *see also* Teratoma, embryonic stem cell gene identification,
　gene expression assays, 452, 453, 457
　materials, 450, 451, 456
　NIH3T3 cell tumor-like property analysis,
　　colony formation assay, 454
　　stable transfection, 454, 457
　　transformation assay, 455
　primary mouse embryonic fibroblast retroviral transduction, 453, 454, 457
　supertransfection, 451, 452, 457
　teratoma formation assay, 455–457
ERas-deficient cells, 450
monkey embryonic stem cell tumor formation,
　allogeneic monkey fetus teratoma formation, 462, 463, 465, 466
　immunodeficient mice and teratoma formation, 461, 462, 465
　materials, 460, 461, 465
　overview, 459, 460
　transplanted cell progeny detection *in situ*, 463–466
PTEN-null cells, 449, 450

U–V

Ultraviolet-induced apoptosis, *see* Apoptosis
Viral vectors, *see* Cytomegalovirus promoter; Herpes simplex virus-1 amplicon vectors; Lentivirus-mediated gene transfer; Simian immunodeficiency virus vector
von Kossa stain, osteoblast staining, 348

W

Western blot,
  cell cycle checkpoint proteins, 323
  Cripto, 162, 166
  p53 induction and target gene analysis in ultraviolet-induced apoptosis, 333, 334
  RNA interference and gene expression analysis, 250–253, 257
  STAT3, 31–33

QH 588 .S83